HORMONE BIOTECHNOLOGY

HORMONE BIOTECHNOLOGY

(Commemorative Volume for *Professor Samir Bhattacharya*, FNA)

Editor
Saumen Kumar Maitra
Professor of Zoology
Visva-Bharati, Santiniketan

2007
DAYA PUBLISHING HOUSE
Delhi - 110 035

ISBN 81-7035-451-X

Published by : **Daya Publishing House**
1123/74, Deva Ram Park
Tri Nagar, Delhi - 110 035
Phone: 27383999
Fax: (011) 23244987
e-mail : dayabooks@vsnl.com
website : www.dayabooks.com

Showroom : 4760-61/23, Ansari Road, Darya Ganj,
New Delhi - 110 002
Phone: 23245578, 23244987

Laser Typesetting : **Classic Computer Services**
Delhi - 110 035

Printed at : **Chawla Offset Printers**
Delhi - 110 052

PRINTED IN INDIA

Preface

There is nothing more powerful than an idea whose time has come. – Victor Hugo

Biotechnology, in recent years, has emerged as a most exciting branch of science that aims at exploitation and application of knowledge of technology in the improvement of the quality of life especially pertaining to economic and clinical benefits of human being. So far genetic engineering has been considered as the most effective tool of biotechnology. But proliferation of information in the field of endocrinology that took place in past few decades could convincingly argue that hormones are equally potential in achieving the goal of biotechnology. The current treatise is an attempt for the first time to present a balanced variety of original and authentic reports on the implication of various hormones in the management of various life processes including reproduction in economically important animals as well as in humans, and thereby to focus the fundamental significance of the study of hormones in biotechnology. The book provides comprehensive, integrated reviews giving sound, critical, and provocative ideas supported by most appropriate illustrations covering diverse areas of hormones and biotechnology, from molecular to organismic levels. One of the notable features is that the articles included herein have been contributed by globally reputed scientists who were responsible for collection of original data from their own long-term research in India and/or abroad, and many of them discussed their findings in a Symposium on Reproductive Biology and Comparative Endocrinology held in February, 2005 in the Department of Zoology, Visva-Bharati University, Santiniketan, India. The common denominator of all the contributors to this treatise is their consensus on providing evidence for the all-pervasiveness of hormones as a hallmark of regulation of all processes that could make life possible to exist and propagate in this earth. There are reasons to believe that the contents of this book would be of particular interest to the readers chiefly as they emphasized implication of hormones in different reproductive and physiological disorders, and molecular basis of their remedies.

This publication is in commemoration of Dr. Samir Bhattacharya, Professor of Zoology, Visva-Bharati, former Director of Indian Institute of Chemical Biology (CSIR), Kolkata, and a Fellow of all the National Academies in India, for his untiring service to science for nearly four decades and unique contribution to the field of hormones and biotechnology.

The proud privilege of the editor is to acknowledge the active participation of all the contributors of the volume, encouragements of both the Department of Biotechnology, Government of India and the authority of Visva-Bharati University, loyal assistance of all the members of the faculty staff and research fellows of the Department of Zoology, Visva-Bharati University, Santiniketan, and especially the efforts of M/S Daya Publishing House, Delhi in nurturing the venture of this novel publication.

Saumen Kumar Maitra

Visva-Bharati, Santiniketan

Contents

— Section B —
Hormones and Reproduction

— Section C —
Comparative Endocrinology and Biotechnology

Professor Samir Bhattacharya: A Profile

Professor Samir Bhattacharya, born in 1942, is a highly reputed scientist known for his break-through research in the fields of molecular mechanisms of hormone action and hormone biotechnology. He started his research career as a Ph.D. student in the Indian Institute of Chemical Biology, CSIR, under the guidance of renowned scientist Dr. A. G. Datta, and subsequently worked with a globally respected comparative endocrinologist Professor Aubrey Gorbman, University of Washington, Seattle, USA, with a most prestigious Senior Fulbright Fellowship. Unlike most of the contemporary scientists, on return to India Professor Bhattacharya selected Visva-Bharati, Santiniketan as the place of academic exercise by setting up a research laboratory at a time when this University was not quite known for its Science departments. Since 1971, he has put his hard work, shown the signs of tremendous tenacity, dedication and devotion to put Santiniketan in the world map of Science. The time has tested his talents, and as a result Professor Bhattacharya became the Fellow of Indian National Science Academy (FNA) in this University in 1990. Subsequently, he was elected as the Fellow of all other National Academies including FNASc., Allahabad (1994) and (FASc), Bangalore (1996), and has been awarded with almost all the coveted national and international awards, a few notable of which are Senior Fulbright Award from Federal Government, USA, (1975 and 1983), J.G. Law Memorial Medal (1992), M.R.N. Prasad Memorial Lecture Award (1998), Meghnad Saha Endowment Memorial Lecture Award, (1999), Rafi Ahmed Kidwai Award (1999), Barclay Memorial Gold Medal Award (2001), Federation of Indian Chamber of Commerce and Industries (FICCI) Award (2001-2002), CSIR Technology Award, (2002), B. C. Guha Memorial Lecture Award, (2004), Professor B. Kizar Ahamath Endowment lecture in Medicine (2004), and Kamal Kumari National Award (2004). As the first Indian, he was elected as the President of "Asia and Oceania Society for Comparative Endocrinology" (an international organization of 17 countries) for a period of 2000-2004, and also elected as the President, Biochemistry, Biophysics and Molecular Biology Section of 89th Indian Science Congress (2002). As an outstanding teacher, Professor Bhattacharya has already produced 30 Ph.D.s and published more than 100 original research papers in globally reputed scientific journals. More than a dozen applications have been filed by him for international patents, of which already been granted a few. In the capacity of an excellent orator, Professor Bhattacharya has delivered numerous lectures in various scientific bodies. In recognition of his unique character of administrative talent blended with scientific temperament, Professor Samir Bhattacharya held the post of Director of Indian Institute of Chemical Biology, a well known CSIR Research Institute, Kolkata over a period of five years (1999 – 2004), and in such capacity he brought the Institute into the light of a number of epochal discoveries. All the faculty staff and students of Visva-Bharati, admirers and well wishers across the world enjoy rare feelings of pride and pleasure in extending their greetings through publication of this commemorative volume for Professor Samir Bhattacharya, who is nearest to their heart and dearest to their mind.

— Section A —
Clinical Endocrinology and Biotechnology

— Section A —
Clinical Endocrinology and Biotechnology

1

Development of Male Contraceptives: Approaches, Associated Problems and Possible Solutions

*D. Chitra Lekha and A. J. Rao**

Department of Biochemistry, Indian Institute of Science, Bangalore–560 012, India

ABSTRACT

With the current problem of population explosion and negligible participation by the male in the family planning programme, the need for a simple and effective male contraceptive need not be overemphasized. One of the major problems in developing a male contraceptive is that it should not interfere with libido and secondary sexual characters and as such the choice of available male contraceptives is extremely limited. In the present review an attempt is made to evaluate the newer approaches which are based on interference at the level of testosterone production and those which do not include interference with sperm maturation, intra-vasal blockade of sperm transport and use of plant products. In addition an attempt is made to assess the current status of these newer approaches.

Introduction

The necessity for control of the population explosion need not be overemphasized particularly in the developing countries. However, it is accepted now that it is not possible to develop a contraceptive

**Corresponding Author.* E-mail: ajrao_2000@yahoo.com

suitable for everybody. The realization, though a disappointing one after decades of research, has in fact been the impetus to develop newer strategies for contraception. Although considerable progress has been achieved in improving the existing as well as developing newer approaches for fertility control usable by females, as of now, there is no reliable method of contraception which is safe and totally reversible for use by males other than vasectomy or condom.

In the present review an attempt is made to provide a brief account of the available methods, approaches for development of newer methods, associated problems, a critical evaluation and possible solutions.

Barrier Methods

Condom

One of the earliest and successful approaches has been the barrier method which essentially involved a method to prevent the access of the sperm when deposited in the female reproductive tract to the egg. Although the use of sheath as a barrier has been referred to as early as 1564, its use became popular in 1930's. Currently, the barrier has been referred to as condom which prevents sperm from entering vagina. Several modifications have been introduced which include spermicidal and microbicidal properties. With the threat of AIDs spreading fast condoms are being used both as contraceptives and for preventing STDs. In spite of the effectiveness and simple to use, condoms are not very popular, and in India the use of condoms is less than 5 per cent. The reasons are not because of failure in protecting from pregnancy but because of lack of awareness, individual dislikes being the most prominent one.

Vasectomy

Vasectomy is a safe, simple and effective method of contraception, which has been used by nearly 42 million couples worldwide. Post-operative complications of vasectomy are minimal and serious side effects are rare if experienced surgical providers perform vasectomy. Since the long-term health consequences of vasectomy have been of concern to the general public and health workers, the relationship between vasectomy and long-term morbidity and mortality have been examined extensively. Concern regarding health risks from vasectomy first rose in 1970's when increased artherosclerosis was reported in vasectomised monkeys which were fed high cholesterol diet (Alexander *et al.*, 1978). However, clinical studies reassured the safety of vasectomy (Walker *et al.*, 1981; Massey *et al.*, 1984). In one of the cohort studies, involving around 22,000 men, vasectomised men had similar or lower rates of 98 diseases, including various types of cancer, autoimmune disorders, and heart disease, compared to control subjects who did not undergo vasectomy (Massey *et al.*, 1984). The only long-term side effect of vasectomy that has been reliably documented is 1.5 to 2.5 times greater risk of genitourinary tract infection or inflammation up to two years after surgery (Goldacre *et al.*, 1978; Walker *et al.*, 1981; Petitti *et al.*, 1982; Massey *et al.*, 1984). These side effects are short-term, could be easily treated, and are not dangerous. Another retrospective cohort study carried out in China in 4596 vasectomised men, with duration of 10-25 years of vasectomy, and in 4340 non-vasectomised men revealed the lack of association between vasectomy and cardiovascular disease (Tang *et al.*, 1988).

Antisperm antibodies were detected in the serum in approximately 75 per cent of vasectomized men (Bronson, 1994). These antisperm antibodies usually appear in the blood in the first six weeks after surgery and peak at one to three years and thereafter the titer either remains constant or declines (Ansbacher, 1981). However, the available clinical evidence suggests that the antisperm immune response elicited following vasectomy does not cause disease (Bullock *et al.*, 1977; Linnet, 1983).

There are two reports indicating insignificant relative risk of 1.56 and 1.66 (95 per cent confidence interval) for prostate cancer following vasectomy (Giovannucci *et al.*, 1993a; Giovannucci *et al.*, 1993b); however, no association was observed between vasectomy and prostate cancer in three studies (Massey *et al.*, 1984; Sidney *et al.*, 1991; Nienhuis *et al.*, 1992). World Health Organization in October 1991 concluded after reviewing all available epidemiological and biologic data that there was no causal link between vasectomy and any type of cancer. Studies also revealed that neither is there a positive relationship for increased risk of testicular cancer after vasectomy nor does vasectomy accelerate the growth or increase the invasiveness of testicular cancer (Farley *et al.*, 1993). The testicular function and serum levels of FSH, LH and testosterone were normal following vasectomy (Devi *et al.*, 1977).

Vasectomy can be reversed by microsurgery with a high degree of success. Spermatozoa were observed in the ejaculate of 89.1 per cent of subjects after reanatomosis (vaso-vasostomy) but pregnancy was reported only in 56.8 per cent of partners (Belker *et al.*, 1991). The lower pregnancy rate after reanastomosis could be due to several factors including dysfunctional changes in the epididymis, testis and vas deferens affecting the fertility potential of spermatozoa, presence of antisperm antibodies in semen, age of the couple and time interval between vasectomy and reanastomosis. Despite these facts vasectomy is still considered as a terminal method due to low rate of reversibility.

Methods, which Interfere with Testosterone Production

It is well established that testosterone is indispensable for normal spermatogenesis in all the species studied so far. Obviously, the first choice is to interfere with testosterone production for fertility control in males. However, testosterone, besides being important for spermatogenesis is also indispensable for maintaining libido and the secondary sexual characters. Thus any method which interferes with testosterone production should not interfere with libido and secondary sexual characters. Although on the face of it appears as unsolvable, problem has been handled by supplementing with testosterone. This has been possible as the concentration of testosterone required for maintaining spermatogenesis is quite high compared to maintenance of secondary sexual characters, which is quite low. Considering this, over the years, attempts have been made to develop fertility control methods for male by interfering with testosterone.

Intereference at the Level of GnRH

The main hormones in regulation of testosterone production are GnRH and LH. GnRH is synthesized in hypothalamus and is responsible for regulation of synthesis and secretion of LH which in turn acts on Leydig cells in testis to stimulate testosterone production. Thus efforts have been made to block the action of GnRH or interfere with release of LH. GnRH agonists and antagonists can be used effectively for this purpose. GnRH agonists act by initially increasing the levels of the gonadotropins due to continuous stimulation of the receptor followed by decline in the level of gonadotropins due to desensitization of the receptors. Agonists, Decapeptyl, Buselerin and Nafarelin at doses of 5-500 mg/day along with testosterone for 10-30 weeks decreased sperm production leading to azoospermia in only 30 per cent of the volunteers (Behre *et al.*, 1999). The low efficacy of GnRH agonist and testosterone combination compared to testosterone alone, in achieving the suppression of spermatogenesis seems to be due to the return of FSH level in the serum to normal range though LH and endogenous testosterone were effectively suppressed. In contrast, FSH remained low in the group receiving androgen alone. The strong positive correlation between FSH and sperm concentration again indicates that the level of both FSH and LH needs to be suppressed.

In contrast to the use of agonists, GnRH antagonists induce a rapid fall in gonadotrophins and hence inhibit spermatogenesis. It has been demonstrated that administration of GnRH antagonist and supplementation of testosterone led to azoospermia in Cynomolgus monkeys (Weinbauer *et al.*, 1987). Our studies have established that spermatogenesis can be effectively suppressed in the bonnet monkey administered GnRH analogues or antagonist via Alzet pumps (Rao *et al.*, 1985; Rao *et al.*, 1990; Ravindranath *et al.*, 1992; Ravindranath *et al.*, 1995). Results from five clinical trials with daily administration of GnRH antagonist Nal-Glu along with TE revealed that 88 per cent of individuals were azoospermic within three months which was better than administering testosterone enanthate (TE) alone (Behre *et al.*, 1999). Studies with newer generation of GnRH antagonist Cetrorelix along with 19-nortestosterone substitution also showed a dose dependent suppression of LH, FSH and testosterone and all volunteers became azoospermic (Behre *et al.*, 1995). Although the results with GnRH antagonist were promising the need for daily subcutaneous administration of 20 mg renders this approach impractical and also continuous administration of GnRH antagonist is necessary to maintain azoospermia. A depot preparation of GnRH antagonist, which is inexpensive, will be necessary to employ it as a male contraceptive.

Another approach to block the action of GnRH is by use of antibodies to it. A vaccine designed to inhibit the function of GnRH alone with testosterone supplementation could serve as a contraceptive. Immunization studies with D-Lysine substituted in the place of glycine at position 6, coupled to diphtheria toxin resulted in arrest of spermatogenesis and induction of infertility along with decline in serum testosterone levels and libido in rodents and monkeys (Talwar *et al.*, 1999). Although active immunization against GnRH combined with androgen supplementation is a promising antifertility vaccine for males (Ladd *et al.*, 1988), the current available preparation requires multiple injections to elicit an adequate level of immunity and frequent booster injections. Even if these problems are overcome, still one has to worry about the use of alternative approach and the side effects, as GnRH has been shown to act on several non-reproductive tissues.

Feed Back Inhibition of Testosterone Production

Just as in the case of females, wherein combination of estrogen and progesterone are used, by the use of testosterone it is possible to inhibit the release of LH by feedback inhibition and suppress 'T' levels. Thus although as early as 1939, McGullah (McGullah *et al.*, 1939) reported that testosterone administration suppresses human testicular functions, it was the work of Reddy and Rao (Reddy *et al.*, 1972), which clearly established for the first time the reversible anti-fertility action of testosterone propionate (TP) in human male. Following this, several studies were carried out using a variety of testosterone derivatives to block hypothamo-pituitary axis and inhibit testosterone production. Among this is a study sponsored by WHO employed 200 mg TE intra muscularly every week which resulted in azoospermia and oligospermia. Although these results provided support for the concept of a male pill, this approach was impractical as it required daily or weekly injections, which was not practical for mass application. In order to overcome this, testosterone derivatives which are long acting, were developed. Among these the most promising one is testosterone buciclate (TB), which is capable of maintaining serum testosterone levels with in normal range for a period approximately 100 days. Studies with multiple site injection of 80mg TB revealed that following the first injection the mean sperm count was reduced by 75 days within the normospermic range. Subsequent injection of TB after 90 days of first injection reduced the mean sperm count to oligospermia (Rajalakshmi *et al.*, 1989).

Following this, a variety of combination of androgen and of estrogen have been tried to improve the efficacy. Thus in a multicentric study wherein a combination of medroxyprogesterone acetate (MPA) and TE was administered, only 53 per cent of subject reached azoospermia. However, a combination of Danazol and 'T' was effective in causing oligospermia in 85 per cent of men (Leonard *et al.*, 1978). Other combinations include Levonorgestrel (LNG) and androgen, Norethinstrone enanthate (NET-EN) and estradiol which were also effective. Yet another progesterone tried was cyproterone acetate (CPA), which is a synthetic steroid with progestational and antiandrogenic properties. Studies with human volunteers administered 100 mg CPA and 100 mg TE resulted in rapid and profound suppression of spermatogenesis (Meriggiola *et al.*, 1996).

Very recently Handelsman and group (Turner *et al.*, 2003) reported that a combination of testosterone plus progestin was highly effective in suppressing fertility and this was hailed as the beginning of the male pill. In a study of 55 healthy men, testosterone (four 200-mg implants, every 4 or 6 months) and 300 mg depot medroxyprogesterone acetate (im) were administered every three months. Once sperm output was suppressed (<1 million/ml for 2 consecutive months), men entered a 12-month contraceptive efficacy period, ceasing other contraception. No pregnancies occurred in 426 person-months (35.5 person-years; 95 per cent confidence limits for contraceptive failure rate, 0-8 per cent/annum). This method was reversible as spermatogenesis was recovered.

Although these various combinations of testosterone or its derivatives are effective to varying degrees, there are several important questions that need to be addressed before they become available for large-scale usage. The most important one is the repeated injections to maintain the effective concentration of steroids, side effects due to high concentration of steroids, its effects on prostate, lipid metabolism and in addition, the ethnic difference in the effectiveness of androgens. Furthermore, the controversy regarding requirement of oligospermia or azoospermia for induction of infertility in male has to be resolved. Finally, with all the combination and dosages, even the best combination needs a minimum of 45 days to decrease the sperm counts. Consequently, during this period one has to use alternative approaches of contraception, which is a major problem on mass application of this approach.

Use of MENT

There is one derivative of testosterone, which overcomes most of the problems mentioned above, excepting the use of alternative approach. This testosterone derivative was originally developed as a substitute in situations where testosterone supplementation is required and is named 7α-methyl-19-nortestosterone (MENT). MENT is a synthetic androgen 10 times more potent than testosterone in stimulating the weight of bulbocavernosus and levator ani muscles and four times more potent based on its ability to maintain the weight of ventral prostate and seminal vesicles in castrated rats (Sundaram *et al.*, 1993). However, MENT cannot be reduced by 5α-reductase but can be aromatized. Our studies using bonnet monkeys which were administered MENT via Alzet pump or silastic tubing could decrease the nocturnal surge of serum testosterone within seven days and sperm counts by 45 days. No pregnancy was recorded during the period of study and the effects of MENT were reversible (Rao, 1998; Ramachandra *et al.*, 2002).

Recently, studies have been carried out in human volunteers who were implanted MENT. For sustained delivery of MENT, ethylene vinyl acetate implants containing MENT acetate (Ac) were used to administer subdermally (von Eckardstein *et al.*, 2003). Thirty-five normal volunteers were recruited in 3 clinics and were randomly assigned to 1 of 3 doses: 1 (12 men), 2 (11 men), or 4 (12 men) MENT Ac implants. The initial average in vitro release rate of MENT Ac from each implant was approximately 400 µg/day. The duration of treatment varied from 6 months for the 2-implant group and to 12 months for the 4-implant group. Dose-related increase in serum MENT levels and decrease in testosterone, LH, and FSH levels were observed. Effects on sperm counts were also dose related. None of the subjects in the 1-implant group exhibited oligozoospermia (sperm count, <3 million/ml). Four subjects in the 2-implant group became oligozoospermic, 2 of whom reached azoospermia. Eight subjects in the 4-implant group reached azoospermia, with 1 exhibiting oligozoospermia, whereas 2 were non-responders. Changes in lipid parameters were moderate and transient. Liver enzymes showed small changes. This suggests that MENT can inhibit spermatogenesis over a prolonged period and has the potential to be used as a male contraceptive.

The main advantage is the relative low dose at which it is effective, lack of effect on prostate and cannot be reduced by 5α-reductase and no supplementation with testosterone is needed. If MENT can be developed into a patch or orally active derivative it will be the first answer to the long quest for a male contraceptive pill.

Approaches which do not Interfere with Testosterone

Blockade of FSH Action

Spermatogenesis is a complex process and it is well established that it is dependent on FSH and LH that regulates testosterone production. The use of FSH or FSH receptor based vaccine originated from the observation that neutralization of endogenous FSH with specific antibodies in the adult male bonnet monkeys resulted in disruption of spermatogenesis leading to production of poor quality sperms, oligospermia, and infertility. It was observed that there was 50 per cent reduction in transformation of spermatogonia to spermatocytes with, reduction in fertility index but no change in serum testosterone levels by 242- 368 days of immunization against ovine FSH (Moudgal *et al.*, 1992). Similar results were also obtained upon immunization against the extracellular region of the FSH receptor peptide representing the extracellular domain (Moudgal *et al.*, 1997). Active immunization of adult bonnet monkey against phage expressed FSH-receptor peptides of the extracellular region also resulted in azoospermia and infertility (Rao *et al.*, 2004). Studies extended in human volunteers revealed that immunization against FSH resulted in a defective spermatogenesis leading to altered sperm quality resulting in a condition of oligospermia and teratospermia which cause infertility in men (Moudgal *et al.*, 1998). Based on these results it has been suggested that FSH could be employed as a candidate contraceptive vaccine in humans. Although based on studies using FSH-β knock out mice and FSH receptor knock out mice it has been suggested that FSH is not required for fertility, it is very clear from the above studies that FSH is required for quantitative restoration of spermatogenesis in immature rats and primates. The differences in the results obtained with immunization studies and knock out models perhaps could be due to differences in these species including the period between birth and puberty which is long in primates and is minimal in rodents, and the duration of spermatogenic cycle (Sriraman *et al.*, 2000). Although the approach of immunocontraception is appealing, one of the major problems encountered is the lack of uniform response in subjects due to genetic variability. However, with the development of synthetic adjuvants, better understanding of the mechanism of immune response, and orally active antigens, it should be possible to make this approach a practical one, although the need for use of alternate approaches until sperm count decreases still persists with this approach.

Interference with Sperm Maturation

Active Immunization Against Sperm Antigen

It is well established that testicular sperms do not have forward motility and are incapable of fertilizing the egg. Testicular sperms have to pass through the rete tubules and epididymis to undergo the process of maturation which involves acquisition of forward motility and fertilizing capacity. In view of this, intensive studies are in progress all over the world to understand the process of sperm maturation. These efforts though not completely successful, have provided important clues. It has been demonstrated that sperm membrane undergoes important changes during its transit through epididymis. Epididymal secretory proteins are associated with the sperm surface during maturation and participate in the development of fertilizing capacity of the sperms. Their role as mediators of sperm maturation has also been supported by the fact that addition of epididymal proteins promotes the maturation of immature sperms by induction of fertilizing capacity and exposure of mature spermatozoa to antibodies that are directed towards epididymal protein impede the fertilizing capacity of the spermatozoa. Thus induction of antibodies that interfere with the association of the protein to the sperm surface or mask these proteins on the sperm surface in the male tract and reduces the functional capacity of sperm could serve as potential contraceptive. One such protein in rats called DE 37 kDa is synthesized by the epithelium of the proximal epididymis originally localized in the dorsal region of sperm head and is involved in sperm egg fusion process. Active immunization with this protein in both male and female rats led to infertility (Cuasnicu *et al.*, 1990). Indirect immunofluorescence assays of spermatozoa indicated that the immune serum was capable of recognizing DE on the fresh spermatozoan. This suggests that the circulating antibodies have access in the reproductive tract, to interfere with the function of the immunized protein. Protein DE fulfils many of the requisites of candidate antigen for an immunocontraceptive and research focused towards identification of the human functional analogues revealed that acidic epididymal glycoprotein related product (ARP) was its functional human counterpart. Further studies using differential cDNA screening procedure to target gene products of post-testicular origin using testis as a negative control and after secondary screening with various tissues led to the cloning of six human epididymal proteins HE1-HE6 (Kirchhoff *et al.*, 1999). With exception of HE5 (CD52) all of these represent completely novel gene products whose expression is highly restricted to epididymis. HE1 accumulates in the caudal region of epididymis, functions as cholesterol transfer protein and plays a major role in capacitation of the sperm which involves change in cholesterol/phospholipid ratio. HE2 is of caput in origin and plays a role in sperm-egg fusion. HE4 is secreted by distal human epithelium and is a decapacitation factor. HE5 is found to be highly concentrated in the caudal epithelium and caudal luminal fluid and found to be GPI anchored protein. Finally HE6 belongs to the novel member of the seven transmembrane domain receptor family whose ligand is not known. The large N-terminal extracellular domain of HE6 resembles that of glycoprotein hormone receptor suggesting a glycoprotein originating from testicular secretion might be a probable ligand. These novel target molecules unique to epididymis may serve as a target for contraception either by developing specific drugs which interfere with its function or by developing neutralizing antibodies by active immunization.

Other sperm specific proteins that have been used include SP-10, Lactate Dehyrogenase (LDH-4), PH-20 and FA-I (Shaha 1999). FA-1 has been tested for its contraceptive efficacy wherein active immunization with these proteins in monkeys and rabbits led to reduction in fertility. Studies involving 15 baboons immunized with the immunodominant epitope of the human LDH-4 conjugated to diphtheria toxin revealed a 75 per cent reduction in fertility compared to control groups of animals (O'Hern *et al.*, 1995). The other antigens tested in experimental animals include RSA-1 human SP17, Tcle-1, βH20 and SP10 (Harrison *et al.*, 1996). However, none of the studies have reached a stage where it can be inferred that a particular antigen can be the candidate for use as a contraceptive vaccines.

Another sperm bound antigen is Riboflavin carrier protein (RCP). Riboflavin is a water-soluble B-complex vitamin, which is essential for the embryonic growth and is transported in the system by RCP. RCP is also involved in intratesticular vitamin transport to germ cells. *In vitro* studies demonstrated that the addition of RCP antibody had profound effects on sperm motility. Sperms were progressively immobilized with time which was further aggravated by addition of guinea pig serum as a source of complement, with 70 per cent decline in penetration capacity as assessed by zona-denuded hamster egg penetration test. Active immunization of male rats led to significant reduction in fertility and anti-RCP antibodies could be identified in the voided spermatozoa (Sridhar *et al.*, 1996). Bonnet monkeys actively immunized with RCP revealed that anti-RCP antibodies are indeed associated with acrosomal caps of the ejaculated spermatozoa and had impaired motility parameters. There was no change in the serum testosterone and mating studies revealed that in repeated mating with fertile females during days 10-16 of ovulatory cycles, no pregnancies were recorded (Adiga *et al.*, 1997).

However, very recently studies employing Eppin, a testis/epididymis specific protein, as a candidate vaccine have reported some success in bonnet monkeys. Seven out of nine males (78 per cent) employed in the study developed high titers to Eppin (O'Rand *et al.*, 2004). It was observed that and all of these high-titer monkeys were infertile. Five out of seven (71 per cent) anti- Eppin titer males recovered fertility when immunization was stopped. This study demonstrates that effective and reversible male immunocontraception is an attainable goal.

While the approach of interfering with the sperm maturation process is very attractive due to its non interference with testosterone, even this approach has its share of problems. Majority of the proteins which were found to be involved in sperm maturation are androgen regulated and to inhibit the production of these proteins involves interference with testosterone. Alternatively one can use the most promising protein as a candidate vaccine and to raise antibodies to interfere with action of these proteins. Also, it is accepted now that more than one antigen need to be used to produce a successful vaccine and further studies are required to delineate the type of immune response and its augmentation required to increase the contraceptive efficacy. However, as in the case of use of FSH as a contraceptive vaccine, the response to the antigen is highly variable due to genetic variability sufficient titers, which can reach the epididymal lumen, have to be maintained by periodic injections. The availability of sufficient quantity of antigen and the need for frequent injections and need to use alternative approaches during early periods of immunization are major hurdles in making this approach a reality.

Blockade of Estrogen Action

Estrogen is no longer considered as a female 'only' hormone. In the adult testis estrogen is synthesized by the Leydig cells and germ cells, and a relatively high concentration of estrogen is found in rete testis fluid. Estrogen receptors are present the testis, efferent ductules and epididymis of most species. Disruption of the α-receptor, either in the knockout (αERKO) or by treatment with a pure antiestrogen, results in dilution of cauda epididymal sperm, disruption of sperm morphology, inhibition of sodium transport and subsequent water reabsorption, increased secretion of Cl^-, and eventual decreased fertility. In addition to this primary function of regulation of luminal fluid and ion transport, estrogen is also responsible for maintaining a differentiated epithelial morphology. This strongly suggests that estrogen or its α-receptor is absolute necessity for fertility in the male (Hess, 2003).

The observation that estrogen has an important role in sperms maturation has suggested the possibility of interfering with estrogen action at epididymal level. Accordingly one can use estrogen receptor antagonists to block the action of estrogen. In fact, our own studies which involved chronic administration ICI 182870, a specific estrogen receptor antagonist, in adult male bonnet monkey, revealed a drastic decrease in sperm motility (Shayu *et al.*, 2005) and the following TMX administration, another E2 receptor antagonist, the male bonnet monkeys were infertile (Rao *et al.*, 1998). While the results are very encouraging one has to remember that estrogen has several sites of action in the male including brain and unless a specific ER blocker which is targeted only to epididymis, this approach is not going to be practical.

Intravasal Male Contraception

The observation that vasectomy is still the most simple, popular and reliable method of contraception for male, though reversibility in all cases is not assured. However, the simplicity of this approach prompted investigators to explore the possibility of developing a male contraceptive by interfering with passage of sperms in the vas deferens. The clips applied over the vas deferens were quite effective but led to fibrous adhesions and most of the time it became irreversible. The percutaneous injection of sclerosing chemicals such as quinacrine, ethanol, silver nitrate and formaldehye into the lumen of vas deferens have been explored. All compounds acted as sclerosing agents and produced inflammation and occlusion, but were invariably irreversible. Currently efforts are in progress to develop a reversible intravasal occlusion approach. The method consists of the injection of polymer, styrene maleic anhydride into the lumen of vas deferens in a vehicle of dimethylsulfoxide and Phase II clinical trials are in progress and the results obtained so for have demonstrated safety and efficacy of the injection of styrene maleic anhydride in humans (Guha, 1999).

Use of Plant Products

Tripchlorolide is a plant product which is isolated from a traditional Chinese medicine plant *Tripterygium wilfordii* as triptolide. It has been used to treat autoimmune disease and has a contraceptive effect which is reversible (Qian *et al.*, 1995). Tripchlorolide primarily affects the spermatid and maturation of sperm function in epididymis thus providing a potential male contraceptive without side effects on male sex hormones. Another plant, the seeds of which were reported to have antifertility effects in male is *Carica papaya*. It was reported that administration of partially purified extract of the seeds of *Carica papaya* to male rats resulted in a drastic decrease in cauda epididymal sperm motility without any effect on spermatogenesis (Lohiya, 2001).

Concluding Remarks

It is evident that out of the newer approaches which are being developed there is none which is ready for practical application. One of the major problems in developing male contraceptive is the fact that unlike in the female where a single oocyte is released during each menstrual cycle, million of sperms are produced in the male continuously. Consequently any successful method should be capable of inactivating the million of sperms. The controversy whether achieving azoospermia is a prerequisite for a successful male contraceptive approach still exists. In addition, the female reproductive system has more points of interference to control without interfering with libido.

It is paradoxical that the simplest and earliest method interfering with sperm production *i.e.* inhibiting testosterone production is not a practical one due to the problems associated with inhibition of libido. Although this problem has been overcome by supplementing with testosterone derivatives, this approach still suffers from several disadvantages such as frequency of administration, effect on prostate, lipid metabolism and more importantly, the need for alternate approaches which haunts all male contraceptive methods. The approach of interfering with sperm maturation at the epididymal level though very attractive, because of the fact that all sperms in the normal case must pass through epididymis, is still not any where near practical application. Our knowledge of exact process of maturation or factors involved is still very meagre. Moreover even in this approach one has to target an androgen independent step. Although use of FSH as a candidate vaccine is very promising it also suffers from problems of availability of antigen in sufficient quantity and frequency of injection for maintenance of sufficient titers.

It is very essential to realize that even after a candidate antigen is identified, produced in enough quantities by recombinant methods and methods are developed to sustain an effective uniform response in 100 per cent of the subjects (which is a daunting task); the main drawback in the application of immunocontraception is the need for periodic injections (which will be a major practical problem in country like ours with a huge rural population) and also need for alternate contraceptive approaches required until sufficient titers are obtained in the subjects. This is all the more so in developing an immuno-contraceptives for use in males, since theoretically, even one sperm is adequate for fertilization and an absolute need for 100 per cent efficacy is required. However, one should realize that immunocontrception is an important tool in identifying the potential target and the candidate antigens. The approach has the power to provide insights into fundamental, immunological and physiological structure and mechanism that could enrich development of the applications. Following the identification of the candidate molecule using the immunological approach, it is the modern approach of understanding the molecular basis of interaction of the hormone with its receptor, which will provide the basis for developing non-protein/non-peptide based orally active mimetics to interfere with the action of hormone. Thus, the knowledge gained out of the immunological approach will be indispensable in developing future contraceptives, although immunocontraception itself may not have a place of its own in fertility control programme at the moment.

It is in this connection the research in male reproduction employing the modern technology has provided important clues. Thus, it has been demonstrated that one of the isozyme in sperm which is very specifically expressed only during spermatogenesis is Glyceraldehyde 3-phosphate dehydrogenase-S (GAPDS). It is tightly bound to the fibrous sheath, a cytoskeletal structure that extends most of the length of the sperm flagellum. Disruption of Gapds expression by gene targeting revealed that the Gapds(-/-) males were infertile and had profound defects in sperm motility, exhibiting sluggish movement without forward progression (Miki *et al.*, 2004). Although mitochondrial oxygen consumption was unchanged, sperm from Gapds(-/-) mice had ATP levels that were only 10.4 per cent of those in sperm from wild-type mice. These results imply that most of the energy required for sperm motility is generated by glycolysis rather than oxidative phosphorylation. Furthermore, the critical role of glycolysis in sperm and its dependence on this sperm-specific enzyme suggest that GAPDS is a potential contraceptive target, and that mutations or environmental agents that disrupt its activity could lead to male infertility. Similarly, the fact that the proteins of vas deferens also have an important role in sperm function was revealed by knocking out P2X1 receptors. P2X1 receptors for ATP are ligand gated ion channels are present on many excitable cells including vas deferens smooth muscle cells. A substantial component of the contractile response of the vas deferens to sympathetic nerve stimulation which propels into the ejaculate is mediated through P2X1 receptors. Male fertility is reduced by ~ 90 per cent in mice with a targeted deletion of the P2X receptor gene. There was neither sperm dysfunction nor effect on testosterone levels (Mulryan *et al.*, 2000).

Thus, these newer leads may permit as to develop simpler orally active compounds which can be used as male contraceptive pills without interfering with testosterone. However, until these become available, one has to rely of the age old approaches of barrier like condoms and vasectomy.

Acknowledgements

Financial assistance from Mellon Foundation, CONRAD U.S.A., IISc., Bangalore, CSIR (Emeritus Scientist Fellowship to AJ Rao), CSIR, DST, ICMR and DBT Government of India, is gratefully acknowledged.

References

Adiga, P.R., Subramanian, S., Rao, J., Kumar, M. (1997). Prospects of riboflavin carrier protein (RCP) as an antifertility vaccine in male and female mammals. Hum. Reprod. Update, 3, 325-334.

Alexander, N.J., Clarkson, T.B. (1978). Vasectomy increases the severity of diet-induced atherosclerosis in *Macaca fascicularis*. Science, 201, 538-541.

Ansbacher, R. (1981). Humoral sperm antibodies: a 10-year follow-up of vas-ligated men. Fertil. Steril., 36, 222-224.

Behre, H.M., Kliesch, S., Lemcke, B., Nieschlag, E. (1995). Suppression of spermatogenesis to azoospermia by combined administration of GnRH antagonist and 19-nortestosterone alone in men. Proc. 77th Ann. Meeting of the Endocrine Society, Washington D. C., USA, pp. OR29-26

Behre, H.M., Nieschlag, E. (1999). GnRH agonist/antagonist-androgen combinations for male contraception. *In*: Rajalakshmi M, and Griffin PD. (Eds.), *Male Contraception: Present and Future*, New Age International (P) Ltd, New Delhi, India, pp. 235-249.

Belker, A.M., Thomas, A.J., Jr., Fuchs, E.F., Konnak, J.W., Sharlip, I.D. (1991). Results of 1,469 microsurgical vasectomy reversals by the Vasovasostomy Study Group. J. Urol., 145, 505-511.

Bronson, R. (1994). Immunologic Infertility. *In*: Whitehead ED, and Nagler HM. (Eds.), *Management of Impotence and Infertility*, Harper and Row, Philadelphia PA, USA, pp. 390-408

Bullock, J.Y., Gilmore, L.L., Wilson, J.D. (1977). Autoantibodies following vasectomy. J. Urol., 118, 604-606.

Cuasnicu, P.S., Conesa, D., Rochwerger, L. (1990). Potential contraceptive use of and epididymal protein that participates in fertilization. *In*: Alexander NJea. (Ed.), X Gamete Interaction. Prospects for Immuno-contraception., Wiley- Liss, New York, pp. 143-153

Devi, P.K., Joshi, U.M., Moodbidri, S.B., Naik, V.K., Susheela, P.S., Sheth, A.R. (1977). Long term effect of vasectomy on pituitary-gonadal axis. Indian J. Med. Res., 66, 591-596.

Farley, T.M., Meirik, O., Mehta, S., Waites, G.M. (1993). The safety of vasectomy: recent concerns. Bull World Health Organ, 71, 413-419.

Giovannucci, E., Ascherio, A., Rimm, E.B., Colditz, G.A., Stampfer, M.J., Willett, W.C. (1993a). A prospective cohort study of vasectomy and prostate cancer in US men. Jama, 269, 873-877.

Giovannucci, E., Tosteson, T.D., Speizer, F.E., Ascherio, A., Vessey, M.P., Colditz, G.A. (1993b). A retrospective cohort study of vasectomy and prostate cancer in US men. Jama, 269, 878-882.

Goldacre, M.J., Clarke, J.A., Heasman, M.A., Vessey, M.P. (1978). Follow-up of vasectomy using medical record linkage. Am. J. Epidemiol., 108, 176-180.

Guha, S.K. (1999). Current status of development and clinical testing of male intravasal contraceptives. *In*: Rajalakshmi M, and Griffin PD. (Eds.), *Male Contraception: Present and Future*, New Age International (P) Ltd, New Delhi, India, pp. 265- 274

Harrison, P.F., Rosenfield, A. (1996). Contraceptive technology and the State of the Science: New Horizons. *In*: Harrison PF, and Rosenfield A. (Eds.), *Contraceptive Research and Development*, National Academy Press, Wasgington, D.C., pp. 125-165

Hess R.A. (2003). Estrogen in the adult male reproductive tract: a review. Reprod. Biol. Endocrinol., 1, 52.

Kirchhoff, C., Osterhoff, C., Kasahara, M., Ivell, R. (1999). Molecular Biology of Sperm maturation-new strategies for male contraception? *In*: Rajalakshmi M, and Griffin PD. (Eds.), *Male Contraception: Present and Future*, New Age International (P) Ltd, New Delhi, India, pp. 103-123

Ladd, A., Prabhu, G., Tsong, Y.Y., Probst, T., Chung, W., Thau, R.B. (1988). Active immunization against gonadotropin-releasing hormone combined with androgen supplementation is a promising antifertility vaccine for males. Am. J. Reprod. Immunol. Microbiol., 17, 121-127.

Leonard, J.M., Paulsen, C.A. (1978). Contrceptive development studies for males: Oral and parenteral steroid hormone administration. *In*: Pantanelli DJ. (Ed.), *Hormonal Control and Male Fertility*, Department of Health, Education and Welfare, National Institute of Health, Bethseda, p. 223

Linnet, L. (1983). Clinical immunology of vasectomy and vasovasostomy. Urology, 22, 101-114.

Lohiya, N.K. (2001). Prospects of Developing a Plant Based Male Contraceptive Pill. In: Chowdhury, S.R., Gupta, C.M., Kamboj, V.P. (Eds.), Current Status in Fertility Regulation: Indigenous and Modern Approaches., Army Printing Press, Lucknow, India, pp. 99-120

Massey, F.J., Jr., Bernstein, G.S., O'Fallon, W.M., Schuman, L.M., Coulson, A.H., Crozier, R., Mandel, J.S., Benjamin, R.B., Berendes, H.W., Chang, P.C., *et al.* (1984). Vasectomy and health. Results from a large cohort study. Jama, 252, 1023-1029.

McGullah, E.P., McGurl, F.J. (1939). Further observations on the clinical use of testosterone propionate. J. Urol., 42, 1265-1273.

Meriggiola, M.C., Bremner, W.J., Paulsen, C.A., Valdiserri, A., Incorvaia, L., Motta, R., Pavani, A., Capelli, M., Flamigni, C. (1996). A combined regimen of cyproterone acetate and testosterone enanthate as a potentially highly effective male contraceptive. J Clin Endocrinol Metab., 81, 3018-3023.

Miki, K., Qu, W., Goulding, E.H., Willis, W.D., Bunch, D.O., Strader, L.F., Perreault, S.D., Eddy, E.M., O'Brien, D.A. (2004). Glyceraldehyde 3-phosphate dehydrogenase-S, a sperm-specific glycolytic enzyme, is required for sperm motility and male fertility. Proc Natl Acad Sci U SA, 101, 16501-16506.

Moudgal, N.R., Ravindranath, N., Murthy, G.S., Dighe, R.R., Aravindan, G.R., Martin, F. (1992). Long-term contraceptive efficacy of vaccine of ovine follicle-stimulating hormone in male bonnet monkeys (*Macaca radiata*). J. Reprod. Fertil., 96, 91-102.

Moudgal, N.R., Sairam, M.R. (1998). Is there a true requirement for follicle stimulating hormone in promoting spermatogenesis and fertility in primates? Hum. Reprod., 13, 916-919.

Moudgal, N.R., Sairam, M.R., Krishnamurthy, H.N., Sridhar, S., Krishnamurthy, H., Khan, H. (1997). Immunization of male bonnet monkeys (*Macaca radiata*) with a recombinant FSH receptor preparation affects testicular function and fertility. Endocrinology, 138, 3065-3068.

Mulryan, K., Gitterman, D.P., Lewis, C.J., Vial, C., Leckie, B.J., Cobb, A.L., Brown, J.E., Conley, E.C., Buell, G., Pritchard, C.A., Evans, R.J. (2000). Reduced vas deferens contraction and male infertility in mice lacking P2X1 receptors. Nature, 403, 86-89.

Nienhuis, H., Goldacre, M., Seagroatt, V., Gill, L., Vessey, M. (1992). Incidence of disease after vasectomy: a record linkage retrospective cohort study. Bmj, 304, 743-746.

O'Hern, P.A., Bambra, C.S., Isahakia, M., Goldberg, E. (1995). Reversible contraception in female baboons immunized with a synthetic epitope of sperm-specific lactate dehydrogenase. Biol. Reprod., 52, 331-339.

O'Rand, M. G., Widgren, E.E., Sivashanmugam, P., Richardson, R.T., Hall, S.H., French, F.S., VandeVoort, C.A., Ramachandra, S.G., Ramesh, V., Jagannadha Rao, A. (2004). Reversible immunocontraception in male monkeys immunized with eppin. Science, 306, 1189-1190.

Petitti, D.B., Klein, R., Kipp, H., Kahn, W., Siegelaub, A.B., Friedman, G.D. (1982). A survey of personal habits, symptoms of illness, and histories of disease in man with and without vasectomies. Am. J. Public Health, 72, 476-480.

Qian, S.Z., Ye, X., Wei, Z.J. (1995). Recent progress in research of Tripterygium: A male antifertility plant. Contraception, 51, 121- 129.

Rajalakshmi, M., Ramakrishnan, P.R. (1989). Pharmacokinetics and pharmacodynamics of a new long-acting androgen ester: maintenance of physiological androgen levels for 4 months after a single injection. Contraception, 40, 399-412.

Ramachandra, S.G., Ramesh, V., Krishnamurthy, H.N., Kumar, N., Sundaram, K., Hardy, M.P., Rao, A.J. (2002). Effect of chronic administration of 7alpha-methyl-19-nortestosterone on serum testosterone, number of spermatozoa and fertility in adult male bonnet monkeys (*Macaca radiata*). Reproduction, 124, 301-309.

Rao, A.J. (1998). Presented in Gordon Conference. In: Plymouth, NH, USA,

Rao, A.J., Chakraborti, R., Kotagi, S.G., Ravindranath, N. (1990). Effect of constant infusion of gonadotropin releasing hormone (GnRH) agonist buserelin and antagonist CDB 2085 A using osmotic minipumps on testicular function in adult male bonnet monkey (*Macaca radiata*). Andrologia, 22, 567-573.

Rao, A.J., Chakraborti, R., Kotagi, S.G., Ravindranath, N., Moudgal, N.R. (1985). Effect of LHRH agonists and antagonists in male and female bonnet monekys (*Macaca radiata*). J. Steroid Biochem., 23, 807-809.

Rao, A.J., Ramachandra, S.G., Ramesh, V., Couture, L., Abdennebi, L., Salesse, R., Remy, J.J. (2004). Induction of infertility in adult male bonnet monkeys by immunization with phage-expressed peptides of the extracellular domain of FSH receptor. Reprod. Biomed. Online, 8, 385-391.

Rao, A.J., Ramachandra, S.G., Ramesh, V., Krishnamurthy, H.N., Jayaraman, S., Gopalakrishnan, K., Juneja, H.S. (1998). Effect of chronic administration of Tamoxifen on fertility in male bonnet monkeys (*Macaca radiata*). Andrologia, 30, 129-132.

Ravindranath, N., Ramesh, V., Krishnamurthy, H.N., Rao, A.J., Moudgal, R.N. (1992). Chronic suppression of testicular function by constant infusion of gonadotropin-releasing hormone agonist and testosterone supplementation in the bonnet monkey (*Macaca radiata*). Fertil. Steril., 57, 671-676.

Ravindranath, N., Rao, A.J., Prasad, M.R.N., Moudgal, N.R. (1995). Effect of chronic treatment of GnRH agonist (Buserelin) on testicular function of the adult male bonnet monkey (*Macaca radiata*). *In:* Singh V, and Muralidhar K. (Eds.), *Neuroendocrine Regulation in Fertility Control,* Vol. 1. NEHU Publications, Shillong, pp. 129-133

Reddy, P.R., Rao, J.M. (1972). Reversible antifertility action of testosterone propionate in human males. Contraception, 5, 295-301.

Shaha, C. (1999). The sperm Vaccine- an utopian goal? *In:* Rajalakshmi M, and Griffin PD. (Eds.), *Male Contraception: Present and Future,* New Age International (P) Ltd, New Delhi, India, pp. 319-328.

Shayu, D., Kesava, C.C., Soundarajan, R., Rao, A.J. (2005). Effects of ICI 182780 on estrogen receptor expression, fluid absorption and sperm motility in the epididymis of the bonnet monkey. Reprod. Biol. Endocrinol., 3, 10.

Sidney, S., Quesenberry, C.P., Jr., Sadler, M.C., Guess, H.A., Lydick, *E.G.*, Cattolica, E.V. (1991). Vasectomy and the risk of prostate cancer in a cohort of multiphasic health-checkup examinees: second report. Cancer Causes Control, 2, 113-116.

Sridhar, L., Kumar, M., Adiga, P.R. (1996). Active immunization with riboflavin carrier protein suppress fertility in male rodents and monkeys. *In*: *Discussion Meeting on Perspective in Biochemistry and Molecular Biology*, Indian Institute of Science Bangalore, India, p. 334

Sriraman, V., Rao, A.J. (2000). Role of FSH in spermatogenesis. In: *ISSRF News Letter*, pp. 15-17

Sundaram, K., Kumar, N., Bardin, C.W. (1993). 7 alpha-methyl-nortestosterone (MENT): the optimal androgen for male contraception. Ann. Med., 25, 199-205.

Talwar, G.P., Diwan, M., Dawar, H., Frick, J., Sharma, S.K., Wadhwa, S.N. (1999). Counter GnRH vaccine. *In*: Rajalakshmi M, and Griffin PD. (Eds.), *Male Contraception: Present and Future*, New Age International (P) Ltd, New Delhi, India, pp. 309-318

Tang, G.H., Zhong, Y.H., Ma, Y.M., Luo, L., Cui, K., Luo, J., Zhang, G.H., An, I.M., Luo, D.C., Qiu, S.H., *et al.* (1988). Vasectomy and health: cardiovascular and other diseases following vasectomy in Sichuan province, People's Republic of China. Int. J. Epidemiol., 17, 608-617.

Turner, L., Conway, A.J., Jimenez, M., Liu, P.Y., Forbes, E., McLachlan, R.I., Handelsman, D.J. (2003). Contraceptive efficacy of a depot progestin and androgen combination in men. J. Clin. Endocrinol. Metab., 88, 4659-4667.

von Eckardstein, S., Noe, G., Brache, V., Nieschlag, E., Croxatto, H., Alvarez, F., Moo-Young, A., Sivin, I., Kumar, N., Small, M., Sundaram, K. (2003). A clinical trial of 7 alpha-methyl-19-nortestosterone implants for possible use as a long-acting contraceptive for men. J. Clin. Endocrinol. Metab., 88, 5232-5239.

Walker, A.M., Jick, H., Hunter, J.R., Danford, A., Rothman, K.J. (1981). Hospitalization rates in vasectomized men. Jama, 245, 2315-2317.

Weinbauer, G.F., Surmann, F.J., Nieschlag, E. (1987). Suppression of spermatogenesis in a nonhuman primate (*Macaca fascicularis*) by concomitant gonadotropin-releasing hormone antagonist and testosterone treatment. Acta Endocrinol. (Copenh), 114, 138-146.

2

Menopause: Hormonal Imbalance and Herbal Management of Clinical Symptoms

*Priya Kapur and B.M.J. Pereira**

Reproductive Biology Laboratory, Department of Biotechnology, Indian Institute of Technology Roorkee–247 667, Uttaranchal, India

ABSTRACT

Estrogens are among the most ubiquitous hormones in the female body. They affect almost every vital organ by interacting with tissues by different mechanisms. The importance of estrogens cannot be viewed in isolation since the level and action of hormones is dependent on one another. Thus when the estrogen levels in the body become deficient the normal balance gets disturbed and several short as well as long-term effects are manifested in the body. Menopause, the irreversible cessation of the menstrual cycle is one such clinical condition in which primary follicles in the ovary starts depleting and hence secretes low amounts of estrogen. This decrease causes variety of unpleasant physical and emotional symptoms like hot flashes, mood swings, memory loss, anxiety, depression, insomnia, cardiac disorders and osteoporosis. There are innumerable reports about the use of HRT (Hormone replacement therapy) for ameliorating the adverse changes associated with menopause but the risks like vaginal discharges, nausea, weight gain and especially the endometrial, breast and ovarian cancers arising after this regime

**Corresponding Author*: E-mail: benmjfbs@iitr.ernet.in; Telephone: +91-1332-285790; Fax: +91-1332-273560

strengthen the desire to focus on other therapies. Estrogenic herbs, acupuncture, mineral and vitamin supplements, exercise, homeopathy are some of the suggested alternatives for effective management of these clinical conditions. All of them help to relieve symptoms but the herbal treatment is the most favorite choice. They appear to boost the good effects while minimizing the bad ones to the maximum. The estrogenic effects of the phytochemicals present in the herbs used for treating menopausal symptoms are beginning to be known and the current state of our knowledge will be reviewed.

The chapter will begin with description on the physiology of menopause and highlight various clinical symptoms often encountered. An attempt would then be made to provide an overview of all the practice especially the ones concerning hormone replacement therapies that are practiced to deal with menopause. Side by side the merits and demerits of the existing modes of treatment will be addressed. The final focus of this chapter will be to explain how herbs together with their reactive components tone up and regulate the endocrine system of the body and how they could be effectively used to improve the quality of life in menopausal women.

Keywords: Menopause, Clinical symptoms, Phytoestrogens, Herbal management.

Introduction

Reproductive system of the female consists of the ovaries (the organs that produce the eggs, or ovum), fallopian tubes, uterus, vagina, accessory glands, and the external genital organs. The ovaries contain follicles, in which the ovum develops. Each month during ovulation, the ovaries not only release the egg into the fallopian tubes but also produce the hormone estrogen which prompts the uterus to build up a lining that will allow the egg if it becomes fertilized, to implant and grow. But if fertilization fails to occur within the specified time, the hormone levels drop, which leads to the washing away of the egg along with the shedding of uterine lining, a process termed menstruation. This cyclic event that comprises of three major phases: the follicular phase, the luteal phase, and the menstruation phase are shown in Figure 2.1. During the follicular phase, which lasts from days 1 to 13, the hypothalamus secretes follicle hormone releasing factor (FHRH) and leutinizing hormone releasing factor (LHRH), which induce the anterior pituitary to secrete follicle stimulating hormone (FSH) and leutinizing hormone (LH). FSH and LH stimulate the development of ovum, and induce the ovaries to release estradiol, a form of estrogen. This presence of estradiol acts on the hypothalamus and pituitary to cause the release of more estradiol. When the ovary begins to release progesterone in the luteal phase, the anterior pituitary halts secretion of LH and FSH. When estrogen and progesterone levels fall on day 28, menstruation begins. The cycle and the interrelationship between hypothalamus, anterior pituitary and the ovary in regulating the production of estrogen is shown in the Figure 2.1.

Estrogen that chiefly regulates the menstrual cycle is not a single hormone but refers to a group of chemically similar substances that belong to gonadocorticoids, one of the classes of steroid hormones (Gruber, 2002). They are primarily C-18 compounds synthesized in the ovary and adrenal cortex from cholesterol and to a lesser extent, by peripheral conversion in adipose tissues from androstenedione. The three forms of endogenous estrogens are β-estradiol, estrone, and estriol (Figure 2:2). β-Estradiol (E_2), the predominant estrogen secreted by the ovaries, is the most potent among all due to its high affinity for the estrogen receptors. Estrone (E_1) secreted by peripheral tissues is 12 times less potent than β-estradiol and estriol (E_3), which is the by-product of β-estradiol and estrone metabolism is 80 times less potent than β-estradiol.

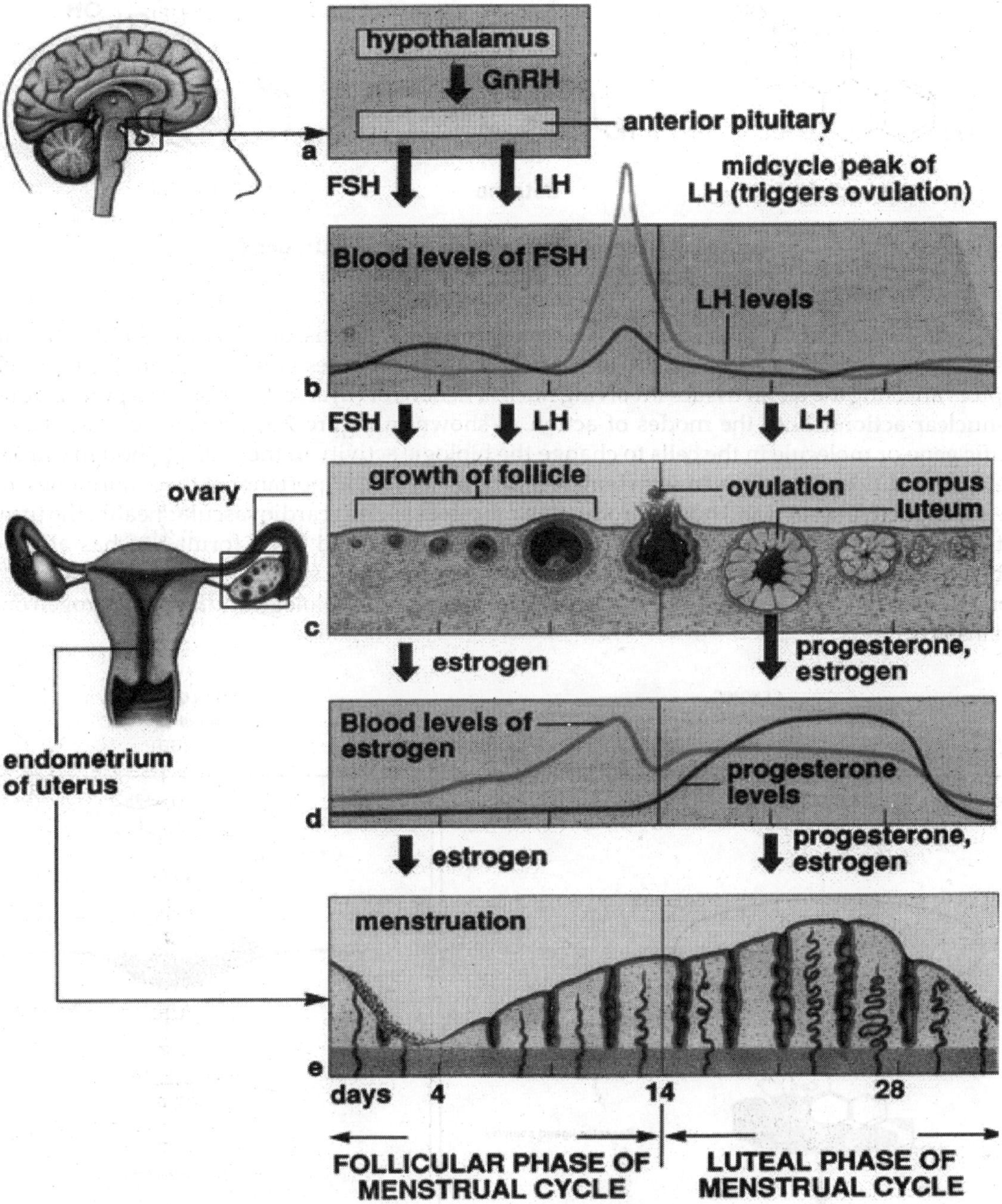

Changing Hormone Level During the Menstrual Cycle

Figure 2.1: Menstrual Cycle
(www. sunnydutchess.edu/scala/bio102/)

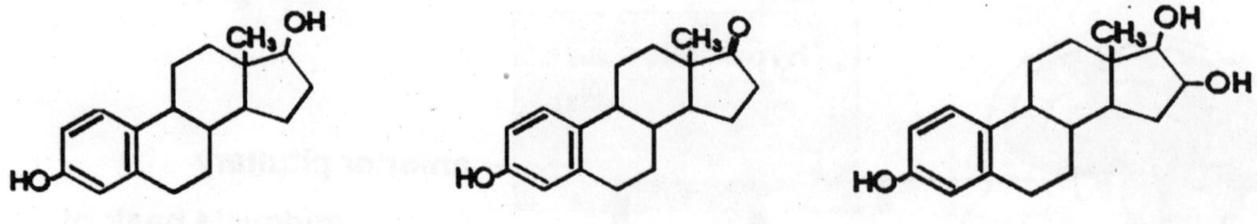

Estradiol Estrone Estriol

Figure 2.2: Three Forms of Endogenous Estrogens

Estrogens have wide spread biological effects on mammalian tissues that are essential for many physiological processes to occur. Estrogens affect the target tissues either through (A) *i.e.* nuclear receptors affecting the target tissues involving nuclear actions or (B) *i.e.* cell surface receptors involving non-nuclear actions. Both the modes of action as shown in Figure 2.3, activate or deactivate the specific gene or molecule in the cells to change the biologic activity in the cells. Although estrogens have been traditionally connected with female reproduction, the importance of these hormones in the male reproductive system and non-reproductive processes such as cardiovascular health, the urinary tract, skin, hair, mucous membranes, pelvic muscles, brain and bone formation has also been established. In addition, the gastrointestinal tract and the immune system are now considered to be among the "estrogen targets". Figure 2.4 clearly shows the various biological effects of estrogen on the human body.

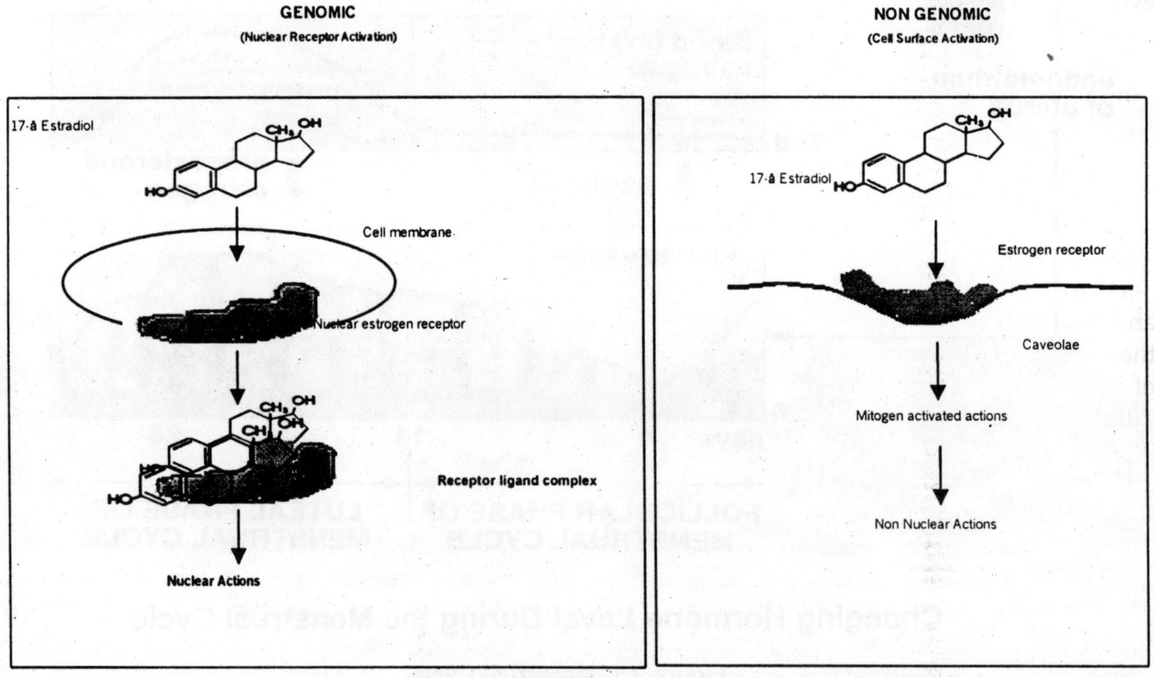

Figure 2.3: Mode of Activation by Estrogen

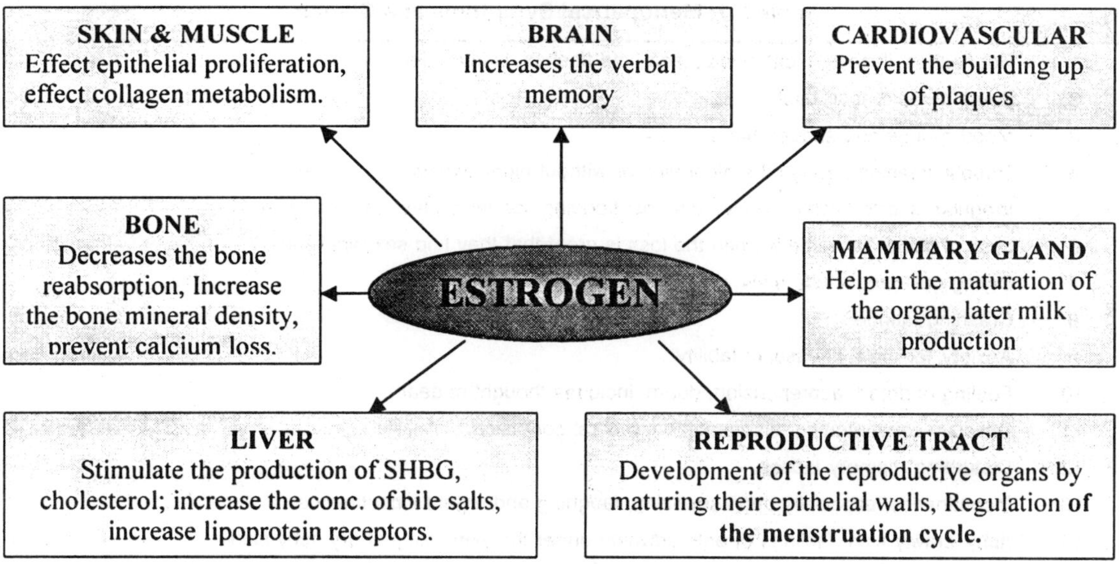

Figure 2.4: Biological Functions of Estrogen

The complicated interrelationships in the body for production, action and effects of estrogen make it difficult to view the biological importance of estrogens in isolation. The integration is such that even a little deficiency in the estrogen levels profusely disturbs the normal balance and results in several short as well as long term effects. Menopause, the irreversible cessation of the menstrual cycle is one such clinical condition. It is a natural part of aging, and occurs when the ovaries stop making estrogens. Actually the primary ovarian follicles start depleting and the remaining follicles become less responsive to gonadotrophins (FSH and LH). Since the follicles produce much of the estrogen in the body, decline in their number therefore results in the low secretion of estrogen (generally one tenth the normal). On the other hand, FSH to LH ratio tends to increase significantly, a positive indication of the menopausal status. There are other conditions where menopause is induced in women by surgical intervention (hysterectomy). In fact, women that have a hysterectomy literally go to sleep with estrogen being produced in their ovaries only to wake up with nothing. The effects are particularly dramatic and immediate when the hysterectomies involve oopherectomy (removal of one or both ovaries) since these organs are the primary source of estrogens. It must be emphasized that although the symptoms of surgical menopause are the same as natural menopause the rate at which they appear is much different.

Clinical Symptoms Associated with Menopause

The disturbance in the regulating hormone levels during menopause causes a variety of unpleasant physiological and psychological symptoms which are consolidated in Table 2.1.The effects vary among individuals and therefore it is not necessary that all the listed symptoms are manifested in every case of menopause. Some notice little physical changes in their bodies and mood while others complain about the change as bothersome and disruptive. Dantas (1999) showed that depending upon the influence, the associated symptoms could be classified into short term and long-term symptoms.

Table 2.1: Menopausal Symptoms at a Glance

1	Hot flashes–flushes, night sweats and cold flashes, clammy feeling.
2	Bouts of rapid heart beat.
3	Mood swings and sudden tears
4	Trouble sleeping through the night with or without night sweats.
5	Irregular, shorter, lighter, heavy periods; flooding; longer cycles; shorter cycles.
6	Loss of libido- for some women the loss is great that they find sex repulsive.
7	Drying out–can include eyes, vagina, mouth, joints and skin.
8	Crashing fatigue.
9	Anxiety, feeling ill at ease, irritability
10	Feeling of dread, apprehension, doom, includes thought of death.
11	Difficulty concentrating, disorientation, mental confusion.
12	Disturbing memory lapses.
13	Incontinence, especially upon sneezing, laughing and urgent incontinence.
14	Itchy, crawly skin- feelings of ants crawling under the skin, not just dry itchy skin.
15	Aching, sore joints, muscles and tendons may include problems as tunnel syndrome.
16	Increased tension in muscles.
17	Breast tenderness.
18	Headache change: increase or decrease.
19	Gastrointestinal distress, indigestion, flatulence, gas pain, nausea, bloating.
20	Depression- the inability to cope is overwhelming. Feeling of loss of self.
21	Increase in allergies, changes in body odor
22	Weight gain–often around the waist and thighs resulting in disappearing waistline
23	Hair loss or thinning, head, pubic, or whole body increase in facial hair.
24	Dizziness, light-headedness, episodes of loss of balance.
25	Electric shock sensation-the feeling of rubber band snapping the layer of tissue between skin and muscle. It is the precursor to a hot flash.
26	Tingling in the extremities. Alternations in the flexibility of blood vessels or a depletion of K or Ca.
27	Gum problems, increased bleeding.
28	Burning tongue, burning roof of mouth, bad taste in mouth, dry mouth, change in breath odor
29	Osteoporosis after several years.
30	Change in fingernails: softer, crack or break easier.
31	Tinnitus; ringing in ears, bells, " whooshing", buzzing etc.

Short-term effects comprise of hot flashes, physical changes, and psychological symptoms. Hot flashes, the feeling of sudden sensation of flushing and extreme warmth often followed by profuse sweating and sometimes shaking or tremor is the hallmark symptom of menopause. They are caused due to the hypothalamic response to the declining ovarian estrogen production. The declining estrogen state induces hypophysiotropic neurons in the arcuate nucleus of the hypothalamus to release gonadotropin-releasing hormone (GnRH) in a pulsatile fashion, which in turn stimulates the release of luteinizing hormone (LH). Extremely high pulses of LH occur during the period of declining estrogen production. The LH has vasodilatory effects, which leads to flushing, leading to destabilization of the body temperature. They cause serious discomfort and sleepless nights for women. Physical changes about which women generally complain are the reproductive and urinary tract changes. Reproductive tract mainly illustrates the shrinkage of vaginal, cervical, uterine and ovarian walls. These walls become thinner, dryer, less elastic (due to reduction in vaginal and cervical secretions) and more vulnerable to infection. Changes in urinary tract often reports about the problem of incontinence *i.e.* the involuntary loss of urine. Thinning of the body hairs, wrinkling and loss of skin tone, slowing of metabolic rate and weight gain are also some other minor changes that are included in short term effects. Psychological response during the menopausal period ranges from little or none to profound alteration of personality. Symptoms include mood swings, lack of concentration, difficulty with memory, tension, anxiety, over reaction to minor upsets, depression, and insomnia.

In the long term, some women experience problems linked to the low levels of estrogen. These may include osteoporosis and increased risk for cardiovascular disorders. Osteoporosis, which means porosity of bones, is a condition in which bones become thin, fragile and highly prone to fracture. Various genetic, medical and life style factors like family history, low calcium diet, lack of exercise, smoking, alcohol use are associated with this condition but the biggest culprit in the process of bone loss is estrogen deficiency like we observe in menopause. Estrogen plays important role in the manufacture of new bone tissue and when its levels drop at the onset and during menopause enough calcium is not deposited and the bone loss quickens.

Cardiovascular disease (CVD), the disorders of the heart and circulatory systems include thickening of arteries, high blood pressure, angina and stroke. For reasons unknown, estrogen helps to protect women against these disorders. But during and after menopause this protection disappears due to estrogen deficiency posing a greater risk for CVD's. Low-density cholesterol (LDL) increase and the high-density cholesterol (HDL) decrease as a direct result of estrogen deficiency, can also lead to stroke, heart attack and even death.

Conventional Therapies for Menopause

To ameliorate the adverse changes associated with menopause suffering women are generally provided with the deficient hormone that is estrogen which their ovaries stop making around the time of menopause. One such hormone therapy that was prescribed by health care providers for coping with menopause was called as estrogen replacement therapy (ERT). However, administration of estrogen raised the risk of endometrial cancer (the cancer of the lining of the uterus) and therefore at first only those women who had undergone hysterectomy were able to take ERT safely. In the later stages for the women who had an intact uterus, the therapy was modified by adding progesterone along with estrogen in the medication to lower the risk of cancer. This was called as HRT or Hormone replacement therapy (Staren, 2004). As a part of hormonal regime artificial/synthetic estrogens are usually administered in the form of oral pills, skin patches, intramuscular injections or vaginal creams. Conjugated equine estrogens (CEE) are the most commonly prescribed formulations *e.g.* Permarin. Other combination estrogen products available are esterified estrogens like Menest and Estratab. They are derived from the sulfate esters of one or more of the estrogens in CEE. 17-β estradiol and ethinyl-estradiols are also included in the list of prescriptions.

Hormone replacement therapy relieves both the short term and the long-term symptoms associated with menopause. The benefits provided by this therapy (HRT) are the mental tonic effect, improved sleep, improved short term memory, prevention of heart attack, elimination of hot flashes, slowering of the overall body deterioration, protection against colon cancer, increased longevity, protection against cardiovascular disorders by raising the levels of high density cholesterol and protection against osteoporosis by reducing calcium loss.

List of the menopausal symptoms being alleviated by hormone replacement therapy (HRT) is so large that anybody can undoubtedly say that it surely enhances the quality of women's life but the irony lies in the fact that only 10-15 per cent of women worldwide are current users of HRT. They are reluctant to use hormone replacement therapy because of the undesirable side effects like irregular vaginal discharges, bleeding, headaches, nausea, fluid retention, weight gain and especially the risk of endometrial, breast, and the ovarian cancers associated with it. Women's increased realization that hormone replacement therapy is not safe strengthens the desire of the researchers worldwide to collaborate and explore the possibilities of alternate therapies.

Alternative Therapies for Menopause

Nowadays there are tremendous ongoing global efforts to develop such therapies that will yield the benefits of conventional hormone replacement therapy but with minimal associated inconveniences or risks. The developed alternatives to HRT are desired mainly to focus on strengthening the female body, encouraging it to balance, regulate and normalize itself naturally using non-drug therapies. Some of the suggested alternatives for the effective management of stated clinical condition are dietary, nutritional and life style changes supplemented by therapies such as acupuncture, estrogenic/ progestenic herbs, homeopathy, mineral and vitamin supplements, and exercise (Kessel, 1998). All of them have proved their worth in improving the quality of women's life but the herbal alternatives is the most appealing.

Herbal Alternatives for Managing Menopause

Herbal alternatives are accepted worldwide as Mother Nature's gift because they appear to boost the good effects of estrogen while minimizing the bad ones to the maximum as compared to the other alternatives. Both estrogenic and progesteronic herbs (Zava, 1998) are found beneficial in treating menopausal patients. *Cimicifuga recemosa, Angelica Sinensis, Mentha pulegium, Alchemilla vulgaris, Trillium erectum, Artemisia vulgaris, Glycine max, Trifolium pratense, Achillea millifolium, and Asparagus officinalis* (Figure 2.5) are few of the herbs that have been exploited commercially.

Phytoestrogens: The New Frontiers

The common reactive components present in all these herbs that contribute to the alleviation of the symptoms are collectively known as phytoestrogens. Phytoestrogens is a general term used to define a part of the wider class of polyphenolic compounds found in all plants (Duncan *et al.*, 2003; Knight and Eden, 1996). They are non-steroidal and are either of plant origin or derived from the *in-vivo* metabolism of precursors present in several plants eaten by human beings. They have a chemical structure very similar to the mammalian estrogen, estradiol, and therefore it comes as no surprise when they possess estrogenic properties. However, their estrogenic activity is generally 1/500 to 1/1000 of the activity of human estradiol. As far as their metabolism is concerned these compounds after entering the body undergo hydrolysis yielding aglycones. These aglycones are either excreted in the urine or bile or absorbed from the gut. On absorption they undergo conjugation in the liver with glucuronic acids or sulphate and follow the pathways of further metabolism.

Types of Phytoestrogens

Phytoestrogens are found in the seeds, stems, roots or flowers of plants, serving as natural fungicides and forms as part of the plant's defense mechanism against microorganisms. The amount

Figure 2.5: Medicinal Herbs (www.google-images.com)

A: *Cimicifuga racemosa*; B: *Angelica sinensis*; C: *Mentha pulegium*; D: *Alchemilla vulgaris*; E: *Trillium erectum*; F: *Artemisia vulgaris*; G: *Glycine max*; H: *Trifolium pratense*; I: *Achillea millefolium*; J: *Asparagus officinalis.*

of phytoestrogens in plants varies with genetic differences between plant species or varieties, influence of environmental factors during growth, and differences in harvesting or subsequent processing. It has been found that they are the actual components that make the herbs useful as alternative for the hormone replacement therapy. Phytoestrogens are broadly classified into Isoflavones, Lignans, Stilbenes and Coumestanes. Isoflavones and Lignans are the most extensively studied among all the known classes.

Isoflavones

Genistein, Daidzein, Biochanin and Formononetin are some of the main types of isoflavones that has been characterized King by (1999) [Figure 2.6(a)]. They are found almost exclusively in legumes including soy, lentils and beans, the soybean being the most abundant source. The highest concentrations of isoflavones are found in the whole food, but to a lesser extent in food products like tofu, soymilk, soy meal and soy flour. Red clover contains all the major four isoflavones types.

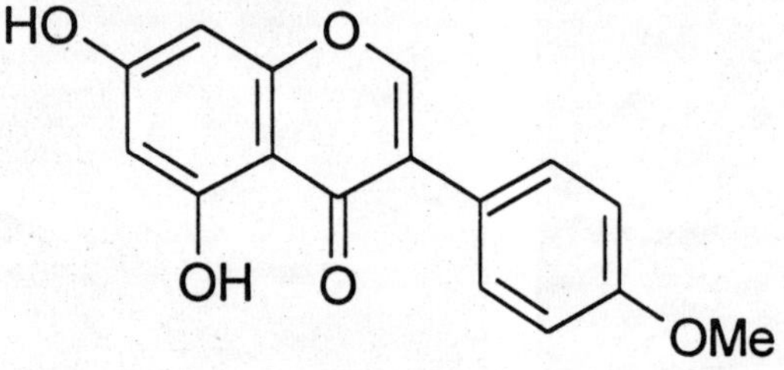

Biochanin A

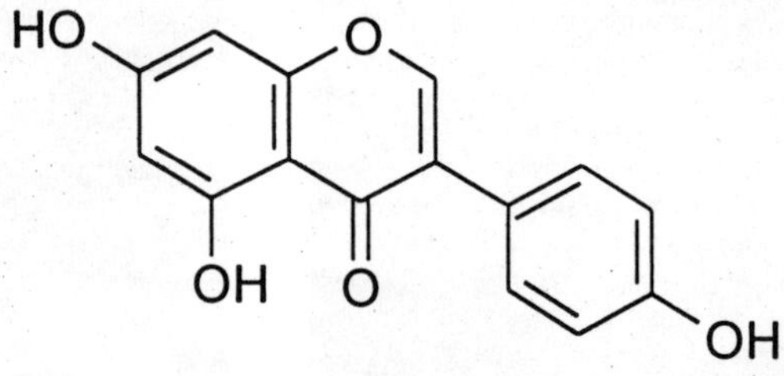

Genistein

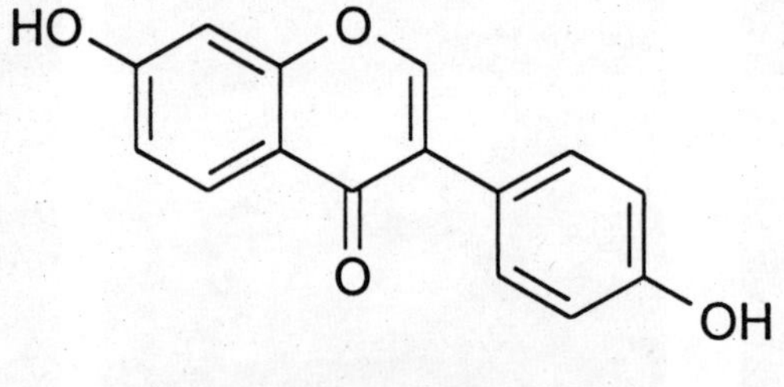

Daidzein

Figure 2.6(a): Isoflavones

Genistein is the most extensively studied among isoflavones to date. It possesses the potent anticancer property due to its ability to block the signals that trigger growth of the cancer cell. Daidzein is also reported to have anti cancer properties but along with this it also provides bone health by regulating the balance between the formation and resorption of the bone mass. Biochanin and Formononetin are two recently discovered forms of isoflavones, which help to lower the body fats thereby protecting the menopausal patients against the cardiovascular disorders. Biochanin is metabolized to genistein. Formononetin is metabolized to daidzein, which is further metabolized to equol the more potent phytoestrogen.

Lignans

Lignans are the diverse class of phenyl propanoid dimers and oligomers. They are found mostly in fiber rich foods, the highest concentration being in flaxseed. They also have wide distribution in cereals, vegetables and fruits; and a lesser extent in wheat, barely, hops, rye, rice and bran. Secoilariciresinol and Matairesinol are the two-lignan dimers that are not estrogenic by themselves but are readily converted to the mammalian lignans enterodiol and enterolactone [Figure 2.6(b)].

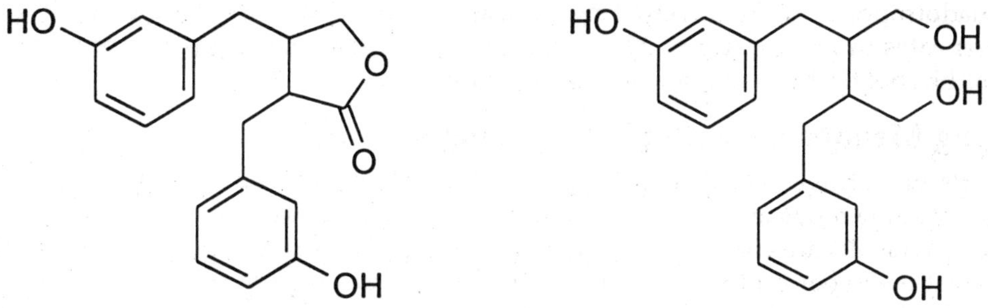

Enterolactone **Enterodiol**

Figure 2.6(b): Lignans

Coumestanes exist in large number of forms but only a small number of them show the estrogenic activity. Coumesterol is one such example shown in Figure 2.6 (c). Main dietary sources for coumestanes are legumes however low levels are also found in Brussels, sprouts and spinach.

Resveratrol is one of the main dietary sources of stilbenes derived from red wine and peanuts (Figure 2.6d).

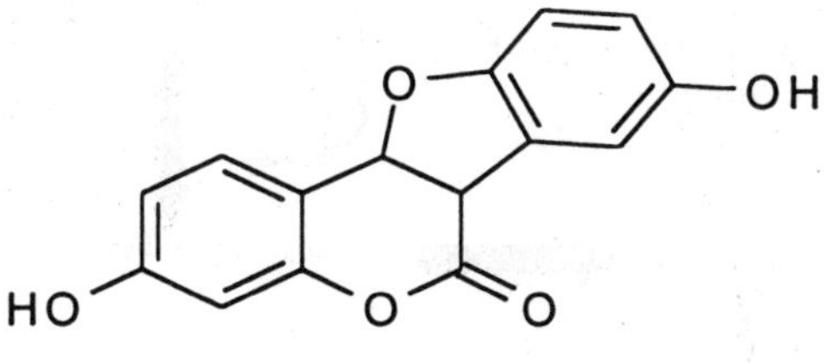

Coumesterol

Figure 2.6(c): Coumestane

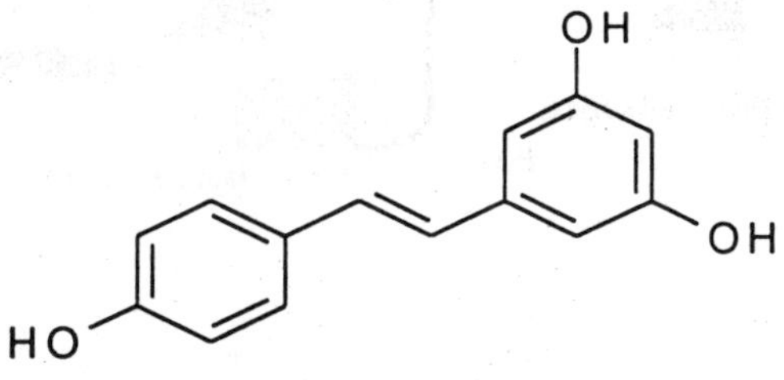

Resveratrol

Figure 2.6(d): Stilbene

Mode of Action of Phytoestrogens

Phytoestrogens, which have properties similar to estrogens, appear to exert both estrogenic and antiestrogenic effects. They interact with numerous molecules, carrier proteins, enzymes, membrane and nuclear receptors directly or indirectly involved in the transfer of estrogen signals (Benassayag *et al*, 2002). Anderson (2005) reported that after entering into the body, they generally exert their effects by an adaptogenic mode (Figure 2.7). This mode of action is beneficial both with increased or decreased estrogen levels in the body. In this mode on metabolization phytoestrogens bind to the same cellular sites as estrogens does. The only difference is that their binding and the subsequent outcomes in terms of biological potency is only a fraction of the results achieved by estrogen binding. Sometimes phytoestrogens also interferes with the endogenous production of estrogen. Normally pituitary gland releases gonadotrophins to stimulate estrogen production in ovaries. Phytoestrogens appear to lower these gonadotrophins levels, thereby lengthening the menstrual cycle. Women with longer cycles have lower rates of breast cancers. Moreover low gonadotrophins levels also reduce menopausal symptoms like hot flashes and other climacteric symptoms.

Managing Menopause with Phytoestrogens

Reports of Cornwell (2004) and Brzezinski (1999) showed that in menopausal women who regularly consume foods and drinks like soya, linseeds, red clover etc rich in naturally occurring estrogens, the results are comparable to women taking hormone replacement therapy. It has been reported that 80-90 per cent of the women who took to this type of treatment claimed about the withdrawal of the short-term symptoms like hot flashes, night sweats, dry vagina and insomnia. Physical ailments like fatigue, weight gain, headaches and mental symptoms like anxiety, confusion and irritability were also reported to be maximally cured in these suffering women. Not only this, the long-term effects like development of cancer and cardiovascular problems were also managed easily (Brandi, 1999).

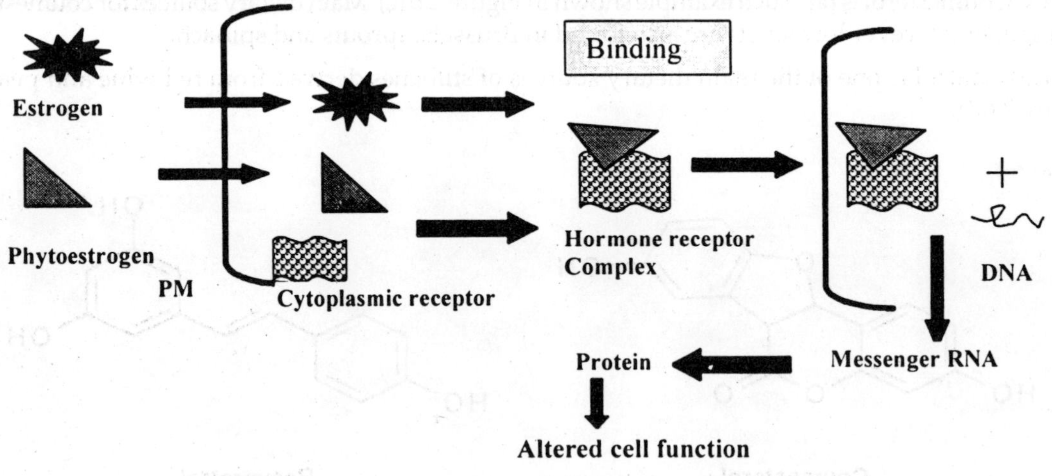

Figure 2.7: Adaptogenic Mode of Action of Phytoestrogens

Cancer Prevention

Phytoestrogen can protect menopausal patients against cancer by more than one route. For example adaptogenic nature of phytoestrogens as explained earlier competitively inhibit estrogen binding to the native estrogen receptor thereby preventing the onset of cancer due to the continuous exposure of potent estrogen. It has been suggested that phytoestrogen being less potent might partly suppress or inhibit normal estrogenic activity in estrogen-responsive tissues such as breast tissue and may reduce the risk of breast cancer. Some studies on soya products reported that they block the formation of nitrites, one of the chemical compounds that are known to cause cancer. Inhibition of several key enzymes like tyrosine kinase by phytoestrogens also helps in controlling cell proliferation and carcinogenesis.

Cardiovascular Disorders (CVD)

With age, arteries, which were earlier clear and flexible, begin to narrow due to the development of plaques made up of fatty deposits of cholesterol and fibrous tissue. These plaques can cause heart attacks as they reduce the internal diameter of blood vessels. Phytoestrogens protect against CVD's by preventing the cholesterol deposition, by reducing the LDL cholesterol levels and raising HDL cholesterol levels. As LDL transports the cholesterol to arteries and HDL takes it away to liver so as to be removed from the body, the increase in HDL and decrease in LDL proves to be beneficial in treating cardiovascular diseases.

Osteoporosis

Osteoporosis or porous bone is a condition that is characterized by low bone mass and structural deterioration of bone tissue which leads to increased bone fragility. An acceleration of bone loss occurring after the menopause, increase the risk of osteoporosis among the suffering women. It has been observed that there is a lower incidence of osteoporosis and hip fractures in south East Asian countries where there is a greater dietary intake of phytoestrogens and it has been suggested that phytoestrogens such as genistein may provide protection from bone loss, by acting as an inhibitor of osteoclasts, which break down and absorb bone. Also estrogen maintains the bone density by regulating the function of enzymes like alkaline phosphatase and deposition of calcium during formation and resorption of bone by osteoblasts and osteoclasts, respectively (Compston 2001). Due to lower circulatory estradiol levels present in menopausal women, calcium is lost from the bones into plasma leading to osteoporosis. Phytoestrogens restore BMD (Bone mass density) thereby preventing against bone loss. Although positive skeletal effects of isoflavones have been demonstrated from studies in rats, there are no reports regarding the clinical trials of the effects of natural isoflavones on bone mass changes in humans.

Ipriflavone, a synthetic isoflavone derivative, has been studied extensively with regard to the prevention of osteoporosis. It appears to inhibit bone resorption slowing the bone loss, thereby improving bone density. These positive findings cannot be directly extrapolated to natural isoflavones found in the diet, but they could form the basis for further research. Daidzein (one of the two principal Soya Isoflavones) is one of the main metabolites of ipriflavone. Hence at present there is only tentative evidence to indicate that naturally occurring phytoestrogens have similar effects in maintaining bone density to those of the related ipriflavone, although emerging evidence suggests these substances are biologically active in experimental models.

Antioxidant Efficacy

The poly-phenolic nature of phytoestrogens give them the ability to act as antioxidants. The inhibition or delay in the oxidation of other molecules prevents the initiation or propagation of oxidizing chain reactions. They also help to reduce the risk of cancer by protecting cells against lipid peroxidation and modulating the production of prostaglandins and leukotrines in carcinogenesis. Some phytoestrogens like genistein has been found as potent inhibitor of hydrogen peroxide production and the activator of antioxidant enzymes like catalase, superoxide dismutase, glutathione peroxidase and glutathione reductase.

Determination of Phytoestrogens

Principal analytical methods used for the analysis of phytoestrogens are divided into two categories the chromatographic and the non-chromatographic procedures. The chromatographic methods involve preliminary chromatographic separations. For example Gas chromatography-mass spectrometry (GC-MS) and High performance liquid chromatography (HPLC). Non-chromatographic methods include UV, IR spectroscopy and immunoassays like radioimmunoassays and enzyme linked immuno sorbent assay (RIA and ELISA). Between the two available kinds, the chromatographic methods are laborious but still useful since they allow determination of physical properties of each phytoestrogen. They permit their identification, confirmation and quantitation very accurately. On the other hand, the non-chromatographic methods are rapid forms of analyses with high degree of specificity.

To date various *in-vitro* and *in-vivo* test systems are also available for screening estrogenic and antiestrogenic activity of phytoestrogens (Wang *et al.*, 2002). The potency of phytoestrogens determined using *in vitro* assays are not always definitive and therefore it is advisable to interpret the data in conjunction with the effects observed *in vivo*. *In vitro* assays include E-screen assays and the competitive ligand-binding assay (Mueller 2002). In the E- screen assay the proliferative potential of the sample is analyzed on breast cancer cell lines. In the competitive ligand-binding assay the adaptogenic nature of phytoestrogen is tested. *In vivo* bioassays most extensively utilize the rodent uterotrophic assay (Diel *et al.*, 2002). In this assay, the ability of the chemical to stimulate uterine growth is determined. The experiments are conducted on the immature, hypophysectomized or ovariectomized rats or mice administered with the test compared either orally or subcutaneouly for a period of three days and absolute weights along with other parameters screened on fourth day.

Korach and McLachlan (1995) also reported about some of the estrogen inducible biomarkers whose expression can also be taken into account to test estrogenic or antiestrogenic property of phytoestrogens. Accordingly, lactoferrin., sex hormone binding globulin (SHBG), ceruloplasmin, uterine protein pattern are some of the estrogen inducible markers that have been utilized for the identification of phytoestrogen.

Conclusion

With large number of women entering the perimenopausal and menopausal phase symptoms like osteoporosis, cardiovascular decline, increased cancer risks, sleep deprivation, hot flashes, mood swings, difficulty concentrating and memory loss deteriorate the overall quality of their life. The many physical and emotional issues involved have necessitated women to seek professional help. Hormone replacement therapy, which provides women with the conjugated estrogens like equine estrogens, esterified estrogens, conjugated estradiol, Ethinyl estradiol etc in the form of oral pills, subcutaneous patches, intramuscular injections to cover up the deficiency state alleviates most of the symptoms like hot flashes and other climacteric symptoms. Unfortunately, this course of action has increased the risk of cancer (breast, endometrial, cervical) in women. Taking into account the risks associated with currently available hormone replacement therapy, health care providers worldwide have now directed their attention to search for alternate therapies for menopause. The vast majority of the symptoms can be treated or improved with an all-natural product. Some of them are phytoestrogens, natural progestins, mineral and vitamin supplements, homeopathy, exercise and acupuncture. Among all these therapies that are currently available in the market, the phytoestrogens, the polyphenolic, nonsteroidal, weakly estrogenic compounds derived from plant sources are accepted maximally as Mother Nature's gift. They appear to boost the good effects of estrogen while minimizing the bad ones to the maximum as compared to other available alternatives. As more and more plants are being screened the estrogenic effects of phytochemicals that have potentials for treating menopausal symptoms are just beginning to be known. Concerted efforts would be needed to specifically land these in the market for managing menopausal symptoms. In order to achieve this, several hormone balancing formulations consisting of herbs and nutritional supports have to be worked out.

Acknowledgement

Priya Kapur thanks the UGC for supporting her with a fellowship to conduct research on herbal alternatives to management of menopausal symptoms

References

Anderson BA. (2005). *Reproductive Health: Women and Men's Shared Responsibility*. Jones and Bartlett Publishers, Boston, Toronto, London, Singapore, pp. 1-350.

Benassayag C, Applanat MP, and Ferre F. (2002). Phytoestrogens as modulators of steroid action in target cells. J. Chrom. B., 777, 233-248.

Brandi ML. (1999). Phytoestrogens and menopause. Environ. Toxicol. Pharmacol., 7, 213-216.

Brzezinski A, and Debi A. (1999). Phytoestrogens: the natural selective estrogen receptor modulators. Eur. J. Obst. Gynecol. Reprod. Biol., 85, 47-51.

Compston JE. (2001). Sex steroids and bone. Physiol. Rev., 81, 420-447.

Cornwell T, and Cohick W. (2004). Dietary phytoestrogens and health. Phytochem., 65, 995-1016.

Dantas S. M. (1999). Menopausal symptoms and alternative medicine. Honorable mention. Elsevier. Prim care update/Obstetrics and Gynecology., 6, 212-220.

Diel P, Schmidt, and Vollmer G. (2002). In vivo tests systems for the quantitative and qualitative analysis of the biological activity of phytoestrogens. J. Chrom. B., 777, 191-202.

Duncan AM, Phipps WR, and Kurzer MS. (2003). Phyto-oestrogens. Best Practice and Research Clinical Endocrinology and Metabolism, 17, 253-271.

Gruber CJ, Tschugguel W, Schneeberger C, and Huber J.C. (2002). Production and actions of estrogens. N. Engl. J. Med., 346, 340-352.

Kessel B. (1998). Alternatives to estrogen for menopausal women. Proc. Soc. Exptl. Biol. Med., 217, 38-44.

King A, and Young G. (1999). Characteristics and occurrence of phenolic phytochemicals. J. Am. Diet Assoc., 99, 213-218.

Knight DC, and Eden JA. (1996). Review of the clinical effects of phytoestrogens. Obstet. Gynecol., 87, 897-904.

Korach KS, and McLachlan JA. (1995). Techniques for the detection of estrogenicity. Environ. Health Perspect., 103 Supp 7, 5-8.

Mueller SO. (2002). Overview of the *in vitro* tools to assess the estrogenic and antiestrogenic activity of phytoestrogens. J. Chrom. B., 777, 155-165.

Murkies AL, Wilcox G, and Davis SR. (1998). Phytoestrogens. J. Clin. Endocr. Metab., 83, 297-303.

Staren ED, and Omer Shuab. (2004). Hormone replacement therapy in postmenopausal women. Am. J. Surg., 188, 136-149.

Wang CC, Prasain JK, and Barnes S. (2002). Review of the methods used in the determination of phytoestrogens. J. Chrom. B., 777, 3-28.

Wuttke W, *et al.* (2003). Phytoestrogens for hormone replacement therapy. J. Steroid Biochem. Mol. Biol., 83, 133-147.

Zava DT, Dollbaum CM, and Blen M. (1998). Estrogen and progestin bioactivity of foods, herbs and spices. Proc. Soc. Exptl. Biol. Med., 217, 369-378.

3

Additive Reaction of Catecholamines and Their Effects upon Outcome and Predictive Diagnosis in Liver Surgery

S. Porta, H. Bacher, H-J. Mischinger, and S. Wurzinger*

Institute for Applied Stress Research, Bad Radkersburg, Austria; Institute for Pathophysiology, Medical University of Graz, Austria; University Hospital, Department for Surgery, Graz, Austria

ABSTRACT

Patients experiencing a post surgical liver failure show a three times higher catecholamine concentration than successfully regenerating patients already during surgery. Even pre-operatively they are distinctly different from non liver failure patients, discernable by metabolic exhaustion markers like triglycerides, cholesterol and magnesium (Mg). Increased catecholamine effects pre- and peri-surgical seem to add up to deplorable post surgical results. Thus knowledge derived from animal experiments about the idiosyncrasy of the catecholamine system of additional secretion on top of already high levels proves beneficiary for treatment of risk patients.

Introduction

Knowledge about additive catecholamine secretion first seen in rats, brought about by absence of a direct feedback system (Porta *et al.*, 1979), has practical consequences in diagnosis and treatment of surgery patients.

**Corresponding Author*: E-mail: stresscenter@netway.at

Catecholamine secretion as a consequence of stress during an extraordinary situation helps the organism to retain physiological balance as quickly as possible. Stressors like surgery and anesthesia do transport the organism into such an extraordinary situation from which it wants to recover as quickly as possible. For this achievement an unusual amount of energy is needed, which is helped alone by increased levels of stress hormones. High energy substrates like triglycerides, cholesterol and glucose are mobilized because energy turnover (ATP-ADP) may be several 100 fold increased by catecholamine induced calcium (Ca) influx into tissues. Thereby also Mg turnover should be increased accordingly.

One should by all means be aware of the fact that similar, though less expressed processes do go on in mentally less stable patients already during a considerable time before surgery, including increased catecholamine turnover already draining the resources of the high energy substrates mentioned above.

Again we would like to state that increased catecholamine secretion which exhausts energy reserves of less mentally stable patients already before surgery may add up with peri-surgical catecholamine increase due to a lack of adreno-medullary feedback mechanism (Porta *et al.*, 1979; Porta *et al.*, 1993). Therefore the pre-operative stress state of a patient should be of importance not only for peri-surgically behaviour but also, in consequence of the additive nature of catecholamine secretion, for post-operative recreation.

Since the worst case for patients as well as for surgeons concerning recreation failure after liver surgery obviously is a cessation of liver function, we subdivided in a post hoc manner patients who underwent major liver surgery into a group experiencing liver failure and into a second successfully recreating group.

Objectivation of pre-, peri- and post-operative stress load is classically done by catecholamine determination, *e.g.* free plasma norepinephrine (NE) and free plasma epinephrine (E). However, the test system is complicated, expensive and last but not least time consuming and therefore not recommendable for quick routine determinations. Therefore, we correlated catecholamine effects like changes in blood gases, changes in the quantity of buffer systems, changes in serum electrolyte differences, changes in fat- and carbohydrate metabolites to actual changes in plasma NE levels (Porta *et al.*, 1997). It turned out, that changes in those plasma and serum constituents closely and linearly correlate with the changes in plasma NE and could therefore be taken as faster and cheaper screening parameters for the assessment of actual stress state. This method, including a pattern recognizing system for automatic calculation, we called Clinical Stress Assessment (CSA) to discern it from psychological stress tests. The aim of the present investigation was to see if pre-surgical exhaustive strain influences peri-operative stress and if this again is subsequently connected with post-operative liver failure.

Material and Methods

Twenty six patients underwent main liver surgery (21 segmental resection, 3 hemihepatectomia, 2 hydatid cysts). Patients who show positive cirrhosis- and hepatitis markers were excluded.

All patients were subjected to a so called "Pringle manoeuvre"–ischemia followed by reperfusion– during surgery. Afterwards the patients were subdivided into 2 groups: the first consisting of those patients who post-operatively experienced liver failure (LF), the second group with those patients without liver failure (nLF). Blood samples were taken before, during and after surgery. Out of those samples the following parameters have been determined: free NE and free E (HPLC with electrochemical detection), GOT, quick test, cholesterol, triglycerides, Mg and Ca (Nova CCX). All data were collected and processed in a software immanent pattern recognition system, including semi automatic calculations of means, standard errors of mean, significance tests and correlations (CSA software, PLK, Austria).

Results

General approach, common to all results and figures.

Ordinate: plasma concentration of hormones, metabolites or electrolytes.

Abcissa: sampling time

1. Two days before surgery
2. cut into the cutaneous layer
3. start of ischemia
4. end of ischemia
5. reperfusion
6. end of surgery

Hormones

1. Catecholamines: free epinephrine levels are 3 fold higher in the LF group than in the nLF group during surgery (Figure 3.1).
2. Metabolic parameters and electrolytes: Significantly higher levels of GOT (Figure 3.3) during surgery in the LF group than in the nLF group exist. Before surgery serum GOT levels do not correlate with plasma concentration of free E but at the end of ischemia and surgery GOT values are significantly linear proportional to free plasma E ($p< 0,05$).

Also in the LF group, serum cholesterol (Figure 3.5) is decreasing linearly proportionally to catecholamine effects, whereby serum Mg (Figure 3.4) increases inversely proportional during surgery (at least $p <0,05$). Serum levels of cholesterol, triglycerides (Figure 3.6) and Mg have been significantly lower in the LF group already before surgery.

Discussion and Conclusion

In previous studies we could show that a number of patients who underwent liver surgery developed a tendency to peri-operative tachycardia (Bacher *et al.*, 2000). The very same patients tended to post-operative complications and showed significantly elevated peri-operative plasma catecholamine levels. Since the worst–case scenario for both patients and surgeons is the post surgical development of liver failure, we investigated early pre- or perioperative idiosyncrasies of patients who experienced post surgical liver failure. Determination of pre- and peri-operative catecholamine changes showed that patients who run the highest post operative risk of liver failure developed 3 times higher catecholamine levels than the other patients already during surgery. Since catecholamine effects on metabolic parameters–which are connected with liver function–are more readily to determine than the catecholamine changes themselves, we took those catecholamine effects as additional predictors for immanent liver failure.

It turned out that routine parameters like Quick Test (Figure 3.2) or GOT are significantly influenced by the catecholamine level during surgery, therefore being useful indicators for the evaluation of surgery stress. A fall in serum cholesterol in LF patients as well as an marked increase in serum

Epinephrine

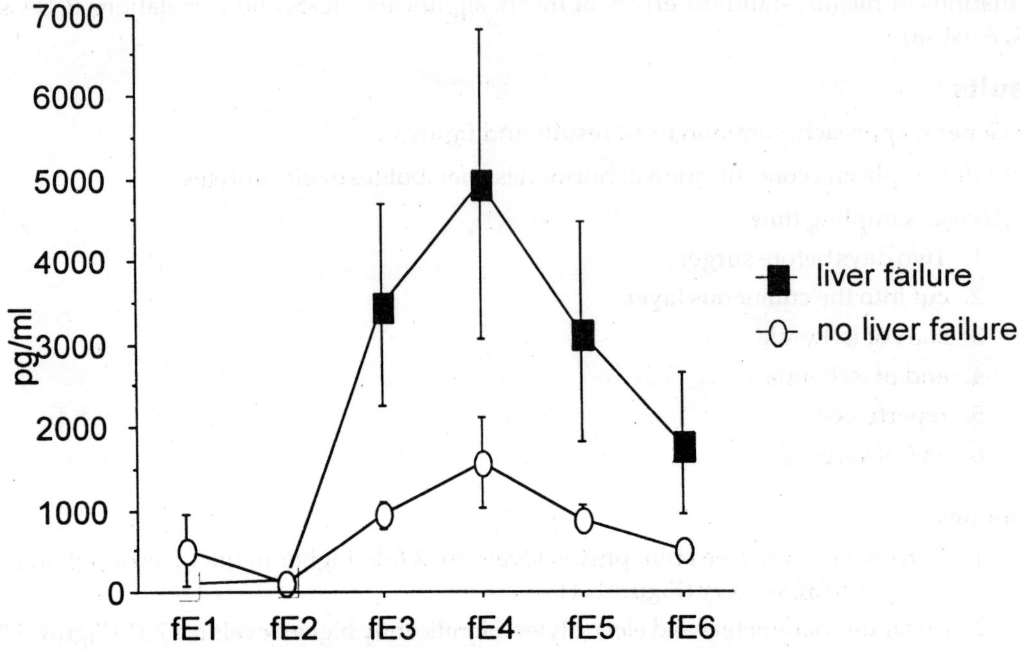

Figure 3.1: Epinephrine Levels Before and During Surgery

Quick Test

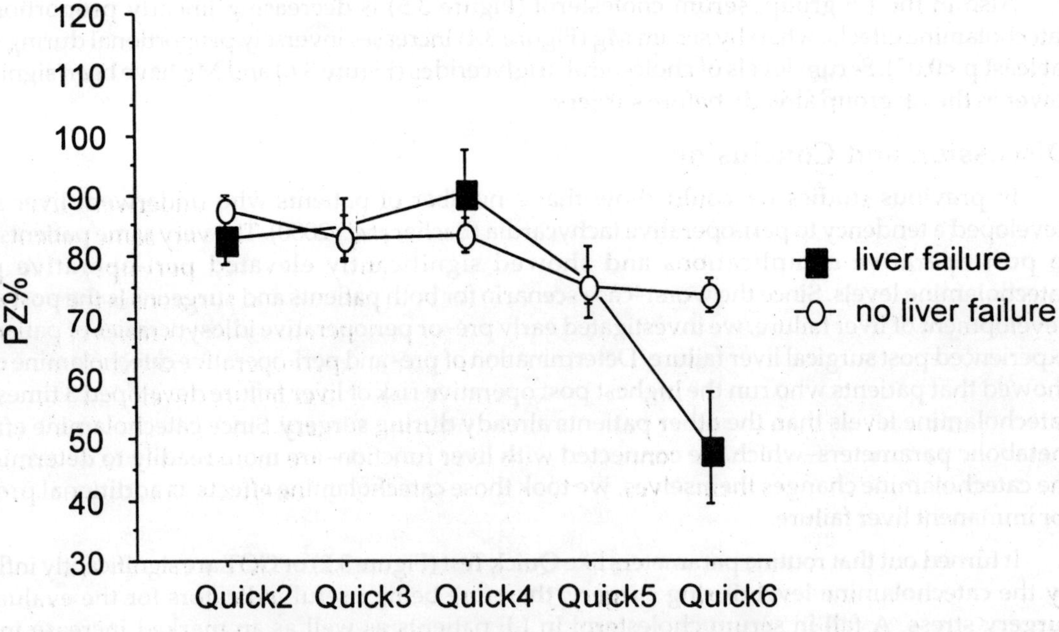

Figure 3.2: Quick Test Before and During Surgery

GOT

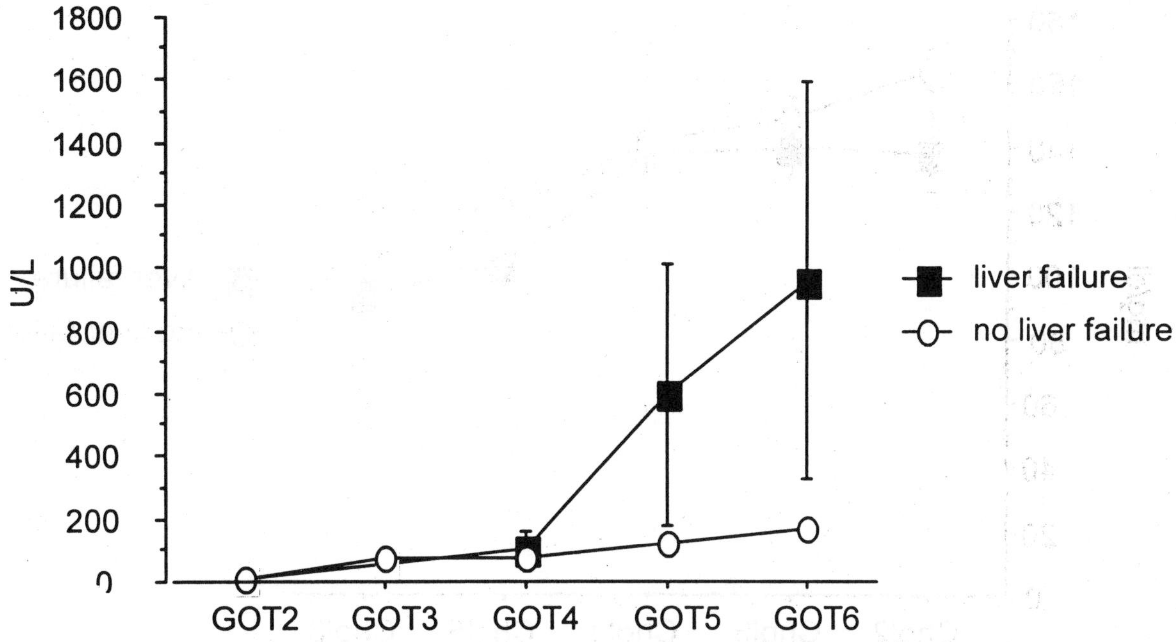

Figure 3.3: GOT Before and During Surgery

Total Magnesium

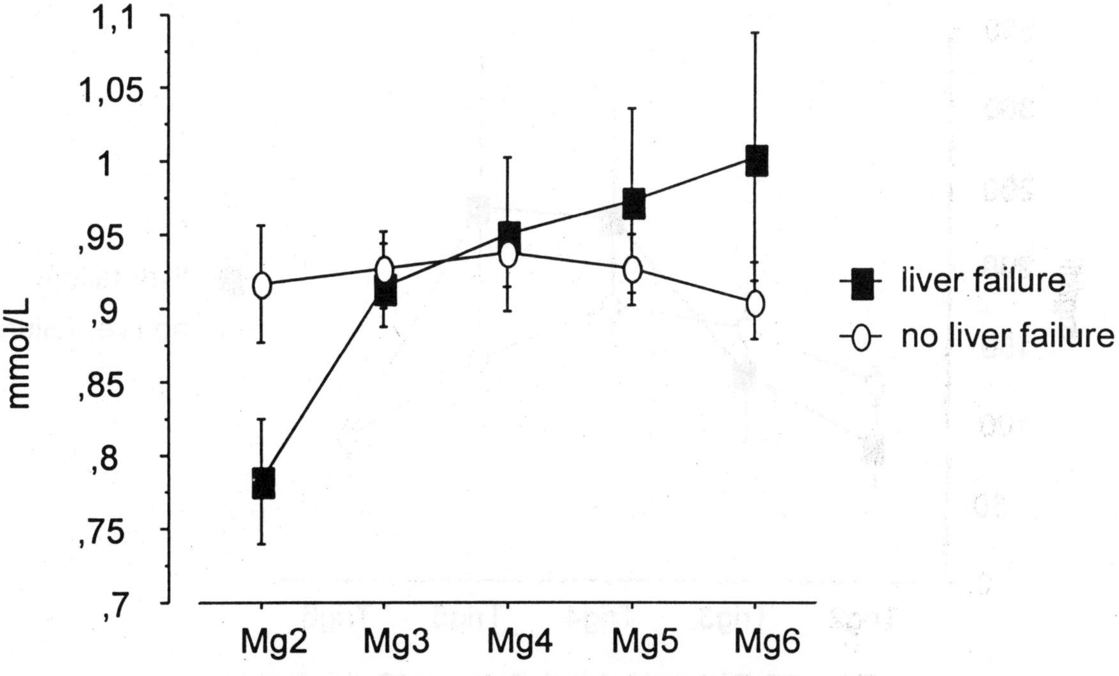

Figure 3.4: Ionized and Protein Bound Mg Before and During Surgery

Cholesterol

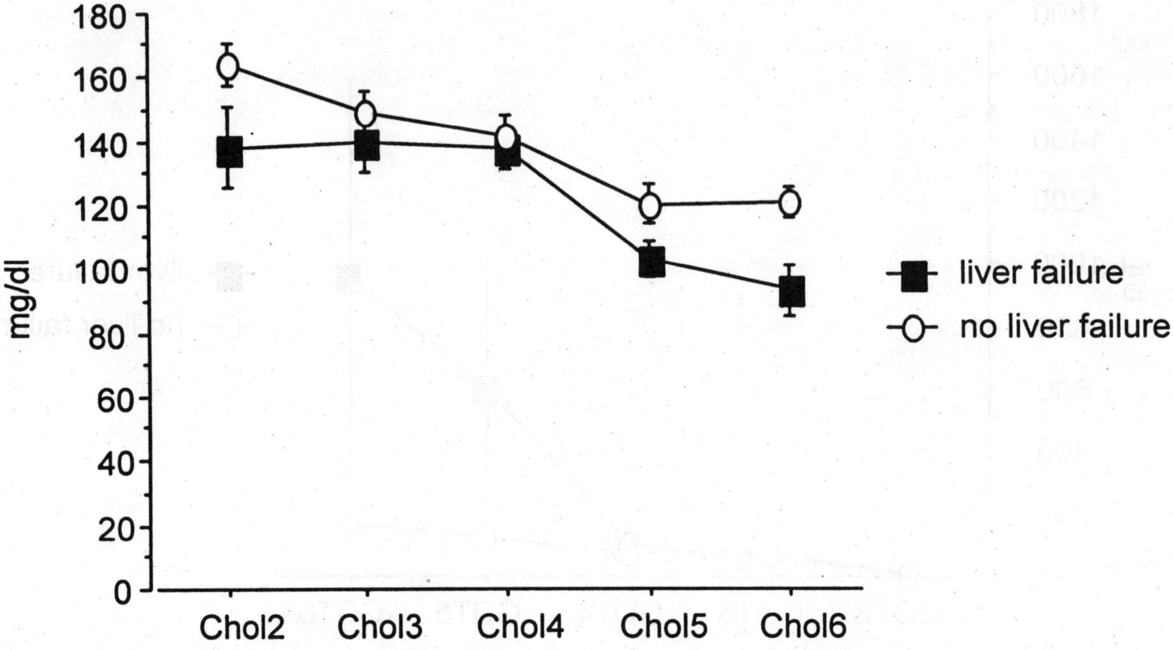

Figure 3.5: Cholesterol Levels Before and During Surgery

Triglycerides

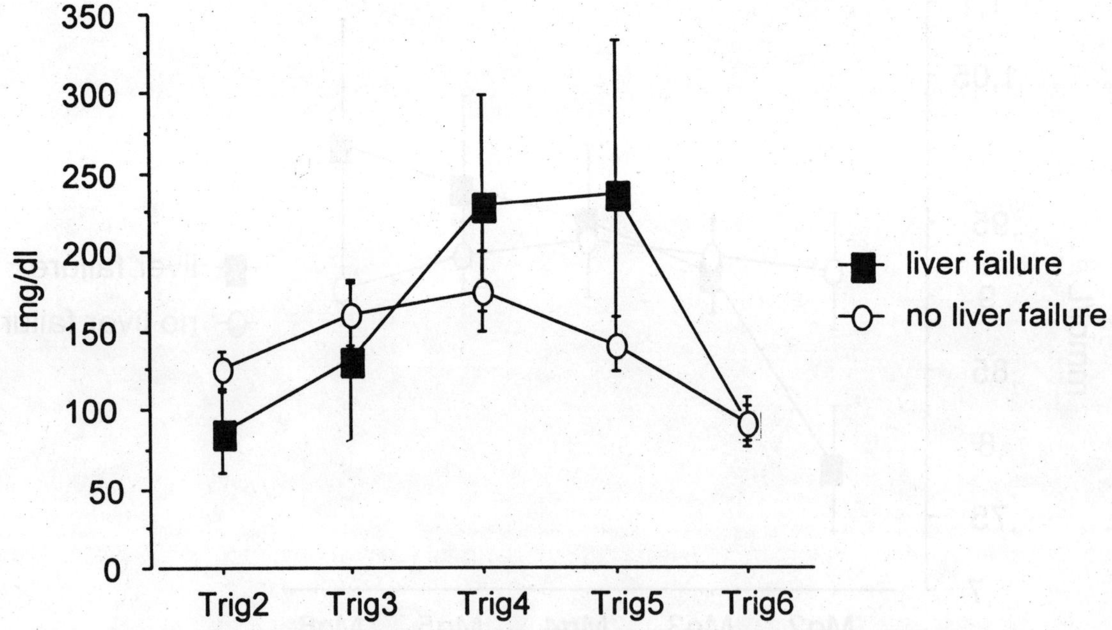

Figure 3.6: Triglyceride Levels Before and During Surgery

triglycerides–are probably brought about by more clearcut effects of liver ischemia–in LF patients.

Increasing serum Mg concentrations together with decreasing Ca levels (Zimmermann *et al.*, 2000) are typical for a classical stress reaction. This quotient therefore renders a suitable indicator for stress quantitification. In our opinion, pre-operative decreased serum concentrations of triglycerides and cholesterol are–in this specific situation–not at all satisfactory, as they woud be under normal circumstances. On the contrary, together with the telling decrease in serum Mg they indicate an already limited or even exhausted metabolic reserve. Most probably they are early indicators for a limited peri-operative stress compatibility. High serum Mg levels may indicate both stress in the immediate past and therefore increased efflux from tissue Mg into the blood or, in low stress patients, a satisfactory Mg state. Low Mg on the other hand always points towards chronical exhaustion. Low cholesterol and "normal" triglyceride serum levels are rare in patients during the second half of their life span. Significant triglyceride decreases were *e.g.* seen during after a week of exhausting mountain training (Porta *et al.*, 1997). Thus "pseudonormal" values of cholesterol and triglycerides together with low serum Mg levels clearly point towards exhaustion.

Increased surgery stress–characterized by notable increased plasma catecholamines before surgery and significant changes of directly catecholamine–influenced metabolic parameters indicate a significant increase of danger of postoperative complications. Along with pre-surgical catecholamine dependent indicators like Mg (Bacher *et al.*, 1998), cholesterol and triglycerides they point to already deficient resources brought about by presurgical catecholamine effects. Due to a lack of feedback ability those early catecholamine effects are topped by perioperative catecholamine increase, which together pave the way for postoperative liver failure.

Thus basic knowledge won by comparative studies in animal experiments about the additive possibilities of catecholamine secretion turned out to be of practical value to prevent liver failure after surgery.

References

Bacher H, Mischinger HJ, Supancic A, Leitner G, Porta S. (1998). Dopamin infusions after liver surgery prevents hypomagnesemia. Magnesium-Bulletin, 20, 2, 49-50.

Bacher H, Mischinger HJ, Cerwenka Werkgartner HG, El-Shabrawi A, Supancic A, andc Porta S. (2000). Liver ischemia, catecholamines and preoperative condition influencing postoperative tachycardia in liver surgery. Life Sciences, 66, 11-18.

Porta S, Englmayer G, Kubat R, Egger G, and Sadjak T. (1979). Evidence of adrenaline secretion regardless of high extraglandular levels. Exp. Path., 17, 530-532.

Porta S, Emsenhuber W, Petek W, Pürstner P, Vogel W, Schwaberger G, Slawitsch G, and Korsatko W. (1993). Detection and evaluation of persisting stress induces hormonal disturbances by a Post Stress Provocation Test in humans. Life Sciences, 53, 1583-1589.

Porta S, Leitner G, Heidinger D, Lang T, Weiß U, Smolle KH, and Hasiba K. (1997). Magnesium während der Alpinausbildung bringt um 30 per cent bessere Energieverwertung. Magnesium-Bulletin, 19, 2, 59-61.

Zimmermann P, Weiß U, Claasen HG, Wendt B, Epple A, Zollner H, and Porta S. (2000). The impact of diets with different magnesium contents on magnesium and calcium in serum and tissues of the rat. Life Sciences, 67, 949-958.

4

Influence of Thyroid Hormones on Cardio Vascular System: An Overview

Peranaidu Govindarajulu

56/64 First Avenue, Indira Nagar, Chennai – 600020, Tamil Nadu, India

ABSTRACT

Thyroid hormones have specific and target oriented influence on heart and cardio vascular system.Tri iodothyronine (T-3) and Tetraiodothyronine (T-4) have potentiating effect on myocytes and vascular smooth muscle cells, by their binding with the receptors which activate thyroid hormone response elements in the target genes and initiate transcription and translation. These genes encode structural as well as regulatory proteins of several kinds. In addition thyroid hormones have extra nuclear actions also. This is how, thyroid hormones bring about their tremendous influence on the heart and blood vessels.

Thyroid hormones have differential effect on myocardial heavy chain gene expressions to regulate myocardial contractility. Thyroid hormones have been implicated in the normal cardiac development and post natal maturation of myocardium. Intra cellular calcium play a critical role in the central regulation of cardiac contractility and in this respect, thyroid hormones have an important influence on the calcium transport mechanisms. In hyperthyroidism, palpitations, increased heart rate, tachycardia due to hyper contractility of cardiac muscles are commonly witnessed. This results in irregular rapid heart rate.

Former Senior Professor of Endocrinology, Dr. A.L.M.P.G. Institute of Basic Medical Sciences, Taramani, Chennai – 600 113, Tamil Nadu, India; E mail: govinaidu@gmail.com

In thyrotoxicosis, a rapid and forceful sensation of heart beat is exhibited. The process of cardiac contractions is mediated through the cardiac contractile proteins, myosin and actins which are modulated by thyroid hormones. There is 50 to 300 per cent in cardiac output due to systemic vascular resistance.

The action of thyroid hormones on the cardiac functional status is measured through cardiac contractility, cardiac size and cardio vascular haemodynamics. In hypothyroidism, broadycardia, systemic hypertension, narrow pulse pressure and slightly increased mean arterial pressure with increased peripheral vascular resistance are often encountered. The cardiac contractility is impaired with uniformly prolonged isovolumic contractile time and pre ejection period.

The detailed impact of thyroid hormones on the cardio vascular system and the underlying mechanisms are presented in the overview.

Introduction

Thyroid hormones have their influence on almost all the tissues and organ systems in our body (Heimberg, 1985; Brent, 1989; Larsen and Brent, 1991). For a proper metabolic activities and functioning of almost every organ and system in our body, thyroid hormones are very essentital.Even mild thyroid disturbances can cause serious health problems. These influences can be identified by means of quantitative changes, basal metabolic rates and clinical assessment in general.However, there are tissue and organ specific influences/actions which are reflected through underlying molecular, biochemical, genetic and physiological mechanisms. Since this review is confined to cardio vascular system, the emphasis is on the specific influences of thyroid hormones on the heart and cardio vascular system. An increase or decrease in thyroid hormone action can influence certain molecular pathways in the heart and vasculature which may result in relevant cardio vascular derangements (Fazio *et al.*, 2004).

Heart is a major target organ for thyroid hormone (Klein and Levey, 1996). The active thyroid hormone which acts profoundly on the heart is triiodothyronine (T-3). Most of the actions of thyroxin (T-4) depend on its conversion to T-3 by 5'-deiodination (5'D) (Oppenheimer *et al.*, 1987). It has been shown that the heart has about one tenth of 5'D activity of the liver (Hesch and Kochle, 1986). In addition, Mori *et al.* (1990) have clearly established that this conversion is essential for the synthesis of Atrial Natriuretic Peptide (ANP) by T-4 in cultured rat myocytes. It has been demonstrated by Yonemoto *et al.* (1991) by using RT-PCR that both T-3 and Ang-11 have stimulated 5' deiodinase activity in the cardiac myocytes,which is necessary for the conversion of T-4 to T-3.In addition to this, these investigators have already shown that T-3 and Ang-11 additively stimulated 1'deiodinase mRNA accumulation in cardiac myocytes.

T-3 has been shown to elevate 5'D activity through its nuclear receptor in myocytes (Mori *et. al.*, 1991). Thyroid stimulating hormone (TSH) also increases 5'D activity and T-3 synergistically increases the TSH induced increase in 5'D activity (Tyoda *et al.*, 1990). It has also been demonstrated in rat myocytes that Angiotensin-II(Ang-II)increases 5'D activity but it is not known whether such an action is independent or by means of any inter-relational pathways with T-3 (Mori *et al.*, 1991). Ang-11 couples with the Ang-11 receptor in cardiac myocytes and activate phospholipase-C and the Voltage sensitive Ca2+ channel. The activated Phospholipase-C acts on Phosphotidylinositol 4, 5 bisphosphate to liberate inositol 1, 4, 5 triphosphate and diacyl glycerol. This results in the release of intra cellular calcium from inositol triphosphate sensitive Ca2++ storage and activation of Protein kinase-C (Griendting and Alexander, 1994; Inagami and Kitami, 1994).

Since T-3 and T-4 are lipophilic in nature, the circulating T-3 and T-4 easily diffuse across the cytoplasmic memberane.The cell membrane has a T-3 specific transport protein and conversion of T-4 to T-3 does not occur significantly in cardiac myocytes (Everts *et al.*, 1996).Once T-3 enters into the myocytes,it gets into the nucleus and binds to its nuclear receptors which then binds to the thyroid hormone response elements in the target genes. The binding of the hormone with the receptor enables the receptor to get activated which either homodimerizes or heterodimerizes with 9-cis-retinoic acid receptor(Brens *et al.*,1994; Long *et al.*,1994).This T-3/THR/THR or T-3. THR/RXR complex recognizes one of the several DNA consensus sequences, the thyroid hormone response element located in the enhanced region of the target genes (Riberio *et al.*, 1995). With the binding of the protein complex to the response element, the promoter of the target gene is activated, which results in the initiation of transcription. In the absence of T-3, the receptor represses genes which are positively regulated by thyroid hormones (Ojamaa *et al.*, 1996).

A number of target genes have been recognized as targets for transcriptional activation by thyroid hormones. The T-3 responsive genes which encode both structural and regulatory proteins in the heart are:

1. Myosin heavy chains alpha and beta which are myofibrillar proteins that make up the thick filament of the contractile apparatus of cardiac myocytes.
2. In animals, transcription of the alpha myosin heavy chain is activated by T-3 but the transcription of the beta myosin heavy chain is repressed (Morkin, 1993; Ojamaa *et al.*, 1996). Whereas in humans, the beta myosin heavy chain is predominantly activated by T-3 (Magner *et al.*, 1988).
3. Sarcoplasmic reticulum proteins
4. Calcium activated ATPase ($Ca2^{++}$ ATPse)
5. Phospholamban
6. Alpha and beta adrenergic receptors
7. Guanine nucleotide regulatory proteins
8. Protein kinase-C
9. Variety of plasma membrane ion transporters such as: Na^+/K^+ ATPase, $Na^+/Ca2^+$ exchanger, voltage gated potassium channels (Kv 1.5, Kv 4.2 and Kv 4.5).
10. Cardiac Troponin-1 and
11. Atrial Natriuretic Peptide (Fullerton *et al.*, 1990; Oriowski and Lingrel, 1990; Tsika *et al.*, 1990; Bahouth, 1991; Averyhart-Fullard and Salero, 1994; Zarain-Horzbeg *et al.*, 1994).

Apart from the modulations and transcriptions of multiple genes as detailed above, thyroid hormones have certain extra nuclear actions in cardiac myocytes (Feldman *et al.*, 1986) which are:

1. Sodium, Potassium and Calcium channels in the heart
2. Intra cellular levels of calcium and potassium for the increase in inotrophy and chronotrophy (Walker *et al.*, 1994).

Release of calcium and its reuptake into the sarcoplasmic reticulum are critical determinants of systolic contractile function and diastolic relaxation (Dillmann, 1990; Mintz *et al.*, 1991; Kiss *et al.*, 1994)

Phospholamban regulates Ca2+ATPase which is involved in the active transport of calcium into the lumen of sarcoplasmic reticulum. The phospholamban activity is modified by its level of phosphorylation (Kiss *et al.*, 1994). Phospholamban plays a key role as a mediator of thyroid hormone action related to changes in cardiac contractility (Murphy *et al.*, 1991). The state of adrenergic activity mediates the influence of T-3 on heart rate, pulse pressure and cardiac output (Levey and Klein, 1990).

Thyroid hormones improve myocardial contractility and such an effect is multifactorial, partially due to differential effect of thyroid hormones on myocardial heavy chain (MHC) gene expression. Three myosin isoforms have been identified in the ventricular muscle; V-1 composed of MHC alpha/alpha, V-2 composed of MHC alpha/beta and V-3 composed of MHC beta/beta. Thyroid hormones by its transcriptional activity on MHC alpha and MHC beta causes a shift in the rate of synthesis of V-1 and decreasing the rate of synthesis of V-3 (Izumo *et al.*, 1986; Morkin, 1993).

Thyroid hormone receptors (TRS) are required to transducer thyroid hormone signals. In the heart there are two isoforms of TRS identified, TR alpha-1 and TR beta-1.Both the receptors have similar affinities to thyroid hormone binding. Another isoform of TR alpha-2 is also present in the heart and this is a splice variant of TRS-alpha-1, which does not bind thyroid hormone. However, this can act as a dominant negative protein for thyroid hormone receptor signaling. A low level expression of another receptor isoform, TRS-beta-2 is also detected in the heart by using reverse transcription polymerase chain reaction (RT-PCR). Even under normal thyroid status diminished TRS signaling can precipitate changes in gene expression consistent with cardiac failure (Sussman, 2001)

It has been shown in failing human hearts, higher levels of TRS-alpha-2 and low levels of TRS alpha-1 and unchanged levels of TRS beta-1 (Kinugawa *et al.*, 2001). However, Sylven *et al.* (1996) and d'Amati *et al.* (2001) have reported elevated levels of TRS mRNA expression of TRS beta-1 and low levels of TRS alpha-1 and alpha-2 and increased expression of TRS alpha-1,alpha-2 and beta-1 in failing human hearts, respectively. The in vitro and in vivo experimental findings indicated that physiological stimulation induced by thyroid hormone treatment in vitro or voluntary exercise in vivo elevated alpha MHC and SERCA with the repression of beta MHC.In pathological stimulation of hypertrophy by treatment with phenylephrine in vitro or by pressure overload after aortic constriction in vivo, decreased all the three isoforms of thyroid receptors (Kinugawa *et al.*, 2001). Therefore, it could be seen that TRS isoform expression modulates thyroid hormone responsive gene expression. These investigators further attempted to establish a direct relationship between TRS isoforms expression and thyroid hormone responsive gene expression. The cultured cardiac myocytes were cotransferred with plasmids encoding specific isoform of human TRS protein together with other plasmids carrying thyroid hormone response promoter-MHC-alpha, MHC-beta or SERCA linked to a receptor gene, by comparing reporter construct activity with each over expressed TRS isoforms, have clearly shown that different TRS isoforms exhibit distinct transcriptional activity profiles. For example, TRS alpha-1 activated the MHC-alpha promoter, TRS beta-1 repressed the MHC beta promoter and the TRS alpha-2 had no effect. Thus it is evident that TRS dependent regulation of thyroid hormone responsive genes and thyroid hormone responsive promoters are differently regulated by the various TRS isoforms.The interdependence between TRS expression and myocardial thyroid hormone signaling requires further in depth exploration. It is believed that though the cardiac contractile function is markedly altered in subjects with thyroid disease, the changes in the expression of myosin heavy chain isoforms may be of insufficient magnitude to account for the functional changes (Levey and Klein, 1996; Hoit *et al.*, 1997).

During the process of normal cardiac development, postnatal changes and in the maturation of myocardium, the importance of thyroid hormones has been clearly implicated. Increasing thyroid hormone signaling after birth induces growth of myocardium and transcriptional reprogramming which leads to the characteristic gene expression profile of the adult heart (Klein *et al.*, 2001; Sussman, 2001). When the thyroid hormone stimulation is relatively low, the gene expression profile of the fetal heart was implicated in heterotrophic and/or failing myocardium (Sussman, 2001). Clinical investigations have shown that subjects suffering from heart failure do possess low thyroid hormone levels and such subjects respond favorably to short term thyroid hormone supplementation with increased cardiac output and exercise performance (Moruzzi *et al.*, 1996; Hamilton *et al.*, 1998).

The central regulator of cardiac contractility is the intracellular calcium. Therefore, alterations in myocyte calcium regulation are critically essential in not only dysfunction but also in arrhythymogenesis associated with congestive heart failure (O'Rourke *et al.*, 1999; Pogwizd *et al.*, 1999). During the cardiac action potential L-type calcium channels are activated and the calcium enters into the cell via calcium current (I ca) and also much lesser amount of calcium enters via Na+-Ca2+ exchange. The calcium influx triggers calcium release from the sarcplasmic reticulum and to some extent can also contribute to the activation of the myofilaments directly. The calcium entry plus the amount of calcium released from the sarcoplasmic reticulum via Ca2+ induced Ca2+ release raises cytosolic free calcium. As a result of which calcium binds to multiple cytosolic calcium buffers. One of the most functionally important cytosolic calcium buffers is the thin filament protein, Troponin-C.Binding of calcium with Troponin-C, switches on the myofilaments in a co-operative manner which activate contractions. Disassociation of calcium from Troponin-C due to the decline in calcium level switches off the contractile machinery. This results in relaxation and diastolic filling (Bers *et al.*, 2000). The four calcium transporters which remove calcium from the cytosol are:

1. Sarcoplasmic reticulum Ca2+ATPase
2. Sarcolemmal Na+-Ca2+ exchange
3. Sarcolemmal Ca2+ ATPase and
4. Mitochondrial Ca2+uniporter

Therefore, it can be seen that calcium in cardiac myocytes is in a dynamic but delicate balance related to cardiac contractile and relaxation mechanism. In this context there is interaction of numerous cellular processes which are subject to many regulatory influences. Thyroid hormones play a critical role in such processes.

Thyroid hormones have significant effects on the peripheral vascular system as well. Thyroid hormones directly act on vascular smooth muscle cells as vasodilator and relaxation of coronary arteries (Kleperer *et al.*, 1995; Ojamaa *et al.*, 1996). Type-1 iodothyronine deiodinase (D-1) is widely distributed and maintains circulating T-3 level. However, Ype-11 iodothyronine deiodinase (D-11) is present only in a limited number of tissues, mainly to provide the local intra cellular T-3. Mizuma *et al.* (2001) have demonstrated the presence of D-11 activity in the cultured human coronary artery smooth muscle cells and human aortic smooth muscle cells, by using Northern analysis and RT-PCR.They further have shown that D-11 mRNA levels and D-11 activities in the human vascular smooth muscle cells were regulated by C-AMP mediated mechanism. A functional C-AMP response element has been reported in the human D-11 promoter site (Bartha *et al.*, 2000). The activity of D-11 is controlled by thyroid hormones at two levels. The D-11 activity is suppressed by T-3 and T-4 mainly at the post translational level by accelerating the degradation of D-11 protein (Leonard *et al.*, 1984). Since rT-3 and T-4 are quite potent in their effect than T-3, this might suggest that such an effect does not require thyroid hormone receptors. Steinsapir *et al.* (1998) have shown that proteasomal degradation is involved in the post translational regulation of D-11 activity by thyroid hormones. Contrary to this finding, T-3 was shown to suppress D-11 activity by decreasing D-11 mRNA without affecting its half life, suggesting that this effect is by the suppression of transcription of D-11 gene through thyroid hormone receptors (Kim *et al.*, 1998).

In hypothyroidism D-11 expression in human vascular smooth cells may be elevated and this may play a significant role in protecting human blood vessels from local T-3 deficiency in hypothyroidism (Mizuma *et al.*, 2001). Human D-11 c DNA contains inframe TGS triplets which are not for the termination codons but for the rare amino acid selenocysteine, which contains selenium (Croteau *et al.*, 1996; Salvotore *et al.*, 1996). Selenium is required to exert for the full D-11 activity. Therefore, it is contemplated that selenium deficiency is related to local T-3 production by D-11 in human vascular smooth muscle cells which may increase the risk of ischemic heart disease but this needs further elucidation.

Hyperthyroidism

Palpitation is one of the most common symptoms associated with hyperthyroidism (Nordyke *et al.*, 1989). Heart rate is constantly increased in day time and exaggerated in response to exercise, even though its circadian rhythm is usually preserved (Von Olshausen *et al.*, 1989; Cacciatori *et al.*, 1996). It was found on analysis of the heart rate variability that the sympathetic–vagal imbalancing with a relative increase in sympathetic tone (Cacciatori *et al.*, 1996). Since beta blockers normally attenuate tachycardia but still heart rate remains slightly higher in hyperthyroid patients, it is derived that thyroid hormones can directly affect sinus node firing (Suu *et al.*, 2001). 5 to 10 per cent of hyperthyroid patients have atrial fibrillation.(Sawin *et al.*, 1994; Aner *et al.*, 2001) which returns to normal rhythm with the normalization of thyroid status. (Nakazawa *et al.*, 1982). Atrial arrthymias including atrial fibrillation or flutter can occur occasionally. This may present an exaggerated sense of an irregular rapid heart rate (Sandler *et al.*, 1959; Ikram, 1985). The characteristic of thyrotoxicosis is the elevated resting heart rate and it is observed that as many as 90 per cent of the thyrotoxicosis patients having resting tachycardia and the degree of sinus tachycardia may be more-than 120 beats/min. (Delit *et al.*, 1961; Olshausen *et al.*, 1989; Magner *et al.*, 1990; Valcavi *et al.*, 1992; Greattinger *et al.*, 1999).

In majority of thyrotoxicosis patients, there is a rapid and forceful sensation of heart beat (Delit *et al.*, 1961). Most often such patients describe the feeling of "exercising while at rest" and occasionally the forcefulness of cardiac contractions can be disturbing and is the presenting clinical manifestation of thyrotoxicosis (Klein *et al.*, 1988). The heart rate declines slowly after exercise, resting and exercise heart rates are exaggerated and diurnal variation in heart rate is blunted. By continuous cardiac monitoring, alterations in para-sympathetic tone is found to underlie the loss of normal heart rate variability (Cacciatori *et al.*, 1996)

Subjects with hyperthyroidism have 50 to 300 per cent increases in cardiac output, compared to normal subjects. Such an increase is due to decrease in systemic vascular resistance, increase in resting heart rate. Increase in left ventricular contractility and ejection fraction, and an increase in blood volume. (Gaettinger *et al.*, 1959; Feldman *et al.*, 1986; Klein and Ojamaa, 1998, 2000). These responses are analogous to the physiological response to exercise as these changes occur in the setting of reduced systemic vascular resistance (Klein and Ojamma, 1995). The increase in left ventricular systolic and diastolic contractile function is consistent with the changes in the expression of contractile and calcium regulatory proteins (Kiss *et al.*, 1994; Ojamma *et al.*, 2000). The rate of increase in intra ventricular pressure during systole, left ventricular ejection fraction and the rate of blood flow across the aortic valve are increased in hyperthyroidism (Feldman *et al.*, 1986). Similarly, the rate of chamber relaxation (isometric relaxation) and left ventricular filling, measured as a flow across the mitral valve during diastole are increased (Mintz *et al.*, 1991). These effects suggest that thyroid hormones act directly on cardiac muscle as it was found that beta adrenergic receptor antagonists slows the heart rate but does not alter systolic and diastolic contractile performance in hyperthyroid subjects (Dillmann, 1990; Morkin *et al.*, 2000). It has been noticed that 13 per cent subjects with unexplained atrial fibrillation have evidence of hyperthyroidism (Forfar *et al.*, 1979). In 5 to 15 per cent of hyperthyroid patients, the presenting problem is atrial fibrillation. Actually sinus tachycardia is the most common rhythm disturbance in hyperthyroidism but it is overshadowed by the changes posed by atrial fibrillation (Nordyke *et al.*, 1986; Forfar *et al.*, 1979).

Varying degrees of chest pain which in some patients is angina pectoris is associated with hyperthyroidism (Polikar *et al.*, 1993).Such a pain is believed to arise from myocardial ischemia due to mismatch between cardiac oxygen supply and demand or from coronary artery spasm. As a result of which increased cardiac work in patients with thyrotoxicosis,there is naturally an increased oxygen demand and this may unmask previously unsuspected coronary artery disease (Klein, 1990; Ross, 1996; Sandler and Wilson, 1959). The process of cardiac contraction is mediated through two major contractile proteins, myosin and actin, which require calcium and the hydrolysis of high energy phosphates in the form of ATP (Dillmann, 1990; Ojamaa *et al.*, 1996). The maximum rate at which muscle fibre shortens in one determinant of systolic contractility is correlated with the inherent ATP hydrolysis activity of the myosin molecule (Dillmann, 1990; Ojamaa *et al.*, 1996). There are two isoforms of cardiac myosin proteins, each characterized by their rate of ATP hydrolysis. The myosin isoform alpha and beta, each one is encoded by a distinct gene. The alpha myosin isoform is associated with a higher level of ATPase activity than the beta isoform (Dillmann, 1990; Klein and Ojamma, 1992).

In smaller mammals, thyroid hormone is the major determinant of the contractile velocity by directly preferring the expression of alpha myosin and not beta myosin, which is suppressed(Chang *et al.*, 1997). Whereas in humans and sub-human Primates. beta myosin is the predominant isoform expressed in the adult heart which is not influenced by thyroid hormones, significantly (Klein and Ojamaa, 1992; Magner *et al.*, 1990). Therefore, in humans, the enhanced cardiac contractility due to hyperthyroidism is by the increased expression of SERCA and other cardiac genes and also due to alterations in haemo dynamic load caused by changes in systemic vascular resistance (Change *et al.*, 1997; Dillmann,1990; Mintz *et al.*, 1991; Ojamaa *et al.*, 1996).

The systemic vascular resistance is reduced from 40 to 60 per cent in hyperthyroidism and the reduction is rapid, occurs before changes in heart rate or cardiac contractility (Dipierro *et al.*, 1996; Klemperer *et al.*, 1996). When systemic vascular resistance declines, the diastolic blood pressure also declines, which results in a reflex increase in heart rate stroke volume and cardiac output. The blood flow to the skin increases with enhanced capillary flow (Graettinger *et al.*, 1959).

The renal physiology is substantially altered due to changes in cardio vascular haemodyanmics.There is a decline in renal perfusion pressure, which stimulates the release of renin, resulting in the increased angiotensin production.(Klein and Ojamaa, 1995). This may account, in part, for the high levels of angiotensin converting enzyme (ACE) in hyperthyroidism (Levy *et al.*, 1994). These effects augment renal reabsoption and expand the total body sodium content, plasma and blood volumes (Anthinisen *et al.*, 1960).

In hyperthyroidism, the resulting heart rate is fast and measured pulses bounding due to widened pulse pressure and an elevated systolic and lowered diastolic pressure. The mean blood pressure is either normal or slightly decreased (Klein, and Ojamma, 1995). However, in some percentage of older patients, have an exaggerated rise in systolic blood pressure due to the loss of normal elastic components of the capacitance arteries (Sandler and Wilson, 1959). Under this circumstances, the mean blood pressure may get elevated. The enhanced cardiac contractility is reflected in the rapid and brisk upstroke of carotid arteries (Klein *et al.*, 1988; Polikar *et al.*, 1989). Cardiac auscultations reveals loud "crisp" heart sounds and soft to moderately loud systolic murmurs may be heard as a result of increased blood flow across the aortic outflow tract (Klein and Ojamma, 1998). Systolic murmurs owing to regurgitant flow across the mitral valve may arise from mitral valve prolapse, left ventricular dilation or functional changes of the mitral valve apparatus (Marks *et al.*, 1985; Olshausen *et al.*, 1989).

Nonspecific ECG changes may occur in hyperthyroidism as shortening of the PR interval is common (Valcavi *et al.*, 1992). Patients with severe. long standing hyperthyroidism have poor cardiac contractility, low cardiac output, symptoms and signs of heart failure which includes a third heart sound and pulmonary congestion. Commonly these things occur with persistent tachycardia or atrial fibrillation, which is related to the so called rate related heart failure (Magner *et al.*, 1988; Klein and Ojamaa, 1998; Sandler and Wilson, 1959). Such subjects are at a higher risk for the development of heart failure (Graettinger *et al.*, 1999; Ikram, 1985; Sandler and Wilson, 1959). In elderly subjects with heart disease, the increased work load due to hyperthyroidism may result in further impairment of cardiac function (Polikar *et al.*, 1993).

A variety of cardiac specific functional parameters have been characterized in thyroid disorders (Biondi *et al.*, 1993; Brucker-Davis *et al.*, 1995; Klein and Ojamma, 2000; Ladenson *et al.*, 1982; Ventrella and Klein, 1993). These parameters are to measure cardiac contractility, cardiac size and cardio vascular haemodynamics (Klein and Ojamma, 2000). To define a variety of non invasive systolic and diastolic performance characteristics, measurement of surface electrocardiogram, phonocardiogram, carotid artery pressure and electrocardiogram are employed.

The standardized measure of cardiac systolic function are, the time from the onset of the QRS complex of the electrocardiogram to the opening of the aortic valve, the pre-ejection period (PEP) and the time from the opening of the aortic valve to the end of ventricular ejection time (LVET) (Bough *et al.*, 1978). It is possible to measure the time from the mitral valve closure to aortic valve opening, the iso volumic contraction time (ICT), and the time from the mitral valve opening and the iso volumic relaxation time (IVRT) with the help of echo cardiography (Biondi *et al.*, 1993; Klein and Ojamma, 2000). The Doppler electrocardiographic measurements of blood flow across the mitral valve, allow for accurate, reproducible and relevant clinical measures of systolic and diastolic function.

Subjects with Grave's disease hyperthyroidism and multi nodular goiter were found to have enhanced measures of PEP, LEVT and ICT and low systemic vascular resistance (Mintz *et al.*, 1991; Ventrella and Klein, 1993). In thyrotoxicosis the diastolic function, including the IVRT and the rate of blood flow across the mitral valve are abnormal (Montz *et al.*, 1991). Thyrotoxicosis involves two sites of action of thyroid hormones, directly on the heart and also the systemic vasculature which in turn, alters cardiac function (Dillmann, 1995; Klein and Ojamma, 1995; Klemperer, *et al.*, 1996).

Subclinical Hyperthyroidism

In subclinical hyperthyroidism, the serum levels of thyroid hormones are within the normal range but TSH levels are subnormal or low. Attimes it may be due to an intrinsic pathology of thyroid gland, which is normally known as endogenous subclinical hyperthyroidism. In clinical practice, exogenous subclinical hyperthyroidism is frequently seen as a result of supportive or replacement of thyroxin therapy (Biondi *et al.*, 2002c). Cardiac function and structure are affected in subclinical hyperthyroidism. Increased heart rate, prevalence of supraventricular arrhythmias and enhanced LV mass are often encountered with subclinical hyperthyroidism (Biondi *et al.*, 1993, 1994, 1996, 1999b, 2000, 2000c; Fazio *et al.*, 1995; Ching *et al.*, 1996; Shapiro *et al.*, 1997; Mercuro *et al.*, 2000).

This is associated with slightly enhanced systolic function and almost always with impaired diastolic function due to slowed myocardial relaxation. The mechanism responsible for the increase in LV mass is not clear and it is assumed to develop in response to a chronic haemodynamic overload due to the mild hyperkinetic cardio vascular state (Klein,1988; Fazio *et al.*, 1995).This is very much akin to overload induced cardiac hypertrophy and the LV mass increase is associated with slowed myocardial relaxation and impaired ventricular filling (Fazio *et al.*, 2004). This diastolic dysfunction results from altered intracellular calcium handling due to reduced expression of sarcoplasmic reticulum calcium ATPase and/or increased expression of phospholamban, with a consequent delayed sarcoplasmic reuptake of calcium.

The increase in LV mass is not much corresponds to an actual LV hypertrophy and is related to the duration of sub clinical hyperthyroidism than the circulating thyroid hormone levels. The increase in LV mass is due to increased wall thickness without changes of the cavity dimension (concentric remodeling) (Fazio *et al.*, 2004).In the long term, the effects of chronically increased cardiac workload on calcium metabolism would prevail on those, promoted by thyroid hormone. Abnormalities of LN morphology and function promptly normalize with the achievement of euthyroidism and beta blocker medicines (Biondi *et al.*, 1994, 1995, 1999b, 2000).

The increase in LV mass as such without a clear hypertrophy and increase of heart rate are associated with an increased risk of sudden death (Haider *et al.*, 1998; Greenland *et al.*, 1999). Diastolic dysfunction may precede development of more severe LV dysfunction and in particular elderly people, it may precipitate cardiac decompensation and congestive heart failure (CHF) (Fazio *et al.*, 2004). This is how subclinical hyperthyroidism is associated with increased cardio vascular mortality (Parle *et al.*, 2001). Therefore, it is preferable to look for subclinical hyperthyroidism in all subjects with a history of arterial fibrillation, particularly in elderly and in those with underlying cardiac disease who complain of worsening of angina pectoris or cardiac decompensation.It is suggested that in elderly subjects with systolic hypertension, a widened pulse pressure, recent onset of angina, trial fibrillation or underlying ischemic heart disease, it is mandatory to assess serum TSH levels (Polikar *et al.*, 1993; Klein and Ojamma,1998; Nordyke *et al.*, 1988; Sandler and Wilson,1959).

Hypothyroidism

In overt hypothyroidism bradycardia and systemic hypertension, with narrow pulse pressure and slightly increased mean arterial pressure, and some degree of exercise impairment are reported as quite usual (McAllister *et al.*, 1995; Klein and Ojamma, 2000). The increase in peripheral vascular resistance (Klein and Ojamma, 2000) and the increase in arterial stiffness which is likely due to myxedema of arterial wall, systemic hypertension is common in overt hypothyroid patients (Dernellis and Panareton, 2002; Obuobie *et al.*, 2002). It is also seen that systemic hypertension associated with hypothyroidism is poorly controlled by conventional treatments but prompt improvement is achieved by correcting hypothyroidism to euthyroid status (Dernellis and Panareton, 2002). Therefore, it better to assess thyroid function in all patients with pre existing systemic hypertension that is resistance to pharmacological treatments (Fazio *et al.*, 2004).The lower cardiac performance and the abnormalities in peripheral and vascular function may contribute to the poor exercise tolerance in hypothyroidism (McAllister *et al.*, 1995).

Compared to hyperthyroidism, the haemodynamic changes are just opposite with fewer signs and symptoms in hypothyroidism (Klein and Ojamma, 2000). The most commonly encountered signs are bradycardia, mild hypertension, a narrowed pulse pressure and attenuated activity on the pre cordial examination (Klein and Ojamma, 2001). In some long standing hypothyroidism, pericardial effusion and non pitting edema are seen (Ladenson *et al.*, 1992; Klein and Ojamma, 2000). The increase in LV mass does not correspond to myocardial hypertrophy but it is interstitial myxedema (Aber, 1964). Increase in wall thickness, cardiac myxedema may further compromise LV mechanics, which contributes to reduced cardiac output (Faizer *et al.*, 2004).The reduced cardiac pre load, in combination with bradycardia and slightly depressed myocardial contractility, accounts for a subnormal cardiac output in hypothyroidism (Wieshammer *et al.*, 1989; Crowley *et al.*, 1997). A 50 per cent increase in systemic vascular resistance is reported in hypothyroid patients with a slowing of diastolic relaxation and filling (Ojamma *et al.*, I996; Klein and Ojamma, 2000; Wieshammer *et al.*, 1989). Since cardiac output is usually sufficient, heart failure is rare (Ladensen, 1992).10 to 20 per cent of hypothyroid subjects have diastolic hyperetension,which in combination with the increase in vascular resistance raises cardiac after load and cardiac work (Klein and Ojamma, 2000; Bengel *et al.*, 2000). The increase in peripheral vascular resistance and arterial stiffness in hypothyroidism contributes to increased cardiac after load (Bengel *et al.*, 2000).This is considered as one of the major factors determining myocardial oxygen consumption. The disproportionate increase in myocardial oxygen uptake with respect to the level of cardiac performance in hypothyroid subjects, may explain at least in part that hypothyroidism may precipitate or worsen angina in patients with suspected or known ischemic heart disease (Keating *et al.*, 1960).This is further supported by the fact that after the initiation of thyroid hormone supplementation, an improvement in anginal symptoms are evident.

Impairment of LV diastolic function which is characterized by slowed myocardial relaxation and impaired early ventricular filling is most consistent abnormality in hypothyroid patients.LV systolic function is marginally subnormal as seen from the reduced values of ejection fraction and stroke volume (Crowley *et al.*, 1997; Wieshammer *et al.*, 1989). Hypothyroidism prolongs the cardiac action potential and the QT interval (Ojamaa *et al.*, 1999). This condition predisposes the patient to ventricular irritability (Davis and Davis, 1993). Although atrial arrhythmias are common and ventricular ectopy is rare in hypothyroid patients, it is an opposite state in hypothyroid patients (Polikar *et al.*, 1993; Klein and Ojamma, 2000). Thyroxin therapy in most of the hypothyroid patients has potentially improved the efficiency of myocardial oxygen consumption (Ojamma *et al.*, 1999) and simultaneously lowered the systemic vascular resistance (Park *et al.*, 1997; Ojamma *et al.*, 1996).

There are also alterations found in thyroid hormones related to certain cardiac problems. In subjects with uncomplicated acute myocardial infraction, serum T-3 was found to decrease by 20 per cent and serum free T-2 by 40 per cent but with a nadir on day 4 after the infarction (Franklyn *et al.*, 1984). Low serum T-3 levels were also found in patients with heart failure and the decrease is proportional to the degree of heart failure as assessed by the New York Heart Association (NYHA) functional classification (Hamilton *et al.*, 1990). However, it is not ascertained whether changes in thyroid hormone metabolism contribute to the impairment of cardio vascular function in patients with heart failure. At the same time, one can not ignore the finding that a single intravenous dose of 58mg of T-3 in 23 patients with advanced heart failure, caused an increase in cardiac output and a decrease in systemic vascular resistance, 2 hours after the administration, without any evidence of myocardial ischemia, rhythm disturbances or other untoward effects (Hamilton *et al.*, 1998). Similarly 20 patients with chronic heart failure, treated with 0.1 mg of T-4 daily for 12 weeks, improved the cardiac index and decreased the systemic vascular resistance (Moruzzi *et al.*, 1996).

Experimental hypothyroidism in animals caused diminished cardiac contractility and altered the expression of cardiac genes (Ojamma *et al.*, 1999). Replacement with T-3 increased the left ventricular function and normalized the expression of T-3 responsive genes (Hamilton *et al.*, 1990; Moruzzi *et al.*, 1998). Serum T-3 is reported to fall by 60 per cent in children who underwent bypass surgery for the correction of complex congenital heart disease and such low levels of T-3 remained up to 8 days after surgery (Bettendort *et al.*, 1997; Samvedt *et al.*, 1996). If the surgical procedures were more complex, the decrease in serum T-3 was more prolonged (Mainwaring *et al.*, 2000). Children with congenital heart diseases who were given T-3 to restore normal serum concentration after surgery, resulted in more than 20 per cent increase in cardiac output and more than 25 per cent decrease in vascular resistance when compared to untreated children (Choudhary *et al.*, 1999).

In hypothyroidism, cardiac contractile functions are impaired. The isovolumic contractile time (ICT), the pre ejection period (PEP) and the ratio of PEP/LVET (left ventricular ejection time) are uniformly prolonged (Bough *et al.*, 1978). Low doses of L-thyroxin with stepwise increase result in the decrease in systolic and diastolic measures which indicates the improvement in left ventricular function. Since thyroid hormones directly acts on the heart and simultaneously alters vascular resistance, the heart and the cardio vascular systems are responsive to variations in thyroid function (Baent *et al.*, 1994; Heimburg *et al.*, 1985; Klein and Ojamma, 2000). By using Positron Emission Tomography (PET) with acetate clearance as a measure of oxidative metabolism in hypothyroid hearts, it was established that the oxidative metabolism is decreased in hypothyroid hearts. When left ventricular efficiency is defined as substrate metabolism divided by work load was calculated, the hypothyroid heart was found metabolically inefficient and thyroxin therapy reversed the same to normalcy (Amico *et al.*, 1984).

Generally ageing is accompanied by the development of cardiac hypertrophy and interstitial fibrosis, which may be responsible for diastolic dysfunction and reduced cardio vascular performance (Fazio *et al.*, 2004).Hypothyroidism, may occasionally precipitate cardiac decompensation and heart failure in elderly people. Diastolic heart failure is strongly associated with hypothyroidism and is more prevalent in women (McDermott *et al.*, 1995). A routine assessment of thyroid function in older patients with newly diagnosed or worsening heart failure, may of significance.

Subclinical Hypothyroidism

Subclinical hypothyroidism is a common clinical problem and is defined as elevated levels of TSH with normal levels of T-3 and T-4. It is generally believed that it affects nearly 10 per cent of the general population and 20 per cent in women who are more than 60 yrs. There are several alternate names for subclinical hypothyroidism, preclinical hypothyroidism, mild hypothyroidism, mild thyroid failure and compensated hypothyroidism.

In subclinical hypothyroid patients, the most consistent cardiac abnormality recognized is LV diastolic dysfunction, characterized by sowed myocardial relaxation and impaired early ventricular filling at rest as well as in exercise (Biondi *et al.*, 1999a; Mouzani *et al.*, 2001; DiBello *et al.*, 2002; Vitale *et al.*, 2002). It is more obvious at exercise (Foldes *et al.*, 1987; Kahaly, 2000; Brenta *et al.*, 2003).Impairment of several exercise related cardio pulmonary responses were found to be associated with subclinical hypothyroidism (Kahaly, 2000). Although majority of sub clinical hypothyroid subjects do not exhibit much variation in cardiac contractility and systolic time intervals, compared to normals,yet some investigators do found impairment in systolic and diastolic functions. Such subjects had positive inotropic response to thyroxin therapy (Klein and Ojamma, 2000).

In a large cross sectional survey of post menopausal women with subclinical hypothyroidism, atherosclerotic cardio vascular disease was noticed and it was independent of the traditional risk factors, such as hypercholestremia, hypertension, diabetes and smoking (Hak *et al.*, 2000). He has studied 1149 post menopausal women in Netherlands with subclinical hypothyroidism and these were likely to have an history of myocardial infarction and had a higher frequency of calcification of aorta. Even though there is no direct evidence of hypothyroidism resulting in accelerated atherosclerosis and coronary artery disease which is associated with hypercholestermia and hypertension, yet there are some hypothyroid subjects who have the same (Klein and Ojamma, 2000; Bengel *et al.*, 2000; Keating *et al.*, 1961). In this context, it is to be noted that subclinical hypothyroidism is found to be associated with an atherogenic lipid profile, characterized by increased serum levels of LDL (Duntas, 2002; Duntas, 2002). In addition subclinical hypothyroidism may affect the haemostatic profile, thereby it may promote hyper coagulable state (Muller *et al.*, 2001). A reduced endothelium dependent flow mediated vasodilation, which is considered as an early marker of atherosclerosis, has been seen in patients with subclinical hypothyroidism (Lekakis *et al.*, 1997). Even though there is a disagreement whether such patients to be treated with thyroid hormones (Chu and Crapo, 2001; McDermott and Ridgway, 2001) yet treatment with L-thyroxin caused an improvement in most metabolic and cardio vascular abnormalities (Danese *et al.*, 2000; Biondi *et al.*, 2002c).

Conclusion

It appears over the years there is preponderance of hypothyroid symptoms and an epidemic of thyroid dysfunction which is more so in women. Further, a serious disturbance in the composition and transport of lipoproteins are associated with thyroid dysfunction/diseases. Hypothyroidism is characterized by hypercholesterolemia and a significant increase in LDL and apolipoprotein-B (Apo-A). This is due to decreased fractional clearance of LDL by a reduced number of LDL receptors in the liver. The HDL levels are either normal or sometimes elevated due to decreased activity of cholesterol ester transfer protein (CETP) and hepatic lipase (HL).These enzymes are regulated by thyroid hormones. The decrease in these enzyme activities.reduce the transfer of cholesterol ester from HDL-2 to VLDL and IDL (Intermediate low density lipoprotein) and reduce the transport of HDL-2 to HDL-3 (Duntas, 2002).In hypothyroidism there is an increased oxidation of plasma cholesterol mainly due to an altered pattern of binding and increase in cholesterol, which presents a substrate for oxidative stress. Hypothyroidism is frequently associated with diastolic hyper tension, which in conjunction with dyslipidemia can promote aterosclerosis. With thyroxin therapy, considerable improvement in lipid profiles are seen.

Hyperthyroidism exhibits an enhanced excretion of cholesterol and an increased turn over of LDL thereby a decrease in total and LDL cholesterol. However, HDL is either decreased or unaffected. Subclinical hypothyroidism is associated with lipid disorders. Total cholesterol may be normal or slightly elevated with increase in LDL and decrease in HDL. Subclinical hypothyroidism is also closely connected with endothelial dysfunction, aortic atherosclerosis and myocardial infarction (Duntas, 2002). This may enhance the risk for development of atherosclerosis and coronary artery disease (CAD) (Kahaly, 2000).

References

Amico J, Richardson V, Alpert B, and Klein I. (1984). Clinical and chemical assessment of thyroid function during therapy with Amidorane. Arch. Intern. Med., 144, 487-498.

Anthonisen P, Holst E, and Thomsen AA. (1960). Determination of cardiac output and other hemodynamic data in patients with hyper-and hypothyroidism, using dye dilution technique: Scan. J. Clin. Lab. Unvest., 12, 472-480.

Auer J, Scheibner P, and Mische T. (2001). Subclinical hypothyroidism as a risk factor for atrial fibrillation. Am. Heart J., 142, 838-842.

Averyhart-Fullard V, Fraker JD, Murphy AM, and Solaro KJ. (1994). Differential regulation of slow skeletal and cardiac troponin–1-mRNA during development and by thyroid hormone in rat heart. J. Mol. Cell. Cardiol., 26, 609-616.

Bengel FM, Nekolla SG, and Ibrahim T. (2000). Effect of thyroid hormones on cardiac function, geometry and oxidative metabolism assessed non-invasively by positron emission tomography and magnetic resonance imaging. J. Clin. Endocrinol. Metab., 85, 1822-1829.

Bettendorf M, Schmidt KG, and Grulich-Henn J. (2000). Triiodothyronin treatment in children after cardiac surgery: a double blind randomized placebo controlled study. Lancet, 356(9229), 529-534.

Betterndorf M, Schmidt KG, and Tiefenbacher U. (1997). Transient secondary hypothyroidism in children after cardiac surgery. Pediatrics, 41, 375-379.

Bindels AJ, Westendrop AG, and Frolich M. (1999). The prevalence of subclinical hypothyroidism at different plasma cholesterol levels in middle aged men and women: A need for case finding? Clin. Endocrinol., 50, 217-220.

Biondi B, Fazio S, and Carella, C. (1993). Cardiac effects of long term thyrotrophin suppressive therapy with levothyroxine. J. Clin. Endocrinol. Metab., 77, 334-337.

Biondi B, Fazio S, and Carello C. (1994). Control of adrenergic over activity by beta blockade improves quality of life in patients receiving long term supportive therapy with levothyroxine. J. Clin. Endocrinol. Metab., 78, 1028-1033.

Biondi B, Fazio S, and Cuocolo A. (1996). Impaired cardiac reserve and exercise capacity in patients receiving long term thyrotrophin suppressive therapy with levothyroxine. J. Clin. Endocrinol. Metab., 81, 4224-4228.

Biondi B, Fazio S, and Palmieri EA. (1999a). Left ventricular diastolic dysfunction in patients with subclinical hypothyroidism. J. Clin. Endocrinol. Metab., 84, 2064-2067.

Biondi B, Fazio S, and Palmieri EA. (1999b). Effects of chronic subclinical hyperthyroidism on cardiac morphology and function. Cardiologia, 44, 443-449.

Biondi B, Palmieri EA, and Fazio S. (2000). Endogenous subclinical hyperthyroidism affects quality of life and cardiac morphology and function in young and middle aged patients. J. Clin. Endocrinol. Metab., 85, 4702-4705.

Biondi B, Palmieri EA, and Lombardi G. (2002a). Effects of thyroid hormone on cardiac function: the relative importance of heart rate, leading conditions, and myocardial contractility in the regulation of cardiac performance in human hyperthyroidism. J. Clin. Endocrinol. Metab., 87, 968-974.

Biondi B, Fazio S, and Plamieri EA. (2002b). Mortality in elderly patients with subclinical hyperthyroidism. Lancet, 359, 799-800.

Biondi B, Palmieri EA, and Lombardi G. (2002c). Effects of subclinical thyroid dysfunction on the heart. Am. Intern. Med., 137, 904-914.

Bahaa M, Fadil Ellahham E, Ringel M, Linday J, Wartofsky L, and Burman KD. (2000). Hyperthyroid heart disease. Clin.Cardiol., 23, 402-408.

Bahouth SW. (1991). Thyroid hormone transcriptionaly regulates the beta-1-adrenergic receptor gene in cultured ventricular myocytes. J. Biol. Chem., 266, 15863-15869.

Brauman A, Algour M, Gilboa Y, Ramoi Y, Golik A, and Stryfer D. (1985). Mitral valve prolapse in hypothyroidism of two different origins. Br. Heart J., 53, 374-379.

Brent GA. (1994). The molecular basis of thyroid hormone action. N. Eng. J. Med., 331, 847-853.

Brenta G, Mutti LA, and Schnitman M. (2003). Assessment of left ventricular diastolic function of radio nucleotide ventriculography at rest and exercise in subclinical hypothyroidism and its response to L-thyroxine therapy. Am. J. Cadiol., 91, 1327-1330.

Cacciatri V, Bellavere F, and Pezzarossa A. (1996). Power spectral analysis of heart rate in hyperthyroidism. J. Clin. Endocrinol. Metab., 81, 2828-2835.

Canaris GJ, Manowitsy NR, Mayor, M. (2000). The Colorado thyroid disease prevalence study. Arch. Intern. Med., 100, 526-534.

Chang KC, Figueredo VM, Schreur SH, Kartiya K, Weiner MW, Simposson PC, and Camacho SA. (1997). Thyroid hormone improves function and Ca^{++} handling in pressure overload hypertrophy: Association with increased sarcoplasmic reticulum Ca^{++} ATPase and myosin heavy chain in rat hearts. J. Clin., Invest., 100, 1742-1749.

Ching GW, Franklyn JA, and Stellard TJ. (1996). Cardiac hypertrophy as a result of long term thyroxine therapy and thyrotoxicosis. Heart, 75, 363-368.

Chu JW, and Crapo, LM. (2001). Treatment of subclinical hypothyroidism is seldom necessary. J. Clin. Endocrinol. Metab., 86, 4591-4599.

Cimochowski GE, Harostock M, Foldes PG. (1997). Minimal operative mortality in patients undergoing coronary artery bypass with significant left ventricular dysfunction by maximization of metabolic and mechanical support. J. Thoracic. Surg., 113, 655-666.

Crowley, WF Jr., Ridgeway EC, and Bough EW. (1977). Non-invasive evaluation of cardiac function in hypothyroidism: Response to gradual thyroxine replacement. N. Eng. J. Med., 296, 1-6.

Chowdhury D, Parnell V, Ojamma K, Boxer R, Cooper R, and Klein I. (1999). Usefullness of tri iodothyronine (T-3) treatment after surgery for complex congenital heart disease in infants and children. Am. J. Cardiol., 84, 1107-1109.

Croteau W, Davey JC, Galton VA, and St.Germaine DL. (1996). Cloning of the mammalian type-11 iodothyronine deiodinase,a selenoprotein differentially expressed and regulated in human and rat brain and other tissues. J. Clin. Invest., 98, 405-417.

Channick BJ, Adlin EV, Marks AD, Denenberg SBS, McDougoh MT, Chakko CS, and Spann JF. (1981). Hypothyroidism and Mitral valve prolapse. N. Eng. J. Med., 305, 497-500.

Daness MD, Ladensen FW, Meinert CL. (2000). Effect of thyroxine therapy on serum lipoproteins in patients with mild thyroid failure-a quantitative review of the literature. J. Clin. Endocrinol. Metab., 85, 2993-3000.

Davis, PJ, and Davis FB. (1993). Cellular actions of thyroid hormone and myocardial function. Am. Thoracic. Surg., 56(Suppl), 516-523.

DeGroot WJ, and Leonard JJ. (1970). Hyperthyroidism as a high cardiac output state. Am. Heart. J., 79, 265-275.

Dillmann WH. (1990). Biochemcial basis of thyroid hormone action in the Heart. Am. J. Med., 88, 629-630.

Duntas LH, Mantzou E, Kontras DA. (2002). Circulating levels of oxidized low density lipoprotein in overt and mild hypothyroidism. Thyroid, 12, 1003-1007.

Duntas LH. (2002). Thyroid disease and lipids. Thyroid, 12, 287-293.

D'Amati G, diGioia CBT, Menticcia Pistilli D, Proietti-Pannunzi L, Miraldi F, Gallo P, and Celi FS. (2001). Increased expression of thyroid hormone receptor isoforms in end stage human congestive heart failure. J. Clin. Endocrinol. Metab., 85, 2080-2084.

Everts ME, Verhoeven FA, and Bezstarosti K. (1996). Uptake of thyroid hormone in neonatal rat cardiac myocytes. Endocrinology, 137, 4235-4242.

Faber J, Petersen L, and Wiinberg N. (2002). Hemodynamic changes after levothyroxine treatment in subclinical hypothyroidism.Thyroid, 12, 319-324.

Fazio S, Biondi B, and Carello C. (1995). Diastolic dysfunction in patients on thyroid stimulating hormone suppressive therapy with levothyroxine: beneficial effects of beta blockade. J. Clin. Endocrinol. Metab., 80, 2222-2226.

Feldman T, Borow KM, and Sarne DH. (1986). Myocardial mechanics in hyperthyroidism: Importance of left ventricular loading conditions, heart rate and contractile state. J. Am. Coll. Cardiol., 7, 967-974.

Foldes J, Istavanfy M, and Halmagyi H. (1987). Hypothyroidism and the heart: Examination of left ventricular function in subclinical hypothyroidism. Acta. Med. Hung., 44, 337-347.

Fommei E, and Iervasi G. (2002). The role of thyroid hormone in blood pressure homeostatis:Evidence from short term hypothyroidism in humans. J. Clin. Endocrinol. Metab., 87, 1996-2000.

Franklyn JA, Gammage MD, and Ramsden DS. (1984). Thyroid status in patients after acute myocardial infarction. Clin. Sci(Lond)., 67, 585-590.

Fredlund BD, and Olsson SB. (1983). Long QT interval and ventricular tachycardia of "torsadede Pointe" type in hypothyroidism. Acta. Med. Scand., 213, 231-235.

Friedman MJ, Okada BD, and Ewy GA. (1982). Left ventricular systolic function in hyperthyroidism. Am. Heart. J., 104, 1303-1308.

Fazio S, Palmieri A, Lombardi G, and Biondi B. (2004). Effects of thyroid hormone on the cardio vascular system. Recn. Prog. Horm. Res., 59, 31-50.

Forfar JC, Muir AL, Sawers SA, and Toft AD. (1982). Abnormal left ventricular function in hyperthyroidism: evidence of possible reversible cardio myopathy. N. Eng. J. Med., 307, 1165-1170.

Forfar JC, Miller HC, and Toft AD. (1979). Occult thyrotoxicosis: a reversible cause of "Idiopathic" atrial fibrillation. Am. J. Cardiol., 44, 9-12.

Fullerton MJ, Stuchbury S, Krozoswski ZS, and Funder JW. (1990). Altered thyroidal status and in vivo synthesis of atrial natriuretic peptide in the rat heart. Mol. Cell. Endocrinol., 69, 227-233

Gibson JG, and Harris AW. (1939). Cinical studies on the blood volume. V. Hyperthyroidism and Myxedema. J. Clin. Invest., 18, 59-65.

Gick GG, Melikian J, and Ismail-Beigi F. (1990). Thyroidal enhancement of ratio myocardial Na-K ATPase:preferential expression of alpha 2 activity and mRNA abundance. J. Membr. Biol., 115, 273-282.

Graettinger JS, Muenster JJ, Silverstone LA. (1959). A correlation of clinical and hemodynamic studies in particular with hyperthyroidism with and without congestive heart failure. J. Clin. Invest., 39, 1316-1327.

Gilligan DM, Ellenbagen KA, Epstein AE. (1996). The management of atrial fibrillation. Am. J. Med., 101, 413-421.

Greenland P, Daviglus ML, and Dyer AR. (1999). Resting heart rate is a risk factor for cardio vascular and non cardio vascular mortality. The Chicago heart association detection project in industry. Am. J. Epidemiol., 149, 853-862.

Haider AW, Larson MG, Benjamin EJ. (1998). Increased left ventricular mass and hypertrophy are associated with increased risk for sudden death. J. Am. Coll. Cardiol., 32, 1454-1459.

Hak AE, Pols HAP, and Visser TJ. (2000). Subclinical hypothyroidism is an independent risk factor for atherosclerosis and myocardial infarction in elderly women: The Rotterdam study. Ann. Intern.Med., 132, 270-278.

Hamilton MA, Stevensen LW, and Lun M. (1990). Altered thyroid hormone metabolism in advanced heart failure. J. Am. Coll. Cardiol., 15, 91-95.

Hamilton MA, Stevensen LW, and Fonarow GC. (1998). Safety and hemodynamic effects of intravenous tri iodothyronine in advanced congestive heart failure. Am. J. Cadiol., 81, 443-447.

Holland FW, Brown PSJr., Weintrant BD, and Clarke RE. (1991). Cardiac pulmonary bypass and thyroid function, a "euthyroid sick syndrome". Ann. Thoracic. Surg., 52, 46-50.

Hammond HK, White FC, Buxton IL, Saltzstein P, Brunton LL, and Longhurst JC. (1987). Increased myocardial beta receptors and adrenergic responses in hyperthyroid pigs. Am. J. Physiol., 252, 11283-11290.

IshikawaT, Chijiwa T, Hagiwara M, Mamiya S, and Hidaka H. (1989). Thyroid hormone directly interact with vascular smooth muscle strips. Mol. Pharmacol., 35, 760-765.

Izuma S, Nadal-Ginard B, and Madhavi V. (1986). All members of the MHC multigenic family respond to thyroid hormone in highly tissue specific manner. Science, 231, 597-600.

Kahaly G, (2000). Cardio vascular and atherogeneic aspects of subclinical hypothyroidism. Thyroid, 10, 665-679

Klein I, and Ojamma K. (2001). Thyroid hormone and the cardio vascular system. N. Engl. J. Med., 344, 501-509.

Kinugawa K, Minobe WA, Wood WM, Ridgway EC, Baxter JD, Riberio RCI, Tawadrous MF, Lowes BA, Long CS, and Bristow MR. (2001). Signalling pathways responsible for fetal gene induction in the failing human heart. Evidence for altered thyroid hormone receptor gene expression.Circulation, 103, 1089-1094.

Kinugawa K, Yonekura K, Ribeiro RCI, Eto, Y, Aoyagi T, Baxter SD, Camacho SA, Bristow MR, Long CS, and Simposn PC. (2001). Regulation of thyroid hormone receptor isoforms in physiological and pathological cardiac hypertrophy. Cir. Res., 89, 591-598.

Kahaly GJ, Nieswandt J, and Wagner S. (1998). Ineffective cardio respiratory function in hyperthyroidism. J. Clin. Endocrinol. Metab., 83, 4075-4078.

Kahaly GJ, Wagner S, and Nieswandt J. (1999). Stress echocardiography in hyperthyroidism. J. Clin. Endocrinol. Metab., 84, 2308-2313.

Katzeft HL, Powell SR, and Ojamma K. (1997). Alterations in cardiac contractility and gene expression during low T-3 syndrome, prevention with T-3. Am. J. Physiol., 273, E951-E956.

Keating FR, Parkin TW, and Selby JB. (1960). Treatment of heart disease associated with myxedema. Prog. Cardiovasc., 3, 304-381.

Kiss E, Jakab G, and Kranias EG. (1994). Thyroid hormone induced alterations in phospholamban protein expression, regulatory effects on sarcoplasmic reticulum Ca^{++} transport and myocardial relazation. Circ. Res., 75, 245-251.

Kiss E, Brittsan AG, and Elds I. (1998). Thyroid hormone induced alteration in phspholamban deficient mouse hearts. Circ. Res., 832, 608-613.

Klein I. (1988). Thyroxine induced cardiac hypertrophy; time course of development and inhibition by propanolol. Endocrinology, 123, 203-210.

Klein I, and Levey GS. (1984). Unusual manifestation of hyperthyroidism. Arch. Intern. Med., 144, 123-128.

Klein I. (1989). Thyroid hormone and high blood pressure. *In:* Laragh JH, Brenner BM, and Kalan MM. (Eds.), *Endocrine Mechanisms in Hypertension*. Vol-2, Raven Press, New York, pp. 61-80.

Klein I, and Ojamaa K. (1998). Thyroid hormone treatment of congestive heart failure. Am. J. Cardiol., 81, 490-491.

Klein I, and Ojamma K. (2000). The cardio vascular system in hypothyroidism. *In:* Braverman IE, and Utiger RD. (Eds.), *The Thyroid: A Fundamental and Clinical Text*. Lippincott-Williams and Witkins Philadelphia, pp. 777-782.

Klein I, and Ojamma K. (2001). Thyroid hormone and cardio vascular system. N. Engl. J. Med., 344, 501-509.

Klemperer JD, Klein I, and Gomez M. (1995). Thyroid hormone treatment after coronary artery bypass surgery. N. Engl. J. Med., 333, 1522-1527.

Krahn AD, Klein I, and Kerr CR. (1982). How useful is thyroid function testing in patients with recent onset atrial fibrillation. Arch. Intern. Med., 72, 903-906.

Kantharia BK, Richards HB, and Battaglia J. (1995). Reversible dilated cardiomyopathy: An unusual case of thyrotoxicosis. Am. Heart. J., 129,1030-1032.

Kontos HA, Shapiro W, and Mauch PJr. (1965). Mechanisms of certain abnormalities of the circulation to the limbs in thyrotoxicosis. J. Clin. Invest., 44, 947-950.

Ladenson PW. (1993). Thyrotoxicosis and heart, something old and something new. J. Clin. Endocrinol. Metab., 77, 332-333.

Ladenson PW, Sherman SI, and Baughman RI. (1992). Reversible alterations in myocardial gene expression in a young man with dilated cardio myopathy and hypothyroidism. Proc. Natl. Acad. Sci., USA, 89, 5251-5255.

Lekakis J, Papamicheal C, and Alevizaki H. (1997). Flow mediated endothelium dependent vasodilatation is impaired in subjects with hypothyroidism, borderline hypothyroidism, and high normal serum thyrotrophin (TSH) values. Thyroid, 7, 411-414.

Levey GS, and Klein I. (1990). Catecholamine-thyroid hormone interactions and the cardio vascular manifestation of hyperthyroidism. Am. J. Med., 88, 642-646.

Lewis BS, Ehrenfold EN, and Lewis H. (1979). Echocardiographic IV function in thyrotoxicosis. Am. Heart. J., 97, 460-468.

Luboshitsky R, Aviv A, and Herer P. (2002). Risk factors for cardio vascular disease in women with subclinical hypothyroidism. Thyroid, 12, 421-425.

Liggett SB, Shah SD, and Cryer PE. (1989). Increased fat cell and skeletal muscle beta adrenergic receptor densities but unattended lipolytic, glycemic, and cardiac chronotropic in vivo in experimental human thyrotoxicosis. J. Clin. Invest., 83, 803-810.

Mainwaring RD, Capparelli E, Schell K, Acosta K, and Nelson JC. (2000). Pharmacokinetic evaluation of tri iodothyronine supplementation in children after modified Fontan procedure.Circulation, 101, 1423-1429.

Mainwaring RD, Lamberti JJ, Carter TIJr., and Nelson JC. (1994). Reduction in tri iodothyronine levels following modified Fontan procedure. J. Card. Surg., 9, 322-331.

Marker AD, Channick BJ, and Adlin ED. (1985). Chronic thyroiditis and mitral valve prolapse. Ann. Intern. Med., 102, 479-463.

Marquesee E, Hadsen ST, and Utiger RD. (1998). Subclinical thyrotoxicosis. Endocrinol. Metab. Clin. North. Am., 27, 37-49.

Magner J, Clark W, and Allenby P. (1988). Congestive heart failure and sudden death in a young woman with thyrotoxicosis. West. J. Med., 110, 759-760.

McAllister RM, Delp MD, and Longhtin MH. (1995). Thyroid status and exercise tolerance: Cardiovascular and metabolic consideration. Sport. Med., 20, 189-198.

McDermutt MJ, and Ridgway EC. (2001). Clinical hypothyroidism is mild thyroid failure and should be treated. J. Clin. Endocrinol. Metab., 86, 4585-4590.

McDermutt MM, Feinglass J, and Sy J. (1995). Hospitalised congestive heart failure patients with preserved versus abnormal left ventricular system function: clinical characteristics and drug therapy. Am. J. Med., 99, 629-655.

Mercure G, Panzuto MG, and Bina A. (2000). Cardiac function, physical exercise capacity and quality of life during long term thyrotrophin supportive therapy with levothyroxine:effect of individual dose tailoring. J. Clin. Endocrinol. Metab., 85, 159-164.

Mintz G, Pizzarelle R, and Klein I. (1991). Enhanced left ventricular diastolic function in hyperthyroidism: noninvasive assessment and response to treatment. J. Clin. Endocrinol. Metab., 73, 146-150.

Monzani F, DiBello V, and Caraccio N. (2001). Effects of levothyroxine on cardiac function and structure in subclinical hypothyroidism: a double blind, placebo controlled study. J. Clin. Endocrinol. Metab., 86, 1110-1115.

Morkin E. (1993). Regulation of myosin heavy chain genes in heart. Circulation, 87, 1451-1460.

Morkin E, Pennock GD, and Spooner PH. (2002). Clinical and experimental studies on the use of 3,5-di iodothyropropionic acid, a thyroid hormone analogue in heart failure. Thyroid, 12, 527-533.

Moruzzi P, Doria E, and Agostoni PG. (1996). Medium term effectiveness of L-thyroxine treatment in idiopathic dilated cardiomyopathy. Am. J. Med., 101, 459-460.

Muller B, Tsakiris DA, and Roth CB. (2001). Haemostatic profile in hypothyroidism as potential risk factor for vascular thrombosis disease. Eur. J. Clin. Invest., 31, 131-137.

Mullis-Janson SL, Argenziaano M, and Corwin S. (1999). A randomized double blind study of the effect of tri iodothyronine effect on cardiac function morbidity. Cardiovasc. Surg., 117, 1128-1134.

Montereggi A, Marconi P, Olivotti I, Castelli G, Dolara A, Luisi ML, and Ghen RG. (1996). Signal averaged P wave duration and risk of paroxysmal atrial fibrillation in hypothyroidism. Am. J. Cardiol., 77, 266-269.

Nakazawa HK, Sakuri K, and Humada N. (1982). Management of atrial fibrillation in the past thyrotoxic state. Am. J. Med., 72, 903-906.

Napoli R, Biondi B, and Gyardsole V. (2001). Impact of hyperthyroidism and its correction on vascular reactivity in humans. Circulation, 104, 3076-3080.

Nordyke RA, Gilbert FIJr., and Harada AS. (1988). Grave's disease: influence of age on clinical findings. Arch. Intern. Med., 148, 626-631.

Obuobie K, Smith J, and Evans LM. (2002). Increased central arterial stiffness in hypothyroidism. J. Clin. Endocrinol. Metab., 87, 4662-4666.

Ojamma K, Kleperer JD, and Klein I. (1996a). Acute effects of thyroid hormones on vascular smooth muscle. Thyroid, 6, 505-512.

Ojamma K, Klemperer JD and MacGilvray SS. (1966b). Thyroid hormones and hemodynamic regulation of beta myosin heavy chain promoter in the heart. Endocrinology, 137, 802-808.

Ojamma K, Sabet A, Kennessey A, and Shenoy R. (1999). Regulation of rat cardiac kVL-5 gene expression by thyroid hormone is rapid and chamber specific. Endocrinology, 140, 3170-3176.

Orlowski J, and Lingrel JB. (1991). Thyroid and glucocorticoid hormones regulate the expression of multiple Na, K-ATPase genes in cultured neonatal rat cardiac myocytes. J. Biol. Chem., 266, 15863-15869.

Ojamma K, Klein I, Sabet A, and Steinberg SF. (2000). Changes in adenyl cyclase isoforms as a mechanism of thyroid hormone modulation of cardiac beta adrenergic receptor responsiveness.Metabolism, 49, 275-279.

Park KW, Dai HB, Ojamma K, Lowenstein E, Klein I, and Sellke FW. (1997). The direct vaso motor effect of thyroid hormone on rat skeletal muscle resistance arteries. Anaesth. Analg., 85, 734-738.

Parle JV, Maisonneuve P, and Sheppard MC. (2001). Production of all cause and cardiovascular mortality in elderly people from one low serum thyrotrophic result: a 10 year cohort study. Lancet, 358, 861-865.

Peterson P, and Hansen JM. (1988). Stroke in thyrotoxicosis with arterial fibrillation. Stroke, 19, 15-18.

Polikar R, Burger AG, Scherrer V, and Nicol P. (1993). The thyroid and the heart. Circulation, 87, 1435-1441.

Portman MA, Fearneyhough C, Ning XH, Duncan BW, Rosenthal GL, and Lupinetti FM. (2000). Triiodothyronine repletion in infants during cardiac pulmonary bypass for congenital heart disease. J. Thorasc. Cadiovas. Surg., 120, 604-608.

Perk M, and O'Neil BJ. (1997). The effect of thyroid hormone therapy on angiographic coronary artery disease progress. Can. J. Cardiol., 13, 273-276.

Reinhardt W, Mocker V, Jocken hovel F, Olbricht T, Reinwein D, Mann K, and Sadony V. (1997). Influence of coronary bypass surgery on thyroid hormone parameters. Horm. Res., 47, 1-8.

Resuick LM, and Largh JH. (1982). Plasma renin activity in syndrome of thyroid hormone excess and deficiency. Life Sci., 300, 585-586.

Ross DS. (1996). Subclinical thyrotoxicosis. *In*: Braverman L, and Utiger R. (Eds.), *The Thyroid*, Lippincott-Raven, Philadelphia, pp. 1016-1064.

Sachs RN, Vignot M, Modigliani E, Perenec-Cardinelli J, Kempeny J.L, Amouroax J, Hatt PY, and Lanfranci J. (1986). Absence of ultra structural histological lesion of the myocardium in cardiac insufficiency of hyperthyroidism. Ann. Med. Interne (Paris), 137, 375-378.

Sagaguchi V, Cui G, and Sen L. (1996). Acute effects of thyroid hormone on inward rectifier potassium channel currents in guinea pig ventricular myocytes. Endocrinology, 137, 4744-4749.

Staub JJ, Althus BU, Engles H, Rtfi R, Trabacca P, Marquaradt K, Berkhardt D, Girrard J, and Weinbraub BD. (1992). Spectrum of subclinical and overt hypothyroidism: effect on thyrotropin, prolactin and thyroid reserve and metabolic impact on peripheral target tissues. Am. J. Med., 92, 631-642.

Salvatore D, Tibor B, Harney SW, and Larsen PR. (1996). Molecular biology and biochemical characterization of the human type-2 selenodeiodinase. Endocrinology, 137, 3308-3315.

Saito I, Ito K, and Saruta T. (1983). Hypothyroidism as a cause of hypertension. Hypertension, 5, 112-115.

Sawin CT, Geller A, and Wolf PA. (1994). Low serum thyrotropin concentrations as a risk factor for atrial fibrillation in older patients. N. Engl. J. Med., 332, 1249-1252.

Shapiro LE, Sievert R, and Ong I. (1997). Minimal cardiac effects in asymptomatic athyroreotic patients, chronically treated with thyrotropin suppression doses of L-thyroxine. J. Clin. Endocrinol. Metab., 82, 2592-2595.

Sun ZQ, Ojamma K, and Nakauma TV. (2001). Thyroid hormone increases pace maker activity in rat neonatal atrial myocytes. J. Mol. Cell. Cardiol., 33, 811-824.

Sudicani P, Hein HO, and Gyntelberg F. (1992). Serum selenium concentration and risk of ischemic heart disease in a prospective cohort study of 3000 males. Atherosclerosis, 96, 33-42.

Suzuki T, Nagai R, and Yazaki Y. (1998). Mechanism of transcriptional regulation of gene expression in smooth muscle cell. Clin. Res., 82, 1238-1242.

Sussman MA. (2001). When the thyroid speaks, the heart listens. Cir. Res., 89, 472-480.

Sylven C, Jasson E, Sotonyi P, Waagstein F, and Bronnegard M. (1992). Cardiac nuclear hormone receptor mRNA in heart failure in man. Life Sci., 59, 1917-1922.

Theilan E, and Wilson WR. (1967). Hemodynamic effects of peripheral vasoconstriction in normal and thyrotoxic subjects. J. Appl. Physiol., 22, 207-210.

Tsai MJ, and O'Malley BW. (1994). Molecular mechanisms of steroid / thyroid super family members. Ann. Rev. Biochem., 63, 451-486.

Tsika RW, Bahl JJ, Leinwand LA, and Makin E. (1990). Thyroid hormone regulators expression of a transfected human alpha myosin heavy chain fusion gene in fetal rat heart cells. Proc. Natl. Acad. Sci., USA, 87, 379-383.

Tunbridge WMG, Evered DC, and Hall R. (1977). The spectrum of thyroid disease in a community: The Whickham survey. Clin. Endocrinol. (Oxford), 7, 481-493.

Tseng KH, Walfish PG, and Persand JA. (1989). Concurrent aortic and mitral valve echocardiography permits measurements of systolic time intervals as an index of peripheral tissue thyroid functional status. J. Clin. Endocrinol. Metab., 69, 633-638.

Valcavi R, Menozz C, and Roti E. (1992). Sinus node function in hypothyroid patients. J. Clin. Endocrinol. Metab., 75, 239-243.

Vanderpump MPJ, Tunbridge WMG, and French JM. (1995). The incidence of thyroid disease in the community: 20 year follow up of the Whickam survey. Clin. Endocrinol., 43, 55-69.

Vanderpump MPJ, and Tunbridge WMG. (2000). The epidemiology of thyroid disease. *In:* Braverman LE, and Utiger RD. (Eds.), *The Thyroid: A Fundamental and Clinical Text.* 8th Edition, Williams and Witkins, Philadelphia, Lippincott, pp. 467-473.

Von Oishansen K, Bischoft J, and Kahaly G. (1989). Cardiac arthymias and heart rate in human hyperthyroidism. Am. J. Cardiol., 69, 930-933.

Vitale G, Galderisi M, and Lupoli BA. (2002). Left ventricular impairment in subclinical hyperthyroidism assessed by a new ultrasound tool pulsed tissue Doppler. J. Clin. Endocrinol. Metab., 87, 4350-4355.

Walker JD, Crawford FA, and Kato S. (1994). The novel effects of 3.5,5tri iodoL-thyronine on myocyte contractile function and beta adrenergic responsiveness in dilated cardio myopathy. J. Thorasc. Cardiovasc. Surg., 108, 672-679.

Waldstein SS. (1966). Thyroid and catecholamine interactions. Ann. Rev. Med., 17, 123-132.

Wieshammer S, Keok FS, and Waitzinger J. (1989). Acute hypothyroidism slows the left ventricular diastolic relaxation. Can. J. Physiol. Pharmacol., 67, 1007-1010.

Yamagishi T, Yanagisawa T, Satoh K, and Taira N. (1994). Relaxant mechanism of cyclic AMP inducing agents in porcine coronary artery. Eur. J. Pharmacol., 251, 253-262.

Yen PM. (2001). Physiological and molecular basis of thyroid hormone Action. Physiol. Rev., 81, 1097-1142.

Zarin-Herzberg A, Marques J, Sukovich D, and Periasamy M. (1994). Thyroid hormone receptor modulates the expression of the rabbit cardiac sarco(endo)plasmic reticulum Ca^{2+} ATPase gene. J. Biol. Chem., 269, 1460-1467.

Zwaveling J, Paffendorf M, and Van Zwicten PA. (1997). The direct effects of thyroid hormones on the rat mesenteric resistance arteries. Fundam. Clin. Pharmacol., 11, 41-46.

5

Glycodelin A: Protector of Fetus from Maternal Immune Attack

Anjali A Karande* and Debaditya Mukhopadhyay

Department of Biochemistry, Indian Institute of Science, Bangalore – 560 012, India

ABSTRACT

Graft rejection is an immunological response displaying the attributes of specificity, memory and self/non self recognition. The rejection is caused by an elaborate and very effective cell mediated immune response to allo-antigens (primarily MHC molecules) expressed on grafts. The fetal genome is half paternal yet under normal conditions the mother does not reject it. Tolerance of fetus, the semi-allograft is a medical enigma that has stimulated research for more than five decades now. Of the different mechanisms that have been reported so far to explain this tolerance, one is the expression of a progesterone regulated protein named glycodelin, earlier known as placental protein 14 (PP14). Glycodelin is a 162 amino acid glycoprotein that bears ~50 per cent identity with bovine β-lactoglobulin because of which it has been placed in the Lipocalin superfamily. The protein is synthesized by the endometrium and decidua in women and has been implicated in the process of endometrial preparation for blastocyst implantation under progesterone regulation. In normally cycling women, glycodelin levels peak during the late luteal phase and remain high till the first few days of the next cycle. In fertile cycles, the glycodelin level remains elevated until about 14 weeks of gestation, declining thereafter. The most significant finding in this context is the presence of subnormal concentrations of circulatory glycodelin in women experiencing habitual abortions. Substantial evidence exists in literature to

**Corresponding Author*: E-mail: anjali@biochem.iisc.ernet.in suggest that glycodelin exhibits immunosuppressive properties and is an indispensable macromolecule in the maternal system for the establishment, maintenance and progression of pregnancy. Recent data from our laboratory suggest that glycodelin induces apoptosis in activated peripheral blood T cells. To understand the ways by which this phenomenon is exhibited has been our research objective that will be discussed in the article.

Glycodelin: An Introduction

The mechanism of tolerance of the fetal 'semi allograft' by the maternal immune system is still unclear. Over the years at least ten different hypotheses have been put forth to explain this phenomena and it is believed that all ten mechanisms elaborated by both mother and fetus work in synergy for successful pregnancy and involve several factors (Thellin et al, 2000). Some of the effects occur on the whole maternal immune system, principally mediated through soluble endocrine factors, others occur locally at the feto-maternal interface, principally through cellular contact, or paracrine and autocrine soluble factors. Many hormones and cytokines are present in the placenta, of which some modulate immune functions. The most important molecule is progesterone which can induce synthesis of regulatory molecules. It is also reported to displace the Th1/Th2 balance towards Th2. Apart from progesterone, human placental growth hormone may also play a role in fetal tolerance (Thellin et al, 2000). One molecule the expression of which is regulated by progesterone is the protein Glycodelin A (GdA; A for amniotic fluid), also known as placental protein 14 (PP14) that belongs to the family of hydrophobic molecule transporter proteins known as lipocalins, secreted by the late secretory phase endometrium during the menstrual cycle (Julkunen *et al.*, 1986) and has been proposed as a biochemical marker of endometrial function in women (Check *et al.*, 1993). It has been well documented in literature that GdA exhibits immunosuppressive properties in that it inhibits the clonal expansion of activated T cells (Bolton *et al.*, 1987). Data from our laboratory suggest that GdA induces apoptosis in human activated peripheral blood lymphocytes (Mukhopadhyay *et al.*, 2001) and thereby modulates the maternal immune response to the fetal allo-antigens.

Molecular Characteristics of Glycodelin

Gd has been placed in the lipocalin superfamily by the virtue of its ~54 per cent identity with bovine beta lactoglobulin (βLG). The protein is synthesized as a 180 amino acid polypeptide, 18 amino acids of which get cleaved during its entry into the endoplasmic reticulum. Gd is dimeric and the extent of glycosylation is ~17 per cent w/w. The monomeric form of Gd is ~24 kDa. The two main isoforms GdA (amniotic fluid) and GdS (seminal plasma) are identical in their primary structure and are glycosylated on the same two asparagines but the glycan compositions vary. The oligosaccharide chains of GdA terminate in sialic acid residues while those of GdS do not carry any sialic acid residues but are heavily fucosylated (Dell *et al.*, 1995). The N-terminal amino acid sequence of the protein was reported in 1987 (Huhtala *et al.*, 1987) where it was noted that the protein is similar to βLG and plasma retinol binding protein (RBP). The complete amino acid sequence was reported in 1988 (Julkunen *et al.*, 1988) after cloning the glycodelin gene from endometrial complementary DNA (cDNA) library

Isoforms of Glycodelin

There are two kinds of isoforms possible for glycodelin, splice variants and differentially glycosylated full length protein. Though splice variants are reported at the mRNA level for hematopoietic glycodelin (Morrow *et al.*, 1994) and GdS (Koistinen *et al.*, 1997), they are not detected at the protein level for GdS. More remarkable are the isoforms generated by differential glycosylation of the protein as in the case of GdA, GdS and zona inhibitory factor (ZIF), a recently identified variant of GdA (Chiu *et al.*, 2003a; Chiu *et al.*, 2003b). The nature of complex glycosylation present on glycodelin A and S has been studied extensively (Dell *et al.*, 1995; Koistinen *et al.*, 1996). Native glycodelin is glycosylated at N28 and N63 (another putative glycosylation site N85 is not glycosylated in the native protein). Both the glycosylation sites in GdA have complex glycosyl groups whereas only N63 in GdS has complex glycosylation. N28 glycosylation in GdS contains branched 5–7 mannose residues. All the different forms of complex glycosylation on GdA are sialylated and contain 1–2 fucose residues, whereas, GdS glycosylation (N63) does not contain any sialic acid. Instead it is rich in fucose residues. Koistinen *et al.*, in their 1999 review on glycodelin, proposed that native folding of glycodelin-A and glycodelin-S is not influenced by the differences in glycosylation or by ligand binding, based on thermodynamic parameters of reversible denaturation. This may not be true as the thermodynamic parameters essentially indicate similarity in the secondary structures and gross folding. There may be slight difference in tertiary structure detectable by other methods which may lead to differences in biological function.

Gene Sequence and Chromosome Localization of Glycodelin

Glycodelin is localized on the chromosome 9q34 (Van-Cong *et al.*, 1991) together with other immunocalins. Structure of the glycodelin gene was reported in 1990 (Vaisse *et al.*, 1990). The gene is 5.05 kb long having seven exons separated by six introns, the exact organization found in βLG. Sequencing of the promoter region of the glycodelin gene revealed that there is a consensus glucocorticoid response element (Mueller *et al.*, 2001) adding support to the progesterone inducible expression of the protein. Further, there is a short gene duplication of 400 base pair (bp) lying at position 2,660. This duplication is homologous to 100 bp of exon 4 and 300 bp of intron 4, including 180 bp corresponding exactly to the right arm of an Alu element lying on the complementary strand. This homology suggests that this duplication may have arisen through a retrotransposition event.

Glycodelin Gene Expression

Glycodelin is synthesized by the endometrium under the control of progesterone. The protein is also detectable in the serum during these periods. The circulatory level of glycodelin is two orders lower than the endometrial or amniotic fluid level, thus, circulatory Gd level is not physiologically relevant. Molecular details of the gene expression by progesterone are still unclear. A recent study has reported that two of the three putative Sp1 sites present in glycodelin promoter region (-1900 bp to +30 bp) are important for human progesterone receptor (hPR) mediated GdA expression (Gao *et al.*, 2001). There is a glucocorticoid response element in the glycodelin promoter which can be modulated directly by progestins and antiprogestins (Taylor *et al.*, 2000). Estrogen has no effect on the synthesis of glycodelin (Gao *et al.*, 2001) and the effect of relaxin on the production of this protein is controversial (Taylor *et al.*, 2000; Stewart *et al.*, 1997). There is a report on regulation of glycodelin gene expression by chorionic gonadotrophin (CG) [Hausermann *et al.*, 1998]. Higher level of GdA expression in case of endometrial malignancies can be a useful marker of the disease (Horowitz *et al.*, 2001). Glycodelin gene expression in the endometrium known today, to some extent is based on data on endometrial cell lines. The regulation of gene expression in case of seminal vesicles however is completely unknown, presumably because of lack of a cell culture model, though mRNA splicing has been reported for GdS (Koistinen, 1997). How and why glycodelin (GdM) is expressed in hematopoietic cells of megakaryocytic lineage (Morrow *et al.*, 1994) and platelets (Tykocinski *et al.*, 1996) is not known.

Tissue Distribution of Glyodelin

Tissue glycodelin was isolated from endometrial tissue by all the research groups except Bohn's laboratory where it was isolated from the placenta (Bohn *et al.*, 1982). Julkunen *et al.* (1986) demonstrated that glycodelin is synthesized by the endometrium and not by placenta which was confirmed by *in situ* hybridization (Julkunen *et al.*, 1990). The same group reported presence of the protein in seminal plasma (Julkunen *et al.*, 1984). The protein has been reported to be synthesized by the fallopian tube (Julkunen *et al.*, 1986), ovary (Seppala *et al.*, 1985), uterine cervix (Pockley *et al.*, 1989) and seminal vesicle (Seppala *et al.*, 1985) tissues. It was widely assumed that glycodelin is a reproduction specific protein until Morrow *et. al.* (1984) demonstrated in 1994 that the protein is expressed by the hematopoietic cells of the megakaryocytic lineage. It has been reported to be expressed by other non reproductive tract tissues like breast, sweat glands, parabronchial glands, hidradenoma, and pancreatic cystadenoma (Kamarainen *et al.*, 1997). From the tissue distribution of the protein it is clear that the protein has its major role in reproductive processes.

In females, glycodelin is synthesized by the endometrium under the control of progesterone. As expected the protein is expressed by the endometrium during the secretory phase of the cycle and during pregnancy in concert with progesterone secretion (Seppala *et al.*, 1997).

Functions of Glycodelin

GdA has been studied in more depth than GdS. To date four different functions have been ascribed to GdA: immunosuppression, contraception, angiogenesis and morphogenesis. Immunosuppression (Bolton *et al.*, 1987) by glycodelin is known for over a decade now, and is the most well studied function of the protein. Its major role is believed to locally down-modulate the maternal immune system to prevent rejection of the fetal allograft. Glycodelin has been shown to have immunosuppressive properties in that it inhibits both phytohemagglutinin (Pockley *et al.*, 1990) as well as anti-CD3 antibody-induced lymphocyte proliferation (Rachmilewitz *et al.*, 1999). Besides, GdA also inhibits NK cell activity (Okamoto *et al.*, 1991). Over the 1990s, this protein has been well established as an important immunosuppressive agent at the feto-maternal interface (Tulppala *et al.*, 1995).

Immunosuppression along with angiogenesis (Song et al, 2001), a recently described function of GdA, can be important for proper blastocyst implantation during pregnancy. Contraceptive (Oehninger *et al.*, 1995) property of the protein on the other hand is important during normal menstrual cycle to define the fertilization window. Morphogenesis by glycodelin in breast tissues and cultured breast carcinoma cells is the least known function of the protein. Function/s of the other isoforms of the protein present in seminal plasma and megakaryocytes is unknown.

There is a clinical correlation between glycodelin level in the endometrium and habitual abortion (Tulppala *et al.*, 1995). As the protein is synthesized under the control of progesterone the level of this protein can indicate endometrial sufficiency. Apart from general endometrial health the level of the protein has been reported to be elevated in the serum in case of endometrial malignancies (Tatarinov *et al.*, 1990) and in ectopic pregnancy (Stabile *et al.*, 1994).

Contributions to Glycodelin-Research from our Laboratory

With all the above background information on glycodelin available in literature, we initiated studies to understand the molecular mechanism of the immunosuppressive property of glycodelin. To study the function of a protein, substantial amounts of the purified form is an absolute prerequisite. The only viable source of GdA is human amniotic fluid from the first trimester–collected either during medical termination of pregnancy or amniocentesis carried out for fetal karyotyping. For any human source material there are many ethical questions to be answered and collecting large samples for purification of GdA proved difficult. Therefore, recombinant DNA technology was resorted to.

The first step to synthesize a recombinant protein is to get the cDNA of the protein. From the available glycodelin gene sequence (Vaisse *et al.*, 1990), the first strand was synthesized using a reverse primer which is unique for glycodelin [Dutta *et al.*, 1998]. Using this first strand as template and a pair of primers corresponding to the start of the signal sequence of the gene and the end of the gene including the stop codon, the gene was amplified by PCR. The PCR product was then cloned into bacterial cloning vector pBS SK+ and absence of mutations was confirmed by automatic DNA sequencing. To get the protein expressed in *E. coli*, the cDNA was amplified using pBS SK+ Gd+signal as the template. Here the forward primer was from the start of the sequence present in the mature protein and the reverse primer was the sequence present at the end of the gene including the stop codon. The amplified product was then cloned into the EcoRV site of pBS SK+ and confirmed by sequencing. The glycodelin cDNA without the sequence for the signal peptide was then subcloned into the bacterial expression vector pRSET A at the XhoI and PstI site, in the proper orientation. The protein was overexpressed when induced with IPTG and was of expected Mw. of 22 kDa. The identity of the protein was confirmed by immunoblotting with the mAb B7B10 that was raised earlier to βLG and shown to cross react with glycodelin [Reddy *et al.*, 1992].

As glycodelin is a glycosylated protein, expression of the protein in a eukaryotic system was also attempted. The full length glycodelin cDNA without the stop codon was cloned in yeast expression vector pPicZα for generating recombinant *Pichia pastoris* expressing and secreting mature glycodelin. pPicZα expresses the recombinant protein as a fusion product having hexa histidine tag at the C-terminus to facilitate one step purification of the protein. When induced with 0.5 per cent methanol the protein was expressed and secreted into the culture supernatant by the recombinant yeast and was purified on a Ni-NTA agarose column. The protein was used to obtain mAbs which were used to standardize glycodelin RIA and purify the native protein by immunoaffinity chromatography.

Monoclonal Antibody Generation and Hybridoma Screening

Three different immunogens namely, *E. coli* expressed Gd, *P pastoris* expressed Gd and GdA (purified by conventional purification procedures) were administered to 3 groups of 6-8 weeks old female BALB/c mice, the first two antigens were emulsified with Freund's adjuvant whereas GdA was mixed with Alugel before administration. Hybridoma were established by fusion of Sp2/0 cells with immune splenocytes [Gangatirkar *et al.*, 2002]. Clones secreting antibodies specific to glycodelin were selected using RIA, ELISA and inhibition ELISA and subsequently subcloned to monoclonality. Several antibody secreting clones were obtained as tested for binding to immobilized Gd in ELISA. Of these some recognised the protein in liquid phase as determined by competitive ELISA. The mAbs belonged to different immunoglobulin classes and isotypes and exhibited varying affnities for antigen binding.

Two mAbs, namely B1C2 and A8C12 were obtained to *E coli*-Gd. These mAbs bind the protein in ELISA and also in liquid phase when the protein is partially denatured. MAb B1C2 and A8C12 were of IgG2a subclass type.

Five mAbs, namely B1B8, D9D4, H9F4, F4D10 and H1D8G7 were obtained to Pic-Gd. Inhibition studies showed that all the mAbs bound the native protein in solution phase but with low affinity except for D9D4. Monoclonal antibody B1B8 was of IgM subclass and mAb D9D4 and mAb F4D10 were of IgG1 subclass type.

From the hybridoma generated using the spleens of the third group of mice, five mAbs, namely B10B12C3, D1H3F8, D12C2B5, E9G9 and C1H6F8 were obtained and all these mAbs bound the protein in solution phase only and more interestingly none recognised immobilized GdA. Monoclonal antibodies obtained using GdA were of higher affinity than D9D4 and specifically bound the protein in solution phase. Dilution curve for all the mAbs were obtained in RIA and 50 per cent binding was calculated. Inhibition RIA showed that all the mAbs were of affinities higher than mAb D9D4. However the binding was of such a high affinity that once bound, the interaction could be dissociated only with strong chaotropic agents that resulted in the denaturation of Gd. All the 5 mAbs were of IgG1 subclass type except for mAb C1H6F8, which was of IgG2a. The summary of the characterization is provided in the Table 5.1. None of the mAbs shared epitope with the other closely related members of the lipocalin superfamily *i.e.* plasma retinol binding protein (RBP) and β Lactoglobulin (βLG). MAb D9D4 being of higher affinity than the other mabs of group 2, was chosen to establish an immuno-affinity column to use for purification of Gd and the mAb B1C2 which bound only denatured Gd but with very high affinity was used for identification of Gd in Western blots and immunohistochemistry.

Purification of Glycodelin

Gd was purified from amniotic fluid and from seminal plasma by immunoaffinity chromatography using the D9D4 monoclonal antibody. Purified D9D4 antibody was oxidized by periodate and coupled to AffiGel hydrazide. This method allows both Fab arms of the antibody to be available for antigen binding and therefore is more efficient. Amniotic fluid (collected from women undergoing elective abortion) and seminal plasma (collected from healthy donors through an infertility clinic), were dialyzed against PBS, and chromatographed on the immunoaffinity column. Bound proteins were eluted with

Table 5.1: Characterization of mAbs Obtained against Glycodelin

Immunization protocol	mAbs	Isotype	WB	ELISA	RIA	Epitope
a.	A8C12	IgG2a	+++	+++	–	Sequential
	B1C2	IgG2a	+++	+++	–	
b.	B1B8	IgM	–	+	+	Conformational
	D9D4	IgG1	–	++	++	
	F4D10	IgG1	–	++	+	
	H9F4	IgG2a	–	++	+	
c.	B10B12C3	IgG2a	–	–	+++	
	C1H6G7	IgG2a	–	–	+++	
	D1H3F8	IgG1	–	–	+++	
	D12C2B5	IgG1	–	–	+++	
	E9G4	IgG1	–	–	+++	

mAb A8C12 recognizes an epitope in the sequence TENPKKFKINYTVAN and mAb B1C2 recognizes an epitope in the sequence EATLLDTDYDNFLFL of glycodelin.

Epitope mapping for A8C12 and B1C2 was done using different overlapping deletion constructs of glycodelin.

low pH buffer. The purity of the proteins was determined by gel staining (Figure 5.1A and B) and the identity of the protein was confirmed by Western blotting using mAb B1C2 (Figure 5.1).

Inhibition of Proliferation of T Cells by GdA and not GdS

As reported earlier (Bohn *et al.*, 1982, Rachmilewitz *et al.*, 1934), GdA exhibited immunosuppressive activity in that the protein inhibited the proliferation of human peripheral blood mononuclear cells (PBMCs) activated by anti-CD3 antibody [Figure 5.2Aa] or by the mitogen phytohemagglutinin (data not shown). The experiment was then extended to a human T cell line namely Jurkat that also did not proliferate when cultured in the presence of GdA (Figure 5.2Ab). Interestingly GdS did not exhibit any such activity either on the activated PBMCs (Figure 5.2Ba) or Jurkat cells (Figure 5.2Bb) even when higher doses of the protein were used.

GdA Induces Apoptosis in Activated T Cells

Depletion of the monocytes from PBMCs did not abrogate the Gd-mediated inhibitory effect on T cells suggesting that GdA exerts this effect directly on T cells (Mukhopadhyay *et al.*, 2001). Moreover, GdA inhibited proliferation of Jurkat cells as well. Because cell lines are homogeneous cell populations, these observations confirmed that the inhibitory effect of GdA is direct on T cells. Studies were then initiated to look into the molecular events that are triggered by GdA. The first experiment in this direction was to determine whether GdA action was reversible. Our studies provided conclusive evidence that the direct effect on both cell types was irreversible indicating that the signal could be triggered through specific receptors. To determine whether GdA blocks the cell cycle progression in

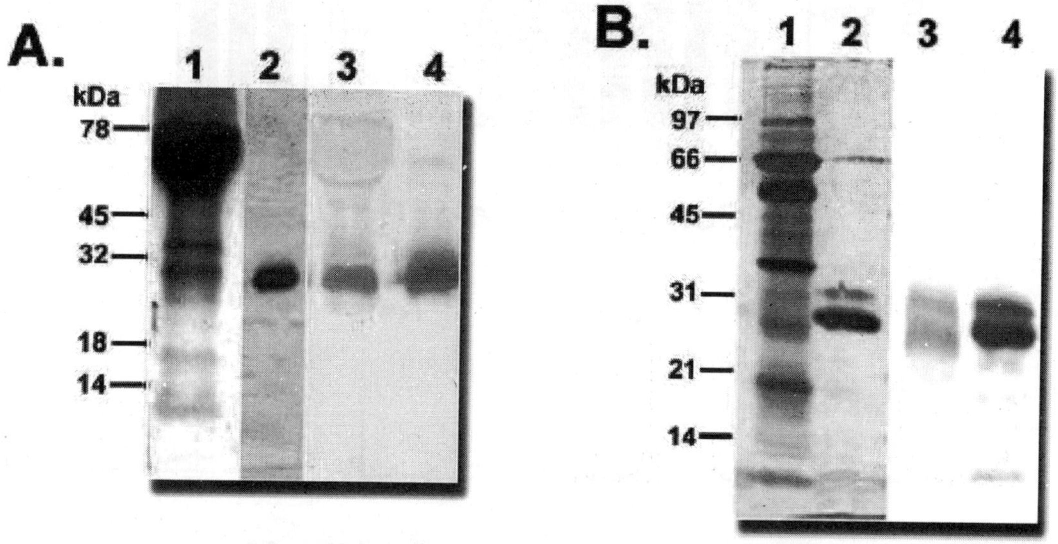

Figure 5.1: Molecular Characterization of Glycodelin
[A.] 5 µl amniotic fluid (lanes 1 and 3), 2 µg GdA purified on mAb D9D4 affigel hydrazide column (lanes 2 and 4) [B] 2 µl seminal plasma (lanes 1 and 3) and 2 µg GdS purified on mAb D9D4 affigel hydrazide column (lanes 2 and 4) were electrophoresed on a 12.5 per cent denaturing polyacrylamide gel under reducing conditions and stained with Coomassie brilliant blue B (lanes 1 and 2). The proteins were transferred to nitrocellulose membrane and subjected to immunoblot analysis with mAb B1C2 (lanes 3 and 4).

any one particular stage, activated PBMCs after culturing with GdA, were fixed and stained with propidium iodide and analysed by FACS (fluorescence activated cell scan). The appearance of the pre G0/G1 cell population in PBMCs after treatment with Gd was the first indication that Gd induces apoptosis in T cells (Figure 5.2C). As could be expected, GdS had no effect. The studies were then extended to Jurkat and another T cell line, namely MOLT-4.

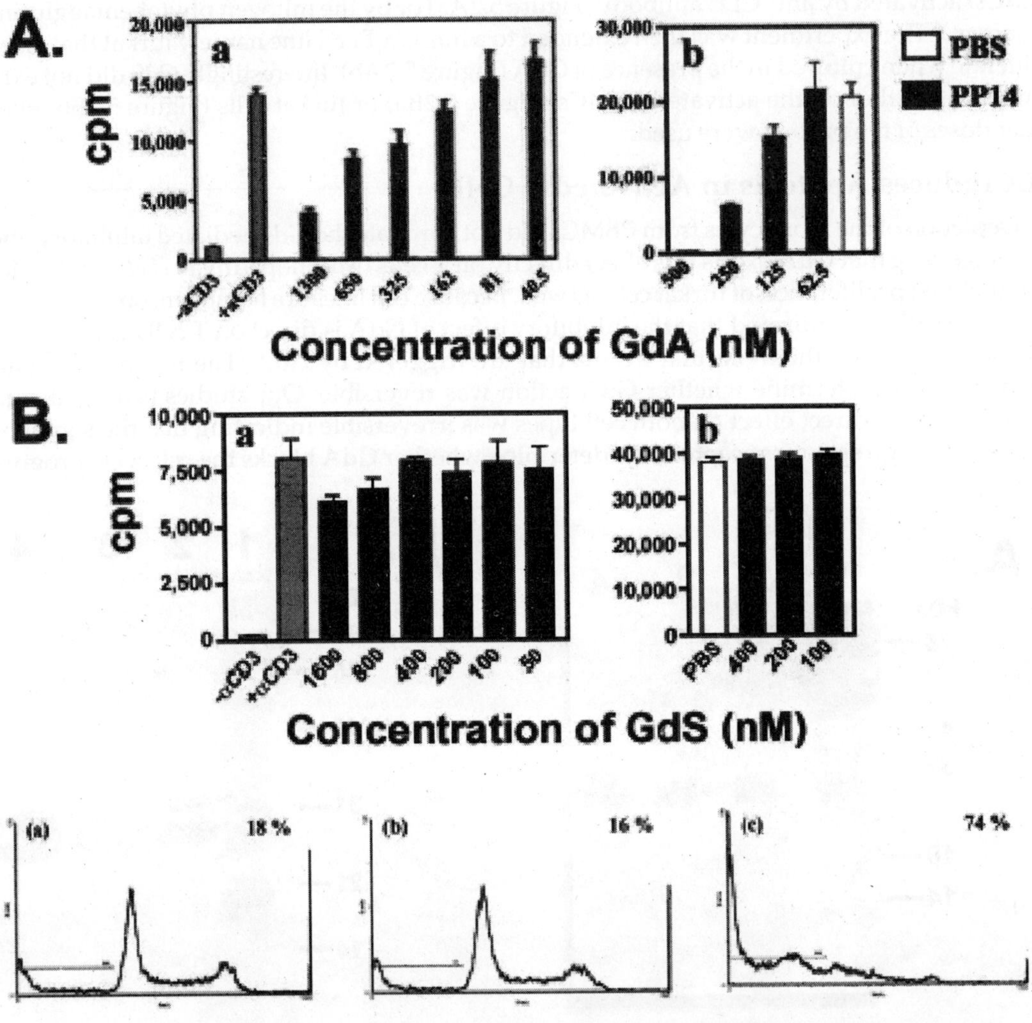

Figure 5.2: Functional Characterization of Glycodelin

[A] and [B] (a) PBMCs were stimulated with OKT3 mAb for 48 h in the presence of varying concentrations of GdA (A) or GdS (B) and later pulsed with [^{3}H]thymidine for 12 h. (b) Jurkat cells were cultured for 24 h in the presence of varying concentrations of GdA (A) or (GdS) and then pulsed with [^{3}H]thymidine. Cells were harvested on glass fibre filters, lysed and the radioactivity measured. [C] Apoptotic effect of glycodelin as determined by EtBr staining and FACS analysis. Jurkat cells were cultured with 200 nM glycodelin for 18 h following which cells were fixed in ethanol, stained with EtBr and analyzed by FACS. (a) Untreated, (b) GdS and (c) GdA

Morphological observation of apoptosis was carried out by dual staining of the cells with EtBr and AcOr acridine orange (AcOr) after exposure to glycodelin for 0, 6, 12, 18 and 24 h. Neither of the cells exhibited any morphological changes until 6 h (data not shown). On culturing with GdA, MOLT-4 cells showed maximum apoptosis at 18 h and Jurkat at 24 h (Figure 5.3A). GdA-induced apoptosis in both the cells was confirmed by DNA fragmentation (Figure 5.3B).

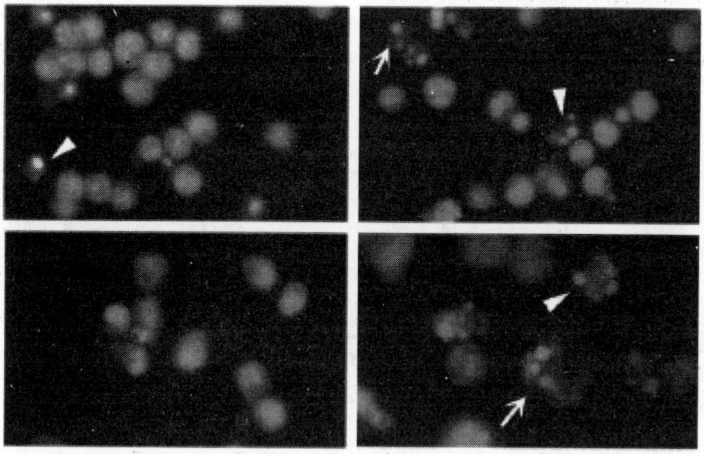

Figure 5.3A: Apoptotic Effect of Glycodelin as Determined by EtBr/AcOr Staining Jurkat (top panels) and MOLT-4 cells (lower panels) were cultured without (left panels) or with purified GdA (right panels) for 18 h, following which the cells were stained with AcOr/EtBr and viewed under the fluorescence microscope using blue filter. Arrowheads indicate cells in early apoptosis, arrows indicate cells in late apoptosis.

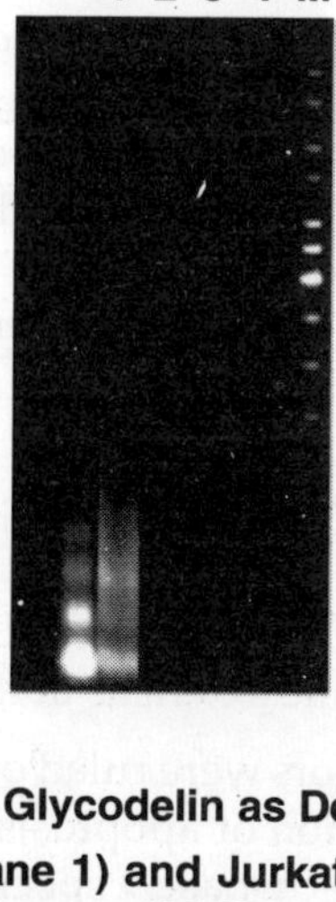

Figure 5.3B: Apoptotic Effect of Glycodelin as Determined by DNA Fragmentation studies. DNA extracts from MOLT-4 (lane 1) and Jurkat cells (lane 2) treated with Gd for 18 hr were electrophoresed on 0.8 per cent agarose gel followed by staining with EtBr. Note the presence of fragmented DNA in lanes 1 and 2. DNA from untreated cells (lanes 3, 4) served as negative control. DNA marker (M) was run alongside.

The above observation lead to the next question ie., is the GdA- induced apoptosis dependant on the activation of caspases. Towards answering this, cells were cultured with GdA in presence of the cell permeable pan caspase inhibitor, Z-VAD-FMK [Calbiochem, Germany] followed by FACS analyses. As can be seen from the data presented in Figure 5.4, the inhibitor rescued cells from Gd-induced apoptosis. Involvement of the caspases in apoptosis triggered by GdA was further confirmed by measuring the level of activated caspase 3 in T cells on culturing with GdA. Caspase 3, a member of the caspase subfamily, is a crucial mediator of apoptosis in many mammalian cells. Our data show that this protease is responsible for Gd-induced apoptosis. That this activity was inhibited by the specific inhibitor Ac-DEVD-CHO (Thornberry *et al.*, 1997) established the specificity of the enzyme assay.

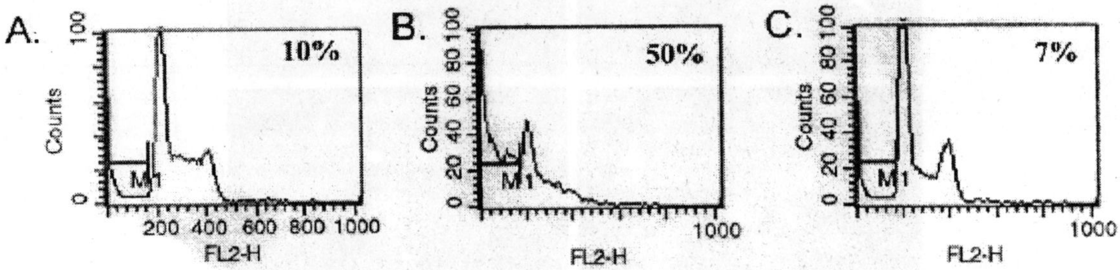

Figure 5.4: GdA Induced Apoptosis is Caspase Dependant

Jurkat (JR4) cells were cultured with 500 nM GdA in presence or absence of pan caspase inhibitor Z-VAD-FMK (25 μM) for 18 h. The cells were harvested, fixed with ethanol, stained with EtBr and analyzed by FACS. A. vehicle control (0.5 per cent DMSO); B. cells treated with GdA; C. cells treated with GdA in presence of 25 μM Z-VAD-FMK.

Elucidation of the Mechanism of Apoptosis Induction by GdA

Our studies confirmed without doubt that.GdA-induced apoptosis is caspase dependent and that the downstream effector enzyme, caspase 3 is activated during GdA induced apoptosis. CD95/ Fas is expressed by Jurkat cells and many apoptotic signalling pathways are known to be Fas dependent (Hildeman *et al.*, 2002). Involvement of Fas in GdA induced apoptosis was therefore assessed using the Fas antagonist mAb CBL537 [Calbiochem, GErmany. This antibody is known to block the signaling of the Fas receptor. The mAb was able to reduce Fas ligand induced cell death by 30 per cent but had no effect on the GdA induced apoptosis (Figure 5.5). Hence it was concluded that GdA induced apoptotic signalling is not initiated by CD95/Fas.

Apoptotic signalling initiated at the cell membrane by the different death receptors (DR) as well as Fas and TNF receptor, activates caspase 8. To determine if the GdA-induced apoptotic signalling is initiated by caspase 8, the caspase 8–/– cell line Jurkat (A3-I9.2, ATCC) was used. Though Fas ligand was unable to induce apoptosis in this cell line as expected (Figure 5.6B), GdA-induced apoptosis in the caspase deficient cell line was to the same extent as in the wild type Jurkat (JR4) (Figure 5.6).

As the involvement of the death receptors were ruled out by the above observations, we explored the involvement of caspase 9 in the initiation of apoptosis by GdA. When cells were cultured with GdA in the presence of the cell permeable caspase 9 specific inhibitor Z-LEHD-FMK [Calbiochem, Germany], the GdA-induced cell death was partially rescued (30 per cent) (Figure 5.7). It is well documented that caspase 9 gets activated only when the mitochondria are damaged [Li *et al.*, 1997;

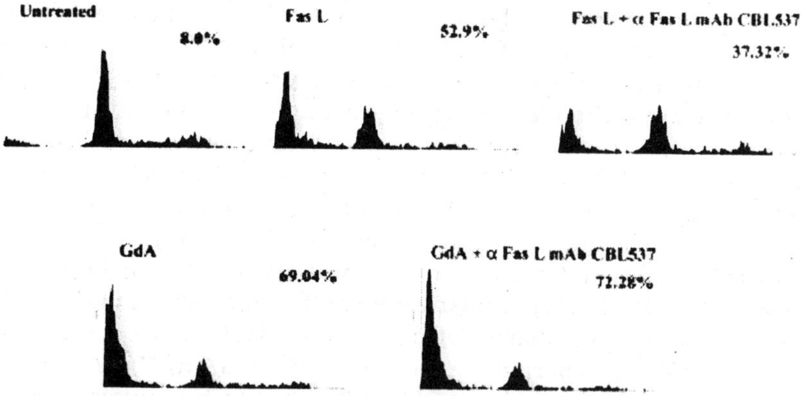

Figure 5.5: GdA Induced Apoptosis is CD95/Fas Independant

Jurkat (JR4) cells were pre-incubated with anti Fas antagonistic mAb CBL537 for 1 h at 37°C. The treated or untreated cells were then cultured with 1 μM pure GdA or 5 ng/ml FasL for 8 h. The cells were harvested, fixed with ethanol, stained with EtBr and analyzed by FACS.

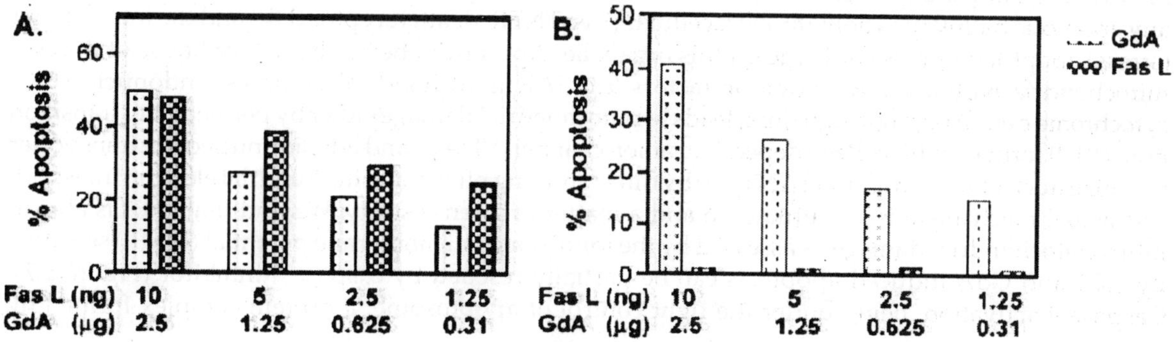

Figure 5.6: GdA Induced Cell Death is Caspase 8 Independant

A. Jurkat (JR4) cells; B. Jurkat (A3-I9.2, caspase 8–/–) cells. Cells were cultured with different concentrations of pure GdA or FasL for 8 h and then stained with EtBr and analyzed by FACS. The percentage of cells in the pre G1/G0 region were plotted over the untreated control for each treatment.

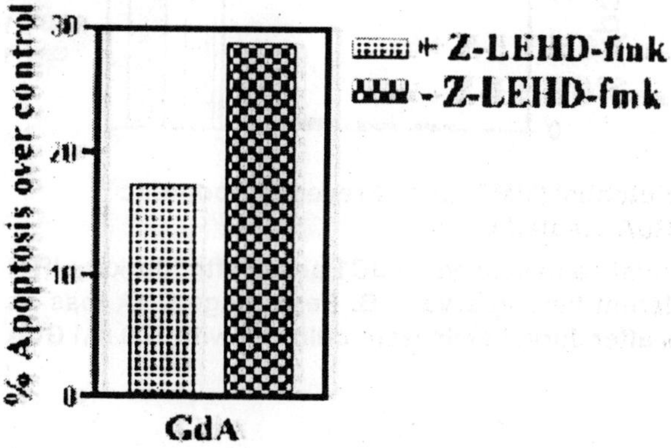

Figure 5.7: GdA Induced Apoptosis is Caspase 9 Dependant

Jurkat (JR4) cells were cultured with 1 μM GdA in presence or absence of 25 μM caspase 9 inhibitor Z-LEHD-FMK for 8 h and then stained with EtBr and analyzed by FACS. The percentage of cells in the pre G1/G0 region were plotted over the untreated control for each treatment.

Cain *et al.*, 2002]. Involvement of caspase 9 in GdA-induced cell death indicated that most probably GdA initiates the apoptotic program at the mitochondria. To study the events taking place at the mitochondria in Jurkat cells in response to GdA, the cationic dye JC1 was used (Cossarizza *et al.*, 1993). The dye accumulates inside the mitochondrial matrix depending on the mitochondrial inner membrane potential. Intact mitochondria accumulate enough dye inside the matrix where the local concentration of the dye crosses the threshold value of 5 μM and the dye forms J aggregates that emit light at 590 nm (Red) instead of 525 84 nm (Green). The ratio of the mean fluorescence intensity (red over green), as measured by flow cytometer, is independent of the number of mitochondria present in different cells and represents the mitochondrial membrane potential (MMP) (Cossarizza *et al.*, 1993) Permanent loss of MMP is closely linked to initiation of apoptotic pathway which is essentially irreversible (Zamzami *et al.*, 1995). Hyperpolarization of the mitochondria just before the complete loss of mitochondrial membrane potential is well documented (Waterhouse *et al.*, 2002). As can be seen in Figure 5.8A, there was a sharp hyperpolarization after 8 h of GdA treatment after which the MMP decreased in a linear fashion with time. Figure 5.8B shows the FACS analyses of the cells at different time periods after GdA treatment. The nuclear DNA loss could be observed only after 10 h of GdA treatment, that is after the mitochondrial damage had set in.

Mitochondria are the chief sensors of different kinds of stress (*e.g.* oxidative damage DNA damage, protein synthesis inhibition etc.). It is widely accepted that mitochondrial damage and spillage of cytochrome c into the cytosol is the point of no return in apoptosis. Mitochondrion is at the center of any apoptotic pathway as both the extracellular (Fas, TNFR, death receptor (DR) and the intracellular (stress) apoptotic signals converge on this organelle. Apart from being the power house of the cell, mitochondria harbor many apoptotic factors (*e.g.* IAP, mitochondrial caspases, endonuclease G, cytochrome c etc. Apoptotic signalling leads to mitochondrial damage either by permeability transition pore (PTP) formation by voltage dependent anion channel (VDAC) and adenine nucleotide transporter (adenine nucleotide transporter (ANT), or by direct shutting off of the mitochondrial electron transport. The actual mechanism is still unknown and a matter of intense study. We find that GdA is able to inflict mitochondrial damage as assessed by the mitochondrial membrane potential (MMP) sensitive dye JC1 and GdA induced apoptosis can be partially rescued by caspase 9 inhibitor (Figure 5.7). Caspase 9 activation, being under the tight control of apoptosome, a protein complex formed by

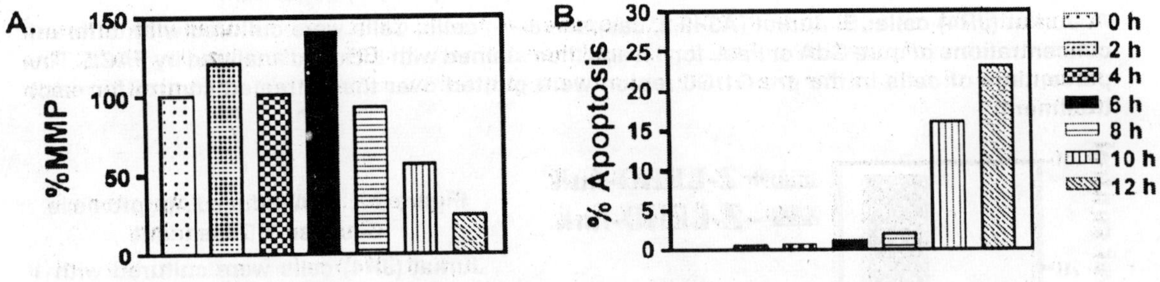

Figure 5.8: Mitochondrial Membrane Potential (MMP) Loss Precedes Apoptotic DNA Loss in GdA Treated Cells

A. Percentage mitochondrial membrane potential as assessed by JC1 assay after Jurkat (JR4) cells were cultured with 100 nM GdA for different time intervals. B. Percentage DNA loss as assessed by EtBr staining and FACS analysis after Jurkat cells were cultured with 100 nM GdA for different time intervals.

Apaf1, ATP and Cytochrome c, is possible only after mitochondrial damage. Hence GdA-induced apoptosis resembles many stress induced situations where the apoptotic signalling is initiated at the mitochondria. The exact pathway by which GdA is able to inflict mitochondrial damage remains to be unraveled and studies have been initiated in our laboratory towards the same.

Activation of the MAP kinase JNK leads to apoptosis as it activates synthesis of Bax, a pro-apoptotic protein (Lei *et al.*, 2002). We decided to look at whether JNK could be activated by GdA for which an inhibitor of JNK was used. As at higher concentration this inhibitor can affect normal cell cycle progression, the concentration of the inhibitor to be used was standardized so that JNK activity was inhibited without affecting the cell cycle (data not shown). As can be seen from Figure 5.9, GdA-induced Jurkat (JR4) cell death was not affected by blocking the JNK pathway, ruling out the involvement of JNK in the GdA induced apoptotic signalling.

GdA and GdS inspite of being identical in their amino acid sequence exhibit differences in their immunomodulatory activities. When GdA was digested with trypsin the apoptotic activity of the same was destroyed suggesting that the activity lies in the protein backbone (Mukhopadhyay *et al.*, 2004). Therefore we looked at the differences in the glycan groups of the two molecules. The most prominent of the differential glycosylation on the two isoforms of glycodelin is the differential sialylation, GdA glycan being rich in the terminal sialic acid residues and GdS being totally devoid of the same (Dell *et al.*, 1995). GdA was desialylated using neuraminidase and the effect of desialylation on the apoptotic activity was tested on Jurkat cells. From the results obtained (Figure 5.10) it is clear that neuraminidase treatment reduced the apoptotic activity of GdA. Hence we speculated that the charged sialic acid residues present on GdA maybe interacting with the protein back bone, leading to an apoptotically active conformation which is absent in case of GdS. We postulate that in the case of GdS, the apoptogenic region could be masked by the glycans in the absence of sialic acid molecules. In this regard, it is pertinent to mention that fl-Pic-Gd is also not apoptotically active (data not shown). It is well known that in the *Pichia pastoris* system glycans comprise of upto 25 mannose but unable to add sialic acid residues on the expressed proteins (Chotigeat *et al.*, 2000).

Conclusions

The fetus is an allograft and has the potential to trigger the maternal immune system strong enough to reject it. Glycodelin, a progesterone regulated glycoprotein exhibits immunomodulatory activity that can protect the fetus from mother's immune attack. Lymphocytes appear in the decidua of the first trimester of pregnancy and disappear toward the third trimester (Bulmer *et al.*, 1987). Those cells that recognize non-self paternal antigens of the trophoblast would need to be deleted from the

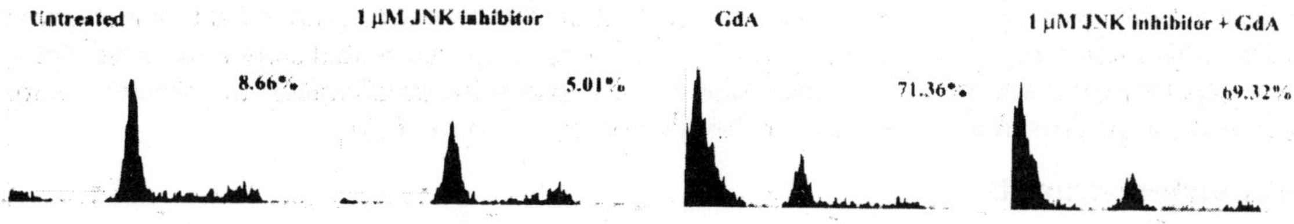

Figure 5.9: JNK Pathway is not Involved in GdA Induced Apoptosis

Jurkat (JR4) cells were cultured with 1 μM GdA in presence or absence of 1 μM JNK inhibitor for 8 h. The cells were harvested, fixed with ethanol, stained with EtBr and analyzed by FACS.

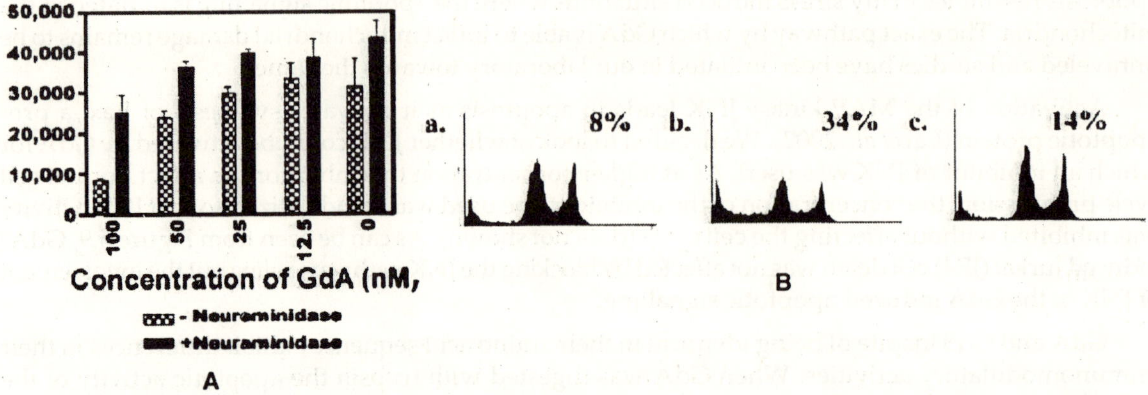

Figure 5.10: A. Neuraminidase Treated GdA Exhibits Lower Inhibition of Jurkat Ceil Proliferation: Jurkat cells were cultured for 24 h in the presence of varying concentrations of neuraminidase treated or untreated GdA and then pulsed with [^{3}H]thymidine. Cells were harvested on glass fibre filters, lysed and the radioactivity measured. B. Neuraminidase treated GdA exhibits lower apoptotic activity: Jurkat cells were cultured for 12 h in presence of 100 nM neuraminidase treated or untreated GdA (300 nM), cells were fixed with ethanol, stained with EtBr and analysed by FACS (a) cells incubated with neuraminidase alone, (b) cells incubated with GdA alone and (c) cells incubated with neuraminidase treated GdA.

local environment, should the embryo survive. Glycodelin, secreted by the glandular epithelium of the endometrium in response to progesterone would be up-regulated at the time of implantation. Upon binding to activated T cells but not naive T cells GdA triggers apoptosis or programmed cell death in these very cells that have the potential to destroy the allograft. Death by apoptosis would be the preferred choice as apoptotic cells are quickly phagocytosed by macrophages via a receptor-mediated process (Fadok *et al.*, 2000) ensuring that there is no release of intracellular contents (containing inflammatory components) in the surrounding tissue, which would be harmful to the developing fetus. This hypothesis is well supported by the observations that women undergoing habitual abortion harbour low levels of GdA (Tulppala *et al.*, 1995).

Any function exhibited by a biological molecule needs to be modulated. An interesting offshoot emerging from our research on glycodelin is that though identical in the primary structure, GdS is not apoptotically active (Mukhopadhyay *et al.*, 2004), one of the reasons being that this molecule lacks sialic acid residues in its glycans. We have recently shown that unglycosylated but refolded recombinant Gd does exhibit apoptotic activity thus confirming that the functional domain of Gd lies in its protein backbone (Jayachandran *et al.*, 2004). Thus it is interesting to speculate that sialylation might play a role in regulating the activity of GdA and research is underway in our laboratory towards elucidating the role of the glycans in the expression of the apoptogenic activity of Gd.

Acknowledgements

The grant provided by the Department of Biotechnology, Government of India is gratefully acknowledged.

References

Bohn H, Kraus W, and Winckler W. (1982). New soluble placental tissue proteins: their isolation, characterization, localization and quantification. Placenta Suppl., 4, 67–81.

Bolton AE, Pockley AG, Clough KJ, Mowles EA, Stoker RJ, Westwood OM, and Chapman MG. (1987). Identification of placental protein 14 as an immunosuppressive factor in human reproduction. Lancet, 1(8533), 593–595.

Bulmer J, Johnson P, and Bulmer D. (1987). *Immunoregulation and Fetal Survival*, Oxford University Press, NY., pp. 111–134.

Cain K, Bratton SB, and Cohen GM. (2002). The Apaf–1 apoptosome: a large caspase–activating complex. Biochimie, 84, 203–214.

Check JH, Vaze MM, and Vetter B. (1993). Measurement of placental protein 14 (PP14) not helpful in predicting infertility or spontaneous abortion. Int. J. Fertil Menopausal Stud., 38, 34–36.

Chiu PC, Koistinen R, Koistinen H, Seppala M, Lee KF, and Yeung WS. (2003). Binding of zona binding inhibitory factor–1 (ZIF–1) from human follicular fluid on spermatozoa. J. Biol. Chem., 278, 13570–13577.

Chiu PC, Koistinen R, Koistinen H, Seppala M, Lee KF, and Yeung WS. (2003). Zona binding inhibitory factor–1 from human follicular fluid is an isoform of glycodelin. Biol. Reprod., 69, 365–372.

Chotigeat W, Chayanunnukul W, and Phongdara A. (2000). Expression of a mammalian alpha 2,6–sialyltransferase gene in Pichia pastoris. J. Biotechnol., 81, 55–61.

Cossarizza A, Baccarani–Contri M, Kalashnikova G, and Franceschi C. (1993). A new method for the cytofluorimetric analysis of mitochondrial membrane potential using the J–aggregate forming lipophilic cation 5,5′,6,6′–tetrachloro–1,1′,3,3′–tetraethylbenzimidazolcarbocyanine iodide (JC–1). Biochem. Biophys. Res. Commun., 197, 40–45.

Dell A, Morris HR, Easton RL, Panico M, Patankar M, Oehniger S, Koistinen R, Koistinen H, Seppala M, and Clark GF. (1995). Structural analysis of the oligosaccharides derived from glycodelin, a human glycoprotein with potent immunosuppressive and contraceptive activities. J. Biol. Chem., 270, 24116–24126.

Dutta B, Mukhopadyay D, Roy N, Das G, and Karande AA. (1998). Cloning, Expression, Purification and Immunocharacterisation of Placental Protein–14. Protein Expression and Purification, 14, 327-334.

Fadok V, Bratton D, Rose D, Pearson A, Ezekewitz R, and Henson P. (2000). A receptor for phosphatidylserine–specific clearence of apoptotic cells. Nature, 405, 85–90.

Gangatirkar P, Gangadharan S, Narendranath A, Nagpal S, Salunke DM, and Karande AA. (2002). Monoclonal Antibodies to Gonadotropin Releasing Hormone (GnRH) Inhibit Binding of the Hormone to its Receptor. Hybridoma Hybridomics, 21, 281-286

Gao J, Mazella J, Seppala M, and Tseng L. (2001). Ligand activated hPR modulates the glycodelin promoter activity through the Sp1 sites in human endometrial adenocarcinoma cells. Mol. Cell. Endocrinol., 176, 97–102.

Hausermann HM, Donnelly KM, Bell SC, Verhage HG, and Fazleabas AT. (1998). Regulation of the glycosylated beta–lactoglobulin homolog, glycodelin [placental protein 14:(PP14)] in the baboon (*Papio anubis*) uterus. J. Clin. Endocrinol. Metab., 83, 1226–1233.

Hildeman DA, Zhu Y, Mitchell TC, Kappler J, and Marrack, P. (2002). Molecular mechanisms of activated T cell death in vivo. Curr. Opin. Immunol., 14, 354–359.

Horowitz IR, Cho C, Song M, Flowers LC, Santanam N, Parthasarathy S, and Ramachandran S. (2001). Increased glycodelin levels in gynecological malignancies. Int. J. Gynecol. Cancer, 11, 173–179.

Huhtala ML, Seppala M, Narvanen A, Palomaki P, Julkunen M, and Bohn H. (1987). Amino acid sequence homology between human placental protein 14 and beta–lactoglobulins from various species. Endocrinology, 120, 2620–2622.

Jayachandran R, Shaila MS, and Karande AA. (2004). Analysis of the role of oligosaccharides in the apoptotic activity of glycodelin-A. J Biol Chem., 279, 8585-8591

Julkunen M, Wahlstrom T, Seppala M, Koistinen R, Koskimies A, Stenman UH, and Bohn H. (1984). Detection and localization of placental protein 14–like protein in human seminal plasma and in the male genital tract. Arch. Androl., 12 Suppl, 59–67.

Julkunen M, Koistinen R, Sjoberg J, Rutanen EM, Wahlstrom T, and Seppala M. (1986a). Secretory endometrium synthesizes placental protein 14. Endocrinology, 118, 1782–1786.

Julkunen M, Wahlstrom T, and Seppala M. (1986b). Human fallopian tube contains placental protein 14. Am. J. Obstet. Gynecol., 154, 1076–1079.

Julkunen M, Seppala M, and Janne OA. (1988). Complete amino acid sequence of human placental protein 14: a progesterone–regulated uterine protein homologous to beta–lactoglobulins. Proc. Natl. Acad. Sci. USA., 85, 8845–8849.

Julkunen M, Koistinen R, Suikkari AM, Seppala M, and Janne OA. (1990). Identification by hybridization histochemistry of human endometrial cells expressing mRNAs encoding a uterine beta–lactoglobulin homologue and insulin–like growth factor–binding protein–1. Mol. Endocrinol., 4, 700–707.

Kamarainen M, Seppala M, Virtanen I, and Andersson LC. (1997). Expression of glycodelin in MCF–7 breast cancer cells induces differentiation into organized acinar epithelium. Lab. Invest., 77, 565–573.

Koistinen H, Koistinen R, Dell A, Morris HR, Easton RL, Patankar MS, Oehninger S, Clark GF, and Seppala M. (1996). Glycodelin from seminal plasma is a differentially glycosylated form of contraceptive glycodelin–A. Mol. Hum. Reprod., 2, 759–765.

Koistinen H, Koistinen R, Kamarainen M, Salo J, and Seppala M. (1997). Multiple forms of messenger ribonucleic acid encoding glycodelin in male genital tract. Lab. Invest., 76, 683–690.

Koistinen H, Koistinen R, Seppala M, Burova TV, Choiset Y, and Haertle T. (1999). Glycodelin and beta–lactoglobulin, lipocalins with a high structural similarity, differ in ligand binding properties. FEBS Lett., 450, 158–62.

Lei K, Nimnual A, Zong WX, Kennedy NJ, Flavell RA, Thompson CB, Bar–Sagi D, and Davis RJ. (2002). The Bax subfamily of Bcl2–related proteins is essential for apoptotic signal transduction by c–Jun NH(2)–terminal kinase. Mol. Cell. Biol., 22, 4929–4942.

Li P, Nijhawan D, Budihardjo I, Srinivasula SM, Ahmad M, Alnemri ES, and Wang X. (1997). Cytochrome c and dATP-dependent formation of Apaf-1/cas pase-9 complex initiates an apoptotic protease cascade. Cell, 91, 479–489.

Morrow DM, Xiong N, Getty RR, Ratajczak MZ, Morgan D, Seppala M, Riittinen L, Gewirtz AM, and Tykocinski ML. (1994). Hematopoietic placental protein 14. An immunosuppressive factor in cells of the megakaryocytic lineage. Am. J. Path., 145, 1485–95.

Mukhopadhyay D, Sundareshan S, Rao C, and Karande AA. (2001). Placental protein (PP-14) induces apoptosis in T cells but not in monocytes. J. Biol. Chem., 276, 28268-28273.

Mukhopadhyay D, SundeRaj S, Alok A, and Karande AA. (2004). Glycodelin A, not glycodelin S, is apototically active: relevance of sialic acid modification. J. Biol. Chem., 279, 8577-8584.

Mueller MD, Vigne JL, Vaisse C, and Taylor RN. (2001). Glycodelin: a pane in the implantation window. Semin. Reprod. Med., 18, 289–298.

Oehninger S, Coddington CC, Hodgen GD, and Seppala M. (1995). Factors affecting fertilization: endometrial placental protein 14 reduces the capacity of human spermatozoa to bind to the human zona pellucida. Fertil. Steril., 63, 377–383.

Okamoto N, Uchida A, Takakura K, Kariya Y, Kanzaki H, Riittinen L, Koistinen R, Seppala M, and Mori T. (1991). Suppression by human placental protein 14 of natural killer cell activity. Am. J. Reprod. Immunol., 26, 137–142.

Pockley AG, Barratt CL, and Bolton AE. (1989). Placental protein 14 (PP14) content and immunosuppressive activity of human cervical mucus. Symp. Soc. Exp. Biol., 43, 317–323.

Pockley, AG, and Bolton AE. (1990). The effect of human placental protein 14 (PP14) on the production of interleukin–1 from mitogenically stimulated mononuclear cell cultures. Immunology, 69, 277–281.

Rachmilewitz J, Riely GJ, and Tykocinski, ML. (1999). Placental protein 14 functions as a direct T–cell inhibitor. Cell. Immunol., 191, 26–33.

Reddy BM, Karande AA, and Adiga PR. (1992). A common epitope of beta–lactoglobulin and serum retinol–binding proteins: elucidation of its core sequence using synthetic peptides. Mol. Immunol., 29, 511–516.

Seppala M, Koskimies AI, Tenhunen A, Rutanen EM, Sjoberg J, Koistinen R, Julkunen, M, and Wahlstrom T. (1985). Pregnancy proteins in seminal plasma, seminal vesicles, preovulatory follicular fluid, and ovary. Ann. N. Y. Acad. Sci., 442, 212–226.

Seppala M, Koistinen H, Koistinen R, Dell A, Morris HR, Oehninger S, and Clark GF. (1997). Glycodelins as regulators of early events of reproduction. Clin. Endocrinol., 46, 381–386.

Song M, Ramaswamy S, Ramachandran S, Flowers LC, Horowitz IR, Rock JA, and Parthasarathy S. (2001). Angiogenic role for glycodelin in tumorigenesis. Proc. Natl. Acad. Sci. USA., 98, 9265–9270.

Stabile I, Olajide F, Chard T, and Grudzinskas JG. (1994). Circulating levels of placental protein 14 in ectopic pregnancy. Br. J. Obstet. Gynaecol., 101, 762–764.

Stewart DR, Erikson MS, Erikson ME, Nakajima ST, Overstreet JW, Lasley BL, Amento EP, and Seppala M. (1997). The role of relaxin in glycodelin secretion. J. Clin. Endocrinol. Metab., 82, 839–846.

Tatarinov YS, Shcherbakova LA, Olefirenko GA, and Petrunin DD. (1990). Protein factor of fertility in malignancies. Tumour. Biol., 11, 113–119.

Taylor RN, Vigne JL, Zhang P, Hoang P, Lebovic DI, and Mueller MD. (2000). Effects of progestins and relaxin on glycodelin gene expression in human endometrial cells. Am. J. Obstet. Gynecol., 182, 841–7; discussion 847–9.

Thellin O, Coumans B, Zorzi W, Igout A, and Heinen E. (2000). Tolerance to the foeto–placental 'graft': ten ways to support a child for nine months. Curr. Opin. Immunol., 12, 731–737.

Thornberry NA, Rano TA, Peterson EP, Rasper DM, Timkey T, Garcia-Calvo M, Houtzager VM, Nordstrom PA, Roy S, Vaillancourt JP, Chapman KT, and Nicholson DW. (1997). A combinatorial approach defines specificities of members of the caspase family and granzyme B. Functional relationships established for key mediators of apoptosis. J. Biol. Chem., 272, 17907–17911.

Tulppala M, Julkunen M, Tiitinen A, Stenman UH, and Seppala M. (1995). Habitual abortion is accompanied by low serum levels of placental protein 14 in the luteal phase of the fertile cycle. Fertil. Steril., 63, 792–795.

Tykocinski ML, Xiong N, and Morrow DM. (1996). Platelet immunoregulatory factors. Stem Cells, 14 Suppl 1, 240–245.

Vaisse C, Atger M, Potier B, and Milgrom E. (1990). Human placental protein 14 gene: sequence and characterization of a short duplication. DNA. Cell. Biol., 9, 401–413.

Van–Cong N, Vaisse C, Gross MS, Slim R, Milgrom E, and Bernheim A. (1991). The human placental protein 14 (PP14) gene is localized on chromosome 9q34. Hum. Genet., 86, 515–518.

Waterhouse NJ, Ricci JE, and Green DR. (2002). And all of a sudden it's over: mitochondrial outer–membrane permeabilization in apoptosis. Biochimie, 84, 113–121.

Zamzami N, Marchetti P, Castedo M, Zanin C, Vayssiere JL, Petit PX, and Kroemer G. (1995). Reduction in mitochondrial potential constitutes an early irreversible step of programmed lymphocyte death in vivo. J. Exp. Med., 181, 1661–1672.

6

Prostate Cancer: An Update

J. Arunakaran

Department of Endocrinology, Dr. A.L.M. P.G. Institute of Basic Medical Sciences, University of Madras, Taramani Campus, Chennai – 600 113, India

ABSTRACT

Prostate cancer is a major public health problem and become one of the most common cancers in males affecting 1 in 6 men in developing countries. Prostate cancer is age related and is uncommon before the age of 50. At this age the incidence and mortality rates increase exponentially. Many mechanisms have been postulated to account for this marked prediction of men including, (*i*) high fetal exposure of circulating androgens, (*ii*) increased expression of 5α reductase, which converts testosterone to the more active androgen dihydrotestosterone and (*iii*) increased dietary fat consumption. Some adverse environmental factors, which have been proposed, include decreased exposure to ultra violet radiation, elevated levels of cadmium and decreased levels of zinc. Several molecular and biochemical studies revealed that prostate cancer commonly represents elevated serum prostatic specific antigen (PSA), prostatic specific antigen (PAP) and prostate specific membrane antigen (PSMA) also nuclear androgen receptor. Traditionally the digital rectal examination (DRE) has been the principal method of obtaining objective evidence to support the consideration of a diagnosis of prostate cancer. Radiation therapy, brachy therapy guide for the placement of radio active seeds into the prostate have been used. For metastatic cancers, castration, androgen ablation therapies are widely used. Androgen ablation therapy includes GnRH analogs and androgen receptor antagonists like flutamide, 5α reductase inhibitors like finasteride etc. have been used. Chemotherapeutic agents fumigillan, thalidomide, suramin, estramustine and vinblastin have been used. In the later stages

E-mail: j_arunakaran@hotmail.com

cancer cells does not respond to androgen ablation therapy because of mutation in the androgen receptor and promotor hypermethylation of androgen receptor (AR), ultimately leads to androgen refractory prostate cancer. Growth factors like EGF, IGF, FGF and IL-6 can activate AR in the absence of ligand and induce mitogenic signals. Severe side effects of these therapies include: impotence, cardiac failure, thrombosis, edema, st. matitis, neutropenia, alopecia etc.

Insulin like growth factors (IGF-I, IGF-II) are mitogenic peptides, which have been related to be the pathogenesis of prostate cancer, and their bioavailability is determined by at least six different binding proteins called IGFBPs. Quercetin, a flavonoid found to inhibit tumor progression and can stimulate apoptosis process in colon, leukemic, mouse melonoma and prostate cancer cells. Garlic organosulfur compounds reduced experimentally induced colon, esophagus, mammary, lung and liver cancers suggesting the protection provided by garlic related compounds. In our study, we proved that diallyl disulfide, one of the effective organosulfur compound of garlic may be useful anticancer effect for management of human prostate cancer.

Although dietary onion or garlic (quercetin or diallyl disulfide) is active against prostate cancer, its molecular mechanism involved in chemotherapeutic potency should be tested on the prostate cancer cells. The potential merits should also be tested in animal models, before going to clinical trials. Apart from this, the role of Vitamin D analogs as an antiproliferative, prodifferentiation and apoptotic agent in prostate cancer merits further study. Vitamin D3 regulates prostate cancer cell growth through multiple mechanisms including, cell cycle arrest, induction of apoptosis and regulation of growth factor signaling.

Despite recent advances, the etiology of prostate cancer remains obscure. Curative treatment is available for many patients with organ confined disease. Unfortunately, there is no curative therapy is available for prostate cancer patients. Apart from hormonal therapy, chemoprevention through dietary compounds enhances our insight into the development of gene therapies, whose impact will well tolerated, effective and durable. Until then it is imperative that we provide these to patients with compassionate and insightful care that will translate into an improvement in the quality of their life as well as low fat diet consumption etc.

Introduction

The prostate is one of the male accessory sex organs along with seminal vesicles, epididymis and vas deferens in mammals. It is a compound tubuloalveolar gland whose base abouts the bladder neck and whose apex merges with the membrane urethra to rest on the urogenital diaphragm. It secretes many components of semen (20 per cent). Prostate cancer is the most prevalent cancer and the second leading cause for cancer death among men. It has become the most common cancer of males in several developing countries including India (Quinn and Babb, 2002). Adenocarcinoma of the prostate is the most common cancer in American men. It constitutes a major factor in the health of the male population in the United states and many other countries in the Western Hemisphere (Kozlwski, 1999).

The rate of detection of occult prostate cancer increases progressively with age, tripling by the ninth decade. Prostate cancers which become clinically manifest often acquire features which facilitate tumour invasion and metastasis. About 25 per cent of men with clinically apparent prostate cancer die of the disease. With increasing age, most men will develop microscopic foci of prostate cancer whether they live in a population at a high or low risk for the invasiform of the disease (Yatani *et al.*, 1982). Autopsy studies have shown that approximately 30 per cent of men over the age of 50 have histologic evidence of prostate cancer (Scardino 1989). In 2003, an estimated 220,900 men were diagnosed with this disease and 28,900 men died from prostate cancer, making it the second most common cause of cancer death in men (Jemal *et al.*, 2003). The incidence of prostate cancer varies world wide with the highest rates found in the United States, Canada, Scandinavia and the lowest rates found in China and other parts of Asia (Quinn and Babb, 2002; Gronberg, 2003).

Prostate cancer mortality varies worldwide with the highest rates approved in the Caribbean and Scandinavia and the lowest rates in China, Japan and countries of the former Soviet Union (American Cancer Society 2003). The number of men over 65 yrs is expected to increase fourfold worldwide between the years 2000 and 2050, representing an increase from 12.4 per cent of the population in 2000 to 19.6 per cent in 2030 (Lunenfeld, 2002).

Etiology

Advancing age is the strongest risk factor for the development of prostate cancer. By age 90, this neoplasm is identified in 67 per cent of male cadavers that undergo autopsy. The post puberal presence of an intact hypothalamic–pituitary–testicular axis is critical since castration prior to the onset of puberty markedly diminishes the rest of developing cancer. Being of the black race is another strong etiologic factor. African American men have the highest incidence and mortality from prostate cancer in the world. In contrast, the Japanese and mainland Chinese have the lowest rates of prostate cancer. Many mechanisms have been postulated to account for this marked predilection of African American men, including (a) higher fetal exposure to androgens, increased expression of 5α reductase which converts testosterone to the more active androgens, dehydrotestosterone and (b) increased dietary fat consumption.

Risk of developing prostate cancer increases two or three fold in the presence of a first degree relative affected with this cancer. The risk increases to nine fold for an individual with two first degree relatives with prostate cancer. Hereditary prostate cancer tends to develop prior to the age of 55. Recent linkage studies have identified the hereditary prostate cancer 1 locus on chromosome 1q24-25. In contrast, most cases of sporadic prostate cancer are associated with hypermethylation of the regulating sequences of the glutathione S transferase gene locus.

A diet rich in saturated fatty acids may predispose to the development of prostate cancer and facilitate its subsequent growth. Sexual activity, sexually transmitted diseases/infection and antecedent vasectomy have a very tenuous association with the development of prostate cancer.

Epidemiology the incidence of prostate cancer increased steadily with the advent of widespread Prostate Specific Antigen (PSA) screening, the incidence increased by 85 per cent between 1987 and 1992 to reach an age-adjusted rate 190.1 per 100,000 men (Greenlee *et al.*, 2000). In the current era of widespread PSA testing, approximately 75 per cent of all prostate cancers cases are detected as a result of abnormal PSA findings. Because of this, the proportion of cases presenting with lymph node involvement of advanced disease has decreased considerably. Now, it is estimated that more than 67 per cent of men are diagnosed with localized and early tumors still, one third of patients with apparently local disease and treated with radical prostatectomy relapse biochemically within 5 years (Remzi *et al.*, 2004).

Environmental and dietary factors play a major role in prostate cancer etiology as evidenced by epidemiological data. Global incidence figures vary widely from 1.08 per 100,000 per year in the USA. Some adverse environmental factors which have been proposed include decreased exposure to ultraviolet light, high-fat diet, elevated levels of cadmium and decreased levels of zinc. Differences in ingestion of high fat diets, green tea or soy products provide potential explanations for the divergent rates of prostate cancer among different populations (Gronberg, 2003).

Hormonal factors and particularly circulating androgens probably play a role in the initiation and promotion of the prostate cancer. Prostate cancer rarely develops in men surgically orchidectomised before the age of 30 yrs (Moon and Logethetis, 2001). Men with prostate cancer report a family history of this tumour 3.1 to 4.3 times more commonly than healthy men. Ross *et al.* (2000) suggested that the lower frequency of prostate cancer in Japanese and Chinese populations may be related to a lower level of 5α reductase activity than in their white counterparts, perhaps on genetic basis.

Pathology

Primary adenocarcinoma of the prostate accounts for over 95 per cent of all prostatic malignancies. Tumors are thought to originate from the adluminal (or secretory) cells of the prostatic acinus. The normal prostate gland consists of four regions: (i) Transition Zone (20 per cent); (ii) Central Zone (10 per cent); (iii) Peripheral Zone (70 per cent) and (iv) Anterior fibro muscular stroma.

About 70 per cent cancer originate from peripheral zone. The transition zone and central zone account for 20 and 10 per cent of these tumors, respectively. Multiple tumor foci of variable volume and grade are commonly encountered and support of the concept of a field change effect for this neoplasm. Histologically, the characteristic double layer of cells lining the acinus may be replaced by a single layer or a piling of cells. The basal cell layer is lost. Another histologic hallmark of prostate cancer is the loss of intervening or enveloping stroma which results in a back to back configuration of the acini. Invasion of the intraprostatic perineural spaces has been seen in 85 per cent of incidental and early carcinoma. Most prostatic cancer stain positively for cytokeratins (CKs) 18. CKs 5/15 are not identified because these CKs are typically associated with the basic cell layer, which is no longer present in prostate cancer. Immuno-histochemical studies have also demonstrated that prostate cancer commonly expresses prostatic specific antigen (PSA), prostatic acid phosphatase (PAP), prostate specific membrane antigen (PSMA) and nuclear androgen receptor (Janulis and Lee, 1999).

Grading is defined as an effort to use histologic characteristics of a tumor to predict its biologic activity. The Gleason classification scheme is the most popular and employs low-power magnification (X40-100) in order to evaluate the glandular pattern of the tumor and its relationship to the stromal compartment (Gleason, 1977). Five tumor grades progressing from the most (1) well defined; (2) less defined (3) poorly defined (4) Ragged infiltrating and (5) Ragged infiltrating with central necrosis of margins of tumor areas with specific gland pattern, size and distribution are recognized. The Gleason tumor score correlates with the risk of developing metastatic disease. Patients with poorly differentiated cancers had a 10 fold greater risk of death from prostate cancer than men who had well differentiated disease. Specifically men with poorly differentiated prostatic cancer had a 5 year cause specific survival of only 34 per cent. Men with well differentiated or moderately differentiated prostate disease had a 5 year cause specific survival of 87 per cent (Remzi *et al.*, 2004).

Symptoms

Early stages prostate cancer does not cause symptoms. This is logical given the peripheral zone origin of most of these tumors, locally advanced prostate cancer is commonly associated with urethral and to bladder neck obstructions. These tumors also exhibit evidence of extra prostatic disease with involvement of the periprostatic tissues, seminal vesicles and bladder base. The patients will complain of irritative and obstructive voiding symptoms including increased frequency, dysuria, nocturia, hesitancy, intermittency, urgency and a perception of incomplete bladder emptying. Urinary retention, hematuria and urinary tract, infections may also occur patients with metastatic disease will often complain of constitutional symptoms such as being easily fatigued and weight loss. Bulky lymph node metastases may be associated with low extremity edema. Patients with skeletal metastages with often complain of multifocal bone pain and may exhibit signs of lower extremity weakness due to spinal cord compression.

Before going to diagnosis or screening of prostate cancer, we should know what is prostate specific antigen (PSA). PSA is a serine protease and it is a chymotrypsin like enzyme and gene on chromosome 19q 13.3–q 13.4. It is secreted by prostatic epithelial cells. It is found in huge levels in seminal plasma. It cleaves seminogelins/semen liquefaction. Its expression is regulated by androgens and progestins. It is the best cancer marker ever discovered. It is almost organ specific biomarker and used for diagnosis or screening, monitoring and prognosis. (Beduschi and Oesterling, 1999).

Diagnosis/Screening

Traditionally, the digital rectal examination (DRE) has been the principal method of obtaining objective evidence to support consideration of a diagnosis of prostate cancer. The features on DRE include a firm palpable nodule with indistinct margins, irregular, stony areas of induration, obliteration of the lateral sulci or median furrow; and induration with loss of capsular integrity in the region of the prostatic base, implying seminal vesicle infiltration. The limitations of DRE as an indicator of the presence and extent of prostate cancer are well documented. Prostate cancer will be documented in only 39 per cent of patients who undergo biopsy on the basis of an abnormal DRE alone. Alternate causes of a nodular induration include a prostatic calculus, prostatitis, tuberculosis, focal infarction, a postbiopsy tissue reaction and a spheroid of benign prostatic hyperplasia. DRE will miss many prostate cancers and will often underestimate the volume/extent of existing disease (Kozlowski and Grayhack, 1996).

None of the current physical imaging modalities are sufficiently sensitive/specific to provide great assistance in the diagnosis of early stage prostate cancer. Currently the anatomic information provided by transrectal ultrasound (TRUS) is superior to that obtained by conventional computed tomographic (CT) scanning or magnetic resonance imaging (MRI). Classically, Prostate cancer presents as a hypoechoic area on TRUS. Many cancers do not manifest this typical feature. The relative recent advent of color Doppler TRUS and transrectal MRI may increase the accuracy of these imaging techniques.

The PSA testing has enhanced our ability to scan at risk populations. PSA is a single chain glycoproteins with a molecular weight of about 34 k Daltons. The human PSA gene is located on chromosome 19. PSA is a cytoplasmic serine protease, which shares homology with the human Kallikreins (Oesterling, 1996). PSA has numerous biological functions, which I have explained in the earlier paragraph. It is a regulator of the insulin like growth factors (IGFs). In fact, PSA cleaves IGF from its binding proteins, releasing the free moiety and thus facilitating the mitogenic impact of the IGFs on prostate cells. PSA possesses properties of a weak matrix degrading protease and its release may facilitate the matrix degradation seen in prostate cancer. PSA is also a mitogen for osteoblasts and may play a role in the osteoblastic proliferation seen in metastatic prostatic cancer (Graves, 1995). PSA has been identified in a number of non prostatic tissues, including periurethral glands, perianal glands, salivary glands, benign and malignant breast tissue, the uterus and a number of solid tumors capable of the constitutive expression of the steroid receptor super family. The normal PSA value should be less than 4 ng/ml. Most serum PSA is complexed with other proteins, particularly 21 antichymotrypsin and α-2-macroglobulin (Lilja *et al.*, 1991). A small fraction of PSA occurs in the free uncomplexed molecular form. The percentage of free PSA decreases in Prostate cancer. It is thought that post transactional modifications in cancer increase the affinity of PSA for these binding proteins. (Oesterling, 1996; Young *et al.*, 1996).

Assessment of the percentage of free PSA is commonly performed in men with a total PSA elevation between 4 and 10 ng/ml. If the percentage of free PSA is more than 24 per cent only 10 per cent of such patients have prostate cancer. Utilizing this approach, about 25 per cent of false negative prostate biopsies can be avoided. A PSA density ie (PSA divided by prostate volume) of more than 0.15 is thought to define an at risk population. Similarly the PSA velocity (rate of change overtime) is another useful barometer. A PSA velocity of more than 0.75 documented over a 12 month interval reflects a worrisome degree of proliferative activity. Both of these abnormalities would prompt ultrasound guided prostate needle biopsies.

Prognosis

The pelvic lymph nodes and the skeletal system are the most prominently identifiable sites of initial dissemination. Involvement of lung, liver and other anatomic areas occurs in frequently and such involvement is usually a premorbid event. Prostate cancers are biologically heterogeneous and contain androgen dependent, androgen sensitive and an androgen–independent sub populations. A majority of prostate cancers maintain a relative of absolute dependence on androgen for much of their natural life. Most of the patients that succumb to metastatic (Prostate) cancer die as a result of the uncontrolled proliferation of hormone refractory cells. The time frame for development and progression of the carcinoma of the prostate is highly variable.

Men assigned to radical prostatectomy had a significantly lower risk of death from prostate cancer than patients assigned to the watchful waiting group. Patients with a reason score of more than 7 and a projected life expectancy of at least 10 years may be at a similar risk of dying of prostate cancer as younger patients.

Treatment Options

Organ Confined Disease

The treatment of organ confined prostate cancer must be individualized. Important variables which impact decision making include age, associated medical co morbidities, anticipated life expectancy, the assigned clinical stage, tumor grade, the pretreatment PSA level and the patients' preference after a discussion of the risk/benefits of available options.

Apart from DRE and TRUS, Radiation therapy has been extensively used for treatment of clinically localized prostate cancer. Most commonly, this therapy involves the administration of high energy electrons or photons. Using CAT scans and computers, very precise dosimetry planning is possible (Hartan *et al.*, 2001). An alternative to above radiation therapy is brachy therapy, which utilizes the use of transrectal ultrasonography to precisely guide the placement of ^{125}I or ^{103}palladium seeds into the prostate (Millikan and Logothetis, 1997). For metastatic disease, castration induced androgen ablation therapy is widely used. The therapy commonly consists of a luteinizing hormone–releasing hormone agonist or with antiandrogen or without antiandrogen (Pienta, 1997).

Another one is scrotal orchidectomy. It can also be performed. The average duration of this treatment is 2 to 3 years ultimately many patients will develop hormone-refractory tumor. Some other androgen ablation therapies include GnRH analogues and androgen receptor antagonists like Flutamide and 5α reductase inhibitor Finasteride etc. (Esper and Pienta, 1997). Another potentially curative treatment is radical prostatectomy which implies the removal of the entire prostate gland, the seminal vesicles and ampulla of the deferens. This operation can be performed through a retropubic or perineal approach.

A variety of procedures are available to facilitate the ultimate staging assignment. The DRE itself if properly performed may permit the identification of capsular perforation and or probable seminal vesicle involvement. Additional serum markers such as PAP and bone isoenzyme of alkaline phosphatase are occasionally helpful. Reverse transcriptase polymerase chain reaction for PSA and PSMA mRNA has recently employed to detect the presence of circulating prostate tumor cells and may facilitate the detection of subclinical metastatic disease. The Prosta Scint scan is the most recent addition to our diagnostic/staging armamentarium. Prosta scint is a murine monoclonal antibody directed against PSMA (Carter *et al.*, 1996).

The complications of radical prostatectomy which implies the reconstruction of urinary tract involves modification of the bladder neck and its anastomosis to the membranous urethra. Preservation of the pelvic splanchnic nerves is technically possible but may be imprudent if tumor is located in close proximity to the nerve bundles. The operation can be performed through a retropubic or perineal approach. Following surgery an indwelling urethral catheter is left in place for several weeks until the anastomosis heals. The operational mortality is 0.8 per cent in healthy patients. Incisional pain and blood loss are common generic complications. Impotence can occur following the procedure. About 50–60 per cent of patients will return to base line sexual activity if both neuro vascular bundles can be preserved. It may take 12 to 18 months for this recovery to take place however, patients who are rendered impotent can have sexual function restored using a wide variety of techniques including a vaceum errection device, intraurethral prostaglandin E (MUSE), intracavernosal injection of prostaglandin E or other vasodilators, insertion of penile prosthesis, or the use of a new oral agent (sildenafil) which inhibits type 5 phospho-diesterase. An other potential complication of radical prostatectomy is urinary incontinence and generally require placement of an artificial sphincter or the performance of a sling procedure.

Metastatic Disease

Patients with hormone refractory metastatic prostate cancer have a poor prognosis with a mean survival of less than a year. If the patient has been maintained on an antiandrogen, such therapy should be stopped because a paradoxical enhancement of tumor growth has been observed in about 30–40 per cent of such patients. Alternative approaches to second line hormonal therapy include ketoconazole plus hydrocortisone, amino glutethimide plus hydrocortisone or hydrocortisone alone. Once again about 30 per cent of patients will transiently benefit from such strategies.

Systemic chemotherapy is generally reserved for patients who are overtly symptomatic and have failed first and or second line hormonal therapy. For these patients, chemotherapy is administered. Mitoxantrone and prednisone is a currently popular regimen. Other efficaous two drug regimens involve the use of estramustine with one of the following agents: Vinblastine; Vinovelbine and Paclitaxel or etopside.

In general, these combinations are associated with a favourable PSA response in about 50 per cent of patients. Relief of pain and decrease in measurable disease have been reported in 30 per cent of patients. Unfortunately the duration of benefit is relatively transient. Many other experimental therapies have been or are being evaluated, including the use of antigrowth factor agents "Suramin" differentiation inducing therapy (Phenylbutyrate), the use of antiangiogeneis agents (TNP 470 and linomide), the use of agents which interfere with tumor cell adhesion (citrus pectin), the induction of programmed cell death (Thapsiagargin) and the use of gene therapy (Millikan and Logothetis, 1997; Chaudhary *et al.*, 2001).

The somatic and visceral pain associated with advanced–stage malignancy should be managed utilizing the world Health organization analgesic "Step ladder" Multifocal skeletal discomfort can be effectively managed utilizing the systematic calcium analog strontium 89. Anemia, constipation, persistent neuropathic pain and depression represent some additional problems that require prompt and appropriate management (Kozolowski, 1999).

Data from epidemiological studies has suggested that Vitamin D deficiency may promote prostate cancer although the mechanism is not understood. The presence of Vitamin D receptors have demonstrated in prostatic cancer tissues. Vitamin D3 may be useful as an additional therapeutic option for the treatment of prostate cancer (Skowronski *et al.*, 1995).

Androgen withdrawal is rarely curative because some prostatic cancer do not express androgen receptor due to gene silenced by promotor hypermethylation, this tumor cells does not respond to androgen withdrawal ultimately leads to the hormone–refractory prostate cancer (Kokontis and Liao, 1999).

The normal development and maintenance of the prostate is dependent on androgen acting through the androgen receptor (AR). AR remains important in the development and progression of prostate cancer. AR expression is maintained throughout prostate cancer progression, and the majority of androgen independent or hormone refractory prostate cancers express AR. Mutation of AR, especially mutations that result in a relaxation of AR ligand specificity, may contribute to the progression of prostate cancer and the failure of endocrine therapy by allowing AR transcriptional activation in response to antiandrogens or other endogenous hormones. Similarly, alterations in the relative expression of AR co regulators have been found to occur with prostate cancer progression and may contribute to differences in AR ligand specificity to transcriptional activity (Heinlein and Chang, 2004).

Prostate cancer progression is also associated with increased growth factor production and an altered response to growth factors by prostate cancer cells. The kinase signal transduction cascades initiated by mitogenic growth factors modulate the transcriptional activity of AR and the interaction between AR and AR coactivators. The inhibition of AR activity through mechanisms in addition to androgen ablation, such as modulation of signal transduction pathways may delay prostate cancer progression. IGFs are important growth factors in the development of prostate cancer and its progression. Overexpression of related proteins or changes which raise the balance of IGF/IGF–IR activity versus IGFBP function can potentially contribute to prostate carcinogenesis (Firth and Baxter 2002). IGFBP 3 also can potentiate IGF bioactivity in prostatic cancer cells (Rajah *et al.*, 1997). In our laboratory we found that the biphasic effect of 17β estradiol in PC 3 cells at 48 hrs could be explained by the increased concentration of IGFBP3 at 100 nm concentration which inhibits the IGFI action by decreasing its bioavailability thereby decreasing the cell proliferation (Sharmila, 2003; Kanagaraj *et al.*, 2004).

We are planning to identify candidate genes in the IGF signaling cascade for their putative use as a molecular probe in the development of drug for prostate cancer. Unfortunately the available drugs for the treatment of metastatic prostate cancer are with side effects. The side effects of above therapies includes impotence, cardiac failure, thrombosis, edema, stomatitis, neutropenia, alopecia etc. Therefore it is necessary to invent new drugs with less side effects for the prostate cancer treatment.

Numerous scientific reports imply that high consumption of certain vegetables and vegetable group reduces the risk of cholorectal, stomach, lung and esophageal cancers. The allium genus of vegetables includes garlic, onion, leeks, scallions, chives and shallots. These vegetables are characterized by a composition that is high in flavonols and organosulfur compounds. Further more, animal and *in vitro* studies provide evidence of the anti-carcinogenic potential of several bioactive compounds in allium vegetables (Wargovich *et al.*, 1988).

In general, garlic has been used throughout the world to treat coughs, toothache, earache, dandruff, hypertension, atherosclerosis, hysteria, diarrhoea, dysentery, diptheria; vaginitis and many other conditions (Block, 1985). It has been demonstrated recently diallyl disulfide (DADS) and diallyl sulfide (DAS) which are organoallylsulfur compounds of garlic extract have apoptotic potential on non small cell lung cancer (Hong *et al.*, 2000), bladder tumor (Chung, 1999), colon tumor (Sundaram and Milner, 1996) and breast cancer cells (Nakagawa *et al.*, 2001). In earlier studies carried out in our laboratory it was shown that DADS induces apoptosis through ethidium bromide acridine orange staining and DNA fragmentation on agarose gel electrophoresis in prostate cancer cell line (LNCaP) (Gunadharini, 2004). These results suggest that DADS may be useful for anticancer effect for management of human prostate cancer. Since LNCaP cell line is androgen dependent, further studies are required to investigate the effect of DADS on the expression of AR and AR responsive genes.

Flavanoids are polyphenols found in fruits, vegetables, nuts, seeds, herbs, spices, stems as well as red wine. They are prominent in citrus fruits (Kefford, 1996). Quercetin, a flavonoid commonly present in many vegetables, has been shown to induce apoptosis in various tumor cell lines including HL60, K562 human leukemia cells and HT human colon cancer cells. Quercetin can interact with a broad range of enzymes, specifically receptor kinases, protein Kinase C, Na^{+}-K^{+} ATPase and topoisomerase. It can also regulate oncogenes like ras and myc and induce wild type P^{53} (see Marchand, 2002).

Quercetin could also be an useful substance for the prevention and intervention treatment of prostate cancer cells (LNCaP) (Yuan *et al.*, 2004). Thus, quercetin appears to be a putative candidate, plant flavonoid for its useful in prostate cancer. However, the exact mechanism by which quercetin interferes the development and progression of prostate cancer is not understood. The work is going on in our laboratory to prove quercetin as a cancer chemopreventive agent on prostate cancer (Vijayababu, 2004). Our aim is to investigate the molecular mechanism underlying the chemopreventive role of quercetin on prostate cancer.

Conclusion

Curative treatment is available for many patients with organ-confined disease. Despite recent advances, the etiology of prostate cancer remains obscure. Treatment must be tailored to the individual patient but aggressive cancers should be treated aggressively and patient's age should not be unappropriately considered in treatment and decision making for localized prostate carcinoma. Unfortunately curative therapy is not available for patients with locally advanced prostate cancer. Sophisticated scientific approaches, molecular mechanism of drugs, dietary elements are beginning to yield significant insights into the genetic and biochemical aberrations associated with aggressive prostate cancer. In all probability this enhanced insight will translate into the development of novel therapies whose impact will be well tolerated, effective and durable.

In this context, it is my privilege to conclude that patients diagnosed with prostate cancer who need information to help them decide on treatment, to avoid side effects such as cardiac problem, uncontinence or impotence etc, they can take dietary flavonoids or diallyl disulfide compounds as a drug as well as preventive measure.

References

American Cancer Society, Cancer facts and figures 2003, Atlanda, GA: American Cancer Society 2003.

Beduschi R, and Oesterling JE. (1999). Prostate specific antigen. *In*: Knobil E, and Neill JD. (Eds), *Encyclopedia of Reproduction*, Volume 4, Academic Press, California, pp. 86-95.

Block, E (1985) The Chemistry of garlic and onions. Sci. Am., 252, 114-119.

Carter RE, Feldman AR, and Coyle JT. (1996). Prostate specific membrane antigen is a hydrolase with substrate and pharmacologic characteristics of aneuropeptidase. Proc. Natl. Acad. Sci. USA, 93, 749-753.

Chaudhary KS, Abel PD, Stamp GWH, and Lalani E. (2001). Differential expression of cell death regulators in response to thapsigargin and adiramycin in Bcl-2 transfected DU145 prostatic cancer cells. J. Pathol., 193, 522-529.

Chung, JG (1999) Effects of garlic components diallyl sulfide and diallydisulfide on arylamine N-acetyl transferase activity in human bladder tumor cells. Drug Chem. Toxicol., 22, 343-358.

Esper PS, and Pienta KJ. (1997). Supportive care in the patient with hormone refractory prostate cancer. Sem. Urol. Oncol., 15, 56-64.

Firth SM, and Baxter RC. (2002). Cellular actions of the Insulin like Growth factor binding proteins. Endocrine Rev., 23, 824-854.

Gleason DF. (1977). Histologic grading and clinical staging of carcinoma of the prostate. *In*: Tannenbaum M. (Ed.), *Urologic pathology: The prostate*, Lea and Febiger, Philadelphia, p. 171.

Graves HCB. (1995). Non prostatic sources of prostate specific antigen; A steroid hormone dependent phenomenon. Clin. Chem., 41, 7–17.

Greenlee RT, Murray T, Bolden S, and Wings PA. (2000). Cancer statistics. Cancer J. Clin., 50, 7-33.

Gronberg H. (2003). Prostate Cancer epidemiology. Lancet, 361, 859-864.

Harlan LC, Potosky A, Gilliland FD, Hoffman R, Albertsen PC, and Hamilton AS. (2001). Factors associated with initial therapy for clinically localized prostate cancer: prostate cancer outcome study. J. Natl. Cancer. Inst., 93, 1864-1871.

Heinlain CA, and Chang C. (2004). Androgen receptor in prostate cancer. Endocrine Rev., 25, 276-378.

Hong Y, Ham Y, Choi J, and Kim J. (2000). Effect of allyl sulfur compounds and garlic extracts on the expressions of Bcl 2, Bax and P^{53} in non small cell lung cancer cell lines. Exp. Mol. Med., 32, 127-134.

Janulis L, and Chung L. (1999). Prostate gland. *In*: Knobil E, and Neill JD. (Eds.), *Encyclopedia of Reproduction*, Volume 4, Academic Press, California, pp.77-85.

Jemal A, Murray T, Samuels A, Crhafoor A, Ward E, Thun M J (2003) Cancer statistics C A Cancer. J. Clin., 53, 5-26.

Kanagaraj P, Ilangovan R, Vijayababu MR, Srinivasan N, Aruldhas MM, Govindarajulu P, and Arunakaran J. (2004). Effect of 17β estradiol on th production of IGFBP 4 and proliferation of prostate cancer PC 3 cells. XXIII Symposium on SRBCE, Univ. of Madras, page 81, abstract 34.

Kefford JF. (1966). Citrus fruits and processed citrus products in human nutrition. World Rev. Nutr. Diet, 6, 197-249.

Kokontis JM and Liao S (1999). Molecular action of androgen in the normal and neoplastic prostate. Vitam. Horm., 55, 219-307.

Kozlowski JM (1999) Prostate Cancer. *In*: Knobil E, and Neill JD. (Eds.), *Encyclopedia of Reproduction*, Volume 4, Academic Press, California, pp. 67-76.

Kozlowski JM, and Grayhack JT. (1996). Carcinoma of the prostate. *In*: Gillenwater *et al.* (Eds.), *Adult and Pedeatric Urology*. Volume 2, 3rd edition, Mosby, St. Louis, Mo, pp. 1575-1713.

Lilja H, Christensson a, Dahlam U *et al.* (1991) Prostate specific antigen in human serum occurs predominantly in complex with α1 antichymotrypsin. Clin. Chem., 37, 1618–1624.

Lunenfeld B. (2002). The ageing male: demographics and challenges. World J. Urol., 20, 11-16.

Marchand LL. (2002). Cancer preventive effects of flavonoids–a review. Biomed, Pharmacother., 56, 296-301.

Millikan R, and Logothetics CJ.(1997). Update of the NCCN guidelines for treatments of prostate cancer. Oncology, 11, 180-193.

Moon C and Legethetis CJ (2001) Endocrine aspects of prostate cancer. *In*: Beeker KL. (Ed.), *Principles and Practice of Endocrinology and Metabolism*. 3rd edition, Chapter 225, Lipincott Williams and Wilkins, Philadelphia, pp. 2046-2057.

Nakagawa H, Tsuta K, Kiuchi K, Senzaki H, Tanaka K, Hioki K, and Tsubara A.(2001). Growth inhibitory effects of diallyl disulfide on human breast cancer cell lines. Carcinogenesis, 22, 891-897.

Oesterling JE. (1996). Molecular PSA. The next frontier in PCA screening Contemp. Urol., April, pp. 76-92.

Pienta KJ. (1997). Advances in the treatment of metastatic prostate cancer. Biomedical communications. Univ. of Michigan, Ann Arbor.

Quinn M, and Babb P. (2002). Patterns and trends in prostate cancer incidence, survival, prevalence and mortality. Part I: international comparisons. BJU International 90, 162-164.

Rajah R, Valentinis B, and Cohen P. (1997). Insulin like growth factor (IGF) binding protein 3 induces apoptosis and medialts the effects of transforming growth factor β 1 and programmed cell death through p 53 and IGF–independent mechanism. J. Biol. Chem., 272, 12181-12188.

Remzi M, Waldert M, and Djavan B. (2004). Prostate cancer in the ageing male. J men's health and gender, 1, 47-54.

Ross KS,. Carter HB, Pearson JD, and Guess HA. (2000). Comparative efficiency of prostate-specific antigen screening strategies for prostate cancer detection. J. Am. Med. Assoc., 284, 1399-1405.

Scardino FT. (1989). Early detection of prostate cancer. Urol. Clin. North Am., 16, 635-685.

Sharmila M. (2003). Influence of 17 β estradiol on the production of IGF I and IGFBP–3 in prostatic carcinoma cell line (PC3) M.Phil Dissertation, University of Madras, Chennai, India.

Skowronski RJ, Peehl DM, and Feldman D. (1995). Actions of vitamin D3 analogs on human prostate cancer cel lins comparison with 1,25 dihydroxy vitamin D3. Endocrinology, 136, 20–26.

Sundaram SG and Milner JA (1996) Diallyl disulfide induces apoptosis of human colon tumor cells. Carcinogenesis, 17, 669–673.

Vijayababu MR, Kanagaraj P, Arunkumar J, Srinivsan N, Michael Aruldass M, and Arunakaran J. (2004). Effects of quercetin on IGF system components in PC 3 cells. XXII symposium on SRBCE, Abstract 14, page 30, University of Madras, Chennai, India.

Wargovich MJ, Words C, Eng VW, Stephens LC, and Gray K. (1988). Chemoprevention of N-nitroso methyl benzylamine induced oesophageal cancer in rats by the naturally occurring thioester, diallyl sulfide. Cancer Res., 48, 6872-6875.

7

DNA Microarray: From Technology to Endocrine Research

Arun Bandyopadhyay[1*], *Kakali De*[1], *Goutam Ghosh*[1], *Aditya Konar*[1], *Sangita Maity*[1], *Malabika Dutta*[2], *Samir Bhattacharya*[3]

[1]*Indian Institute of Chemical Biology, 4 Raja S. C. Mullick Road, Kolkata – 700 032, India*
[2]*Institute of Genomic and Integrated Biology, New Delhi, India*
[3]*Department of Zoology, Visva-Bharati, Santiniketan, India*

ABSTRACT

cDNA Microarray allows the simultaneous analysis of gene expression patterns of thousands of genes under a similar set of experimental conditions. Gene expression array printed on nylon filter membranes is extremely sensitive in detecting very less abundant mRNA with the help of radioactive labelling. To understand the global gene expression profile in hyperthyroid rat heart, we have utilized cDNA array comprising 588 genes those are important for cardiac function. Hyperthyroid rat was produced by administering rats with 3,5,3′ triiodo-L-thyronine (T3) for 15 days. The heart weight to body weight ratio (HW/BW) in hyperthyroid rat was increased by about 58 per cent over that of normal rat. Based on a threshold of greater than 10 per cent change, about 37 genes were found to be regulated by T3. Further analyses by Western blotting, Northern blotting and real-time quantitative RT-PCR of some of the genes confirmed the microarray results. The T3-altered genes encode various types of proteins related to metabolism, matrix and cytoskeletal structures, growth factors, transcription receptors, Ca^{2+}-

**Corresponding Author*: E-mail: arunb@iicb.res.in

channels etc. Physiological significance of one of such altered proteins, insulin-responsive glucose transporter (GLUT) type 4 (GLUT4) in hyperthyroid heart was studied in details. The expression of GLUT4 was drastically reduced in the ventricular tissues of hyperthyroid heart. Insulin-induced glucose uptake in hyperthyroid cardiomyocytes was reduced significantly indicating the impaired glucose transport in cardiac cells. Interestingly, a few genes such as GLUT4, cytochrome P450 isoforms, superoxide dismutase (SOD), collagens, matrix metalloproteinases (MMP), tissue inhibitors of matrix metalloproteinases etc which were not reported earlier were found to be altered in hyperthyroid heart. The results show some new aspects of hyperthyroid heart which will be important to assess the pathophysiology of hypertrophied cardiomyocytes.

Introduction

DNA Microarray

In biological system genomic DNA is transcribed into messenger RNA which is translated into protein (the gene product). Thus the flow of biological information described in the so-called central dogma is from DNA, to RNA, to Protein. The protein is the primary molecular machinery that carries out cellular functions. Genes on the other hand, serves as the blueprints from which proteins are constructed, and when accessed by the cell's transcriptional machinery, are transcribed into the scaffolding on which proteins are built: the mRNA. As such the relative abundance and complexity of mRNA transcripts, collectively known as transcriptome, reflects the potential of all cellular pathways. Knowledge of the transcriptome therefore provides clues as to the molecular processes that operative, or not, in a cell.

To understand the molecular logic underlying biological processes and cellular states, scientists have used different methods of gene expression analysis. This has led to the discovery of genes that play important role in cell behaviour and human disease. The techniques used for determining gene expression levels are Northern Blot analysis, and Differential Display PCR. The limitation of this approach was to look for transcripts one or few at a time. The recent invention of genome-wide transcript screening and and DNA microarray technology have overcome this limitation. These techniques are helpful for the determination of the relative abundance and complexity of tens of thousands of transcripts simultaneously.

Types of Gene Expression Arrays

There are generally three types of gene expression arrays; oligonucleotide chips, cDNA arrays printed on glass slides using fluorescent probes and cDNA arrays printed on nylon filter membranes. Each format has associated advantages and disadvantages. cDNA arrays using fluorescent probes utilise typical two coloured microarray hybridisation procedure. Cellular mRNA is extracted from two types of tissues (or cell population) such as 'normal' and 'treated' for which relative gene expression levels are to be compared. The RNAs are then reverse transcribed into cDNA by simple enzymatic reaction and fluorescently labelled with different colour fluorophores *i.e.* Cy3 and Cy5 dyes. Thereby one cDNA target population will display green fluorescence (Cy3) and the other will be displaying red fluorescence (Cy5). These labelled probes are then mixed together and simultaneously hybridised to the same microarray. Using microarray scanner the hybridisation signals are captured. If the magnitude of the average signal intensity for a given gene probe is equivalent in the Cy3 and Cy5 channels than the actual transcript level of that gene in the two RNA samples (normal and treated) are said to be equally expressed. If red signal is more than the green signal or vice versa then the corresponding genes are differentially expressed. The colour with higher intensity corresponds to the sample in which the gene is highly expressed.

cDNA arrays printed on membranes require radioactive labelling where two nylon different membranes are used for the hybridisation with cDNA from two different tissues (or cell population). This system offer multiple advantages as compared to glass/fluorescent system.

1. Higher efficiency of labelling cDNA with radioactive labels as compared to fluorescent labels, which result in higher specific activity probes. In case where RNA is limiting this allows a greater chance for multiple replication leading to statistical significance. (Bertucci, 1999).
2. A single label system as compared to dual label system. This allows direct comparison of two cDNA probes.
3. Multiple use of nylon filters. Filter membranes can be used three to five times. This reduces cost, and allows direct comparison on the same array.
4. Universal protocols and accessibility. Nylon membrane based technology including radioactive labelling is widely used and widely accessible. Using existing reagent, hardware and software.

Gene Expression Profile in Hypertrophied Heart: Influence of Thyroid Hormones

Thyroid hormones (TH) have profound effects on metabolism, growth and development (Oppenheimer *et al.*, 1987). Although the action of TH involves virtually all organ systems, the heart responds to minimal changes in the serum thyroid hormone levels. It was reported that defects in TH receptor expression and the TH signaling occur in the heart of failing/hypertrophied human patients (Kinugawa *et al.*, 2001). Moreover hyperthyroid human patients experience cardiovascular complications with alteration of several cardiac parameters. These include increased resting heart rate, abnormal myocardial contractility and an increased left ventricular muscle mass (Fadel *et al.*, 2000). Therefore, prolonged hyperthyroidism, if left untreated results into abnormal myocardial functions and cardiac hypertrophy that may lead to heart failure (Chien *et al.*, 1993; Cohn, 1996).

Cardiac hypertrophy is an adaptive enlargement of the myocardium characterized by the increased size of the individual myocytes in response to a wide variety of extrinsic as well as intrinsic stimuli. Extrinsic stimuli are pressure overload, volume overload and neurohumoral factors. Intrinsic stimuli include contractile abnormalities due to altered expression or mutations of the sarcomeric proteins. Though hypertrophic growth is initially compensatory in nature, the function in the hypertrophied heart eventually becomes abnormal and a transition from hypertrophy to heart failure occurs (Feldman *et al.*, 1993). Due to abnormal heart functions in the long run, the overall growth process initiating the cardiac disease has been termed as pathological hypertrophy (Scheuer and Battrick, 1987).

Treatment of cardiac myocyte by T3 causes beta adrenergic stimulation, increased protein kinase A (PKA) activity, cAMP production and phosphorylation of proteins related to Ca^{2+} cycling kinetics (Kiss *et al.*, 1994; Ojamaa *et al.*, 2000). Thus, beta-adrenoreceptor stimulation may be responsible for thyroid mediated positive inotropic actions and enhanced atrial contractility (Dilman, 1990). Because TH also enhances cardiac functions characterized by faster myocardial relaxation and a decline in the cytosolic Ca^{2+} transient (MacKinnon *et al.*, 1988, Bing *et al.*, 1994, Kiss *et al.*, 1994), TH treatment has been suggested for improving cardiac performance in the postinfarction model of heart failure (Gay *et* al., 1988; Pennock et al., 1993; Mahaffey et al., 1995). However, the long-term effects of TH on cardiac function and metabolism are incompletely known. Recently, it has been shown that TH induced cardiac hypertrophy is associated with an initial increase in cardiac performance followed by a decline in cardiac function indicating detrimental effects of chronic hyperthyroidism (Degens et al., 2003).

Although it is reported that TH influences cardiac growth and metabolism via the signaling pathway distinct from other hypertrophic stimuli, the molecular mechanisms underlying hyperthyroid-induced versus pathological hypertrophy remain uncertain (Scheuer et al., 1982; Schaible et al., 1984; Izumo et al., 1987). Some of the unique features of TH–induced gene expression in cardiac muscle are induction of alpha myosin heavy chain (MHC-α), sarco-endoplasmic reticulum Ca^{2+} ATPase (SERCA) and repression of beta MHC (MHC-β) (Dillmann, 1990; Klein and Ojamaa 2001; Kinugawa et al., 2001). It is also postulated that TH-induced improvement in myocardial contractility is partly due to the differential expression of MHC gene expression (Fadel et al., 2000). To understand the gene expression programme in TH-induced hypertrophied heart, we have conducted cDNA microarray analysis of heart tissue obtained from rat treated/untreated with T3 for 15 days. In the present study, the alterations of at least 37 genes that are involved with the cardiovascular functions are demonstrated and the physiological significance of one such altered genes has been explored in details. Microarray analysis revealed that a few genes involved with the pathological hypertrophy are also altered in hyperthyroid heart in addition to the genes those are not yet known to be TH responsive in cardiac tissues.

Materials and Methods

Treatment of Rat with T3

Sprague-Dawley (SD) rats were intraperitoneally injected daily with 7.5 mg T3/100g body weight (Hamasaki et al., 2000; Weitzel et al., 2001) for 15 days. The control group was treated with the same volume of saline. The animals were sacrificed on 16th day. In parallel groups, rats treated with T3 for 15 days were maintained untreated for another 15 days to check the effect of T3 withdrawal *i.e.* these rats were sacrificed after one month from the start of the experiment. Animals were anesthetized, and hearts were surgically removed and immersed in ice–cold 0.9 per cent NaCl. The degree of hypertrophy was calculated in terms of heart weight/body weight ratio.

Histology

Cardiac tissue from vehicle, T3 treated or T3-withdrawn rats were fixed in 10 per cent formalin and embedded in paraffin following routine procedure. Tissue sections of 5 mm thickness were prepared and stained with Mayer's hematoxylin and eosin. The sections were mounted and observed under microscope under 400 X magnifications.

Northern Blotting Analysis

Total left ventricular RNA from control and T3-treated rats were isolated using TriReagent (Sigma-Aldrich Co. USA). About 15 µg of RNA was resolved on 1 per cent agarose/formaldehyde gels and transferred to nylon membrane (Millipore). After transfer, RNA was fixed by UV cross linking, hybridized with [α-^{32}P]dATP labeled (Hexalabel Plus DNA labeling kit, MBI Fermentas), Atrial Natriuretic Factor (ANF) cDNA and GLUT4 cDNA (gifted by Dr. Jeffry E Pessin, University of Iowa, Iowa). The membranes were washed twice for 10 min each with 2XSSC, 0.1 per cent SDS at room temperature followed by a single wash with the same solution at 65^{0}C for 30 min. Membranes were exposed to KODAK autoradiographic film.

cDNA Expression Arrays

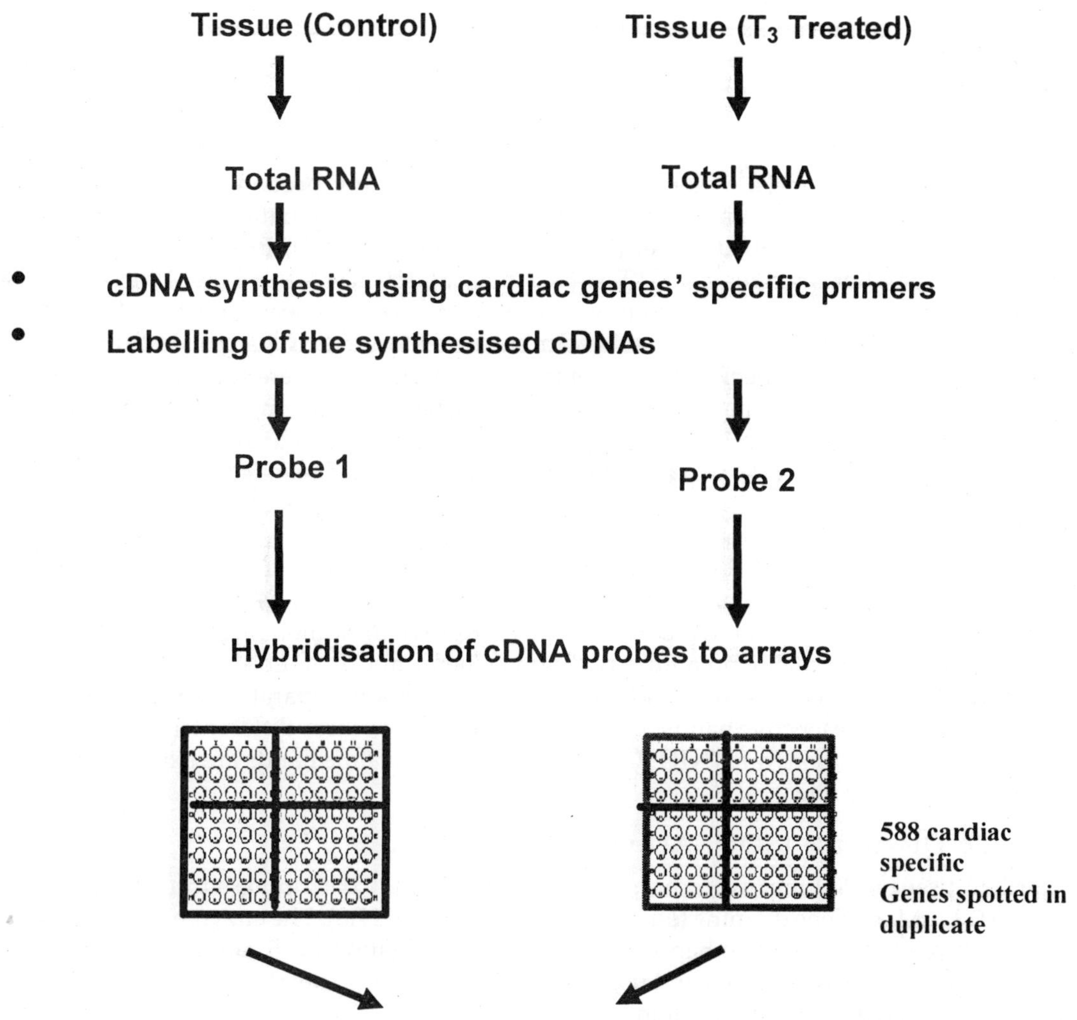

Protocol

Atlas Cardiovascular Array kit was purchased from Clontech (U.S.A). These arrays consisted of 588 cDNAs (organized into broad functional groups), nine housekeeping control cDNAs and negative controls. Membrane hybridization was performed according to the manufacturer's protocol. Briefly, 5 μg of total RNA from the left ventricular tissues of control and T3-treated rat hearts were reverse transcribed in presence of [α-^{32}P]dATP (3000Ci/mmole; Amersham Pharmacia) using MMLV Reverse Transcriptase (100units) and CDS primer mix at 50^{0}C for 25 min. Array membranes were hybridized overnight at 68^{0}C in Express Hybridization solution containing 100μg/ml sheared salmon testes DNA. The array membranes were then washed twice with 2XSSC, 0.1 per cent SDS at room temperature followed by a single wash with the same solution at 65^{0}C for 30 min. The array membranes were exposed to Kodak (X-OMAT, Sigma) autoradiographic film for 48 hours. Spots developed were normalized by equalizing the intensity of signals from a set of house keeping genes provided on the array. The autoradiographs were scanned and analyzed using ImageJ software (NIH, USA). The results presented are the mean ± SEM of three separate microarray experiments using RNA prepared from hearts from three different batches of treatments.

Western Blot Analysis

For Western blot analysis, either whole homogenate (80 mg protein sample) or microsome preparations (50 mg proteins sample) were used. The whole homogenate from the left ventricular tissues was prepared as described earlier (Bandyopadhyay *et al.*, 2000). Briefly, rat heart tissue was homogenized in a buffer containing 2mM Tris-HCl (pH 7.8), 150 mM NaCl, 1 mM PMSF and 100mg/ml pepstatin A. For the microsome preparation, the whole homogenate was centrifuged at 4000 X g for 15 min at 4°C. The post nuclear supernatants were subjected to a centrifugation for 40 min at 20000 X g at 4°C using a fixed-angle Beckman T50 rotor. The supernatants were then ultra-centrifuged at 105000 X g for 90 min (4°C). The pellet was resuspended in the homogenizing buffer and the expression of GLUT1 and GLUT4 was examined in this preparation. Immunoblot analyses with the antibodies against SOD-1, GLUT1, GLUT4, Cox1, collagen IV, MMP2, Na-Ca-exchanger and MHC β were performed as described earlier (Bandyopadhyay *et al.*, 2000). Anti-goat SOD-1, anti-rabbit GLUT1, anti-goat GLUT4, anti-rabbit Cox1, anti-rabbit collagen IV antibodies were purchased from Santa Cruz Biotech (Santa Cruz, CA, USA). Anti-mouse Na-Ca-exchanger and anti-rabbit MMP2 antibodies were purchased from Sigma Chemical Co. (Sigma Chemical Co. St. Louis, MO, USA). Antibody against MHC β was received as gift from Dr. Do Han Kim (S. Korea). Immunoblotting in parallel sets were performed with antimouse β-actin antibody (Sigma Chemical Co, St. Louis, MO, USA) as loading controls both with whole homogenate and microsome preparations separately. For quantification, the pixel intensities of the respective bands of the immunoblots from three different individual rats were calculated using ImageJ software (NIH, USA). The intensities of the bands are expressed as relative pixel density (arbitrary units).

Superoxide Dismutase (SOD) Activity

Cardiac tissues (200 mg) were homogenized in 0.5 ml. 0.05M potassium phosphate buffer containing 0.1 mM ethylene diamine tetra acetic acid (EDTA, pH 7.8). The enzyme was extracted with 0.35 ml mixture of chloroform/ethanol (3:5). After centrifugation for 5 minutes at 12,000 x g, the supernatant was carefully collected and used for the Cu, Zn-SOD assay (Martin *et al.*, 1987). The SOD activity is expressed as units/mg protein.

Real Time Quantitative RT-PCR Analysis

One step quantitative RT-PCR of each sample was performed in triplicates using the QuantiTect™ SYBR® Green RT-PCR Kit (Qiagen, Germany) and the iCycler realtime detection system and software (BioRad, USA) according to the manufacturer's instructions. Sequence-specific PCR primers were designed using the GeneFisher software (http://bibiserv.techfak.uni-bielefeld.de/genefisher/) based on published nucleotide sequences. A standard curve for each cDNA was obtained by performing amplifications from serial dilutions of myocardial total RNA. For all specific mRNA amplified, linear inverse correlations were observed between an amount of RNA and the C_T values (number of cycles needed to generate a fluorescent signal above a pre-defined threshold). A higher C_T value indicates a lower amount of mRNA in the sample and vice versa. Expression level of the 'housekeeping gene' GAPDH was used to normalize for variations in the amount of RNA, purity, etc. The normalized C_T

values of T3-treated samples for each cDNA were expressed as per cent change compared to the untreated control and depicted as histogram in Figure 7.4.

Isolation of Adult Rat Ventricular Cardiomyocytes

Adult rat ventricular cardiomyocytes from control and T3-treated animals were isolated by an earlier described procedure (Piper *et al.*, 1982) with some modifications. The left ventricles were cut, minced thoroughly and the tissue was incubated at 37°C in enzyme solution (0.1 per cent collagenase and 0.1 per cent hyaluronidase in calcium free Krebs-Henseleit (KH) bicarbonate buffer (NaCl 119mM, KCl 4.7mM, $CaCl_2$ 1.25mM, $MgSO_4$ 1.2mM, $NaHCO_3$ 25mM, KH_2PO_4 1.2mM, Hepes 12.5mM, BSA 0.2 per cent, pH 7.4) for 30 min in a CO_2 incubator with intermittent shaking. After digestion the cell suspension was centrifuged at 50 g for 1 min. The pellet obtained was suspended in wash medium (3 parts of KH buffer +1 part of medium 199+10 per cent FBS) and centrifuged at 50 g for 1 min. The pellet contained 90-97 per cent rod shaped viable ventricular myocytes which was resuspended in glucose-free KH buffer. The cells were stored in the CO_2 incubator at 37° C for 30 min before they were used for monitoring glucose uptake.

Glucose Uptake

Glucose uptake was monitored following the protocol as described earlier (Roy *et al.*, 2003) with some modifications. Ventricular cardiomyocytes were incubated for 30 min either in absence (basal) or presence of porcine insulin (100 nM). After 30min of insulin treatment, 2-deoxy-D[^{3}H]-glucose (0.4 nmoles) was added to each incubation and after 5 min, uptake was terminated by washing cells thrice with ice cold KH buffer in presence of 0.3 mM Phloretin to correct the glucose uptake data from simple diffusion and non specific trapping of radioactivity. Cells were pelleted down, washed twice with KH buffer and sonicated in presence of Nonidet P40 (NP40) and acetone. [^{3}H]deoxy glucose was measured in liquid scintillation counter (Packard, Tricarb 2100TR) and the results are expressed as cpm/mg protein.

Statistical Analysis

Data are presented as mean ± SEM. The level of significance was calculated using ANOVA. A p value ≤ 0.05 was considered to be significant.

Results

Development of Left Ventricular Hypertrophy in Hyperthyroid Rat

The histological sections of the left ventricular tissues displayed normal, healthy and compact arrangement of myofibrils in control heart whereas a massive enlargement of individual myofibrils, myocyte disarray occurred in the hypertrophied heart while normal architecture was restored in the heart following T3 withdrawal (Figure 7.1A). Figure 7.1B showed HW/BW ratio was increased with the duration of thyroid hormone treatment which attained maximum and exhibited a 58 per cent increase ($p<0.001$) over that of control rat group after 15 days. When injections of rat with T3 was discontinued for 15 days following 15 days of daily treatments, there was a restoration of HW/BW ratio (3.0 ±0.09) similar to that of control rats (2.88 ± 0.04) indicating the specificity of thyroid hormone action on heart in the present study (Figure 7.1B).

Molecular Marker of Cardiac Hypertrophy

Induction of ANF synthesis in ventricular tissue is a common event in various form of cardiac hypertrophy (Chien *et al.*, 1993). As shown in Figure 7.2, the ANF mRNA level was very high in the left

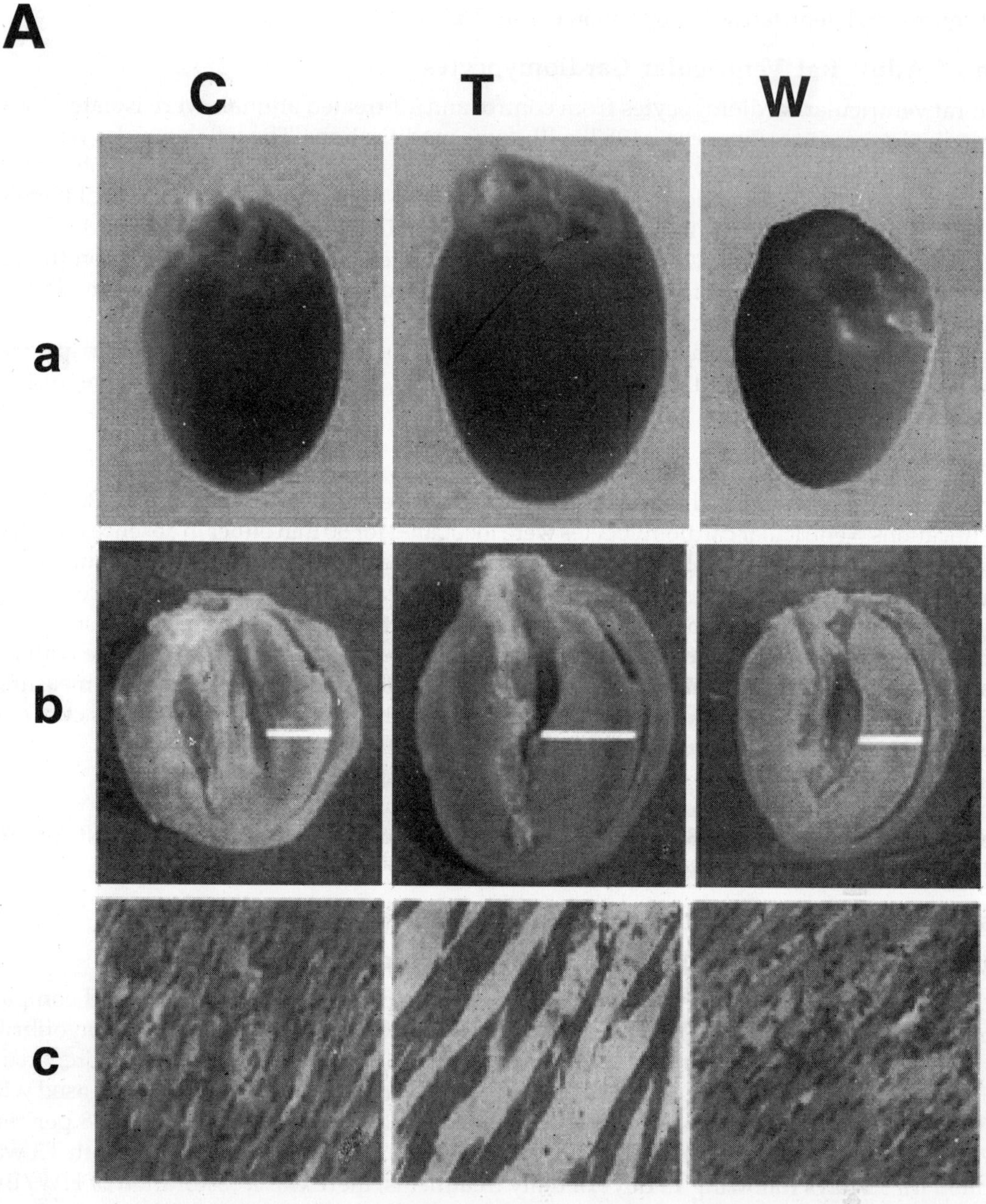

Figure 7.1(A): Left Ventricular Hypertrophy in Hyperthyroid Rat

Morphology (a), longitudinal section (b) and histological examination (c) of hearts (400X magnification) from rats injected with vehicle (C) or T3 (7.5 mg/100 g body weight) for 15 days (T) or 15 days withdrawal after T3 administration (W) showing the increased wall thickness (thick line) and reduced chamber volume in hyperthyroid rat heart.

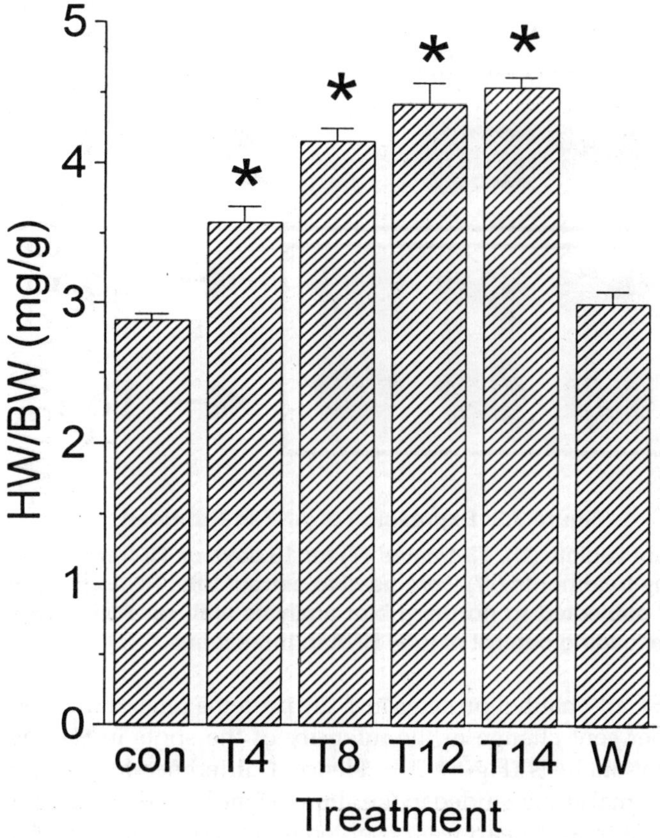

Figure 7.1(B): Left Ventricular Hypertrophy in Hyperthyroid Rat

Heart weight to body weight ratio (mg/g) of rats treated without T3 (Con) or with T3 for 4 days (D4), 8 days (D8), 12 days (D12) or 15 days (D15) or 15 days withdrawal following T3 treatment (W). The data represents mean ± SEM of 12 individual animals. * $p<0.01$ vs. control. ** $p<0.001$ vs. control.

ventricular tissue of T3-treated rat heart compared with saline treated control. When the treatment of rat with T3 was discontinued after 15th day, ANF level was reduced. This data indicated that prolonged hyperthyroid condition also induces severe cardiomyopathy.

Gene Expression Profile in Hyperthyroid Rat Heart

To examine the effect of thyroid hormone treatment on cardiac gene expression, RNA from normal and T3-treated rat hearts was prepared, reverse transcribed, radiolabeled with [α-^{32}P]dATP and the labeled probes thus prepared were used for hybridization of the array membranes separately as described in the materials and methods. The autoradiographs were scanned and analyzed using ImageJ software (NIH, USA). After normalization of the signals with the control, 37 genes were found to be altered in T3-treated heart (Table 7.1). Out of these, 11 genes were upregulated and 26 genes were downregulated in the heart due to T3 treatment. The altered genes can be grouped into 5 categories based on their functions. These are related to general metabolism, cytoskeletal/matrix protein,

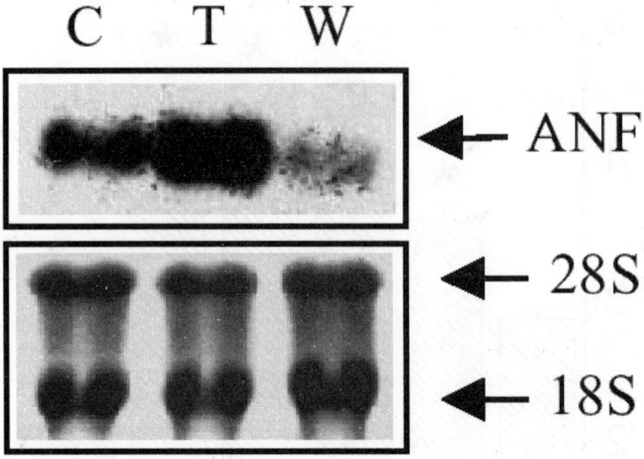

Figure 7.2: Expression of ANF in Rat Heart

Ventricular tissues from vehicle (C), T3-treated (T) for 15 days or T3-treated for 15 days followed by 15 days without any treatments (W) were used to extract total RNA which was transferred to nylon membrane and subjected to northern blot analysis with rat ANF cDNA as a probe. The lower panel shows the loading control of total RNA in three lanes.

hormone/growth factor, Ca^{2+}-channel proteins and the proteins related to receptor-transcription. A difference of above 10 per cent change in the intensity of the spots in the microarray blots were validated by Western blot analysis (Figure 7.3). Hence a difference above 10 per cent change are considered to be the real signal and included in the altered gene list. About 212 per cent increase in the expression of ANF in the T3-treated heart was observed which is consistent with typical hypertrophic growth of cardiomyocytes (Chien *et al.*, 1993). A plethora of metabolic enzyme genes of cytochrome P450 system were upregulated in T3-treated heart. MHC-α was upregulated and the MHC-β was downregulated in hyperthyroid rat heart as reported earlier (Dillmann 1990, Klein and Ojama 2001, Kinugawa *et al.*, 2001, Fadel *et al.*, 2000). Some of the genes important for cardiovascular functions which are downregulated but not yet known to be influenced by TH are Cu,Zn-SOD, GLUT4, tissue inhibitor of metalloproteinase 2 (TIMP2), Collagen XVI α1, collagen IVα4, vascular endothelial growth factor receptor and ERR α. Similarly, MMP2, RAR α and RARg are positively regulated by TH in heart are revealed for the first time.

Validation of Microarray Results

Some of the genes altered in hyperthyroid rat heart as revealed by the cDNA microarray analysis (Table 7.1), were further examined by Western blot analysis using homogenate prepared from the left ventricular tissues of control and hyperthyroid rat heart. Figure 7.3A shows significantly increased ($p<0.05$) level of expression of MMP2, COX 1(cyclooxygenase 1) and a decreased ($p<0.05$) level of expression of GLUT4, SOD, MHC β, collagen IVα4, Na-Ca^{2+} exchanger consistent with the cDNA microarray analysis. Since SOD expression was reduced, we additionally examined the activity of this antioxidant enzyme in the homogenate prepared from control and T3-treated rat hearts. It is interesting that SOD activity was also reduced significantly ($p<0.01$) in hypertrophied heart tissues (Figure 7.3B).

Table 7.1: Gene Profile in Hyperthyroid Heart

Group	*Gene/protein Name*	*GenBank Accession No*	*Induction/ Repression*
Metabolism	Cytochrome P450 IIB6, Cytochrome P450 2B3	M29874; J02864	+ 92.3
	Cytochrome P450 VIIIA1	X56088	+ 33.49
	Cytochrome P450 III A3	M13784	+ 194.6
	HDL binding protein	M83789	+ 120
	Adipocyte fatty acid binding protein	J02874	- 35.5
	Phospholipid transporter protein	L26232	- 56
	Cyclooxygenase 1 (COX 1)	M59979	+ 16.44
	Insulin responsive glucose transporter (GLUT) type 4 (GLUT4)	M20747	- 24
	Cu,Zn-SOD	J02947	-40.4
Cytoskeletal/ Matrix proteins	Cardiac muscle myosin heavy chain α isoform (MHC-α)	D00943	+24
	Cardiac muscle myosin heavy chain β isoform (MHC-β)	M58018	- 30.41
	Matrix metalloproteinase 2 (MMP2)	J03210	+ 15
	Tissue inhibitor of metalloproteinase 2 (TIMP2)	J05593	- 60.86
	Collagen IVα1	M11315	- 12
	Collagen XVI α1	M92642	- 56.28
	L-selectin	M25280	+ 18
	Junction placoglobulin Desmoplakin III	M23410 Z68228	- 42.55
	Erythrocyte adducing alpha subunit	X58141	- 21.4
Hormone/growth factor	Atrial natriuretic factor (ANF)	HS75640	+ 211
	Placental growth factor 1	X54936	- 30
Ca^{2+} Channel protein	Na^{+}-Ca^{2+} Exchanger	M91368	- 15
	Cardiac Phospholamban	M63603	- 77
Receptor/Transcription factors	Estrogen related receptor alpha (ERRα)	X51416	- 45.58
	Retinoid acid receptor alpha (RAR α)	X06614	+ 26
	Retinoic acid receptor gamma (RARg)	M24857	+ 13.74
	V-erb related protein 3, transcription factor COUP2; nuclear receptor subfamily 2 group F member 2 (NR2F2)	X12795; X16155; X58241	- 41.63
	Progesterone receptor	M15716	-26
	Mineralocorticoid receptor	M16801	- 45.6
	Purinoreceptor	U41070	- 41.5
	Vascular Endothelial growth factor receptor	AF016050	- 43.71
	Vesopressin V-3 receptor	D31833	- 23
	Sterol regulatory element-binding transcription factor 1	U00968	- 18.3
	Protein tyrosine phosphatase receptor type–F	Y00815 X69398	+29.16
Functionally unclassified protein	Tafazzin	X92762	- 26.1

Alteration of some of the nuclear hormone receptor was validated by real time quantitative PCR. The real time quantitative PCR analysis revealed that the expression of ERR α and MR was reduced while that of RAR α was increased in hyperthyroid rat heart compared to the control consistent with microarray analysis (Figure 7.4).

Functionality of Microarray Results

The inhibition of GLUT4 expression as evident from Microarray analysis prompted us to investigate the expression of GLUT4 with the progression of cardiac hypertrophy following T3 treatment. Microsomes were prepared from left ventricular tissues after treatment of rats without or with T3 for 4 and 15 days and subjected to Western blot analysis. As shown in Figure 7.5, the expression of GLUT4 level was reduced after 4 days which was further reduced after 15 days of T3 treatments. The level of GLUT4 was restored to normal when T3 treatment was discontinued for 15 days after 15 days of injection. This data indicates that hyperthyroidism repressed GLUT4 transcription in the heart.

Figure 7.3(A): Validation of Microarray Results

Representative immunoblots showing alteration of proteins in ventricular tissues obtained from rat treated with vehicle (C) or T3 (T) for 15 days. The approximate molecular mass of the proteins are given in the parentheses. The levels of the proteins were quantitated using the software Image J (NIH, USA) and the pixel intensities are shown as histograms. Immunoblots with β-actin antibody confirmed the equal amount of protein loading of the whole homogenate and microsome preparations (Figure 7.6B). Each immunoblot was repeated with the samples from three different hearts (n=3). **p*<0.05 vs. control.

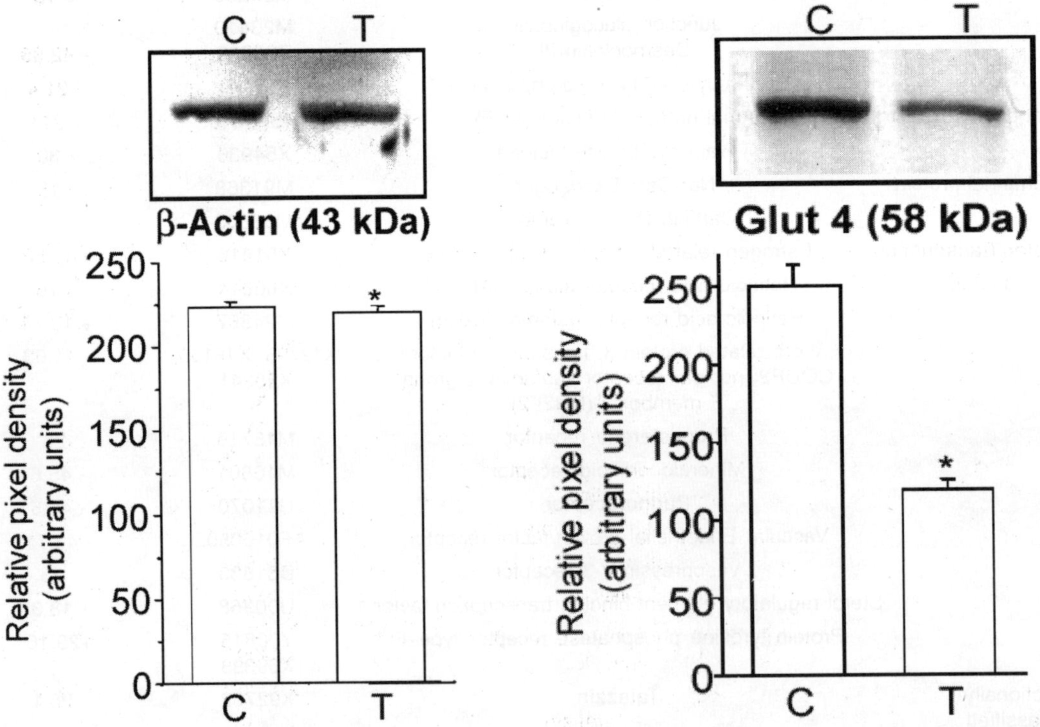

Contd...

Figure 7.3–Contd...

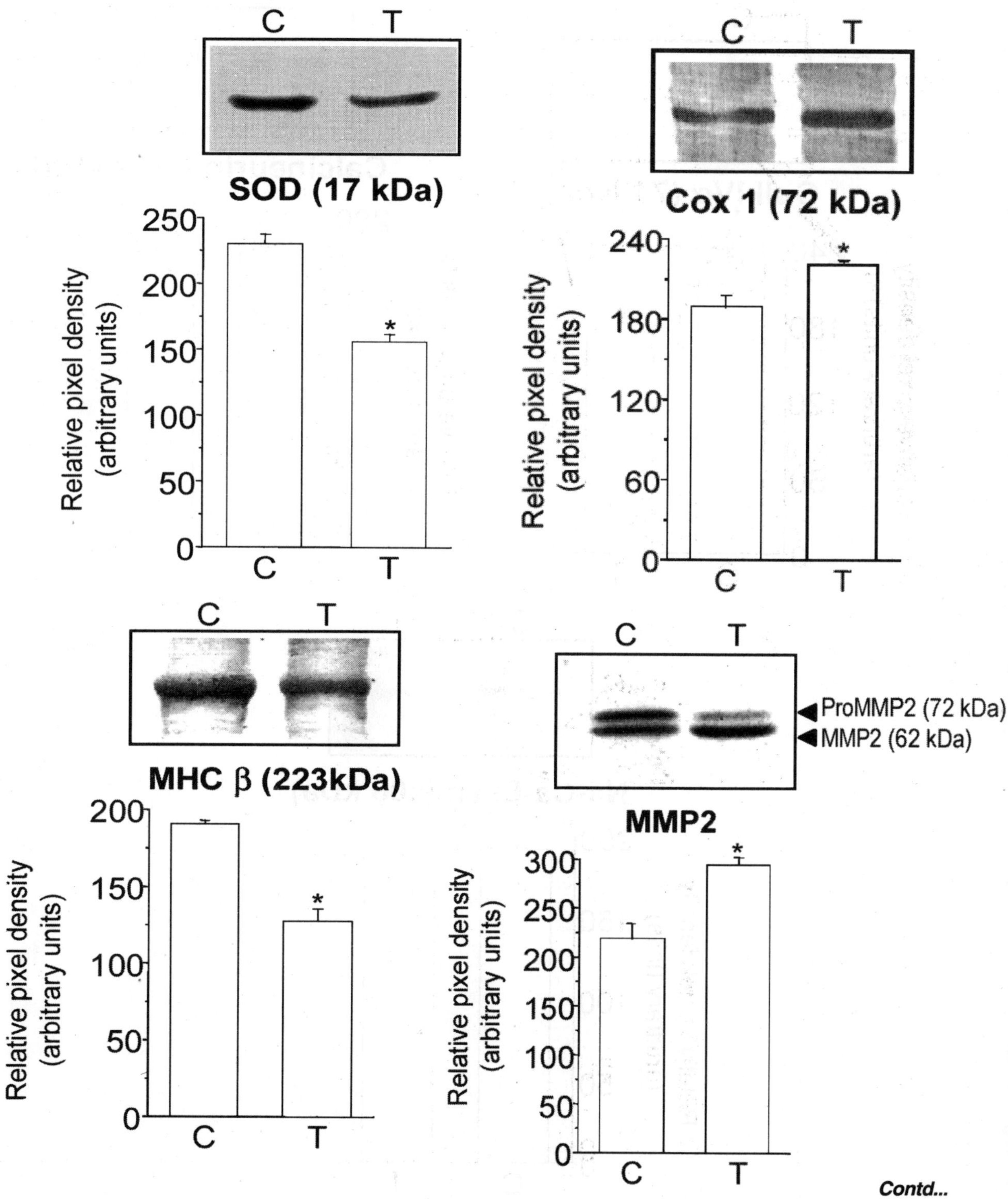

Contd...

Figure 7.3–Contd...

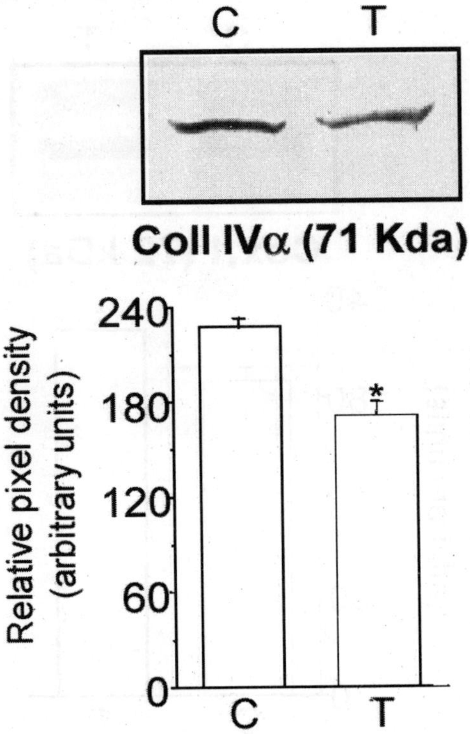

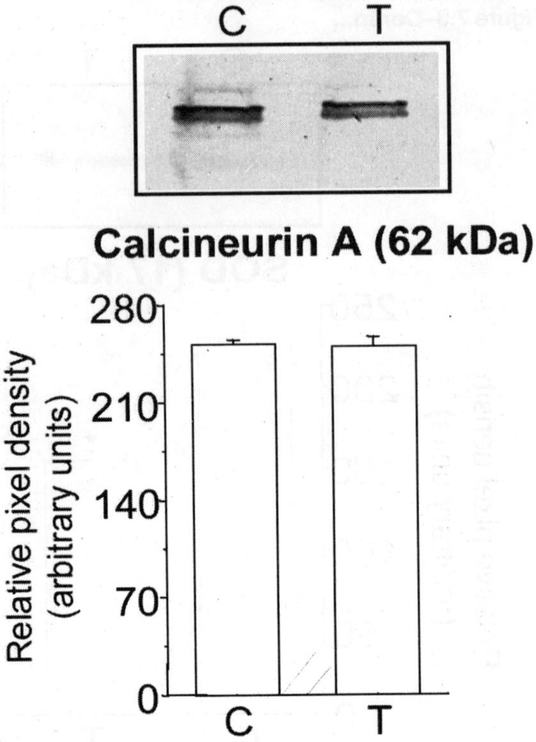

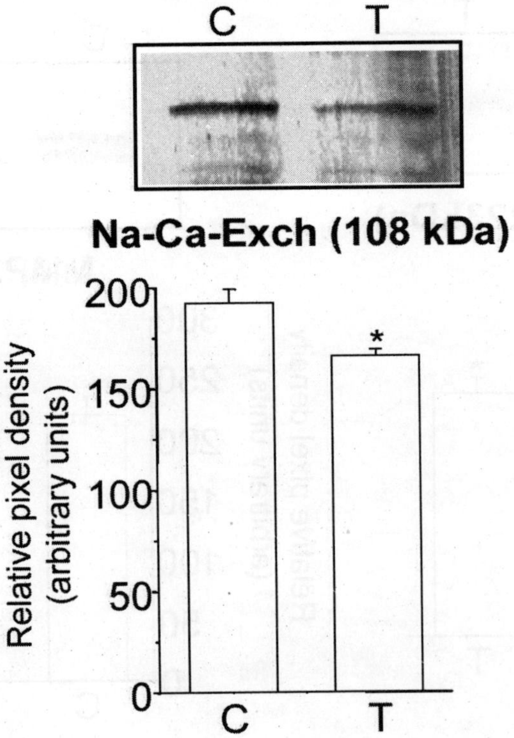

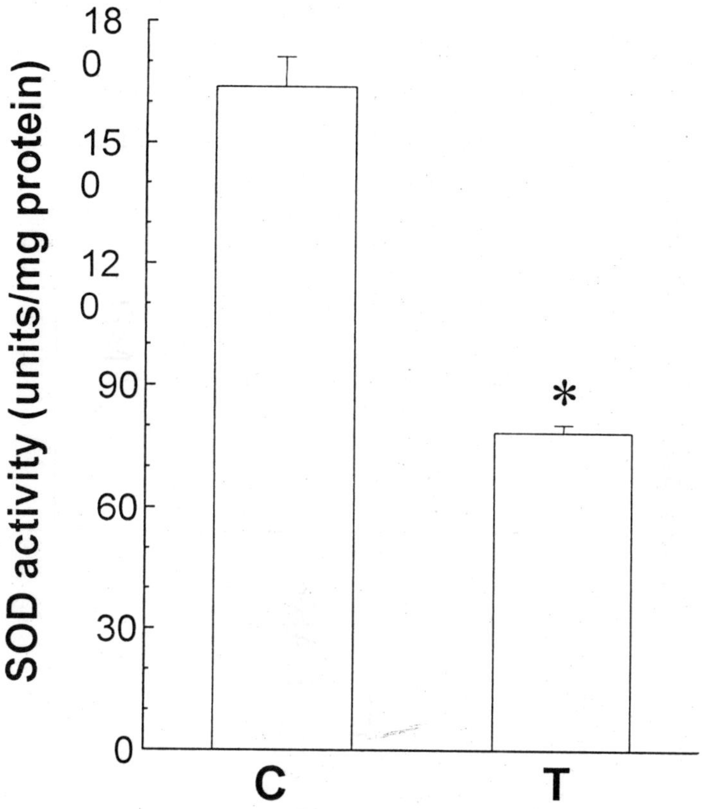

Figure 7.3(B): Validation of Microarray Results

Activity of Cu, Zn-SOD in the heart tissues obtained from control (C) and T3-treated (T) rats. The results are mean ± SEM of 3 individual rats. *p<0.01 vs. control.

Thyroid Hormone Inhibits Glucose Uptake in Cardiac Myocytes

To examine the effect of low level of GLUT4 expression on heart, glucose uptake in cardiomyocytes prepared from normal and T3-treated animals was examined. Cardiomyocytes from the adult rats were prepared and 2-Deoxy–D-[^{3}H]-Glucose uptake was monitored in the absence or presence of insulin. Figure 7.6 shows that insulin stimulated 2-Deoxy–D-[3H]-Glucose uptake was reduced in the cardiomyocytes prepared from T3-treated rat indicating the impairment of glucose transport in hyperthyroid heart. Since, the basal level of 2-Deoxy–D-[3H]-Glucose uptake was also reduced in cardiomyocytes of hyperthyroid rat, the level of GLUT1 in ventricular microsome preparations was examined by Western blot analysis with anti-GLUT1 antibody.

Discussion

Using cDNA microarray analysis we show that thyroid hormone induces changes in gene expression in heart, some of which were previously not known to be thyroid hormone responsive and associated with cardiac hypertrophy in vivo. One of the consequences of long term thyroid hormone treatment results into reduction of GLUT1 and GLUT4 expression and inhibition of glucose entry into

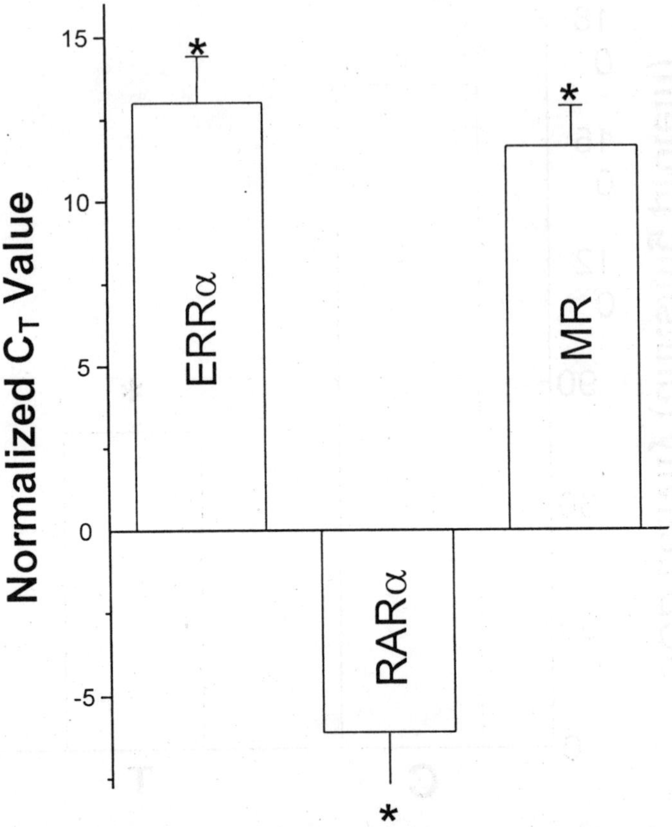

Figure 7.4: Validation of Microarray Results by Real Time Quantitative RT-PCR Analysis

Expression pattern of ERRα, RARα and MR were examined by real time quantitative RT-PCR analysis using RNA samples prepared from left ventricular tissues of rat treated with vehicle (control) or T3 for 15 days. Each bar represents per cent change in C_T value over the control after normalization with the C_T value of GAPDH mRNA as described in the Materials and Methods. Each experiment was repeated with three different batches of samples. *$p<0.05$ vs. control.

the cardiac myocytes. Analysis of gene expression reveals that thyroid hormone signals the hypertrophic growth of the cardiac myocytes through gene expression program distinct from other types of hypertrophic stimuli and altered genes are associated with the pathological hypertrophy. Whether the altered genes are directly regulated by T3 needs further study. Considering the characteristics of altered genes and specifically abnormal glucose handling of the cardiomyocytes, it is reasonable to suggest that in the event of increased excitation-contraction coupling, the hyperthyroid heart could fail due to defect in the energy metabolism.

TH regulates expression of a wide variety of genes after binding with TH receptor on DNA. Using cDNA microarray analysis recently it has been shown that TH regulates the expression of about 55 genes in its main target organ, liver (Feng *et al.*, 2000). Microarray analysis also demonstrated thyroid hormone-induced gene profile in hypothyroid animal (Weitzel *et al.*, 2003). Consistent with hyperthyroid induced cardiovascular complications such as increased resting heart rate, myocardial

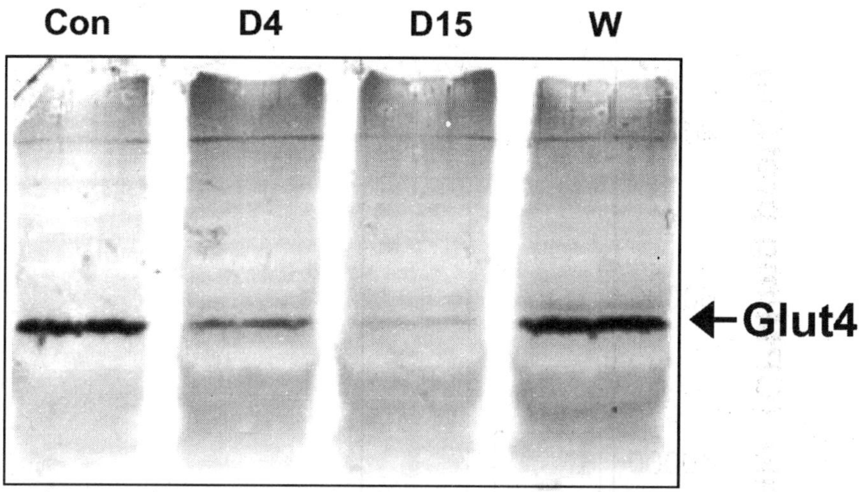

Figure 7.5: Influence of T3 on GLUT4 Expression in Ventricular Tissues

Representative immunoblots showing GLUT4 expression in the ventricular microsome preparations obtained from rat treated without (Con) or with T3 for 4 days (D4) and 15 days (D15) or 15 days treated but left for 15 days without any treatments (W).

contractility and left ventricular muscle mass (Fadel *et al.*, 2000) we show here that T3 resulted in up and downregulation of a diverse range of genes in the heart tissue. These gene products influence metabolism, growth and transcriptions, cytoskeletal structures and Ca^{2+}-signaling (Table 7.1). Interestingly, out of the 37 differentially expressed genes in the left ventricular tissue of hyperthyroid rat, majority (26) are negatively regulated as reported in mouse liver (Feng *et al.*, 2000).

It is well known that ANF is released as an important molecular marker of cardiac hypertrophy, which is also differentially expressed in left ventricular tissue as revealed by our microarray and Northern blot analysis. Similar to other hypertrophic stimuli, the dramatic increase in the level of ANF in left ventricular tissue is associated with hyperthyroid-induced cardiac hypertrophy, which was reduced following withdrawal of T3 treatment after two weeks (Table 7.1, Figure 7.2).

Cytochrome P450 enzymes catalyze oxidation of a wide range of drugs and other xenobiotics, and are responsible for the metabolism of endogenous compounds. Expression of cytochrome P450 mono-oxygenases is reported for extrahepatic tissues including lung, kidney, gastrointestinal tract, endothelium, and heart (Thum and Borlak 2000a). Heart tissue is metabolically competent and disease may influence the expression of cytochrome P450 monooxygenases (Thum and Borlak 2000b). In the hypertrophied heart of spontaneously hypertensive rat, the expression and activities of cytochrome P450 monooxygenases were shown to be induced resulting into the increased lipid peroxidation and oxidative stress (Thum and Borlak 2002). Consistently, we also show a dramatic increase in the expression of cytochrome P450 isoforms in the left ventricular tissues of hyperthyroid rat. This is likely to be due to direct effect of T3 on cardiac metabolism. Furthermore an increased amount of lipid peroxidation (data not shown) and a decreased level and activity of SOD 1 (Table 7.1, Figures 3A, 3B), in TH-induced rat heart indicates the occurrence of severe oxidative stress. Such a decrease in the level and activity of SOD due to increased oxidative stress has also been reported by others in a number of different experimental situations (Indik *et al.*, 2001, Simon-Giavarotti *et al.*, 2002).

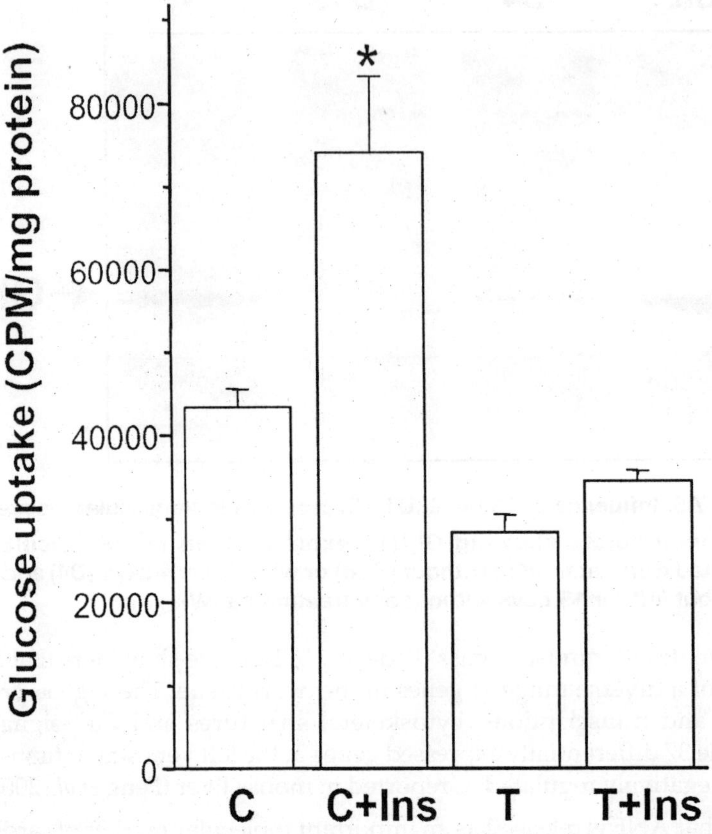

Figure 7.6: Effect of T3 on the Glucose Transport into Adult Rat Ventricular Cardiomyocytes

Cardiomyocytes were prepared from the adult rat heart injected with vehicle (C) or T3 (T) for 15 days. The cells prepared from vehicle or T3-treated rats were stimulated without (C) or with 100 nM insulin (I) for 30 min followed by incubation for 5 min in the presence of 2-Deoxy–D-[3H]-Glucose (0.4 nmoles) as described in the Materials and Methods. [^{3}H]deoxy glucose was measured in liquid scintillation counter and results are expressed as cpm/mg protein. Each bar represents mean ± SEM of three individual animals, each in duplicates. * *p*<0.01 vs. control. *p*<0.05 vs. control.**

Thyroid hormone-induced hypermetabolism, involving all major fuel sources, is a hallmark of hyperthyroidism (Moller *et al.*, 1996). As a direct consequence of hypermetabolism, T3 also alters a series of enzymes of the metabolic pathway in the heart such as HDL binding proteins, adipocyte fatty acid binding protein, phospholipid transporter, COX 1 and glucose transporter (Table 7.1). The lowering of plasma lipid levels by T_3 (Trost *et al.*, 2000) could be linked to the reduced expression of fatty acid and lipid targeting proteins such as adipocyte fatty acid binding protein and phospholipid transporters, respectively. Excess of thyroid hormone affects intermediary metabolism (Moller *et al.*, 1996) and as a result of which COX 1 might have increased in the heart of hyperthyroid rat.

Consistent with the earlier reports, microarray analysis reveals the differential expression of MHC genes *i.e.* upregulation of MHC-α and repression of MHC-β in the heart of hyperthyroid rat (Table 7.1, Figure 7.3A). Cardiac tissues that contain a relatively more MHC-α than MHC-β contract faster than those in which the ratio of MHC-α to MHC-β is low. This could explain the increased heart rate in hyperthyroid condition.

It is interesting to observe that MMP2 is increased as revealed by microarray and Western blot analyses whereas TIMP2 expression is decreased (Table 7.1). We examined the expression of TIMP2, the endogenous inhibitor of MMP2, by semi quantitative RT-PCR. Concomitant to the increased expression of MMP2 by microarray and immunoblotting, the expression of TIMP2 was decreased. Consistent with these effects, expressions of collagen IVα4 and collagen XVIα1 are decreased in the heart of hyperthyroid rat. A possible mechanism involves reduction of collagen synthesis coupled with increased MMP2 activity leading to the low collagen deposition in the heart of hyperthyroid rat.

The present study shows that phospholamban is downregulated in the hyperthyroid heart as reported earlier (Kiss *et al.*, 1994). Downregulation of phospholamban expression has further strengthened our microarray analysis since the regulation of phospholamban in relation to pathophysiology of hyperthryroid heart is well established (Kiss *et al.*, 1998, Holt *et al.*, 1999). However, SERCA is not depicted in the present study since the microarray blot does not include this cDNA. In the cardiac myocytes, Ca^{2+} release and reuptake by the sarcoplasmic reticulum regulate the rate of myocardial fiber contraction and relaxation. Phospholamban phosphorylates SERCA and inhibits it ability to reuptake Ca^{2+} into the lumen of sarcoplasmic reticulum. It is known that thyroid hormone allows an accelerated reuptake of calcium by the sarcoplasmic reticulum by upregulation of SERCA and downregulation of phospholamban protein expression (Fadel *et al.*, 2000). This is a mechanism by which thyroid hormone improves diastolic relaxation properties of the hyperthyroid heart.

Hyperthyroid rat heart shows an altered level of expression of some of the nuclear hormone receptors such as ERR α, RAR α, RARg, mineralocorticoid receptors etc. Retinoid receptors such as specific retinoic acid receptors (RARs) and retinoic acid X receptors (RXRs) hetero- or homodimerize with specific regulatory sequences in gene promoters (Leng *et al.*, 1994). Retinoids have long been recognized as important for cardiac morphogenesis and differentiation (Colbert *et al.*, 1997). The present study shows for the first time that thyroid hormone mediated cardiac hypertrophy is signalled through retinoid signaling pathway specifically via upregulation of RARα (Table 7.1, Figure 7.4). It is likely that hyperthyroid rat heart requires upregulation of its transactivators RARs and thus RARα expression is increased more than 26 per cent than its normal level. Microarray analysis show an altered expression of other nuclear hormone receptors as mentioned in the Table 7.1, however the physiological significance of such alterations require further study.

To meet energy demands for contractile function or increased workload, the heart utilizes a variety of substances including glucose, fatty acids, lactate, ketone bodies and amino acids. The rate of glucose utilization in heart is greater than many other tissues such as skeletal muscle, adipose tissues or lung (James *et al.*, 1985a, 1985b, Zorzano *et al.*, 1997). Thus, glucose transport/utilization is critical for the maintenance of normal function. Glucose uptake in heart is mediated by the facilitative transporter GLUT4 and to a lesser extent by GLUT1 (Young *et al.*, 1997). Insulin stimulates GLUT4 translocation in the heart from the intracellular storage membrane to the cell surface where they facilitate glucose entry into the cell (Young *et al.*, 1997, Sun *et al.*, 1994). Surprisingly, Microarray, Northern and Western blot analyses show reduction of GLUT4 expression in the T3-treated left ventricular tissues in a situation when cardiomyocytes require more energy under increased workload. Initially, the expression of GLUT4 is changed a little which was considerably declined after 15 days of T3 treatment suggesting that GLUT4 expression is affected due to prolonged hyperthyroid condition (Figure 7.5). A recent report shows that reduction of GLUT4 results into development of cardiac hypertrophy in mice (Kaczmarczyk *et al.*, 2003). The cardiac myocytes prepared from adult rat treated with T3 show a decreased glucose uptake in response to insulin compared to the untreated cardiomyocytes (Figure 7.6). Depolarization across the plasma membrane results in Ca^{2+} induced Ca^{2+} release, which is directly coupled to excitation contraction coupling of myofibrils (Hoffmann *et al.*, 1995). It is known that KCl induces depolarization of cardiac myocytes in vitro and this process stimulates glucose entry into the myocytes. It is interesting that KCl-induced glucose uptake was also inhibited in cardiomyocytes of T3-treated rat heart compared to the untreated control (data not shown).

Taken together, we demonstrate the gene expression profile in left ventricular tissues of hyperthyroid rat in which some of the altered genes are common to pathological hypertrophy. Alteration of GLUT4 expression is one of the distinct features of hyperthyroid induced altered gene expressions leading to impaired glucose transport into the myocytes which could be detrimental to cardiac physiology if prolonged hyperthyroidism is unchecked.

Acknowledgement

This work is supported by grants from Department of Science and Technology, New Delhi.

References

Bandyopadhyay A, Shin Dw, Ahn JH and Kim DH. (2000). Calcineurin regulates ryanodine receptor/ Ca release channels in rat heart. Bioch. J., 352, 61-70.

Bertucci F, Bernard K, Loriod B, Chang YC, Granjeaud S, Birnbaum D, Nguyen C, Peck K, and Jordan BR.(1999). Sensitivity issues in DNA chip array-based expression measurements and performance of nylon microarrays for small samples. Hum. Mol. Genet., 9, 1715-1722.

Bing OH, Hague NL, Perreault CL, Conrad CH, Brooks WW, Sen S, and Morgan JP. (1994). Thyroid hormone effects on intracellular calcium and inotropic responses of rat ventricular myocardium. Am. J. Physiol., 267, H1112-H1121.

Chien KR, Zhu H, Knowlton KU, Miller HW, Van BM, O'brien TX, and Emans SM.(1993). Transcriptional regulation during cardiac growth and development. Ann. Rev. Physiol., 55, 77-95.

Cohn JN. (1996). The management of chronic heart failure. New England J. Med., 335, 490-498.

Colbert MC, Hall DG, Kimball TR, Witt SA, Lorenz JN, Kirby ML, Hewett TE, Klevitsky R, and Robbins J. (1997). Cardiac compartment-specific overexpression of a modified retinoic acid receptor produces dilated cardiomyopathy and congestive heart failure in transgenic mice. J. Clin. Invest., 100, 1958-1968.

Degens H, Gilde AJ, Lindhout M, Willemsen PH, Van Der Vusse GJ, and Van Bilsen M. (2003). Functional and metabolic adaptation of the heart to prolonged thyroid hormone treatment. Am. J. Physiol.-Heart and Circulatory Physiol., 284, H108-H115.

Dillmann WH. (1990). Biochemical basis of thyroid hormone action in the heart. Am. J. Med., 88, 626-630.

Fadel BM, Ellahham S, Ringel MD, Lindsay J, Wartofsky L, and Burman KD. (2000). Hyperthyroid heart disease. Clin. Cardiol., 23, 402-408.

Feldman AM, Weinberg EO, Ray PE and Lorell BH. (1993). Selective changes in cardiac gene expression during compensated hypertrophy and the transition to cardiac decompensation in rats with chronic aortic banding. Circulation Res., 73, 184-192.

Feng X, Jiang Y, Meltzer P and Yen PM. (2000). Thyroid hormone regulation of hepatic genes in vivo detected by complimentary DNA microarray. Mol. Endocrinol., 14, 947-955.

Gay RG, Graham S, Aguirre M, Goldman S, and Morkin E. (1988). Effects of 10- to 12-day treatment with L-thyroxine in rats with myocardial infarction. Am. J. Physiol., 255, H801-806.

Hamasaki Y, Shinohara O, Ishida H, Hayashi Y, and Nakazawa H. (2000). Decreased protein kinase C-epsilon expression in hyperthyroid cardiac ventricles induced by triiodothyronine treatment in the rat. Life Sci., 67, 1859-1869.

Hoffmann P, Richards D, Heinroth-Hoffmann I, Mathias P, Wey H, and Toraason M. (1995). Arachidonic acid disrupts calcium dynamics in neonatal rat cardiac myocytes. Cardiovascular Res., 30, 889-898.

Holt E, Sjaastad I, Lunde PK, Christensen G, and Sejersted OM. (1999). Thyroid hormone control of contraction and the Ca(2+)-ATPase/phospholamban complex in adult rat ventricular myocytes. J. Mol. Cell. Cardiol., 31, 645-656.

Indik JH, Goldman S, and Gaballa MA. (2001). Oxidative stress contributes to vascular endothelial dysfunction in heart failure. Am. J. Physiol.-Heart and Circulatory Physiol., 281, H1767-1770.

Izumo S, Lompre AM, Matsuoka R, Koren G, Schwartz K, Nadal-Ginard B, and Mahdavi V. (1987). Myosin heavy chain messenger RNA and protein isoform transitions during cardiac hypertrophy. Interaction between hemodynamic and thyroid hormone-induced signals. J. Clin. Invest., 79, 970-977.

James DE, Jenkins AB, and Kraegen EW. (1985a). Heterogeneity of insulin action in individual muscles in vivo: euglycemic clamp studies in rats. Am. J. Physiol., 248, E567-E574.

James DE, Kraegen EW, and Chisholm DJ. (1985b). Muscle glucose metabolism in exercising rats: comparison with insulin stimulation. Am. J. Physiol., 248, E575-E580.

Kaczmarczyk SJ, Andrikopoulos S, Favaloro J, Domenighetti AA, Dunn A, Ernst M, Grail D, Fodero-Tavoletti M, Huggins CE, Delbridge LM, Zajac JD, and Proietto J. (2003). Threshold effects of glucose transporter-4 (GLUT4) deficiency on cardiac glucose uptake and development of hypertrophy. J. Mol. Endocrinol., 31, 449-459.

Kinugawa K, Yonekura K, Ribeiro RC, Eto Y, Aoyagi T, Baxter JD, Camacho SA, Bristow MR, Long CS, and Simpson PC. (2001). Regulation of thyroid hormone receptor isoforms in physiological and pathological cardiac hypertrophy. Circulation Res., 89, 591-598.

Kiss E, Brittsan AG, Edes I, Grupp IL, Grupp G, and Kranias EG. (1998). Thyroid hormone-induced alterations in phospholamban-deficient mouse hearts. Circulation Res., 21, 608-613.

Kiss E, Jakab G, Kranias EG, and Edes I. (1994). Thyroid hormone-induced alterations in phospholamban protein expression: regulatory effects on sarcoplasmic reticulum Ca^{2+} transport and myocardial relaxation. Circulation Res., 75, 245-251.

Klein I and Ojamaa K (2001) Thyroid hormone and the cardiovascular system. New England J. Med., 344, 501-509.

Leng X, Blanco J, Tsai SY, Ozato K, O'Mally BW, and Tsai MJ. (1994). Mechanisms for synergestic activation of thyroid hormone receptor and retinoid X receptor on different response elements. J. Biol. Chem., 269, 31436-31442.

MacKinnon R, Gwathmey JK, Allen PD, Briggs GM, and Morgan JP. (1988). Modulation by the thyroid state of intracellular calcium and contractility in ferret ventricular muscle. Circulation Res., 63, 1080-1089.

Mahaffey KW, Raya TE, Pennock GD, Morkin E, and Goldman S. (1995). Left ventricular performance and remodeling in rabbits after myocardial infarction. Effects of a thyroid hormone analogue. Circulation, 91, 794-801.

Martin JP, Dailey M, and Sugarman E. (1987). Negative and positive assays of superoxide dismutase based on hematoxylin autooxidation. Arch. Bioch. Biophys., 255, 329-336.

Moller N, Nielson S, Nyholm B, Porksen N, Alberti KG, and Weeke J. (1996). Glucose turnover, fuel oxidation and forearm substrate exchange in patients with thyrotoxicosis before and after medical treatment. Clin. Endocrinol., 44, 453-459.

Oppenheimer JA, Schwatz HL, Mariash CN, Kinlaw WB, Wong NCW, and Freake HC. (1987). Advances in our understanding of thyroid hormone action at the cellular level. Endocr. Rev., 8, 288-308.

Ojamaa K, Klien I, Sabet A, and Steinberg SF. (2000). Changes in adenylyl cyclase isoforms as a mechanism for thyroid hormone modulation of cardiac β adrenergic receptor responsiveness. Metabolism, 49, 275-279.

Pennock GD, Raya TE, Bahl JJ, Goldman S, and Morkin E. (1993). Combination treatment with captopril and the thyroid hormone analogue 3,5-diiodothyropropionic acid: a new approach to improving left ventricular performance in heart failure. Circulation, 88, 1289-1298.

Piper HM, Probst I, Schwartz P, Hutter JF, and Spieckermann PG. (1982). Culturing of calcium stable adult cardiomyocytes. J. Mol. Cell. Cardiol., 14, 397-412.

Roy SS, Mukherjee M, Bhattacharya S, Mandal CN, Kumar LR, Dasgupta S, Bandyopadhyay I, and Wakabayashi KA. (2003). New cell secreting insulin. Endocrinol., 144, 1585-1593.

Schaible TF, Ciambrone GJ, Capasso JM, and Scheuer J. (1984). Cardiac conditioning ameliorates cardiac dysfunction associated with renal hypertension in rats. J. Clin. Invest., 73, 1086-1094.

Scheuer J, and Battrick PM. (1987). The cardiac hypertophic responses to pathologic and physiologic loads. Circulation, 75, 163-168.

Simon-Giavarotti KA, Giavarotti L, Gomes LF, Lima AF, Veridiano AM, Garcia EA, Mora OA, Fernandez V, Videla LA, and Junqueira VB. (2002). Enhancement of lindane-induced liver oxidative stress and hepatotoxicity by thyroid hormone is reduced by gadolinium chloride. Free Radical Res., 36, 1033-1039.

Sun D, Nguyen N, DeGrado TR, Schwaiger M, and Brosius FC. (1994). Ischemia induces translocation of the insulin-responsive glucose transporter GLUT4 to the plasma membrane of cardiac myocytes. Circulation, 89, 793-798.

Thum T, and Borlak J. (2000a). Cytochrome P450 mono-oxygenase gene expression and protein activity in cultures of adult cardiomyocytes of the rat. Br. J. Pharmacol., 130,1745-1752.

Thum T, and Borlak J. (2000b). Gene expression in distinct regions of the heart. Lancet., 18, 979-983.

Thum T, and Borlak J. (2002). Testosterone, cytochrome P450, and cardiac hypertrophy. FASEB J., 16, 1537-1549.

Trost SU, Swanson E, Gloss B, Wang-iverson DB, Zhang H, Volodarsky T, Grover GJ, Baxter JD, Cheillini G, Scanlan TS, and Dilmann WH. (2000). The thyroid hormone receptor-β-selective agonist GC-1 differentially affect plasma lipids and cardiac hypertrophy. Endocrinol., 141, 3057-3064.

Weitzel JM, Radtke C, and Seitz HJ. (2001). Two thyroid hormone-mediated gene expression patterns in vivo identified by cDNA expression arrays in rat. Nucleic Acid Res., 29, 5148-5155.

Weitzel JM, Hamann S, Jauk M, Lacey M, Filbry A, Radtke C, Iwen KA, Kutz S, Harneit A, Lizardi PM, and Seitz HJ. (2003). Hepatic gene expression patterns in thyroid hormone-treated hypothyroid rats. J. Mol. Endocrinol., 31, 291-303.

Young LH, Renfu Y, Russell R, Hu X, Caplan M, Ren J, Shulman GI, and Sinusas AJ. (1997). Low flow ischemia induces translocation of canine heart Glut 4 and Glut 1 glucose transporters to sarcolemma in vivo. Circulation, 95, 415-422.

Zorzano A, Sevilla L, Camps M, Becker C, Meyer J, Kammermeier H, Munoz P, Guma A, Testar X, Palacin M, Blasi J, and Fischer Y. (1997). Regulation of glucose transport, and glucose transporters expression and trafficking in the heart: studies in cardiac myocytes. Am. J. Cardiol., 80, 65A-76A.

— Section B —
Hormones and Reproduction

8

Estrogens and Spermatogenesis

Chandrima Shaha and Durga Prasad Mishra*

National Institute of Immunology, Aruna Asaf Ali Marg, New Delhi – 110 067, India

ABSTRACT

The study of the role of estrogens in spermatogenesis has attracted significant interest because of the increasing awareness that estrogen is essential for spermatogenesis, spermatogenic cells express estrogen receptors and estrogen like chemicals present in the environment adversely affect male reproductive health. Such chemicals have the ability to affect gene expression and cellular function by binding to the hormone receptors and have been implicated in the declining trend of male fertility and an increase in testicular cancers. Effects of endocrine disruptors, a term used to describe agents that mimic hormones have been particularly well documented in wild life populations. Regular apoptosis of spermatogenic cells is required to maintain proper testicular homeostasis, although, increased cell death can result in defective spermatogenesis leading to infertility. Even though it is well established that estrogens induce spermatogenic cell apoptosis, the dissection of complete pathways leading to apoptotic death due to exposure to estrogens have not been worked out. With increasing data accumulating on the role of estrogens in spermatogenesis and the detrimental effects of environmental estrogens on the same process, it has become essential to investigate the mechanism of estrogen induced apoptosis in order to understand spermatogenetic disorders. In that light, the present review provides a discussion on the role of estrogens in spermatogenesis, the effect of environmental estrogens on fertility related events and data from two models, one *in vitro* model using rat spermatogenic cells and one *in vivo* system using rats for estrogen induced rat germ cell apoptosis are presented to show experimental evidence for the proof of concept.

**Corresponding Author.* E-mail: cshaha@nii.res.in

Abbreviations

E_2, estradiol; estrogen receptor α–Erα; estrogen receptor β–Erβ; Luteinizing hormone–LH; Follicle Stimulating Hormone–FSH; estrogen receptor α knockout mice- ERKO; beta estrogen receptor knockout mice- BERKO; activation function-1–AF-1; EGF–epidermal growth factor; cytochrome P450 aromatase–CYP 450 arom; 17 alpha-ethynylestradiol–EE2; Bisphenol A–BPA; p-nonylphenol–p-NP; 4-octylphenol–OP; butyl benzyl phthalate–BBP; 3, 3', 4, 4', 5-Pentachlorobiphenyl–PCB; diethylstilbestrol–DES].

Introduction

Estradiol (E_2) is a key regulator of growth, differentiation, and function in a wide array of target tissues, including the male and female reproductive tracts, mammary gland, and skeletal and cardiovascular systems. The predominant biological effects of E_2 are mediated through two distinct intracellular receptors, estrogen receptor α (ERα) and estrogen receptor β (ERβ), each encoded by unique genes. Within the reproductive tract of the male the levels of estrogens are far higher than in the general blood compartment (Hess, 2000), therefore favoring a testicular source of estrogens (Carreau *et al.*, 2001). It is now very well established that the E_2 is produced by the testis and estrogens are present in the circulation of a normal adult male, and that conversion of testosterone to E_2 is important in androgen signaling particularly in the brain. Production of E_2 by the Leydig cells in response to Luteinizing hormone (LH) and by the immature rat Sertoli cells in response to Follicle Stimulating Hormone (FSH) have been studied in considerable detail. In recent years there has been a resurgence of interest in the study of the role of estrogens in the male. This stemmed primarily from the following developments: (1), Discovery of the second type of estrogen receptor, named ERβ and demonstration that its distribution in the testis and in the male reproductive tract differs from that of the long-known ERα (Kuiper *et al.*, 1997); (2), Development of mice with targeted disruption of ERα or ERβ (Couse and Korach, 1999), which allowed studies of the physiological role of estrogen signaling in the male; and (3), Suggestion that reduction in sperm counts and increase in the incidence of testicular tumors in men in some industrialized areas may be due to estrogenic substances in the environment (Nikula *et al.*, 1999). The role of estrogens in the development and physiology of male reproductive tract of mammals is still a matter of debate, even though there is a growing body of evidence suggesting that estrogens are playing a role via their specific receptors (ERα and ERβ) which are distributed all along the genital tract.

Compelling evidence has accumulated on the detrimental effects of estrogen like chemicals present in the environment on male reproductive function. These chemicals include industrial pollutants like bisphenol A (BPA) and polychlorinated biphenols, and pesticides like DDT, methoxychlor or chlorodecone. The extent of exposure to these chemicals on members of a population differs as occupations in agriculture, petrochemicals and the construction industry entails higher exposure. Estrogen receptors are present in the pituitary and therefore, in spermatogenic cells estrogen like chemicals can act as agonists or antagonists for the hormone and interfere with spermatogenesis.

Overview of Spermatogenesis

The mammalian testis is a complex organ with two functions: synthesis of steroid hormones and production of spermatozoa. Maintenance of spermatogenesis is controlled by gonadotrophins and testosterone whose effects are modulated by locally-produced factors, and among them estrogens are notable.

Distribution of Cells

Spermatogenesis is the process by which immature germ cells undergo division, differentiation, and meiosis to give rise to haploid elongated spermatids. This process takes place within the seminiferous tubules of the testis, in close association with the somatic cells of the seminiferous epithelium, the Sertoli cells. When germ cell development is complete, the mature spermatids are released from the Sertoli cells into the tubule lumen, and proceed through the excurrent duct system, known as the rete testis, until they enter the epididymis via the efferent ducts. During passage through the epididymis, the spermatids undergo a series of biochemical changes to become the motile spermatozoa capable of fertilization.

The testicular parenchyma, consisting of seminiferous tubules and interstitial tissue, is enclosed by a capsule called the tunica. The interstitial tissue contains the blood and lymphatic vessels, which are essential for the movement of hormones and nutrients into, and out of the testis. The most frequently encountered cell type in the interstitium is the Leydig cell (Kurland, 1975), which is primarily involved in the secretion of androgens, notably testosterone, as well as other steroids including estrogen.

Within the seminiferous tubules, the Sertoli cells reside on a basement membrane, under which there are the lymphatic endothelium and the peritubular myoid cells (Dym and Fawcett, 1970). The structure of the Sertoli cell is extremely complex with numerous cup-shaped processes encompassing the various germ cell types (Russell, 1993). Developing germ cells form intimate associations with Sertoli cells, with multiple germ cell types in contact with one Sertoli cell. The various generations of germ cells, are arranged in strictly defined cellular associations (Leblond and Clermont, 1952). It is the unique associations of these germ cells with Sertoli cells that constitute the cycle of the seminiferous epithelium, and each particular association of germ cells is referred to as a stage. The number of stages of spermatogenesis in a particular species is thus defined by the number of morphologically recognizable germ cell associations within the testis; in the mouse there are 12 stages, in the rat there are 14, and in the human there are 6.

Germ Cell Development

Germ cell development involves a complicated series of events, and the various germ cell types can be distinguished on the basis of morphology and the differential expression of proteins. Spermatogonia are present between Sertoli cells, close to the basement membrane of the tubule. They are the most immature germ cells in the testis, and include type A spermatogonia, intermediate spermatogonia (found only in rodents), and type B spermatogonia, the latter of which are considered to be committed to differentiation. The true stem cells of the germ cell population are considered to be a subset of the type A spermatogonial population, although their identity cannot be discerned on the basis of morphology (Couse and Korach, 1999). Spermatogonia undergo numerous mitoses to produce a large number of germ cells available for entry into meiosis. Thus, proliferation of the spermatogonial population provides the source for the millions of sperm that are produced per day (Russell *et al.*, 1990).

After the last mitosis of type B spermatogonia, preleptotene primary spermatocytes are formed (Russel *et al.*, 1990). These cells replicate their DNA and hence initiate meiosis. During the prophase of the first meiotic division, germ cells undergo morphological transitions that can be classified on the basis of nuclear size and morphology. In the zygotene phase, pairing of homologous chromosomes occurs, and cells with completely paired chromosomes are termed pachytene spermatocytes. After the pachytene phase, a brief diplotene phase follows in which the chromosome pairs partially separate, and the cells then undergo the first meiotic division to yield secondary spermatocytes. These cells quickly undergo the second meiotic division to yield the haploid round spermatid.

The differentiation of round spermatids into the mature elongated spermatid form takes place, with no further division, during the process known as spermiogenesis. Briefly, spermiogenesis involves formation and development of the acrosome and flagellum, condensation of the chromatin, reshaping and elongation of the nucleus, and removal of the cytoplasm before release of the spermatid during spermiation. After commencement of spermatid elongation, the highly condensed spermatid nucleus becomes incapable of transcription, and immature round spermatids transcribe high levels of mRNAs that are subject to translational delay until translation is required during elongation. Spermiation is the final step of spermiogenesis and involves the release of the mature elongated spermatid from the Sertoli cell into the lumen of the seminiferous tubule (Dym and Fawcett, 1970).

Estrogens and Spermatogenesis

The Estrogen Receptors

In order to exert a biological role, testicular estrogens should interact with estrogen receptors which in turn modulate the transcription of specific genes. The presence of ERα in the Leydig cells, efferent ducts and epididymis, as well as in the pituitary, has been shown in different species. Expression of ERα is particularly intense in the efferent ducts that transport spermatozoa suspended in the fluid secreted by the Sertoli cells from the testis to the epididymis. In contrast, ERβ is widely expressed in the testis, including Sertoli cells, Leydig cells, and germ cells, as well as in the epididymis, prostate, and seminal vesicles.

Male ERKO (estrogen receptor α knockout mice) mice with targeted disruption of ERα have increased plasma testosterone levels but are infertile due to distension of the rete testis and excurrent ducts and progressive loss of testicular weight, reflecting damage to seminiferous epithelium and exfoliation of germ cells (Kuiper, 1997). Studies in these animals revealed a previously unsuspected important physiological role of endogenous estrogen: the control of fluid reabsorption in the efferent ducts and the initial segment of the epididymis. Male BERKO (beta estrogen receptor knockout mice) mice with selective disruption of ERβ are fertile and exhibit no major alterations in the function of the male reproductive system. This suggests that the effects of endogenous estrogens on the development and function of the male reproductive system identified by treatment with aromatase inhibitors or knockout of the aromatase gene are due primarily, if not exclusively, to signaling via ERα (Carreau *et al.*, 2001). However, several weak estrogens, including plant estrogens and environmental pollutants, were recently shown to bind to ERβ with greater affinity than to ERα. There is also evidence for the inhibitory effects of liganded ERβ on the signaling via ERα. These findings are potentially very important because they may explain the apparent paradox of significant biological effects of phytoestrogens and environmental estrogenic pollutants that are very weakly estrogenic and are usually ingested in very limited amounts. Current progress in elucidating the actions of estrogens and antiestrogens at the molecular level and in development and study of selective ER modulators raises hope for resolving at least some of the controversies surrounding the possible impact of environmental estrogens on male reproductive health. Of particular concern is the suspected importance of prenatal exposure to environmental estrogens and other endocrine disruptors on development of the male reproductive tract and adult reproductive function (Hess, 1998).

Classical Mechanism of Estrogen Action: Ligand-dependent

The ligand-dependent mechanism of ER action is the defining element of the Class I members of the nuclear steroid/thyroid receptor superfamily (Mangelsdorf *et al.*, 1995), of which the ERα and β are members. This model states that in the absence of hormone, the receptor is sequestered in a multiprotein inhibitory complex within the nuclei of target cells. The binding of ligand induces an activating conformational change within the ER and promotes homodimerization and high affinity binding to specific DNA response elements, which are cis-acting enhancers located within the regulatory regions of target genes. The DNA-bound receptors contact the general transcription apparatus either directly or indirectly via cofactor proteins (McKenna *et al.*, 1999), of which several have been identified, including SRC-1, GRIP1, AIB1, CBP/p300, TRAP220, PGC-1, p68 RNA helicase, and SRA. It is generally accepted that the ER-coactivator interactions stabilize the formation of a transcription preinitiation complex and facilitate the necessary disruption of chromatin at the ERE. Depending on the cell and promoter context, the DNA-bound receptor exerts either a positive or negative effect on expression of the downstream target gene.

The ligand-dependent transcriptional activity of ERα is mediated by two separate nonacidic activation domains, a constitutive activation function-1 (AF-1) located within the N terminus (A/B domain) and a hormone-dependent AF-2, located in the ligand-binding domain (Tora *et al.*, 1989). The AF-2 functional domain includes a highly conserved amphipathic α-helix (H12) that is essential for ligand-dependent transcriptional activity and interaction with members of the SRC family of coactivators (McKenna *et al.*, 1999). AF-1 activity is also dependent on recruitment of co-activators, both similar and unique from those utilized by AF-2 (McKenna *et al.*, 1999, Webb *et al.*, 1998). The AFs function in a synergistic manner in most ligand-activated mechanisms but may also function independently in certain cell and promoter contexts (Tzukerman *et al.*, 1994).

Ligand-independent Activation of ER: Cross-talk with Peptide Growth Factors

In addition to hormone-mediated activation, it is now well accepted that ER function can be modulated by extracellular signals in the absence of E_2. These findings focus primarily on the ability of polypeptide growth factors such as epidermal growth factor (EGF) and insulin-like growth factor-1 (IGF-1) as well as the intracellular effector analog 8-bromo-cyclic adenosine monophosphate to activate ER and increase the expression of ER target genes (Smith, 1998). Many of these findings have been corroborated with *in vivo* studies, such as the ability of EGF to mimic the effect of E_2 on the female reproductive tract (Curtis *et al.*, 1996). Although the molecular mechanisms involved in ligand-independent activation of ER are somewhat characterized, the biological role of these processes is not yet clear. It is possible that hormone-independent pathways allow ER activation in the presence of low E_2 levels, as found in males. Alternatively, this phenomenon may serve as a mechanism to amplify growth factor pathways and thereby enhance mitogenesis within ER-positive tissues.

The mechanisms by which the ER and growth factor pathways converge are not entirely clear. However, studies do indicate that each pathway may be dependent on the other for full manifestation of the respective ligand-mediated response. In the mouse uterus, co-treatment with anti-EGF (epidermal growth factor) antibodies was able to attenuate the uterine response to E_2; in turn, administration of the ER antagonist ICI 164, 384 reduced the uterine response to EGF. These studies together with similar observations made in the mammary gland (Ankrapp *et al.*, 1998) led to the proposed model that the mitogenic action of E_2 in these tissues is at least partially mediated by EGF; conversely, the mitogenic effects of EGF require the presence of ERα. More definitive evidence comes from studies of the uteri of ERKO females, which despite expressing wild-type levels of functional EGF and EGF receptor and showing evidence of EGF signaling, remain unresponsive to the mitogenic actions of EGF (Curtis *et al.*, 1996).

Specific receptor domains of the ER are critical to E_2-independent activation. Specifically, the effects of elevated intracellular cAMP are mediated through AF-2, whereas growth factor activation of

ER requires the N-terminal AF-1 domain of the receptor (El-Tanani and Green, 1997). The majority of evidence indicates that modification of the phosphorylation state of the ER by cellular kinases may serve as an important mechanism of ligand-independent activation. The serine 118 residue of the human ERα AF-1 is phosphorylated by the mitogen-activated protein kinase (MAPK) pathways following treatment with EGF or IGF, enabling the receptor to interact with the ERα-specific coactivator p68 RNA helicase and activate target gene transcription (Kato, 2001). Interestingly, the MAPK pathway also enhances the activity of the murine ERβ through stimulating the recruitment of SRC-1 to the N terminus. Recent evidence has emerged that coactivators may also serve as points of convergence between ER and growth factor signaling pathways, as it was shown that SRC-1 and AIB1 are phosphorylated by MAPK, an event thought to enhance their transcriptional activities (Rowan *et al.*, 2000).

The Aromatase

The cytochrome P450 aromatase (CYP 450 arom) is involved in the irreversible transformation of androgens into estrogens. The P450 arom is a microsomal enzymatic complex composed of two proteins: a ubiquitous NADPH-cytochrome P450 reductase and a cytochrome P450 aromatase, which contains the heme and the steroid binding pocket. In humans the P450 arom is the product of a single gene called *Cyp19*, which belongs to the cytochrome P450 gene family, containing more than 500 members. The *Cyp19* gene lies on more than 120 kb length with a coding region of 9 exons plus 9 untranslated exons I. The *Cyp19* gene expression is regulated by tissue-specific promoters producing alternate 5'-untranslated exons I that are then spliced onto a common 3'-splice acceptor site in the exon II upstream of the translation start site (Sebastian and Bulun, 2001).

In the mammalian testis, aromatase is mainly localized in Leydig cells. In rodents, Leydig cells synthesize estrogen in adults, whereas Sertoli cells are the major source in immature animals. However it has been shown that germ cells from adult male rat represent a new source of estrogens (Carreau *et al.*, 2001). The P450 arom transcripts is higher in pachytene spermatocytes than in round spermatids and spermatozoa.

Cellular Apoptosis in the Testis

It is very well documented that both natural and experimentally induced fluctuations in production of spermatozoa can almost always be traced to alterations in the proportion of germ cells that survive particular stages of spermatogenesis. Loss of germ cells in the testis, similarly to the situation in the ovary, occurs primarily, and perhaps exclusively, via programmed cell death or apoptosis (Billig *et al.*, 1995). Germ cell apoptosis in the testis increases in response to seasonal or experimentally induced hormonal deprivation, elevated temperature, ischemia, or toxicants known to suppress sperm production, with both Hypothalamo-Pituitary-Testicular axis and somatotropic axis (growth hormone and insulin-like growth factor-I) acting as germ cell survival factors. Testicular apoptosis involves the Fas system. Sertoli cells produce Fas ligand (FasL), a proapoptotic factor that binds to Fas, a transmembrane receptor protein on germ cells (Boekelheide *et al.*, 2001). A recent analysis of the available data led to the conclusion that survival of germ cells may reflect balance of prosurvival and proapoptotic signals (*e.g.* FasL) originating from the Sertoli cells and concomitant changes in Fas protein in germ cells (Nandi *et al.*, 1999). Studies in mice with targeted disruption of genes involved in the regulation of apoptosis provided evidence that this process is essential for normal sperm production and male fertility (Knudson *et al.*, 1995; Rodriguez *et al.*, 1997). Further studies of the Fas system and the regulation of pro- and anti-apoptotic signals in the testis should produce major advances in the understanding of paracrine interactions between the Sertoli cells and the germ cells and may provide new directions for the treatment of male infertility and for development of novel approaches to contraception. It should be mentioned that apoptotic cell death is also importantly involved in regression of male accessory reproductive glands in response to androgen withdrawal and in the loss of Leydig cells in response to toxicants (Knudson *et al.*, 1995).

Environmental Estrogens

An environmental estrogen is a chemical that acts like endogenous estrogens. These compounds have a molecular shape that allows them to fit like a key into the binding sites of the body's estrogen receptors. Once bound, the chemical can unlock or provoke the same physiological changes induced by endogenous estrogens (Hess, 1998). Pollutants with estrogenic activity bind to the receptor more weakly than does estrogens, so larger doses are needed to elicit the usual physiological responses. Natural compounds capable of producing estrogenic responses, such as the phytoestrogens, occur in a variety of plants and fungi, in addition to the synthetic chemicals that mimic estrogen. The concern arises from the fact that, surprisingly, a number of synthetic chemicals (even some that do not closely resemble the structure of natural estrogens) have estrogen-like activity in animals, including humans. In some cases, these chemicals imitate estrogen in the body, but estrogen-like activity can also result from interference with the action or production of the natural hormones. The environmental estrogens can belong to pesticides (insecticides such as o, p'-DDT, endosulfan, dieldrin, methoxychlor, kepone, dicofol, toxaphene, chlordane; herbicides such as alachlor, atrazine and nitrofen; fungicides such as benomyl, mancozeb and tributyl tin; nematocides such as aldicarb adn dibromochloropropane); products associated with plastics (bisphenol A, phthalates); pharmaceuticals (drug estrogens–birth control pills, DES, cimetidine); ordinary household products (breakdown products of detergents and associated surfactants, including nonylphenol and octylphenol); industrial chemicals (polychlorinated biphenyls (PCBs), dioxin and benzo(a)pyrene); heavy metals (lead, mercury, and cadmium).

Modes of Action

Environmental estrogens, although different in shape from natural hormones, can travel in the bloodstream, enter a cell, bind with a receptor and trigger gene expression. Once bound with the receptor, the mimicker compounds can produce a normal hormone response, cause an abnormal response or elicit no response as it blocks the receptor site and prevents natural hormones from binding. Some environmental estrogens relay molecular messages through a complex array of cellular proteins that indirectly turn genes on and alter cell growth and division. The environmental estrogen, beta-hexachlorocyclohexane (beta-HCH) produces estrogenic responses (cell division and growth) at levels found in human breast cancer tissue (Coosen and Velson, 1989). This compound, which does not bind the estrogen receptor (Coosen and Velson, 1989; Steinmetz *et al.*, 1996), may promote DNA transcription, and thus produce estrogenic responses, by passing signals through a highway of hormone and nonhormone response elements resulting in gene transcription (Steinmetz *et al.*, 1996). Another environmental estrogen p, p'-DDT, at or below levels found in human breast fat tissue, bypasses the estrogen receptor and stimulates a complex mixture of cell signaling proteins (growth factor receptors) and processes that eventually results in cell division (Shen and Novak, 1997). These results suggest that the same chemical can influence the endocrine system in more than one way.

Effects of Environmental Estrogens on Wild Life Populations

Many widely-used synthetic chemicals and natural plant compounds can alter or interfere with the endocrine system. These foreign substances have been associated with health and reproductive problems in wildlife and laboratory animals. For example, abnormal blood hormone levels; reduced fertility; altered sexual behavior; modified immune system; masculinization of females; feminization of males (reduced testes and penis size); undescended testicles (Gill *et al.*, 1976); cancers of the female and male reproductive tract; malformed Fallopian tubes, uterus and cervix (Nollar *et al.*, 1990); altered bone density and structure. As with any potential health risk, many factors such as length of exposure, dose, age and individual differences will influence the kinds and severity of health problems experienced.

Effects of Environmental Estrogens on Human Populations

The adverse effects of environmental estrogens documented in wildlife populations are also believed to be influencing reproductive abnormalities in both male and female populations. In women developmental or chronic lifetime exposure to estrogen mimics are believed to cause breast and reproductive organ tissue cancers, fibrocystic disease of the breast, polycystic ovarian syndrome, endometriosis, uterine fibroids and pelvic inflammatory diseases. While in men poor semen quality (low sperm counts, low ejaculate volume, high number of abnormal sperm, low number of motile sperm), testicular cancer (Fisch and Goluboff, 1996), malformed reproductive tissue (undescended testes, small penis size), prostate disease is speculated to be the result of prolonged exposure to environmental estrogens.

Interference with sex hormone action can result in a variety of developmental and reproductive anomalies (Akingbemi and Hardy, 2001). Although homeostatic mechanisms in the body have the potential to ameliorate or lessen the effects of endocrine disrupting capabilities of these environmental estrogens, these pathways are also potential target sites for the actions of these chemicals (Ben-Jonathan *et al.*, 1999). While it has long been known that in wildlife populations, fertility is impaired by estrogen exposure, it is unclear whether environmental pollutants with weak estrogenic effects are sufficiently potent and prevalent to have biological effects in humans. However there is good evidence for an increase in testicular cancers, and possibly in other conditions that certain spatial characteristics plus evidence on heritability suggesting possible association with a impaired spermatogenesis (Joffe, 2003). Some other environmental estrogens capable of impairing reproductive functions are summarized below.

The synthetic estrogen 17-alpha-ethynylestradiol (EE2) is a commonly used oral contraceptive that has been increasingly detected in sewage effluents. Exposure of sub lethal doses of the synthetic estrogen EE2 caused hermaphroditism, disturbed maturation of the germ cells and spermatogenesis of post-F1-generation males in amphipod *Hyalella azteca* (Vandenbergh *et al.*, 2003). Embryos or very early life-stage in fishes like the fathead minnows (*Pimephales promelas*) exposed to an environmentally relevant concentration of ethinylestradiol for very short durations of time lead to alterations in gonadal development that potentially have reproductive consequences and thus population level effects (van Aerle *et al.*, 2001). Male trouts (*Oncorhynchus mykiss*) were exposed to graded water concentrations of EE2 (10, 100, and 1, 000 ng/L) for 62 days leading up to the time of spawning. This exposure significantly reduced sperm motility, spermatocrit, gonadosomatic and hepatosomatic indices as well as testicular mass.These results suggested that the sexually maturing male rainbow trout are susceptible to detrimental reproductive effects of short-term exposures to environmentally relevant levels of EE2 (Schultz *et al.*, 2003).

The existence of ER α and β in the testis suggests that estrogens can directly affect germ cells during testicular development and spermatogenesis, and differential modulation of ER α and β in the testis could be involved in the effects of environmental estrogens like bisphenol A (Takao *et al.*, 2003). Bisphenol A is a high-volume chemical used to make polycarbonate plastic, epoxy resins and is a global environmental contaminant that has been implicated as a potential endocrine disruptor with estrogenic activity. Neonatal administration of BPA in mice results in abnormalities in the process of spermiogenesis, for example, deformities in the formation of acrosomal granule and nucleus. Ectoplasmic specialization between the Sertoli cell and spermatids are also affected by BPA exposure (Toyama and Youasa, 2004). Fathead minnows (*Pimephales promelas*) when exposed to water concentrations of BPA in the range from 1 to 1280 micrograms/L that acts as a weak estrogen adversely affects germ cell development, egg quality and hatchability (Sohoni *et al.*, 2001).

The effect of maternal exposure of rats to the estrogenic environmental substance p-nonylphenol (p-NP) resulted in lower testicular mass indicating a direct toxic effect on the testis in animals exposed to p-NP during foetal life, the postnatal period and after weaning until termination at 10 weeks of age. p-NP has toxic effects on testis and epididymis and both structures are important in impairing male fertility (de Jager *et al.*, 1999). Because p-NP has an effect on established spermatogenesis in the rat, one could speculate that the same effects might also occur in humans. Bio-accumulation may also enhance the negative effects at even lower p-NP concentrations over longer exposure periods (de Jager *et al.*, 1999). NP had a negative effect on spermatogenesis and sperm quality with a resultant impact on *in vivo* fertility in mice (Kyselova, 2003). Exposure of male rats to two estrogenic, environmental chemicals, 4-octylphenol (OP) and butyl benzyl phthalate (BBP) during gestation or during the first 21 days of postnatal life, adversely affected testicular size or spermatogenesis in adulthood (Sharpe *et al.*, 1995). The 4-tert-octylphenol (OP), is a known environmental estrogen. To examine the relationship between OP exposure and reproductive impairment in male fish, adult male fish medaka were exposed to OP, in concentrations ranging from 20 to 230 ppb for 21 days, under flow-through conditions. Following exposure, male fish were mated in the absence of OP with unexposed females. Breeding groups composed of exposed males and control females produced about 50 per cent fewer eggs than control groups. OP caused a significant increase in the number of abnormally developing embryos, suggesting that OP may be genotoxic as well as estrogenic (Brouwer *et al.*, 1999). Alkylphenols, like the estrogenic 4-tert-octylphenol (OP), are environmental pollutants. The environmental estrogens are often considered to alter the calcium homeostasis in Sertoli cells.

Occupational exposure to environmental estrogens is often speculated to be a prime cause for reproductive impairment in the male. Dibromochloropropane is one of the best known occupational testicular toxins and prolonged exposure to it is known to adversely affect semen parameters. Human semen parameters can serve as valuable indicators of toxic and, in future even genotoxic effects of occupational and environmental factors (Lahdetie, 1995). Male fertility can be impaired by various Sertoli cell specific toxicants as these cells play an essential role in spermatogenesis. Production of lactate and secretion of inhibin B have been used as general and specific cell markers for exploring the potential of Sertoli cell toxicants (Monsees *et al.*, 2000).

There is a growing body of evidence that industrial pollutants can exert potent environmental estrogenic effects *in vivo* through receptor independent mechanisms by directly interacting with cellular enzyme targets. A recent study has shown that the environmental estrogens like alkylphenols directly act and inhibit enzymes like sulfotransferases, which convert active estrogenic steroids to inactive steroid sulfates; and also $Ca(^{2+})$-ATPases, which are responsible for maintaining low, physiological, intracellular $Ca(^{2+})$ concentrations. As a result elevated concentrations of active estrogens and intracellular calcium cause cellular toxicity underlying the potential adverse effects of this widespread industrial pollutant on the male reproductive system causing sexual dysfunction and reduced spermatogenesis (Kirk *et al.*, 2003).

Little information exists on the transfer of endocrine-disrupting effects through the food chain and the transfer of estrogenic chemicals, particularly from the aquatic ecosystem. A study was conducted on rats fed on fish living in water bodies contaminated with environmental estrogens. There was no effect on the body weights or in the male reproductive organ weights between groups except for adult epididymal weight, which was significantly decreased in the environmental estrogen group. Adult sperm concentrations and sperm motility parameters were all significantly decreased in the environmental estrogen group as compared to the control groups. These data indicate that contaminants from environments may pass through the food chain and exert effects on male reproductive functions (Aravindakshan *et al.*, 2004). The 3, 3', 4, 4', 5-Pentachlorobiphenyl (PCB126), a congener with a planar configuration, has been established to have relatively strong toxicities similar to those of 2, 3, 7, 8-tetrachlorodibenzo-p-dioxin via aryl hydrocarbon receptors. A study looking in to the effects of this coplanar PCB on mammalian early spermatogenesis and steroidogenesis in a mouse neonatal testicular organ culture system revealed that PCB does not alter the proliferative activity of spermatogenic cells and Sertoli cells in neonatal testis, but that it directly affects the expression of steroidogenic enzyme genes (Fukuzawa *et al.*, 2003).

Models to Study Estrogen Action

In vivo

DES, a potent estrogenic analogue is known to disrupt spermatogenesis by the suppression of pituitary gonadotropins which leads to the inhibition of testosterone production (Goyal *et al.*, 2001) leading to insufficient circulating and intratesticular concentration of the hormone. This results in abnormal spermatogenesis, as testosterone is vital for spermatogenic cell survival (Akingbemi *et al.*, 2001). We observed a loss of testicular weight in treated animals in a dose and time dependent manner indicating a decrease in the number of testicular cells.

In the regimen of hormone administration that we used, the 1 mg diethylstilbestrol (DES) treated group exhibited a severe depletion of spermatogenic cells by day 22, whereas 0.1 mg DES treatment showed a nearly comparable depletion of spermatogenic cells only by day 31 showing that depletion of spermatogenic cells was dependent on dose as well as time of treatment. Taken together, the above data show the ability of DES to reduce serum testosterone levels and induce a loss of testicular spermatogenic cells in a dose and time dependent manner that is paralleled by a loss in testicular weight. Biochemically, internucleosomal DNA strand breaks, which are detected as a ladder pattern on agarose-gel electrophoresis of DNA are regarded as a hallmark of apoptosis (Stewart, 1994). Genomic DNA prepared from spermatogenic cells of animals after day 3 of DES treatment showed distinct DNA laddering, regarded as a hallmark of apoptosis as compared to controls. Therefore, the above data showing nucleosomal DNA laddering, increase in cells with fragmented DNA and changes in the ploidy of spermatogenic cells clearly established the apoptotic nature of spermatogenic cell death, particularly in the haploid cell type as induced by DES.

Since it was evident that DES induced cell death occurred via apoptosis, the expression of possible apoptosis inducing proteins were studied in order to dissect the apoptotic pathway. Our studies revealed a clear up-regulation of FasL in spermatogenic cells within 24 h of DES treatment as determined by Western blots and RT-PCR with FasL specific antibodies and probes respectively. Similar up-regulation of Fas was also detectable by Western blots of purified spermatogenic cell extracts. Interestingly, in isolated spermatogenic cells from DES treated animals on day 1, FasL and Fas showed differential localization. While Fas was distributed evenly throughout the cell, FasL staining was associated with vesicular bodies. In summary, the above data shows that DES treatment selectively increases FasL and Fas expression in spermatogenic cells indicating an apoptosis modulating capacity of the spermatogenic cells independent of Sertoli cell control and provides a model for the study of Fas-

FasL interaction in the most complex epithelium of the body. In addition, testosterone supplementation clearly downregulates the Fas-FasL upregulation induced by DES showing that testosterone is important for regulation of apoptotic signals to germ cells. The fact that Fas is an absolute necessity in case of DES induced apoptosis was proven by studies with lpr mice where DES was unable to induce changes. There was a temporal relationship between the activation of caspase-8, translocation of Bax to mitochondria, release of cytochrome c and subsequent activation of caspase-3 provide a clear picture that exposure to potent estrogens could induce spermatogenic cell apoptosis in a caspase dependent manner (Nair and Shaha, 2003).

The finding that primarily haploid spermatogenic cells were undergoing apoptosis in response to DES is in contrast to what is found in other toxin induced cell death models where the diploid population of spermatocytes are most affected (Miura *et al.*, 2002). Clearly, in case of DES treatment, the spermatids are mainly affected suggesting that estrogen induced cell death primarily involves apoptosis of the haploid population. The belief that Sertoli cells are the only source of FasL dominated thinking in testicular biology (Xu *et al.*, 1999), however, in the recent past it has been shown that the spermatogenic cells have the ability to express both Fas and FasL (D'Alessio *et al.*, 2001). The presence of Fas-FasL in testicular cells have been linked to the ability of the testis to act as an immune privileged site and maintenance of testicular homeostasis by Sertoli cells eliminating spermatogenic cells. However, the recognition that both the Fas receptor and its cognate ligand FasL can be expressed in spermatogenic cells has led to the speculation that apart from the above functions, the expression of both these proteins on spermatogenic cells may indicate an as yet unidentified function. Here, we demonstrate an important novel function for the Fas-FasL system in the testis. We show that estrogen exposure induces an increase in the expression of spermatogenic cell Fas-FasL leading to the activation of components downstream of Fas that indicate control of cell death within the same lineage. More importantly, the observation that the number of apoptotic cells is highest in seminiferous tubule stages where both Fas-FasL expression were high reinforced the idea that these two proteins expressed in spermatogenic cells are the main modulators of death. Our data clearly shows that the mitochondria is involved in DES induced death as shown by the translocation of Bax from the cytosol to the mitochondria causing the release of cytochrome-c and loss of mitochondrial membrane potential. This identification of spermatogenic cells as type II cells is important as therapeutic strategies that involve modulation of Fas-mediated apoptosis will first have to identify the tissue to be determined as type I or type II.

The intrinsic and extrinsic pathways are linked by the proapoptotic members of the Bcl-2 family. In case of DES treatment, Bax protein translocation to mitochondria occurred after caspase-8 was activated. A role for Bax has been predicted in germ cell apoptosis using different models, but predictions of involvement of Bax have largely been made based on cellular localization (Yu *et al.*, 2001) or from Western blots of whole testis extracts (Yamamoto *et al.*, 2000). Using purified spermatogenic cells our study clearly demonstrated that there is a translocation of Bax from the cytosol to the mitochondria resulting in the release of cytochrome c (Nair and Shaha, 2003). We also postulate that normal testicular homeostasis may involve the Fas-FasL system to maintain proper spermatogenic cell number, which may be under estrogen regulation.

In vitro

Our *in vitro* studies with estrogens (water soluble estradiol-17β) revealed some interesting results (Mishra and Shaha, 2005). (i) 17β-estradiol was able to induce spermatogenic cell death in the absence of testicular somatic cells and (ii) estrogen receptor modulators were able to prevent 17β-estradiol induced death. Genomic DNA prepared from spermatogenic cells after 9 and 10 h of 17β-estradiol treatment showed distinct DNA laddering as compared to 0 h controls. Thus, from the above data, it was evident that many features of apoptotic phenotype were expressed upon 17β-estradiol exposure. Taken together, the data showed that in response to 17β-estradiol following changes occurred (i) several apoptotic phenotypes were expressed by the cells; (ii) FasL upregulation occurred; (iii) there was a release of soluble FasL in the media (iv) caspase-8 cleavage was observed; (iv) release of cytochrome c in the cell cytosol occurred. The above data established that (i) a transient mitochondrial hyperpolarization resulted due to estrogen exposure; (ii) this response was modulated through estrogen receptor interactions, thus linking an estrogen receptor mediated pathway with the mitochondrial changes.

Conclusions

The above review shows that there is compelling evidence for a role of estrogen in testicular function. Germ cells contain ERs as well as aromatase, and thus it is possible that estrogen acts in an intracrine manner in these cells, to control viability/apoptosis and, potentially, acrosome biogenesis. The presence of ER's and aromatase makes the germ cells possible source and target cell for the estrogens. The models of gene disruption and estrogen administration studies have answered many questions as to the role of, and sites of action for estrogen in several aspects of male fertility, from testicular function to sexual behavior. However, a large amount of conflicting data also exist, such as the comparison of the phenotypes of mice deficient in ERα and/or ERß and aromatase. Such conflicting data highlight the complexity of the molecular mechanisms of estrogen action in the male, and serve to remind us that the study of the role of estrogen in male fertility is far from complete.

The evidence from the wildlife population studies and animal studies as well demand a comprehensive study into the importance of estrogens as endocrine disrupters.

References

Akingbemi BT, and Hardy MP. (2001). Oestrogenic and antiandrogenic chemicals in the environment: effects on male reproductive health. Ann. Med., 33, 391-403.

Ankrapp DP, Bennett JM, and Haslam SZ. (1998). Role of epidermal growth factor in the acquisition of ovarian steroid hormone responsiveness in the normal mouse mammary gland. J. Cell Physiol., 174, 251-260.

Aravindakshan J, Gregory M, Marcogliese DJ, Fournier M, and Cyr DG. (2004). Consumption of xenoestrogen-contaminated fish during lactation alters adult male reproductive function. Toxicol. Sci., 81, 179-189.

Ben Jonathan N, Cooper RL, Foster P, Hughes CL, Hoyer PB, Klotz D. *et al.* (1999). An approach to the development of quantitative models to assess the effects of exposure to environmentally relevant levels of endocrine disruptors on homeostasis in adults. Environ. Health Perspect., 107, Suppl 4, 605-611.

Billig H, Furuta I, Rivier C, Tapanainen J, Parvinen M, and Hsueh AJ. (1995). Apoptosis in testis germ cells: developmental changes in gonadotropin dependence and localization to selective tubule stages. Endocrinol., 136, 5-12.

Boekelheide K, Fleming SL, Johnson KJ, Patel SR, and Schoenfeld HA. (2001). Role of Sertoli cells in injury-associated testicular germ cell apoptosis. Proc. Soc. Exp. Biol. Med., 225, 105-115.

Brouwer A, Longnecker MP, Birnbaum LS, Cogliano J, Kostyniak P, Moore J. *et al.* (1999). Characterization of potential endocrine-related health effects at low-dose levels of exposure to PCBs. Environ. Health Perspect., 107, Suppl 4, 639-649.

Carreau S, Bourguiba S, Lambard S, Galeraud-Denis I, Genissel C, Bilinska B. *et al.* (2001). Aromatase expression in male germ cells. J. Steroid Biochem. Mol. Biol., 79, 203-208.

Coosen R, and van Velsen FL. (1989). Effects of the beta-isomer of hexachlorocyclohexane on estrogen-sensitive human mammary tumor cells. Toxicol. Appl. Pharmacol., 101(2), 310-318.

Couse JF, and Korach KS. (1999). Reproductive phenotypes in the estrogen receptor-alpha knockout mouse. Ann. Endocrinol. (Paris), 60, 143-148.

Curtis SW, Washburn T, Sewall C, DiAugustine R, Lindzey J, Couse JF *et al.* (1996). Physiological coupling of growth factor and steroid receptor signaling pathways: estrogen receptor knockout mice lack estrogen-like response to epidermal growth factor. Proc. Natl. Acad. Sci. USA, 93, 12626-12630.

D'Alessio,A, Riccioli,A, Lauretti,P, Padula,F, Muciaccia,B, De Cesaris,P *et al.* (2001). Testicular FasL is expressed by sperm cells. Proc. Natl. Acad. Sci. USA, 98, 3316-3321.

de Jager C, Bornman MS, and van der HG. (1999). The effect of p-nonylphenol, an environmental toxicant with oestrogenic properties, on fertility potential in adult male rats. Andrologia, 31, 99-106.

Dym M, and Fawcett DW. (1970). The blood-testis barrier in the rat and the physiological compartmentation of the seminiferous epithelium. Biol. Reprod., 3, 308-326.

El Tanani MK, and Green CD. (1996). Insulin/IGF-1 modulation of the expression of two estrogen-induced genes in MCF-7 cells. Mol. Cell. Endocrinol., 121, 29-35.

Fisch H, and Goluboff ET. (1996). Geographic variations in sperm counts: a potential cause of bias in studies of semen quality. Fertil. Steril/. 65, 1044-1046.

Fukuzawa NH, Ohsako S, Nagano R, Sakaue M, Baba T, Aoki Y. *et al.* (2003). Effects of 3,3',4,4',5-pentachlorobiphenyl, a coplanar polychlorinated biphenyl congener, on cultured neonatal mouse testis. Toxicol. In Vitro, 17, 259-269.

Gill WB, Schumacher GF, and Bibbo M. (1976). Structural and functional abnormalities in the sex organs of male offspring of mothers treated with diethylstilbestrol (DES). J. Reprod. Med., 16, 147-153.

Goyal HO, Braden TD, Mansour M, Williams CS, Kamaleldin A, and Srivastava KK (2001). Diethylstilbestrol-treated adult rats with altered epididymal sperm numbers and sperm motility parameters, but without alterations in sperm production and sperm morphology. Biol. Reprod., 64, 927-934.

Hess RA. (1998). Effects of environmental toxicants on the efferent ducts, epididymis and fertility. J. Reprod. Fertil. Suppl., 53, 247-259.

Hess RA. (2000). Oestrogen in fluid transport in efferent ducts of the male reproductive tract. Rev. Reprod., 5, 84-92.

Joffe M. (2003). Increased concentrations of polychlorinated biphenyls, hexachlorobenzenes, and chlordanes in mothers of men with testicular cancer. Environ. Health Perspect., 112, A21-A22.

Kato S. (2001). Estrogen receptor-mediated cross-talk with growth factor signaling pathways. Breast Cancer, 8, 3-9.

Kirk CJ, Bottomleym L, Minican N, Carpenter H, Shaw S, Kohli N *et al.* (2003). Environmental endocrine disrupters dysregulate estrogen metabolism and Ca2+ homeostasis in fish and mammals via receptor-independent mechanisms. Comp. Biochem. Physiol. A Mol. Integr. Physiol., 135, 1-8.

Knudson CM, Tung KS, Tourtellotte WG, Brown GA, and Korsmeyer SJ. (1995). Bax-deficient mice with lymphoid hyperplasia and male germ cell death. Science, 270, 96-99.

Kuiper GG, Carlsson B, Grandien K, Enmark E, Haggblad J, Nilsson S. *et al.* (1997). Comparison of the ligand binding specificity and transcript tissue distribution of estrogen receptors alpha and beta. Endocrinol., 138, 863-870.

Kurland,G, Christensen,AK, and Huseby,RA (1975). Changes in fine structure accompanying estrogen-induced tumorigenesis of Leydig cells in the mouse testis. Cancer Res., 35, 1671-1686.

Kyselova V, Peknicova J, Buckiova D, and Boubelik M. (2003). Effects of p-nonylphenol and resveratrol on body and organ weight and in vivo fertility of outbred CD-1 mice. Reprod. Biol. Endocrinol., 1, 30.

Lahdetie J. (1995). Occupation- and exposure-related studies on human sperm. J. Occup. Environ. Med., 37, 922-930.

Leblond CP, and Clermont Y. (1952a). Definition of the stages of the cycle of the seminiferous epithelium in the rat. Ann. N. Y. Acad. Sci., 55, 548-573.

Mangelsdorf DJ, Thummel C, Beato M, Herrlich P, Schutz G, Umesono K. *et al.* (1995). The nuclear receptor superfamily: the second decade. Cell, 83, 835-839.

McKenna NJ, Xu,J, Nawaz,Z, Tsai,SY, Tsai,MJ, and O'Malley,BW (1999). Nuclear receptor coactivators: multiple enzymes, multiple complexes, multiple functions. J. Steroid Biochem. Mol. Biol., 69, 3-12.

Mishra DP, and Shaha C. (2005). Estrogen-induced spermatogenic cell apoptosis occurs via the mitochondrial pathway: role of superoxide and nitric oxide. J. Biol. Chem., 280, 6181-6196.

Miura M, Sasagawa I, Suzuki Y, Nakada T, and Fujii J. (2002). Apoptosis and expression of apoptosis-related genes in the mouse testis following heat exposure. Fertil. Steril., 77, 787-793.

Monsees TK, Franz M, Gebhardt S, Winterstein U, Schill WB, and Hayatpour J. (2000). Sertoli cells as a target for reproductive hazards. Andrologia, 32, 239-246.

Nair R, and Shaha C. (2003). Diethylstilbestrol induces rat spermatogenic cell apoptosis *in vivo* through increased expression of spermatogenic cell Fas/FasL system. J. Biol. Chem., 278, 6470-81.

Nandi S, Banerjee PP, and Zirkin BR. (1999). Germ cell apoptosis in the testes of Sprague Dawley rats following testosterone withdrawal by ethane 1,2-dimethanesulfonate administration: relationship to Fas? Biol. Reprod., 61, 70-75.

Nikula H, Talonpoika T, Kaleva M, and Toppari, J. (1999). Inhibition of hCG-stimulated steroidogenesis in cultured mouse Leydig tumor cells by bisphenol A and octylphenols. Toxicol. Appl. Pharmacol., 157, 166-173.

Nollar KL, O'Brien PC, Colton T, Kaufman R, and Melton LJ.(1990).Medical and surgical diseases associated with *in utero* exposure to diethylstelbisterol. *In:* Nollar KL (Ed.), *Clinical Practice of Gynocology,* Elsevier Science Publishing Co. Inc., New York, pp. 1-7.

Rodriguez I, Ody C, Araki K, Garcia I, and Vassalli P. (1997). An early and massive wave of germinal cell apoptosis is required for the development of functional spermatogenesis. EMBO J., 16, 2262-2270.

Rowan BG, Weigel NL, and O'Malley, BW. (2000). Phosphorylation of steroid receptor coactivator-1. Identification of the phosphorylation sites and phosphorylation through the mitogen-activated protein kinase pathway. J. Biol. Chem., 275, 4475-4483.

Russell LD, Corbin TJ, Borg KE, de Franca LR, Grasso P, and Bartke A. (1993). Recombinant human follicle-stimulating hormone is capable of exerting a biological effect in the adult hypophysectomized rat by reducing the numbers of degenerating germ cells. Endocrinol., 133, 2062-2070.

Russell LD, Ren HP, Sinha Hikim I, Schulze W, and Sinha Hikim AP. (1990). A comparative study in twelve mammalian species of volume densities, volumes, and numerical densities of selected testis components, emphasizing those related to the Sertoli cell. Am. J. Anat., 188, 21-30.

Schultz IR, Skillman A, Nicolas JM, Cyr DG, and Nagler JJ. (2003). Short-term exposure to 17 alpha-ethynylestradiol decreases the fertility of sexually maturing male rainbow trout (Oncorhynchus mykiss). Environ. Toxicol. Chem., 22, 1272-1280.

Sebastian S, and Bulun SE. (2001). A highly complex organization of the regulatory region of the human CYP19 (aromatase) gene revealed by the Human Genome Project. J. Clin. Endocrinol. Metab., 86, 4600-4602.

Sharpe RM, Fisher JS, Millar MM, Jobling S, and Sumpter JP. (1995). Gestational and lactational exposure of rats to xenoestrogens results in reduced testicular size and sperm production. Environ. Health Perspect., 103, 1136-1143.

Shen K, and Novak RF. (1997). Differential effects of aroclors and DDT on growth factor gene expression and receptor tyrosine kinase activity in human breast epithelial cells. Adv. Exp. Med. Biol., 407, 295-302.

Smith CL. (1998). Cross-talk between peptide growth factor and estrogen receptor signaling pathways. Biol. Reprod., 58, 627-632.

Sohoni P, Tyler CR, Hurd K, Caunter J, Hetheridge M, Williams T. *et al.* (2001). Reproductive effects of long-term exposure to Bisphenol A in the fathead minnow Pimephales promelas). Environ. Sci. Technol., 35, 2917-2925.

Steinmetz R, Young PC, Caperell-Grant A, Gize EA, Madhukar BV, Ben Jonathan N. *et al.* (1996). Novel estrogenic action of the pesticide residue beta-hexachlorocyclohexane in human breast cancer cells. Cancer Res., 56, 5403-5409.

Stewart BW. (1994). Mechanisms of apoptosis: integration of genetic, biochemical, and cellular indicators. J. Natl. Cancer. Inst., 86, 1286-1296.

Takao T, Nanamiya W, Nazarloo HP, Matsumoto R, Asaba K, and Hashimoto K. (2003). Exposure to the environmental estrogen bisphenol A differentially modulated estrogen receptor-alpha and–beta immunoreactivity and mRNA in male mouse testis. Life Sci., 72, 1159-1169.

Tora L, White J, Brou C, Tasset D, Webster N, Scheer E. *et al.* (1989). The human estrogen receptor has two independent nonacidic transcriptional activation functions. Cell, 59, 477-487.

Toyama Y, and Yuasa S. (2004). Effects of neonatal administration of 17beta-estradiol, beta-estradiol 3-benzoate, or bisphenol A on mouse and rat spermatogenesis. Reprod. Toxicol., 19, 181-188.

Tzukerman MT, Esty A, Santiso-Mere D, Danielian P, Parker MG, Stein RB. *et al.* (1994). Human estrogen receptor transactivational capacity is determined by both cellular and promoter context and mediated by two functionally distinct intramolecular regions. Mol. Endocrinol., 8, 21-30.

van Aerle R, Nolan TM, Jobling S, Christiansen LB, Sumpter JP, and Tyler CR. (2001). Sexual disruption in a second species of wild cyprinid fish (the gudgeon, Gobio gobio) in United Kingdom freshwaters. Environ. Toxicol. Chem., 20, 2841-2847.

Vandenbergh GF, Adriaens D, Verslycke T, and Janssen CR. (2003). Effects of 17 alpha-ethinylestradiol on sexual development of the amphipod *Hyalella azteca*. Ecotoxicol. Environ. Safety, 54, 216-222.

Webb P, Nguyen P, Shinsako J, Anderson C, Feng W, Nguyen MP. *et al.* (1998). Estrogen receptor activation function works by binding coactivator proteins. Mol. Endocrinol., 12, 1605-1618.

Xu JP, Li X, Mori E, Guo MW, Matsuda I, Takaichi H. *et al.* (1999). Expression of Fas-Fas ligand in murine testis. Am. J. Reprod. Immunol., 42, 381-388.

Yamamoto CM, Sinha Hikim AP, Huynh PN, Shapiro B, Lue Y, Salameh WA. *et al.* (2000). Redistribution of Bax is an early step in an apoptotic pathway leading to germ cell death in rats, triggered by mild testicular hyperthermia. Biol. Reprod., 63, 1683-1690.

Yu X, Kubota H, Wang R, Saegusa J, Ogawa Y, Ichihara G. *et al.* (2001). Involvement of Bcl-2 family genes and Fas signaling system in primary and secondary male germ cell apoptosis induced by 2-bromopropane in rat. Toxicol. Appl. Pharmacol., 174, 35-48.

9

A Need to Study the Hormonal Regulation of Sertoli Cells Isolated from Spermatogenically Inactive and Active Testis: Implications in Determining Causes Underlying Idiopathic Male Infertility

*Kanchan Sarda, Jibananda Mishra, Y. Sangeetadevi and Subeer S. Majumdar**

National Institute of Immunology, Aruna Asaf Ali Marg, New Delhi – 110 067, India

ABSTRACT

At present, half of infertile males remain untreated. This is mainly due to the fact that most of such defects are homed in the testis and our knowledge of intra-seminiferous tubular regulation of spermatogenesis is extremely poor. Only a small percentage of infertile or subfertile males undergo a rational, effective treatment. Infertility caused by primary testicular failure due to cryptorchidism, torsion, varicocele, chemotherapy, ischemic damage etc. is invariably associated with varying degree of damage to the seminiferous tubule, *the major cell type affected being the Sertoli cell (Sc)* (Foresta *et al.*, 2001) and such patients are not amenable to treatment of infertility (Winters, 1990). Idiopathic, oligo, astheno and tetrato zoospermia may be related to defective

**Corresponding Author*: E-mail: hisubeer@yahoo.com

maturation or spermiation of sperm from Sc. In few of them, genetic factors are also responsible(Grimaldi *et al.*, 2002).

In human and monkey, LH, FSH and testosterone (T) are high during infancy (0-4 months of age) as well as puberty but spermatogenesis is initiated only at puberty (Burger *et al.*, 1991; Mann and Fraser, 1996). In rats, although there is no major difference in the gonadotropin levels between day 9 and 60 of age, progression of spermatogenesis is evident only at day 11-15 of postnatal life (Dohler and Wuttke, 1975; Lee *et al.*, 1975; Yang *et al.*, 1990). Such introvert behavior of Sc during postnatal/infant period in presence of sufficient hormonal milieu attracts attention to investigate inability of immature Sc to support spermatogenesis. Several infertile males have normal or high levels of gonadotropins but they do not produce sperms, a situation very much similar to that found during infancy. Appropriate transduction of hormonal (FSH and T) signals is necessary for the initiation and maintenance of spermatogenesis. Hence, it is important to investigate whether the Sc from such infertile males are capable of responding to the robust hormonal drive. Since it is difficult to obtain Sc from infertile individuals, Sc from 9 days old rat or more importantly 3-4 months old monkeys may act as surrogates of infertile adults for investigative research to analyze the status with respect to hormone responsiveness and protein production of such Sc which fail to support spermatogenesis even in the presence of adequate hormonal milieu. Sertoli cells isolated from the spermatogenetically active stage of development may be used as a yardstick of mature hormonally responsive Sc and their functional comparison with infant Sc may help us in resolving the deficits residing in the infant Sc.

With the advent of molecular biology tools and unraveling of the human genome such studies can reveal the differential gene expression by DNA micro array, RT PCR and differential display analyses. Proteomic approaches would help in determining the translation profiles in these two situations. All this would lead to the identification of molecular mechanisms vital to spermatogenesis, which may help in the diagnosis and treatment of idiopathic male infertility and in identifying sites for the development of non-hormonal approaches for contraception.

Introduction

At an alarming rate, population is growing fastest in the poorest countries, those least able to provide for basic needs and create opportunities. At the same time, 61 countries are seeing fertility at or below replacement level, and their populations could decline over the long term (UNFPA: state of world population 1999). As fertility falls in more countries, this phenomenon could affect countries with as many as two thirds of the world's people. Hence, development of acceptable male contraceptives as well as treatments for infertility, both, demand a great attention. Of the infertile couples, approximately 40–50 per cent causes are attributed to males, approximately half of which are idiopathic (causes unknown) in nature. Only a small percentage of infertile or subfertile males undergo a rational, effective treatment. Infertility caused by primary testicular failure due to cryptorchidism, torsion, varicocele, chemotherapy, ischemic damage etc. is invariably associated with varying degree of damage to the seminiferous tubule, *the major cell type affected being the Sertoli cell* (Forestà *et al.*, 2001), and such patients are not amenable to treatment of infertility (Winters, 1990). This is mainly due to the fact that most of such defects are homed in the testis and our knowledge regarding intra-seminiferous tubular regulation of spermatogenesis in the testis of normal adults is extremely poor. It would be ideal to compare Sc functions from spermatogenically active (normal) testis with that of Sc from the spermatogenically inactive testis for determining the deficits in the hormonal regulation of Sc function in the latter.

Male Infertility

A myriad of factors contribute to male infertility. The causes of male infertility are listed in Table 9.1. Idiopathic, oligo, astheno and tetrato zoospermia may be related to defective maturation or spermiation of sperm from Sc. In few of them, genetic factors are also responsible (Grimaldi *et al.*, 2002). Defect in GnRH production is known to be the cause of several endocrine disorders. Isolated GnRH deficiency is a clinical syndrome that results from the failure of normal pattern of episodic GnRH secretion (Seminara *et al.*, 1998). GnRH deficiency due to Kallmann's syndrome (KS) or idiopathic hypogonadotropic hypogonadism (IHH) resulting from a deficient hypothalamic secretion of GnRH (Georgopoulos *et al.*, 1997; Naftolin *et al.*, 1971) culminates into lack of gonadotropins, onset of puberty and hence absence of spermatogenesis. Kallmann's syndrome is usually caused due to migration defect of GnRH neurons from olfactory bulbs. However, a more recent study demonstrated normal olfactory bulbs in 25 per cent of male KS patients (Quinton *et al.*, 1996), suggesting that further studies are required to differentiate KS from IHH. Reports of rodents displaying KS or IHH do not exist. Many questions remain unanswered regarding the fascinating neurobiological model presented by KS. Studies related to human puberty, hence, can only be envisaged using non-human primate models because a well-defined puberty and human-like hypothalamic awakening at the time of puberty is not represented in rats (Plant, 1988; Plant and Marshall, 2001). There are several other causes of infertility which are mentioned in the Table 9.1.

In addition to the known causes of infertility, there are several forms of infertility which are idiopathic in nature. Lack of sperm production in spite of normal or elevated hormones in the circulation of infertile men indicates presence of defects in the seminiferous tubules which are not known to us till date (Winters, 1990). Since hormonal regulation of testicular cells, androgen production and maturation of spermatozoa is similar in man, monkeys and rodents, fertility regulation at the level of testis may be investigated equally well in rats and monkeys, as far as present knowledge is concerned. Species to species differences can be noted in future but non availability of human testis for research compels us to utilize these animal models to increase our understanding about testicular regulation of spermatogenesis.

Necessity of Rodent as well as Primate Studies

The basic framework of male reproductive endocrinology concerning regulation of fertility is more or less similar in all higher mammals. On the basis of this general understanding, most of the research is carried using rodents (mostly rats) as the preferable model of choice because they can be bred easily, they become adult and reproductively active by 8 weeks of age and the gestation is also of a short duration of 21 days or so as compared to that in primates. Also, a large number of animals can be maintained in a small space at an affordable maintenance cost. Although results from studies using rodents have helped in establishing several hypotheses valid to human situation, an in-depth comparison of the endocrine regulation of sexual maturation and spermatogenesis in rodents and primates suggests that all results obtained from rodent studies may not necessarily be extrapolated to human. Since experiments can not readily be carried out in human, it is necessary to have animal models, in addition to rats, which provide a situation resembling the endocrine regulation of fertility in men. The macaques fit very well into the description of the aforementioned framework. In the present chapter, an attempt has been made to justify use of sub-human primate and rodent models for increasing our understanding about idiopathic infertility in men.

Table 9.1: Pre-testicular and Testicular Causes of Infertility

PRE-TESTICULAR CAUSES OF INFERTILITY
Hypothalamic disease
— Isolated gonadotropin deficiency (Kallmann's syndrome)
— Idiopathic hypogonadotropic hypogonadism
— Isolated LH deficiency ("Fertile eunuch")
— Isolated FSH deficiency
Pituitary disease
— Pituitary insufficiency (tumors, infiltrative processes, operation, radiation)
— Hyperprolactinemia
— Hemochromatosis
TESTICULAR CAUSES OF INFERTILITY
Genetic Disorders
— Chromosomal abnormalities (Klinefelter's syndrome, XX disorder (sex reversal syndrome), XYY syndrome)
— Noonan's syndrome (male Turner's syndrome)
— Myotonic dystrophy
Testicular Dysgenesis Syndrome
— Bilateral anorchia (vanishing testes syndrome)
— Sertoli-cell-only syndrome (germinal cell aplasia)
— Gonadotoxins (drugs, radiation)
— Orchitis
— Trauma
— Systemic disease (renal failure, hepatic disease, sickle cell disease)
— Cryptorchidism
— Varicocele
Androgen resistance syndrome
— Defective androgen synthesis or action
— Partial or complete androgen insensititvity
POST-TESTICULAR CAUSES OF INFERTILITY
Disorders of sperm transport
— Congenital disorders
— Acquired disorders
— Functional disorders
Disorders of sperm motility or function
— Congenital defects of the sperm tail
— Maturation defects
— Immunologic disorders
— Infection
Sexual dysfunction

Hypothalamus-Pituitary-Testicular Axis

In human, the neuroendocrine axis, which regulates functions of testis, is known to be active during peri-natal as well as peri- and post pubertal period. In between, there lies a protracted prepubertal juvenile period during which circulating levels of GnRH, LH, FSH and testosterone (T) are barely detectable (Plant and Marshall, 2001). Such ontogenically regulated GnRH and gonadotropin secretion is found in sub-human primates but contrasts with the situation in laboratory rodents, where puberty is manifest within the first few weeks of life without a prepubertal hiatus in gonadotropin secretion (Plant, 1988; Plant and Marshall, 2001). Although different hypothalamus-pituitary-testicular axes exist ontogenically in rats and monkeys, levels of reproductive hormones during infancy is similar to that during puberty in both the species (Figure 9.1).

Developmental Changes in the Endocrine Stimulus to the Testis of Primates and Rodents

Primates

The developmental pattern in higher primates is characterized by three distinct phases, namely neonatal, juvenile, and adulthood (Ramaswamy and Plant, 2001). It is well recognized that the hypothalamic-pituitary-testicular hormonal axis is transiently activated during the first few postnatal months (neonatal period) in primates including humans, where levels of gonadotropins (LH and

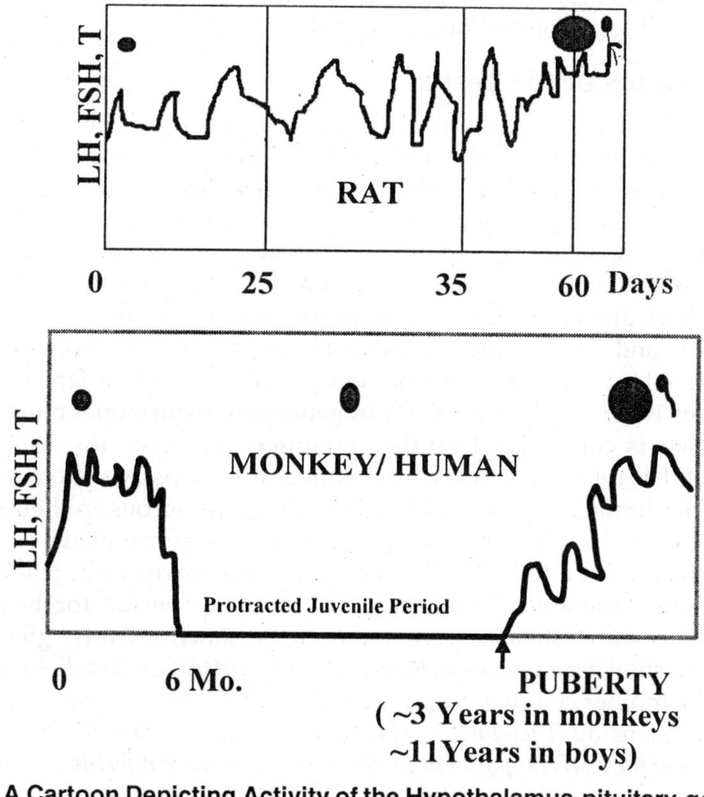

Figure 9.1: A Cartoon Depicting Activity of the Hypothalamus-pituitary-gonadal Axis from Birth to Adulthood in Male Rats and Primates

FSH) and testicular hormones (T and inhibin) are elevated (Bergada *et al.*, 1999; Burger *et al.*, 1991). In human, after birth, inhibin B and gonadotropins start to increase significantly during the first months, with the highest peak of inhibin B, FSH and LH levels found at 3-4 months of age (Chada *et al.*, 2003). Correspondingly, a rise in T levels is observed during the first few months of the postnatal life. After this early increase, serum gonadotropins and T level decrease rapidly to barely detectable levels during the first year and remain low (juvenile period) until puberty. Serum inhibin B concentration decreases gradually over a period from 4 months up to two years of age. In contrast to T, inhibin B is well measurable during childhood. At the onset of puberty, the serum levels of LH, FSH, T and inhibin increase with similar rates of pubertal progression associated with Gc proliferation and differentiation. This developmental pattern of primate is diagrammatically represented in Figure 9.1.

Rodents

In case of rats, unlike primates, there is no distinct phase of hormonal development (Figure 9.1). Serum FSH levels during first 10 days of life are not remarkably different than those during puberty or adulthood (Lee *et al.*, 1975). There is no strict age related deficiency of circulating levels of LH although this varied from one study to other (Dohler and Wuttke, 1975; Lee *et al.*, 1975). The serum T levels are usually around 100ng/100 ml from birth to 50 days. A stepwise rise then occurs to reach levels in excess of 230 ng/100 ml between Days 70 and 80. It is important to note that spermatozoa can be found in the seminiferous tubule by day 45-50 of age. Stereological studies reveal that in one day old rats, the only Gc present are gonocytes; at 5 days, a mixture of gonocytes and spermatogonia are present. Primary spermatocytes can first be appreciably noticed on day 15 of age, the round spermatids appear by approximately 30 days of age(Yang *et al.*, 1990).

Structure and Functions of the Testis

Before investigating the defects in the testicular compartment of reproductive system, it is important to know the structure and function of the testis. Testis is not only responsible for producing the hormones necessary for sexual maturation and the development of the physiologic milieu required for reproduction but also acts as a harboring place for Gc proliferation and differentiation. The testes in all mammals are paired encapsulated ovoid organs consisting of seminiferous tubules separated by interstitial tissue (Figure 9.2). The testis displays two major compartments: the interstitial compartment (interstitium) and the seminiferous tubule compartment. The interstitium contains the blood vessels, lymphatics, nerve fibers and appreciable numbers of mast cells and macrophages. It also contains Leydig cells (Lc), scattered in groups that lie near blood vessels, and seminiferous tubules (de Kretser and Kerr, 1994). This cell is the major source of androgenic steroid hormones most important being T. Most of the testis mass is contributed by the seminiferous tubule inside which the steps of spermatogenesis occur. The tubules are two ended, convoluted loops, with both ends opening at rete testes. Each seminiferous tubule is lined on its inside by the seminiferous epithelium, which contains two kinds of cells–male Gc and Sc. This seminiferous epithelium is surrounded by noncellular material, myoid cells, collagen fibers and endothelial cells. The peritubular myoid cells (PTc) are responsible for the irregular contractions of the seminiferous tubules that are responsible for the propulsion of fluid secreted by the Sc together with testicular spermatids into the lumen and through the tubular network to the region of the rete testis (Clermont, 1958; Ross and Long, 1966). Sertoli cells bear receptors for FSH as well as T and act like mother cells feeding, supporting and allowing Gc to differentiate within its branched cytoplasmic extension. *Therefore, it is logical to consider that lack of sperm production inspite of required hormonal milieu in the testis (as found in some forms of idiopathic infertility) may be a consequence of defect(s) in Sc function.*

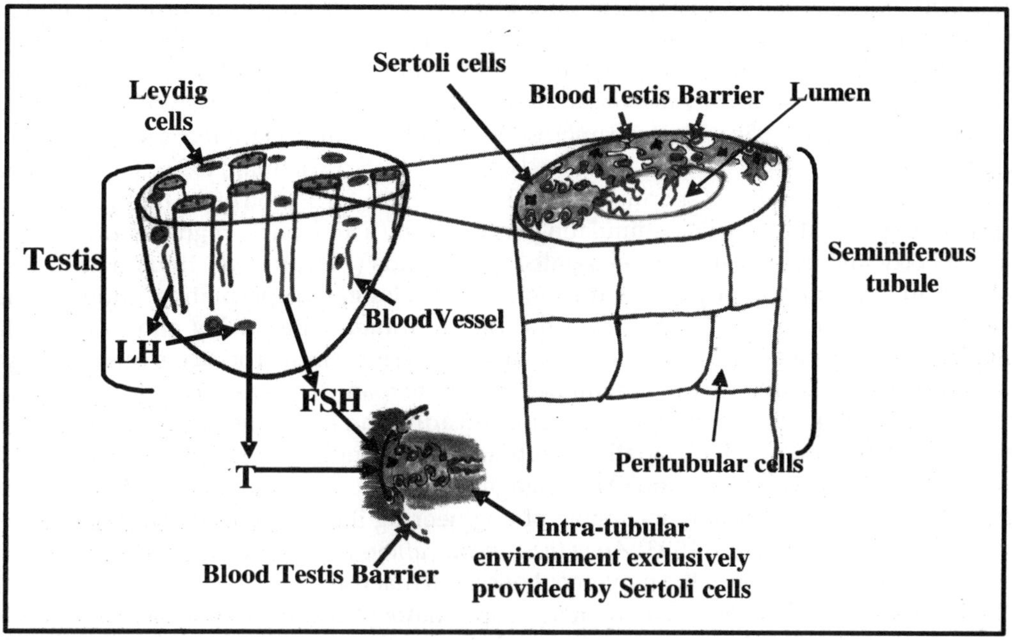

Figure 9.2: A Cartoon Depicting the Organization of Various Components of the Testis

Sertoli Cell and its Functions

The Sc, named after Enrico Sertoli who described the cell in 1865, is essential for spermatogenesis (Sertoli, 1865). This cell is also known by many names such as ramifying cell, sustentacular cell, supporting cell or nurse cell. Within the mammalian testes the Sc is a tall irregular columnar cell extending perpendicularly through the seminiferous epithelium (de Kretser and Kerr, 1994). Sertoli cells participate in translocating, anchoring and positioning of spermatogenic cells within and releasing spermatozoa from the seminiferous epithelium (Russell and Steinberger, 1989). Sertoli cells undergo extensive structural and functional changes during the process of postnatal and pubertal development. The structural evolution of Sc can be directly related to changes in the adjacent Gc (Gondos, 1993). Sertoli cells undergo extensive cytoplasmic differentiation during the first few weeks after birth in rat as well as human (Nistal and Paniagua, 1983). The structural differentiation occurring during the prepubertal period correlates with the appearance of various physiologic and metabolic changes in the developing testis at this time. Numerous cytoplasmic projections extending between maturing Gc and into the seminiferous tubule lumen are seen in fully differentiated Sc at puberty.

The immature Sc differs both in morphology and biochemical activity compared to the fully differentiated cell. At the start of puberty, the tight junctions of the barrier are formed and the Sc display more differentiated biochemical activities manifested by the production of transferrin (Skinner and Griswold, 1980) and androgen binding protein (Tindall *et al.*, 1975). Adjacent Sc form tight or occluding junctions at their baso-lateral surfaces resulting into the formation of blood-testis barrier (Setchell *et al.*, 1988). Establishment of the occlusive junction results in the formation of two compartments within seminiferous epithelium, a basal compartment containing spermatogonia and early spermatocytes and an adluminal compartment containing meiotic spermatocytes and spermatids at various steps of spermatogenesis. This results in sequestering of the meiotic and post-meiotic Gc in an adluminal compartment, which is inaccessible to macromolecules from the serum or lymph. Another consequence of such a barrier is that Sc must be involved in the transport of substances from the basal compartment to the germinal cells present in the adluminal compartment. *Thus, the multiple secretion products of the Sc determine this local environment and can influence meiosis as well as spermatid and spermatocyte development* (Griswold, 1995b). Electron microscopic studies with rodent testis sections have shown that degenerating spermatogenic cells are indeed engulfed by Sc (Miething, 1992). The Sc also normally process a considerable volume of spermatid cytoplasm after the formation of the tubulobular complexes and just before spermiation. Follicle stimulating hormone, by acting on Sc, regulates the internalization of residual body and cytoplasts from elongated spermatids (Pineau *et al.*, 1991). The Sc also play a crucial role in spermiation, a process through which elongated spermatids undergo their final maturation and are released from Sc into the tubule lumen ((Beardsley and O'Donnell, 2003). Spermiation is a protracted, complex process that involves 1) the removal of excess spermatid cytoplasm from around the flagella to form the streamlined spermatozoan; 2) extension of the spermatid into the tubule lumen; 3) progressive removal of specialized adhesion junctions, including the ectoplasmic specialization; 4) formation and degradation of tubulobulbar complexes; 5) retraction of the Sc cytoplasm from around the spermatid head; and 6) disengagement of the spermatid from the Sc into the tubule lumen. The spermatid is not only incapable of engineering its own release but in fact is relatively passive in this process (Russell, 1993) *All of these modifications in preparation for disengagement involve active participation of the Sc*. Since the movement of serum macromolecules from the interstitium to the inner or adluminal compartment, where meiotic spermatocytes and spermatids reside, is barred by the blood–testis barrier, it is generally assumed that nutrients received by Gc of the adluminal compartment must be provided by Sc. The delivery of nutritive substances such as sugars, amino acids, lipids, metallic elements, etc. to the germinal cells is generally considered plausible because of the close association of Sc processes with each and every Gc. Production of various substances, important for Gc development and differentiation, by Sc has been reported. For example, Sc produce cellular retinol-binding protein and lecithin-retinol acyltransferase, which are involved in transport of vitamin A towards the testicular lumen. Vitamin A deficiency in mammals results in the loss of Gc on the adluminal side of the blood-testis barrier, suggesting a major role of Sc in trafficking of some molecules to Gc (Schmitt and Ong, 1993). Similarly, the transport of iron by Sc to Gc via transferrin has also been reported (Gilmont *et al.*, 1990). Lactate, the preferred energy substrate used by Gc, is known to be produced by Sc in response to FSH *in vitro* (Jutte *et al.*, 1983; Mita and Hall, 1982). It is suggested that Sc transport lactate to Gc by monocarboxylate transporters (Goddard *et al.*, 2003). Conversion of T to estradiol, a survival factor for Gc, involves enzyme aromatase which is present in the Sc (Ando *et al.*, 2001; Pentikainen *et al.*, 2000). In addition to transporting nutritive substances to the Gc, the Sc produce and secrete various proteins. These secretory products create the local environment conducive for Gc development. Knowledge regarding most of the secreted products of Sc have been gathered from studies in rats and few other species. Sertoli cells dedicate a reasonably large percentage of their activity to the synthesis and secretion of glycoproteins (Russell, 1993). The first glycoprotein product of Sc to be characterized to any extent was androgen binding protein (ABP, Hagenas *et al.*, 1975). But the biochemical evidence for the proposed "nurse cell" function of Sc was provided by the discovery of transferrin as one of the major products of this cell. A model for iron transport from the basal surface of Sc to spermatocytes and spermatids has been proposed (Gilmont *et al.*, 1990).

The proteins secreted by Sc have been placed in four categories by Griswold according to their known biochemical properties (Griswold, 1995a; Griswold, 1993). The first category includes the transport or bioprotective proteins; the second category of secreted proteins includes proteases and protease inhibitors; the third category includes the basement membrane proteins and finally, the regulatory proteins.

Spermatogenesis

Spermatogenesis represents the process by which immature Gc termed as spermatogonia undergo a complex series of cytological events resulting in the formation of mature spermatozoa (de Kretser and Kerr, 1994). This process takes place within the seminiferous epithelium, where the Sc form an intimate physical relationship with the Gc (Fawcett, 1975; Russell, 1993). The unique structural characteristics of the Sc physically support migratory population of Gc, which proliferate from the base of the tubules and move progressively towards the lumen of the tubules as they differentiate. The physical and biological relationships between Sc and Gc are reflected in observations that have led to the conclusion that the number of Sc in a testis determines the spermatogenic output in a range of mammals, including humans (Sharpe *et al.*, 2003).

Spermatogenesis can be divided into three major phases (1) proliferation and differentiation of spermatogonia, (2) meiosis in which genetic material is recombined and segregated in spermatocytes, (3) spermiogenesis which represents a complex metamorphosis involved in the transformation of round spermatids arising from the final division of meiosis into the complex structure of the spermatozoon. All these phases of spermatogenesis are homed within the Sc cytoplasmic extensions and influenced by the secretions of the Sc.

Hormonal Dependency of Spermatogenesis

The adult testis has two important roles, namely the production of spermatozoa and the secretion of T, which is needed for the expression of secondary sexual characteristics. These functions are dependant on stimulation by pituitary gonadotropins, FSH and LH. Testosterone is secreted by the Lc under LH stimulation and is essential for promoting spermatogenesis (McLachlan *et al.*, 1996). Although there is general agreement that T is essential for the maintenance of normal spermatogenesis, three major areas of dispute and doubt remain; 1) dispute about whether T alone can maintain quantitavely normal spermatogenesis, 2) dispute about the level of T required to maintain spermatogenesis and its correlation with the normal testicular T levels and 3) the exact mode of action of T which is responsible for its pivotal role in spermatogenesis. Effect of T on spermatogenesis has been studied using the T-suppressed model and quantitative Gc counting methods (McLachlan *et al.*, 1994; Meachem *et al.*, 1998). It is known that T acts at several sites during spermatogenesis but, at some sites, it is not possible to conclusively separate the actions of T and FSH.

While there is general agreement that T is essential for promoting spermatogenesis, controversy persists as to whether FSH is absolutely essential for the maintenance of adult spermatogenesis (McLachlan *et al.*, 1996), though its role in the development of the immature testis is well recognized (Orth, 1984). The data obtained from the initiation of testicular function in hypogonadotrophic men indicates that FSH is required for the induction of spermatogenesis in a significant proportion (MacLeod *et al.*, 1966; Santen and Paulsen, 1973). However, the results of studies in rodent models of FSH deficiency have raised the possibility that FSH is not required for the completion of spermatogenesis. In the hypogonadal (hpg) mouse model, where failure of GnRH secretion causes failure of sexual maturation, treatment with T alone restored full spermatogenesis despite smaller than normal testes (Handelsman *et al.*, 1999; Singh *et al.*, 1995). In mice with the targeted disruption of the β subunit of FSH, spermatogenesis proceeds to completion and the mice are fertile despite a significant reduction in testis size (Kumar *et al.*, 1997). The knockout model of the FSHR gene produced a similar phenotype to the FSH β subunit knockout model but also showed impaired progression in spermatogenesis, and defects in sperm chromatin condensation (Dierich *et al.*, 1998; Krishnamurthy *et al.*, 2000). In men, inactivating mutations of the FSH receptor (FSHR) gene can result in a variable reduction of sperm counts, some of which are associated with fertility (Tapanainen *et al.*, 1997). Interestingly, a mutation in the FSHβ subunit gene resulted in azoospermia (Phillip *et al.*, 1998), in contrast to the normal fertility maintained in the FSH β knockout mouse (Kumar *et al.*, 1997).

Whereas FSH and T are capable of acting independently, it is clear that in terms of potentiating spermatogenesis, their actions can be synergistic. For example, FSH may act to potentiate T stimulation with FSH usually leading to increased androgen production and AR expression (O'Shaughnessy *et al.*, 1992; Vihko *et al.*, 1991). Recently, increases in cAMP have been shown to activate androgen receptor (AR) to move to nucleus and induce expression of the prostrate specific antigen (PSA) gene in androgen independent cells (Sadar, 1999). Furthermore, using hypophysectomised animals, the addition of FSH and T together are able to maintain higher levels of Gc compared to addition of either hormone alone (Sharpe, 1994). Recent report using hpg mice expressing transgenic FSH, alone or in combination with T, has shown that FSH actions combine synergistically with androgens to have a marked effect on both meiotic and post meiotic Gc maturation (Haywood *et al.*, 2003).

Role of Sertoli Cells in Spermatogenesis

Sertoli cells are the first cells to differentiate recognizably in the indifferent fetal gonad, an event which enables seminiferous cord formation, prevention of germ-cell entry into meiosis and differentiation and function of the Leydig cells (Mackay, 2000). The Sc also ensure regression of the Mullerian ducts via secretion of anti-Mullerian hormone (Josso *et al.*, 2001; Mackay, 2000). When these events have occurred, and some time has passed, the role of the Sc switches during puberty to the support of spermatogenesis. The majority of Sc differentiated functions appear at the onset of puberty and increase to optimal levels upon the completion of puberty in the adult animal. *Without the physical and metabolic support of the Sc, Gc differentiation, meiosis and transformation into spermatozoa would not occur* (Sharpe, 1994). Therefore, Sc have an essential role in the control and maintenance of the process of spermatogenesis. Moreover, *the number of Sc determine the number of Gc that can be supported through spermatogenesis and hence the Sc numerically determine the extent of sperm production* (Orth et al., 1988; Sharpe, 1994; Sharpe *et al.*, 1999).

Hormonal Regulation of Sertoli Cells

Since Sc possess receptors for both FSH and T, and Gc development is solely depends on Sc, interaction of these hormones (FSH and T) with Sc has a crucial role in the initiation and maintenance of spermatogenesis. Interaction of Sc with FSH leads to coupling of the FSHR to G_{soc}, which, in turn, stimulates adenylyl cyclase and results in the production of cAMP (Griswold *et al.*, 1995; Heckert and Griswold, 2002; Rannikko *et al.*, 1996). Increased concentrations of cAMP release the catalytic subunit of protein kinase A (PKA) from regulatory subunits, allowing for the phosphorylation of numerous cytoplasmic and nuclear proteins including the cAMP responsive element binding protein (CREB). The phosphorylated form of CREB (P-CREB) acts as a transcription factor and directly stimulates many genes (Meyer and Habener, 1993). In addition, CREB stimulates the expression of transcription factors c-fos, jun-D, C/EBP-β and CREB itself, which then modulate other genes. (Figure 9.3). The consequences of FSH action on Sc are different depending on the developmental stages. FSH produced by the fetal and early postnatal pituitary is of critical importance in stimulating the proliferation of Sc. FSH plays a central role in regulating Sc proliferation in the fetus. Specific suppression of fetal FSH dramatically reduces the percentage of rat Sc entering the S phase of the cell cycle (Orth, 1984). In

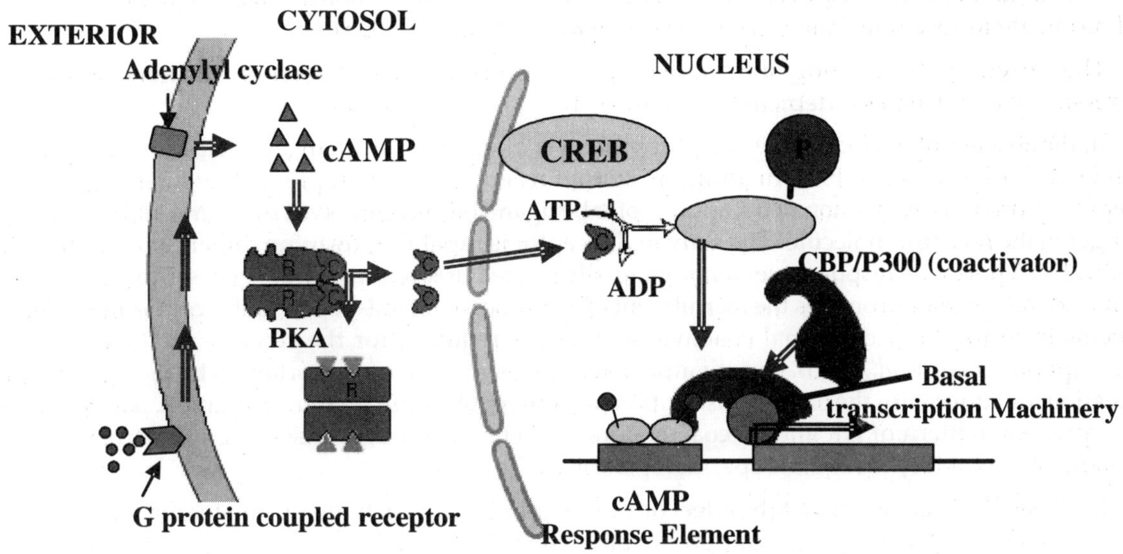

Figure 9.3: Schematic Representation of the FSH Signaling Pathway

higher primates, involvement of gonadotropins in fetal Sc proliferation has been suggested by the findings that anencephaly in man (Baker and Scrimgeour, 1980) and fetal hypophysectomy in the rhesus monkey (Gulyas *et al.*, 1977), result at term in a marked reduction in testicular size. At the time of puberty, primate Sc mitosis is driven by the gonadotropic hormones is confirmed by the finding that proliferation of Sc may be induced precociously in juvenile rhesus monkeys in which the pituitary-testicular axis is prematurely activated by intermittent stimulation with GnRH (Marshall and Plant, 1996).

Two dimensional polyacrylamide gel electrophoresis (2D PAGE) and autoradiography have been used to show that within five minutes of FSH treatment, new phosphoproteins appear in lysates of culture Sc (Ireland *et al.*, 1986). Few of these phosphorylated proteins have been identified. It is known that a structural protein, vimentin, is phosphorylated in culture Sc in response to FSH (DePhilip and Kierszenbaum, 1982; Spruill *et al.*, 1983). As mentioned earlier, FSH, through cAMP, has been shown to increase the Sc production and secretion of a number of metabolites, for example androgen binding protein (Tindall and Means, 1976), transferrin (Skinner and Griswold, 1980), lactate (Jutte *et al.*, 1983), estradiol 17-B and metabolites of androstenedione (Welsh and Wiebe, 1976). Oligonucleotide microarray analysis, to characterize the changes in gene expression in FSH-treated Sc, has revealed several genes which were either up or down regulated by FSH (McLean *et al.*, 2002).

The principle steroidal androgens, T and its metabolite DHT predominantly mediate their biological effects through binding to the AR, a 110-kDa ligand-inducible nuclear receptor that regulates the expression of target genes through binding to an androgen response element (Heinlein and Chang, 2002a). The binding of T or DHT to AR induces receptor dimerization, facilitating the ability of AR to bind to its cognate response element and recruit coregulators to promote the expression of target genes.

However, androgens have also been reported to induce the rapid activation of kinase-signaling cascades and modulate intracellular calcium levels (Heinlein and Chang, 2002b).

The current understanding of the molecular mechanism by which steroid receptors alter gene expression of target genes is depicted in (Figure 9.4).

In the absence of hormone, the receptor protein is maintained in an inactive state by association with heat shock proteins (HSP). In addition, steroid receptors can be repressed by interaction with corepressor proteins. Activation of the aporeceptor by ligand binding involves structural and functional changes in the receptor molecule. The activated receptor is capable of forming dimers and binding to specific DNA hormone response elements, generally located upstream of target genes. In addition, the ligand-bound receptor promotes the recruitment of a number of ligand-dependant receptor-interacting proteins including transcriptional coactivators that are required for the receptor to fully activate transcription (Onate *et al.*, 1995). Coactivators have been shown to be important in bridging receptor activation functions with the basal transcription machinery to stimulate transactivation. Alteration in the expression pattern of AR and its coactivators may change the response of Sc to androgens and consequently its ability to regulate spermatogenesis.

Previously it was shown that the effect of T on Sc proliferation in higher primates may differ from that in rodents, with T exhibiting an antiproliferative effect on the rat Sc. (Buzzard *et al.*, 2003). However, the view that FSH is of prime importance in rodent Sc proliferation has been challenged by new findings that indicate that androgens can also play an important role in determining Sc number in mice (Haywood *et al.*, 2003; Johnston *et al.*, 2004). It has been shown that androgens might be particularly

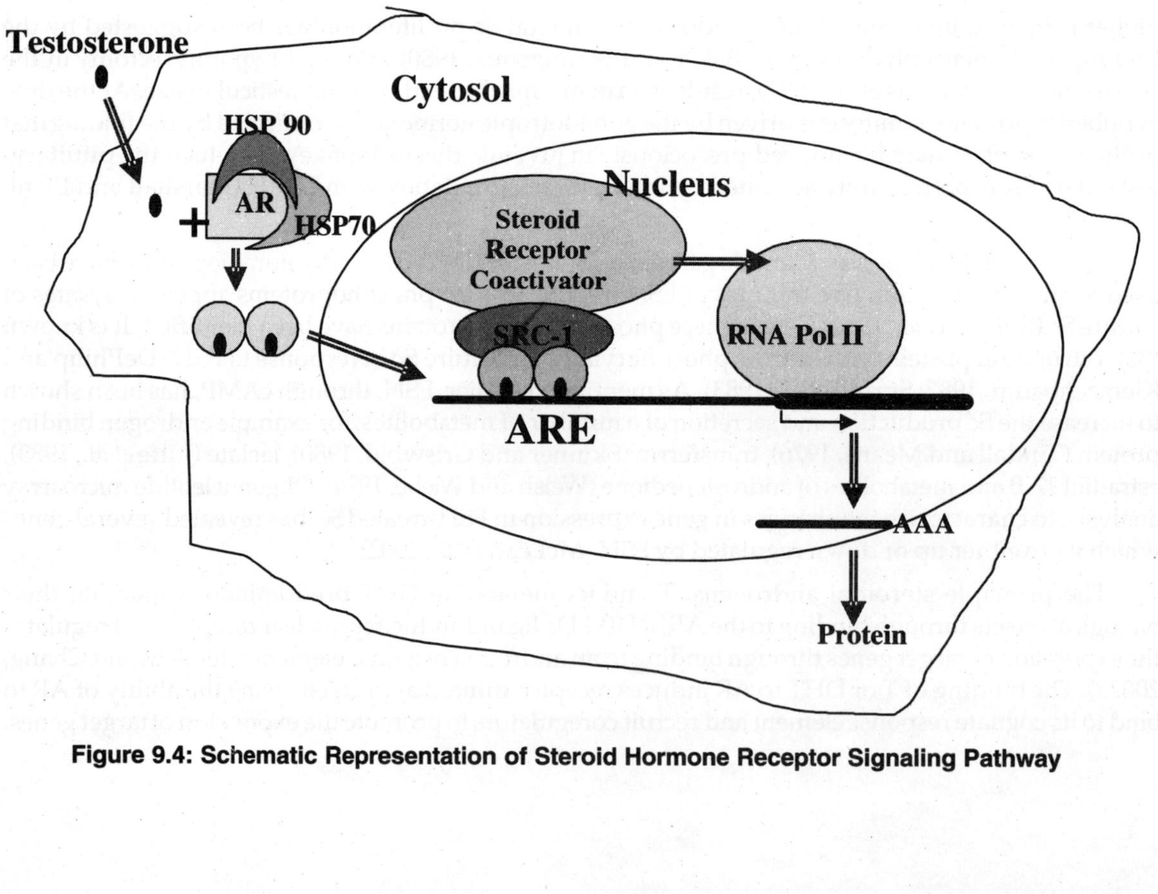

Figure 9.4: Schematic Representation of Steroid Hormone Receptor Signaling Pathway

important in the fetal period when FSH appears unnecessary for Sc proliferation (Baker and Scrimgeour, 1980; Johnston *et al.*, 2004). Johnston *et al.* (2004) have shown that testicular feminized mice (*Tfm*), which lack functional AR, exhibit an approximately 50-70 per cent reduction in Sc numbers. Similar findings were observed in complete AR knock out mice; however, mice with selective knock out of AR in Sc develop apparently normal number of Sc (De Gendt *et al.*, 2004). Studies of Atanassova *et al.* (2005) using various combinations of hormonal manipulations or treatments resulting in selective block of androgen action provide strong support for an effect of neonatal androgens on Sc proliferation in the rat (Atanassova *et al.*, 2005). Despite these findings in the rodents, species specificity in the action of androgens may not be ruled out completely.

The studies of more potent direct effect of T on certain functions of cultured Sc are lacking. This is so because, culture studies generally utilize Sc from immature rats (because of the methodological problems culturing Sc from adult rats) which have relatively low level of AR. Stimulation of ABP and transferrin secretion upon T treatment by Sc has been reported. Expression of AR in Sc is stage-specific, but how this expression pattern and T action correlate with the function of Sc is not well understood. Only a limited number of androgen-regulated genes in Sc have been studied (Benbrahim-Tallaa *et al.*, 2002; Sadate-Ngatchou *et al.*, 2004; Sutton *et al.*, 1998).

Necessity of Sertoli Cell Studies to Delineate Causes Underlying Idiopathic Infertility

As mentioned earlier, in human and monkey, LH, FSH and T are high during infancy, 0-4 months of age (Burger *et al.*, 1991; Mann and Fraser, 1996), as well as puberty (Chada *et al.*, 2003) but spermatogenesis is initiated only at puberty. In rats, although there is no major difference in the gonadotropin levels between day 9 and 60 of age, progression of spermatogenesis is evident only at day 11-15 of postnatal life (Dohler and Wuttke, 1975; Lee *et al.*, 1975; Yang *et al.*, 1990). Such introvert behavior of Sc during postnatal/infant period in presence of sufficient hormonal milieu attracts attention to investigate inability of immature Sc to support spermatogenesis. Several male infertile individuals have normal or high levels of gonadotropins (like during infancy) but they do not produce sperms, a situation very much similar to that found during infancy. It can be hypothesized that the Sc from such individuals either 1) fail to transduce the hormone signals resulting in the ability of these Sc to provide a conducive milieu for Gc development and differentiation or 2) these cells are unable to synthesize the proteins necessary to support spermatogenic progression or produce inhibitors to limit spermatogenesis (Figure 9.5 a, b, c, d). These two hypotheses may not necessarily be mutually exclusive. It is important to investigate whether these hypotheses hold true in Sc from such infertile males in which a robust hormonal drive exists.

It is difficult to obtain Sc from infertile individuals. However, procedures to culture of Sc from the infant as well as adult rodents and monkeys have now been established (Griswold, 1995b; Majumdar *et al.*, 1995; Majumdar *et al.*, 1998; Welsh and Wiebe, 1976). Sertoli cells from 9 days old rats or more importantly 3-4 months old monkeys may act as surrogates of infertile adults for investigative research to analyze status of a) hormone signaling pathways and b) protein secretions of such Sc. Sertoli cells isolated from the spermatogenetically active stage of development may be used as a yardstick of mature hormonally responsive Sc and their functional comparison with infant Sc may help us in identifying the deficits, if any, residing in the infant Sc (a surrogate of infertile male). Preliminary studies in our laboratory have shown that Sc isolated from infant and pubertal rats as well as monkeys differ in their functional response to hormones (Majumdar *et al.*, 2004; Sarda *et al.*, 2004). A differential 2D PAGE profile of Sc secretory proteins from the two developmental groups has also been observed

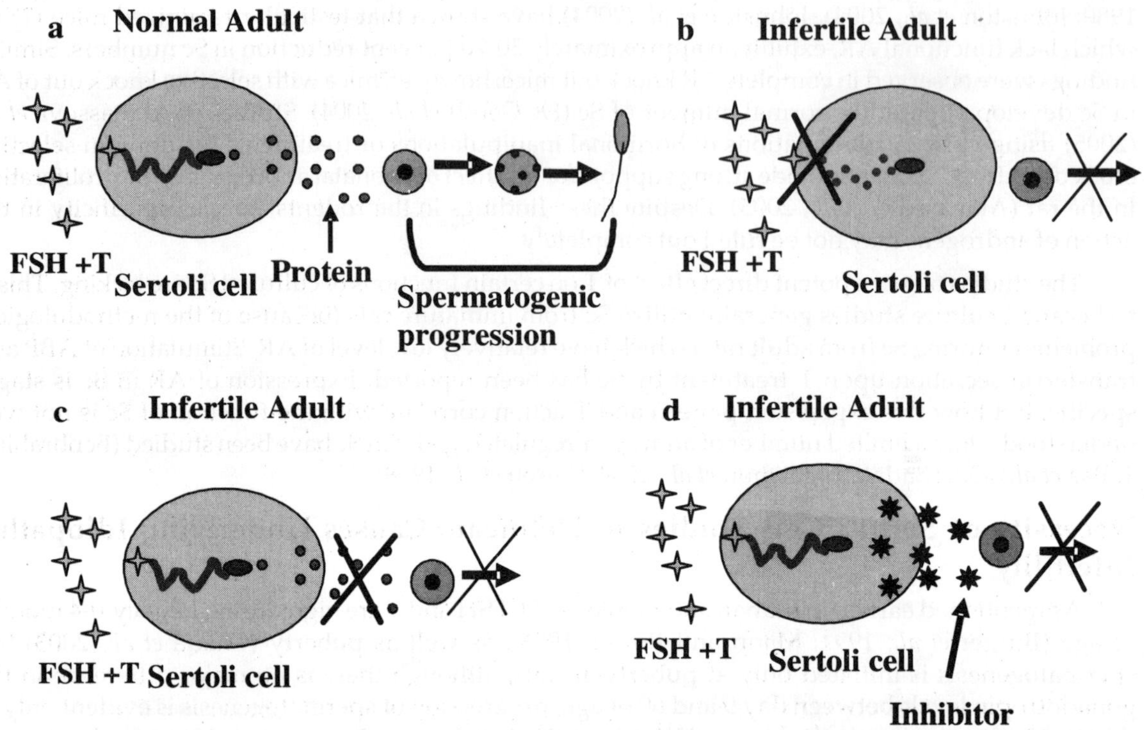

Figure 9.5: Proposed Hypotheses of Sertoli Cell Deficits in the Infertile Male
(a) Normal adult Sc, in the presence of hormone signaling produce factors which are necessary for spermatogenic progression; (b) The Sc from the infertile adult fail to transduce the hormone signals and hence cannot produce factors necessary for spermatogenesis. or (c) In presence of adequate hormonal signaling, the Sc from the infertile male fail to produce protein products necessary for spermatogenesis or (d) they produce inhibitors which prevent spermatogenesis.

(Majumdar *et al.*, 2003; Y.Sangeetadevi and Majumdar, 2004). With the recent availability of high throughput tools, comparative genomic analysis of the Sc from infant and adult may be carried out by techniques like DNA micro array, differential display and RT PCR. Such studies coupled with analysis of the Sc proteome by 2D PAGE followed by MALDI- TOF analysis in the two age groups would result in a two-pronged approach to address the problem of Sc-related infertility.

References

Ando S, Sirianni R, Forastieri P, Casaburi I, Lanzino M, Rago V, Giordano F, Giordano C, Carpino A, and Pezzi V. (2001). Aromatase expression in prepuberal Sertoli cells: effect of thyroid hormone. Mol. Cell. Endocrinol., 178, 11-21.

Atanassova NN, Walker M, McKinnell C, Fisher JS, and Sharpe RM. (2005). Evidence that androgens and oestrogens, as well as follicle-stimulating hormone, can alter Sertoli cell number in the neonatal rat. J. Endocrinol., 184, 107-117.

Baker TG, and Scrimgeour JB. (1980). Development of the gonad in normal and anencephalic human fetuses. J. Reprod. Fertil., 60, 193-199.

Beardsley A, and O'Donnell L. (2003). Characterization of normal spermiation and spermiation failure induced by hormone suppression in adult rats. Biol. Reprod., 68, 1299-1307.

Benbrahim-Tallaa L, Tabone E, Tosser-Klopp G, Hatey F, and Benahmed M. (2002). Glutathione S-transferase alpha expressed in porcine Sertoli cells is under the control of follicle-stimulating hormone and testosterone. Biol. Reprod., 66, 1734-1742.

Bergada I, Rojas G, Ropelato G, Ayuso S, Bergada C, and Campo S. (1999). Sexual dimorphism in circulating monomeric and dimeric inhibins in normal boys and girls from birth to puberty. Clin. Endocrinol. (Oxf.), 51, 455-460.

Burger HG, Yamada Y, Bangah ML, McCloud PI, and Warne GL. (1991). Serum gonadotropin, sex steroid, and immunoreactive inhibin levels in the first two years of life. J. Clin. Endocrinol. Metab., 72, 682-686.

Buzzard JJ, Farnworth PG, de Kretser DM, O'Connor AE, Wreford NG, and Morrison JR. (2003). Proliferative phase sertoli cells display a developmentally regulated response to activin in vitro. Endocrinol., 144, 474-483.

Chada M, Prusa R, Bronsky J, Kotaska K, Sidlova K, Pechova M, and Lisa L. (2003). Inhibin B, follicle stimulating hormone, luteinizing hormone and testosterone during childhood and puberty in males: changes in serum concentrations in relation to age and stage of puberty. Physiol. Res., 52, 45-51.

Clermont Y. (1958). Contractile elements in the limiting membrane of the seminiferous tubules of the rat. Exp. Cell Res., 15, 438-440.

De Gendt K, Swinnen JV, Saunders PT, Schoonjans L, Dewerchin M, Devos A, Tan K, Atanassova N, Claessens F, Lecureuil C, Heyns W, Carmeliet P, Guillou F, Sharpe RM, and Verhoeven G. (2004). A Sertoli cell-selective knockout of the androgen receptor causes spermatogenic arrest in meiosis. Proc. Natl. Acad. Sci. USA, 101, 1327-1332.

de Kretser DM, and Kerr JB. (1994). The cytology of the testis. *In*: Knobil E, and Neill JD. (Eds.), *The Physiology of Reproduction*, Raven Press, New York, pp 1177-1290.

DePhilip RM, and Kierszenbaum, AL. (1982). Hormonal regulation of protein synthesis, secretion, and phosphorylation in cultured rat Sertoli cells. Proc. Natl. Acad. Sci., USA, 79, 6551-6555.

Dierich A, Sairam MR, Monaco L, Fimia GM, Gansmuller A, LeMeur M, and Sassone-Corsi P. (1998). Impairing follicle-stimulating hormone (FSH) signaling in vivo: targeted disruption of the FSH receptor leads to aberrant gametogenesis and hormonal imbalance. Proc. Natl. Acad. Sci. USA, 95, 13612-13617.

Dohler KD, and Wuttke W. (1975). Changes with age in levels of serum gonadotropins, prolactin and gonadal steroids in prepubertal male and female rats. Endocrinol., 97, 898-907.

Fawcett DW. (1975). Gametogenesis in the male: prospects for its control. Symp. Soc. Dev. Biol., 25-53.

Foresta C, Bettella A, Moro E, Roverato A, Merico M, and Ferlin A. (2001). Sertoli cell function in infertile patients with and without microdeletions of the azoospermia factors on the Y chromosome long arm. J. Clin. Endocrinol. Metab., 86, 2414-2419.

Georgopoulos NA, Pralong FP, Seidman CE, Seidman JG, Crowley WF, Jr. and Vallejo M. (1997). Genetic heterogeneity evidenced by low incidence of KAL-1 gene mutations in sporadic cases of gonadotropin-releasing hormone deficiency. J Clin Endocrinol. Metab., 82, 213-217.

Gilmont RR, Coulter GH, Sylvester SR, and Griswold MD. (1990). Synthesis of transferrin and transferrin mRNA in bovine Sertoli cells in culture and in vivo: sequence of partial cDNA clone for bovine transferrin. Biol. Reprod., 43, 139-150.

Goddard I, Florin A, Mauduit C, Tabone E, Contard P, Bars R, Chuzel F, and Benahmed M. (2003). Alteration of lactate production and transport in the adult rat testis exposed in utero to flutamide. Mol. Cell. Endocrinol., 206, 137-146.

Gondos B. (1993). Ultrastructure of developing and malignant germ cells. Eur. Urol., 23, 68-74.

Grimaldi P, Rossi P, Dolci S, Ripamonti CB, and Geremia R. (2002). Molecular genetics of male infertility: stem cell factor/c-kit system. Am. J. Reprod. Immunol., 48, 27-33.

Griswold MD. (1995a). Interactions between germ cells and Sertoli cells in the testis. Biol. Reprod., 52, 211-216.

Griswold MD. (1995b). Interactions between germ cells and Sertoli cells in the testis. Biol. Reprod., 52, 211-216.

Griswold MD, Heckert L, and Linder C. (1995). The molecular biology of the FSH receptor. J. Steroid. Biochem. Mol. Biol., 53, 215-218.

Griswold M. (1993). Protein Secretion by Sertoli cells: General Considerations. *In*: Russell L, and Griswold M. (Eds.), *The Sertoli Cell*, Cache River Press, Clearwater, FL, pp 493-508.

Gulyas BJ, Tullner WW, and Hodgen GD. (1977). Fetal or maternal hypophysectomy in rhesus monkeys (*Macaca mulatta*): effects on the development of testes and other endocrine organs. Biol. Reprod., 17, 650-660.

Hagenas L, Ritzen EM, Plooen L, Hansson V, French FS, and Nayfeh SN. (1975). Sertoli cell origin of testicular androgen-binding protein (ABP). Mol. Cell. Endocrinol., 2, 339-350.

Handelsman DJ, Spaliviero JA, Simpson JM, Allan CM, and Singh J. (1999). Spermatogenesis without gonadotropins: maintenance has a lower testosterone threshold than initiation. Endocrinol., 140, 3938-3946.

Haywood M, Spaliviero J, Jimemez M, King NJ, Handelsman DJ, and Allan CM. (2003). Sertoli and germ cell development in hypogonadal (hpg) mice expressing transgenic follicle-stimulating hormone alone or in combination with testosterone. Endocrinol., 144, 509-517.

Heckert LL, and Griswold MD. (2002). The expression of the follicle-stimulating hormone receptor in spermatogenesis. Recent Prog. Horm. Res., 57, 129-148.

Heinlein CA, and Chang C. (2002a). Androgen receptor (AR) coregulators: an overview. Endocr. Rev., 23, 175-200.

Heinlein CA, and Chang C. (2002b). The roles of androgen receptors and androgen-binding proteins in nongenomic androgen actions. Mol. Endocrinol., 16, 2181-2187.

Ireland ME, Rosenblum BB, and Welsh MJ. (1986). Two-dimensional gel analysis of Sertoli cell protein phosphorylation: effect of short term exposure to follicle-stimulating hormone. Endocrinol., 118, 526-532.

Johnston H, Baker PJ, Abel M, Charlton HM, Jackson G, Fleming L, Kumar TR and O'Shaughnessy PJ. (2004). Regulation of Sertoli cell number and activity by follicle-stimulating hormone and androgen during postnatal development in the mouse. Endocrinol., 145, 318-329.

Josso N, di Clemente N, and Gouedard L. (2001). Anti-Mullerian hormone and its receptors. Mol. Cell. Endocrinol., 179, 25-32.

Jutte NH, Jansen R, Grootegoed JA, Rommerts FF, and van der Molen HJ. (1983). FSH stimulation of the production of pyruvate and lactate by rat Sertoli cells may be involved in hormonal regulation of spermatogenesis. J. Reprod. Fertil., 68, 219-226.

Krishnamurthy H, Danilovich N, Morales CR, and Sairam MR. (2000). Qualitative and quantitative decline in spermatogenesis of the follicle-stimulating hormone receptor knockout (FORKO) mouse. Biol. Reprod., 62, 1146-1159.

Kumar TR, Wang Y, Lu N, and Matzuk MM. (1997). Follicle stimulating hormone is required for ovarian follicle maturation but not male fertility. Nat. Genet., 15, 201-204.

Lee VW, de Kretser DM, Hudson B, and Wang C. (1975). Variations in serum FSH, LH and testosterone levels in male rats from birth to sexual maturity. J. Reprod. Fertil., 42, 121-126.

Mackay S. (2000). Gonadal development in mammals at the cellular and molecular levels. Int. Rev. Cytol., 200, 47-99.

MacLeod J, Pazianos A, and Ray B. (1966). The restoration of human spermatogenesis and of the reproductive tract with urinary gonadotropins following hypophysectomy. Fertil. Steril., 17, 7-23.

Majumdar S S, Plant, T M, and Sangeetadevi Y. (2003). Differential secretion of proteins by Sertoli cells from spermatogenitically active and quiescent testis of the rhesus monkey (*Macaca mulatta*). Biol. Reprod., 66 Suppl 1, 335-336.

Majumdar SS, Sarda K, Onate SA, Friedman RL, and Plant TM. (2004). Limited Sertoli cell expression of androgen and FSH receptors in the rhesus monkey during infancy may underlie the failure of spermatogonial stem cells to differentiate in the presence of an adult-like gonadotropin milieu at this stage of development. 86th Annual Meeting of the Endocrne Society, New Orleans, USA Abstr.# P-3 318.

Majumdar SS, Tsuruta J, Griswold MD, and Bartke A. (1995). Isolation and culture of Sertoli cells from the testes of adult Siberian hamsters: analysis of proteins synthesized and secreted by Sertoli cells cultured from hamsters raised in a long or a short photoperiod. Biol. Reprod., 52, 658-666.

Majumdar SS, Winters SJ, and Plant TM. (1998). Procedures for the isolation and culture of Sertoli cells from the testes of infant, juvenile, and adult rhesus monkeys (*Macaca mulatta*). Biol. Reprod., 58, 633-640.

Mann DR, and Fraser HM. (1996). The neonatal period: a critical interval in male primate development. J. Endocrinol., 149, 191-197.

Marshall GR, and Plant TM. (1996). Puberty occurring either spontaneously or induced precociously in rhesus monkey (*Macaca mulatta*) is associated with a marked proliferation of Sertoli cells. Biol. Reprod., 54, 1192-1199.

McLachlan RI, Wreford NG, Meachem SJ, de Kretser DM, and Robertson DM. (1994). Effects of testosterone on spermatogenic cell populations in the adult rat. Biol. Reprod., 51, 945-955.

McLachlan RI, Wreford NG, O'Donnell L, de Kretser DM, and Robertson DM. (1996). The endocrine regulation of spermatogenesis: independent roles for testosterone and FSH. J. Endocrinol., 148, 1-9.

McLean DJ, Friel PJ, Pouchnik D, and Griswold MD. (2002). Oligonucleotide microarray analysis of gene expression in follicle-stimulating hormone-treated rat Sertoli cells. Mol. Endocrinol., 16, 2780-2792.

Meachem SJ, Wreford NG, Stanton PG, Robertson DM, and McLachlan RI. (1998). Follicle-stimulating hormone is required for the initial phase of spermatogenic restoration in adult rats following gonadotropin suppression. J. Androl., 19, 725-735.

Meyer TE, and Habener JF. (1993). Cyclic adenosine 3',5'-monophosphate response element binding protein (CREB) and related transcription-activating deoxyribonucleic acid-binding proteins. Endocr. Rev., 14, 269-290.

Miething A. (1992). Germ-cell death during prespermatogenesis in the testis of the golden hamster. Cell Tissue Res., 267, 583-590.

Mita M, and Hall PF. (1982). Metabolism of round spermatids from rats: lactate as the preferred substrate. Biol. Reprod., 26, 445-455.

Naftolin F, Harris GW, and Bobrow M. (1971). Effect of purified luteinizing hormone releasing factor on normal and hypogonadotrophic anosmic men. Nature, 232, 496-497.

Nistal M, and Paniagua R. (1983). The postnatal development of the human Sertoli cells. Z. Mikrosk. Anat. Forsch., 97, 739-752.

O'Shaughnessy PJ, Bennett MK, Scott IS, and Charlton HM. (1992). Effects of FSH on Leydig cell morphology and function in the hypogonadal mouse. J. Endocrinol., 135, 517-525.

Onate SA, Tsai SY, Tsai MJ, and O'Malley BW. (1995). Sequence and characterization of a coactivator for the steroid hormone receptor superfamily. Science, 270, 1354-1357.

Orth JM. (1984). The role of follicle-stimulating hormone in controlling Sertoli cell proliferation in testes of fetal rats. Endocrinol., 115, 1248-1255.

Orth JM, Gunsalus GL, and Lamperti AA. (1988). Evidence from Sertoli cell-depleted rats indicates that spermatid number in adults depends on numbers of Sertoli cells produced during perinatal development. Endocrinol., 122, 787-794.

Pentikainen V, Erkkila K, Suomalainen L, Parvinen M, and Dunkel L. (2000). Estradiol acts as a germ cell survival factor in the human testis in vitro. J. Clin. Endocrinol. Metab., 85, 2057-2067.

Phillip M, Arbelle JE, Segev Y, and Parvari R. (1998). Male hypogonadism due to a mutation in the gene for the beta-subunit of follicle-stimulating hormone. N. Engl. J. Med., 338, 1729-1732.

Pineau C, Le Magueresse B, Courtens JL, and Jegou B. (1991). Study in vitro of the phagocytic function of Sertoli cells in the rat. Cell Tissue Res., 264, 589-598.

Plant TM. (1988). Neuroendocrine basis of puberty in the rhesus monkey (*Macaca mulatta*). *In*: Martini L and Ganong WF (Eds.), *Frontiers in Neuroendocrinology*, Raven Press Ltd., New York, pp 215-237.

Plant TM, and Marshall GR. (2001). The functional significance of FSH in spermatogenesis and the control of its secretion in male primates. Endocr. Rev., 22, 764-786.

Quinton R, Duke VM, de Zoysa PA, Platts AD, Valentine A, Kendall B, Pickman S, Kirk JM, Besser GM,

Jacobs HS, and Bouloux PM. (1996). The neuroradiology of Kallmann's syndrome: a genotypic and phenotypic analysis. J. Clin. Endocrinol. Metab., 81, 3010-3017.

Ramaswamy S, and Plant TM. (2001). Operation of the follicle-stimulating hormone (FSH)-inhibin B feedback loop in the control of primate spermatogenesis. Mol. Cell Endocrinol., 180, 93-101.

Rannikko A, Penttila TL, Zhang FP, Toppari J, Parvinen M, and Huhtaniemi I. (1996). Stage-specific expression of the FSH receptor gene in the prepubertal and adult rat seminiferous epithelium. J. Endocrinol., 151, 29-35.

Ross MH, and Long IR. (1966). Contractile cells in human seminiferous tubules. Science, 153, 1271-1273.

Russell LD, and Steinberger A. (1989). Sertoli cells in culture: views from the perspectives of an in vivoist and an in vitroist. Biol. Reprod., 41, 571-577.

Russell L. (1993). Role in spermiation. *In*: Russell L, and Griswold M. (Eds.), *The Sertoli Cell*, Cache River Press, Clearwater, FL, pp 269-302.

Sadar MD. (1999). Androgen-independent induction of prostate-specific antigen gene expression via cross-talk between the androgen receptor and protein kinase A signal transduction pathways. J. Biol. Chem., 274, 7777-7783.

Sadate-Ngatchou PI, Pouchnik DJ, and Griswold MD. (2004). Identification of testosterone-regulated genes in testes of hypogonadal mice using oligonucleotide microarray. Mol. Endocrinol., 18, 422-433.

Santen RJ, and Paulsen CA. (1973). Hypogonadotropic eunuchoidism. II. Gonadal responsiveness to exogenous gonadotropins. J. Clin. Endocrinol. Metab., 36, 55-63.

Sarda K, Shankar B, Dhup S, Nagarajan P, Venkatesan R, and Majumdar SS. (2004). Low androgen binding activity in the Sertoli cells of infant rhesus monkeys may be responsible for quiescence of spermatogenesis in the face of high hormone levels. 14th Annual Meeting of the Indian Society for the Study of Reproduction and Fertility, Bangalore.

Schmitt MC, and Ong DE (1993) Expression of cellular retinol-binding protein and lecithin-retinol acyltransferase in developing rat testis. Biol. Reprod., 49, 972-979.

Seminara SB, Hayes FJ, and Crowley WF, Jr. (1998). Gonadotropin-releasing hormone deficiency in the human (idiopathic hypogonadotropic hypogonadism and Kallmann's syndrome): pathophysiological and genetic considerations. Endocr. Rev., 19, 521-539.

Sertoli E. (1865). Dell'esistenza di particolari cellule ramificate nei canalicoli seminiferi del testicolo umano. Morgagni, 7, 40.

Setchell BP, Pollanen P, and Zupp JL. (1988). Development of the blood-testis barrier and changes in vascular permeability at puberty in rats. Int. J. Androl., 11, 225-233.

Sharpe RM. (1994). Regulation of spermatogenesis. *In*: Knobil E, and Neill JD. (Eds.), *The Physiology of Reproduction*, Raven Press Ltd., New York, pp 1363-1434.

Sharpe RM, McKinnell C, Kivlin C, and Fisher JS. (2003). Proliferation and functional maturation of Sertoli cells, and their relevance to disorders of testis function in adulthood. Reproduction, 125, 769-784.

Sharpe RM, Turner KJ, McKinnell C, Groome NP, Atanassova N, Millar MR, Buchanan DL, and Cooke

PS. (1999). Inhibin B levels in plasma of the male rat from birth to adulthood: effect of experimental manipulation of Sertoli cell number. J. Androl., 20, 94-101.

Singh J, O'Neill C, and Handelsman DJ. (1995). Induction of spermatogenesis by androgens in gonadotropin-deficient (hpg) mice. Endocrinol., 136, 5311-5321.

Skinner MK, and Griswold MD. (1980). Sertoli cells synthesize and secrete transferrin-like protein. J. Biol. Chem., 255, 9523-9525.

Spruill WA, Steiner AL, Tres LL, and Kierszenbaum AL. (1983). Follicle-stimulating hormone-dependent phosphorylation of vimentin in cultures of rat Sertoli cells. Proc. Natl. Acad. Sci., USA, 80, 993-997.

Sutton KA, Maiti S, Tribley WA, Lindsey JS, Meistrich ML, Bucana CD, Sanborn BM, Joseph DR, Griswold MD, Cornwall GA, and Wilkinson MF. (1998). Androgen regulation of the Pem homeodomain gene in mice and rat Sertoli and epididymal cells. J. Androl., 19, 21-30.

Tapanainen JS, Aittomaki K, Min J, Vaskivuo T, and Huhtaniemi IT. (1997). Men homozygous for an inactivating mutation of the follicle-stimulating hormone (FSH) receptor gene present variable suppression of spermatogenesis and fertility. Nat. Genet., 15, 205-206.

Tindall DJ, and Means AR. (1976). Concerning the hormonal regulation of androgen binding protein in rat testis. Endocrinol., 99, 809-818.

Tindall DJ, Vitale R, and Means AR. (1975). Androgen binding protein as a biochemical marker of formation of the blood-testis barrier. Endocrinol., 97, 636-648.

Vihko KK, LaPolt PS, Nishimori K, and Hsueh AJ. (1991). Stimulatory effects of recombinant follicle-stimulating hormone on Leydig cell function and spermatogenesis in immature hypophysectomized rats. Endocrinol., 129, 1926-1932.

Welsh MJ, and Wiebe JP. (1976). Sertoli cells from immature rats: in vitro stimulation of steroid metabolism by FSH. Biochem. Biophys. Res. Commun., 69, 936-941.

Winters S. (1990). Male Infertility. *In*: Seibel M (Ed.), *Infertility, a comprehensive text*, McGraw-Hill Publishing Co., UK, pp 157.

Sangeetadevi Y, and Majumdar SS. (2004). Differential protein secretion profile of infant and adult rat Sertoli cells. XXII annual Symposium on Reproductive biology and Comparative endocrinology, Chennai.

Yang ZW, Wreford NG, and de Kretser DM. (1990). A quantitative study of spermatogenesis in the developing rat testis. Biol. Reprod., 43, 629-635.

10

On the Surveillance of Sperm Auto-Antigens by the Epididymal Epithelium

M. A. Akbarsha[1*] *and M. Michael Aruldhas*[2]

[1]*Department of Animal Science, Bharathidasan University, Tiruchirappalli – 620 024, India*
[2]*Department of Endocrinology, Dr ALM PGIBMS, University of Madras, Chennai – 600 113, India*

ABSTRACT

The epididymis is a novel organ endowed with varied roles in the processing of sperm towards their post-testicular physiological maturation. Recent studies have found evidence for other roles such as modulation of the fluid microenvironment of the sperm, cleansing the debris that arrive from the epididymis, acquisition and processing of the cytoplasmic droplets shed by the sperm during their post-testicular maturation, management of harmful electrophiles which reach the epididymis, etc. One more role attributed to the epididymis is management of the sperm auto-antigens, which escape into the epithelium as a routine and, more so when the luminal sperm are induced into a pathophysiological state. We have found evidences to this effect. The article reviews the knowledge in this regard thus far and also proposes newer directions in research in this connection.

**Corresponding Author:* E-mail: akbarbdu@yahoo.com

Barriers to Sperm Antigens from Reaching Circulation

The testis and the male reproductive tract are considered immunologically privileged sites in respect to sperm auto-antigens. The antigens of the spermatozoa are capable of evoking an auto-immune response since in a healthy person the sperm auto-antigens are never exposed directly to the blood. For reasons like viral infections, trauma, ductal obstruction, etc., the sperm auto-antigens get exposed to the blood either directly or carried by the virus, and antibodies are generated against the sperm auto-antigens. This can lead to pathological states of spermatic granulomas, auto-immune architis and immunological infertility. The mechanisms that prevent exposure of sperm auto-antigens to blood lie in the blood-testis barrier and the blood-epididymis barrier.

Blood-epididymis Barrier is Weak, but Still Sperm Antigens Don't Reach the Circulation

Whereas the blood-testis barrier, consisting of the tough tunic, the peritubular myoid cells and the special tight junctions between the Sertoli cells, is complete such that sperm auto-antigens can never reach the blood under normal circumstances (de Kerster and Kerr, 1994), the blood-epididymis barrier is not always complete and can provide for access of sperm auto-antigens to circulation. This is in view of three factors *viz.*, (*i*) the blood-epididymis barrier can be leaky, allowing entry of the sperm auto-antigens and even sperm themselves into the intercellular spaces between the lining epithelial cells (Pollanen and Cooper, 1994), (*ii*) the principal cells (PCs), the predominant epithelial cell type lining the epithelium, are capable of acquiring the sperm auto-antigens and cell debris through endocytosis (Yeung *et al.*, 1994) and (*iii*) the sperm, during transit from corpus to cauda epididymidis shed the cytoplasmic droplet and, thus, millions of cytoplasmic droplets drain into the lumen of cauda epididymidis, undergo fragmentation and are endocytosed by the clear cells (Robaire and Hermo, 1988; Akbarsha *et al.*, 2000a). Yet, there is no immune response to sperm auto-antigens. It is believed that the lining epithelium has mechanisms for surveillance of sperm auto-antigens and cell debris of testicular origin (Nashan *et al.*, 1989).

Heterogeneous Nature of the Epithelial Lining of Ductus Epididymidis

The mechanism underlying this surveillance is to be sought in the heterogeneous nature of the cells constituting the epithelium of the ductus epididymidis. There are at least six different cell types, *viz.*, principal cell, narrow cell, apical cell, clear cell, basal cell and intra-epithelial leukocytes, differing among themselves in structure, development, distribution along the different segments of the epididymis and role. Among these cell types the intraepithelial leucocytes, *i.e.*, intraepithelial lymphocytes (IE-Lyms), and intraepithelial macrophages (IE-Macs), have been established to be immune-competent. These are migratory cells and arrive at the epithelium from circulation and/or peritubular/interstitial tissue (Seiler *et al.*, 1998, 1999, 2000). The role of the IE-Lyms is not clearly defined. The IE-Macs are phagocytic as well as antigen presenting cells. They can phagocytose spermatozoa or their fragments, and cell debris of testicular origin entering the intercellular spaces between the tall columnar cells through the leaky junctions. But, the role of these immune-competent cells in the handling of sperm auto-antigens and cell debris is highly limited, and sperm/their auto-antigens can escape management by these cells. Thus, existence of a more efficient system of surveillance of sperm auto-antigens was conceived (Nashan *et al.*, 1989).

The Candidate Cell Types in the Surveillance of Sperm Antigens

The basal cells are the second largest cell population present in the epididymal epithelium along the entire length (Figure 10.1). The cell is confined to the basal aspect of the epithelium forming a three-dimensional latticework lying on the top of the basement membrane, sending out processes into the intercellular spaces between the principal cells perhaps even up to the lumen (Veri *et al.*, 1993). Sparse in organelles, the BC may contain lipofuscin inclusions and mitochondria, but Golgi apparatus is sparse or not detectable. Basal cell appears late during post-natal development of the male reproductive tract (Seiler *et al.*, 1998). Recent evidences indicate that they are derivatives of monocyte-macrophage cell lineage and, thus, of extra-epithelial origin (Yeung *et al.*, 1994; Seiler *et al.*, 1998, 1999, 2000; Holschbach and Cooper, 2002).

Basal Cells, Intra-epithelial Macrophages and Principal Cells

Though the basal cells may play a supportive role, in view of their unique location as a three-dimensional latticework at the base of the tall columnar epithelium, it is in the more recent times attributed with novel roles such as scavenging the epithelium, and probably lumen, of harmful electrophiles (Veri *et al.*, 1993; Hermo *et al.*, 1994; Papp *et al.*, 1995) and immune surveillance of sperm and other testicular auto-antigens (Yeung *et al.*, 1994; Seiler *et al.*, 1998, 1999, 2000; Holschbach and Cooper, 2002). In the latter case, the basal cells are believed to acquire the sperm auto-antigens through a not yet clearly known mechanism and process them to produce lipofuscin material. Incidentally, IE-Macs also acquire sperm/sperm fragments and cell debris of testicular origin by phagocytosis and process them to produce lipofuscin material (Yeung *et al.*, 1994). Interestingly, the tall columnar principal cells of the more proximal segments of the epididymis are also believed to acquire sperm auto-antigens and germ cell debris through endocytosis and process them into lipofuscin material. It is believed that the partially processed sperm auto-antigens and germ cell debris into lipofuscin material are discharged from the principal cells along their lateral and/or basal aspects, to be acquired by the IE-Macs and basal cells (Yeung *et al.*, 1994), though definite evidence to this effect is still awaited. It will be pertinent to find i) how the apical membrane of the principal cells interact with the sperm and germ cell debris auto-antigens and translocate them to the interior, ii) if the lipofuscin material present in the principal cells are produced due to the partial processing of these antigens, and iii) if the principal cells discharge the lipofuscin material for acquisition by the IE-Macs and basal cells.

Basal Cells are Key Role Players

Whereas acquisition of sperm/sperm fragments and cell debris of testicular origin by IE-Macs is conceivable, it is not known how the basal cells, believed as derived from monocytes-macrophage lineage and express Mac1 and F4/80 antigens that are only late expression macrophage antigens, acquire the sperm and germ cell debris auto-antigens and/or the lipofuscin material apparently released from the principal cells. It also remains yet to be established that the lipofuscin material in the principal cells and basal cells are derived from sperm auto-antigens and germ cell debris. Yet another riddle in this context pertains to the fate of the apparently sperm auto-antigen and germ cell debris-derived lipofuscin material in the basal cells. Definite evidence of migration of lipofuscin material-ladenend basal cells to leave the epithelium and reach the peritubular and/or interstitial tissue such that new basal cells take residence at the epithelium has not yet been obtained, though it has been suggested. If such recruitment of fresh basal cells occurs, one should find plumby basal cells in the adult and functional male's epididymal epithelium when the pre-exisiting basal cells have left the epithelium. This is hypothesized based on the fact that on their first establishment at the epithelium during post-natal development the basal cells appear plumby (Figure 10.2), with a regular outline, containing clear cytoplasm very sparse in organelles and lacking lipofuscin material and possessing a large clear nucleus (Veri *et al.*, 1993; Hermo *et al.*, 1994; Seiler *et al.*, 1998, 1999, 2000).

A Role for Narrow Cells?

Yet another riddle in this connection relates to cytoplasmic processes of basal cells extending in the intercellular spaces between tall columnar cells (Goyal and Williams, 1991; Arrighi *et al.*, 1993; Yeung *et al.*, 1994) and can extend all the way to the lumen (Veri *et al.*, 1993). The basal cells are, after full establishment, flat cuboidal cells confined to the very basal aspect of the epithelium. It is, therefore, intriguing as to how processes from basal cells reach up to the lumen. This is because the segment where Veri *et al.* (1993) showed this is the initial segment whose epithelium is about 60-100 µm tall. Interestingly, it is this segment that has the most abundant narrow cells. Narrow cells are confined to the initial segment and intermediate zone (Robaire and Hermo, 1988; Akbarsha and Averal, 1999). We have found vague evidence for a direct continuity or at least a physical association between basal cells and narrow cells at these segments of the epididymis (Figure 10.3) (Akbarsha *et al.*, 2000b). Further, in as much as basal cells increase in number in response to pathophysiological state of the sperm and abundant germ cell debris, the narrow cells also undergo hypertrophy and hyperplasia. The question is whether the narrow cell is lumen-ward extension of the basal cell or material could be transferred from narrow cell to basal cell or vice-versa. We found apical portions of the narrow cells, when provoked by certain male reproductive toxicants, protrude deep into the lumen and such protruding ends, with some content, pinch off into the lumen (Akbarsha and Averal, 1999) (Figure 10.4). Is it a mechanism of warding off some material connected with processing of the sperm auto-antigens and germ cell debris into the lumen? A commonality between basal cells and narrow cells is seen in the localization of cathepsin D in both the cell types (Adamali and Hermo, 1996).

Inducing Sperm into Pathophysiological State would Augment Epididymal Immune Surveillance Mechanism

Thus, there are several riddles in the acquisition, processing and disposal of the sperm auto-antigens and germ cell debris in the epididymal epithelium, and the role in it of principal cells, basal cells, IE-Macs and, probably, narrow cells. We have found certain clues, which are leads for newer discoveries to resolve these riddles. Treatment of rat/mouse/monkey with aflatoxin B_1 (Agnes and Akbarsha, 2001) or Cr-VI (Aruldhas *et al.*, 2004), and also several other potential toxicants like vincristine (Averal *et al.*, 1996), phosphamidon (Akbarsha and Sivasamy, 1998), malathion (Akbarsha *et al.*, 2000a), dichlorvas (Akbarsha *et al.*, 2000a), andrographolide (Akbarsha *et al.*, 2000), ursolic acid (Akbarsha *et al.*, 2000), plumbagin (Akbarsha, M.A. and Girija, R., unpublished observation), etc., produces a pathophysiological state of the sperm and availability of plenty of germ cell debris, warranting immunological management of the sperm and the germ cell debris.

Unique Responses of Epididymal Epithelium to Pathophysiological Sperm in the Lumen

Sperm, if obstructed, can find escape through certain weak patches in the epithelium. A unique role to the basal cells in forming into uni- (mouse/rat) (Figure 10.5) or multi- (monkey) nucleate (Figure 10.6) microcanaliculi to seclude such sperm and process them so as to prevent extravasation of sperm has been advocated, suggesting a kind of sperm auto-antigen surveillance very different (Agnes and Akbarsha, 2001; Aruldhas *et al.*, 2004). More importantly, presence of toxicant-induced pathophysiological sperm and germ cell debris in the lumen leads not only to increase in the number of basal cells and IE-Macs in the epithelium but dense accumulation of lipofuscin material, different in electron density between the cell types, in principal cells, basal cells and IE-Macs (Figure 10.7–10.9) in rat, mouse and monkey (Yeung *et al.*, 1994; our unpublished observations). These are all incidental

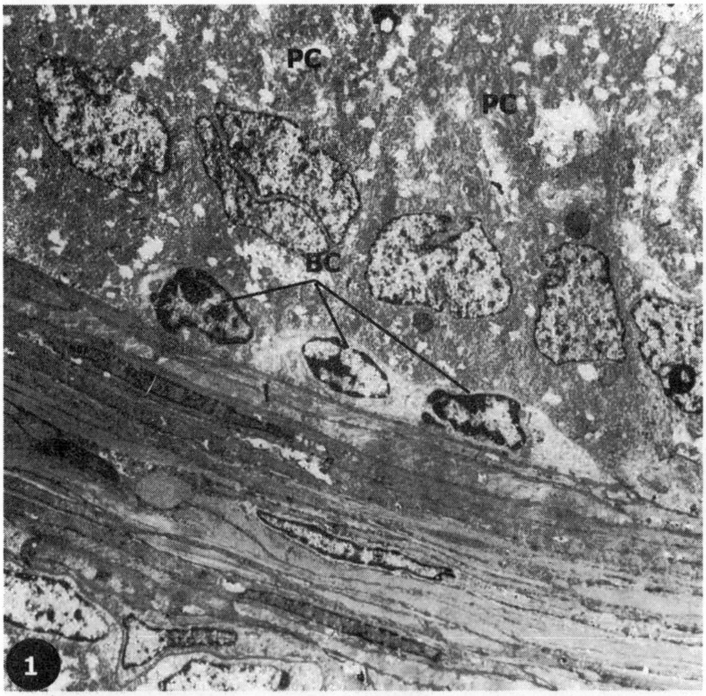

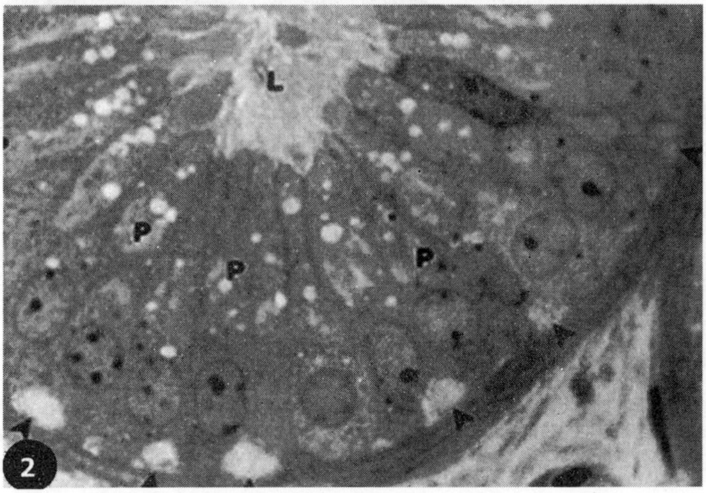

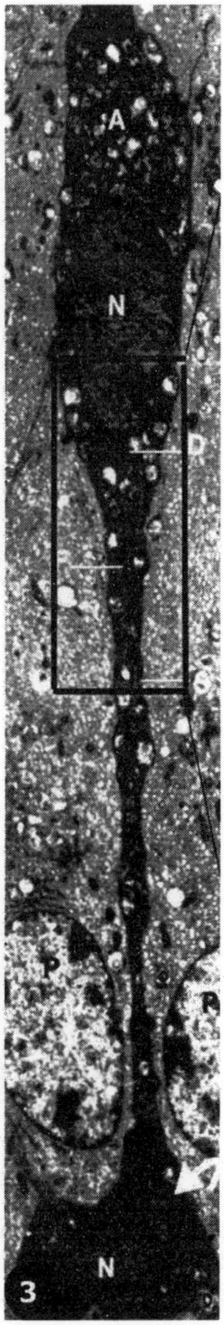

Figure 10.1: TEM: Epithelium of Epididymis of Showing Basal Cells Adherent to the Basal Lamina; BC: Basal cell; PC: Principal Cell

Figure 10.2: LM: Plumby Basal Cells (Arrowhead) in Adult Epididymal Epithelium; (AFB_1: Treated Mouse); P: Principal Cell; L: Lumen

Figure 10.3: TEM: An Apparent Direct Continuity Between Basal Cell and Narrow Cell in Mouse Epididymal Initial Segment (PI's Collection). Curved Arrow, Basal Peduncle; A: Apex; N: Nucleus; S: Stalk; P: Principal Cell

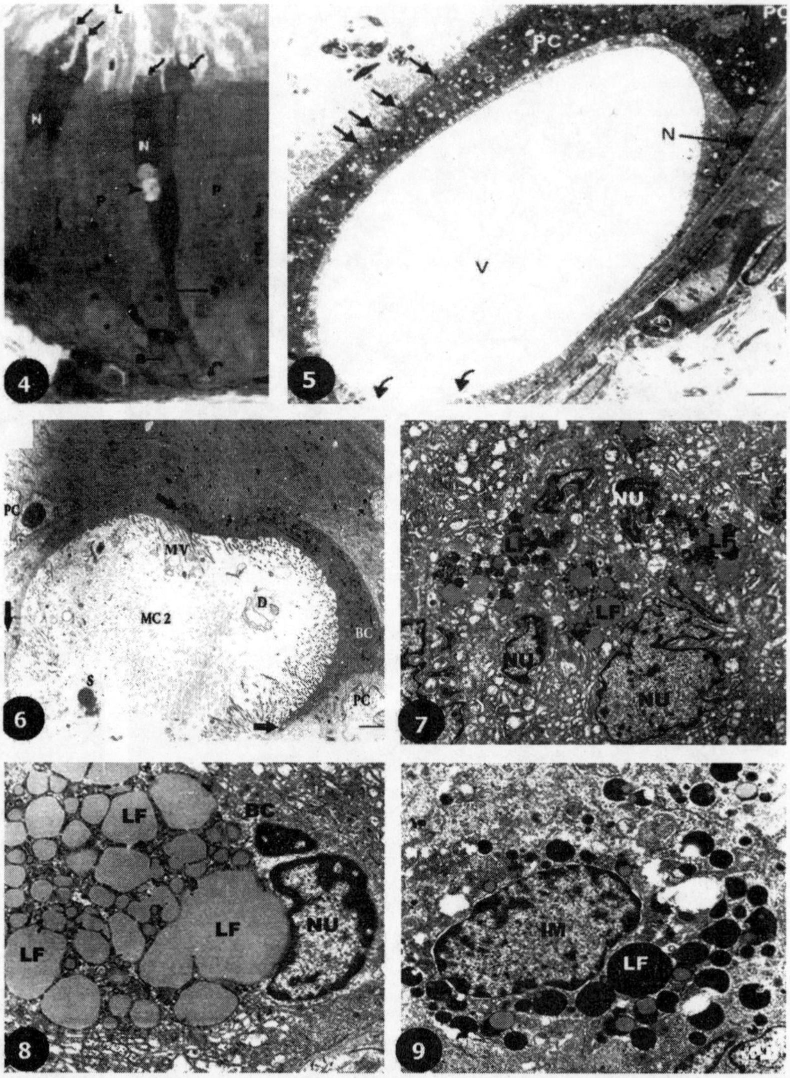

Figure 10.4: LM: Pinching of Apical Protruding Ends of Narrow Cell into the Lumen in Mouse Epididymal Initial Segment (Arrow). Narrow cell connection with basal cell is also indicated. B: Basal Peduncle; S: Stalk; P: Principal Cells; N: Nucleus; L: Lumen; Arrowhead vacuole.

Figure 10.5: TEM: Uninucleate Microcanaliculus in the Epididymal Epithelium, Apparently Derived from Principal Cell which would Prevent Extravasations of Sperm. (AFB_1: Treated Mouse). PC: Principal Cell Cytoplasm; Curved Arrow, Microvilli; V: Vacuole; N: Nucleus; Straight Arrow, Occluding Junctions.

Figure 10.6: TEM: Multinucleate Microcannaliculus [Cr(VI)-treated Monkey]. PC: Principal Cell Cytoplasm; BC: basal Cell; D: Debris; MV: Microvilli of Cell Enclosing the Microcanaliculus; S: Sperm; Arrow: Occluding Junctions.

Figure 10.7: TEM- Lipofuscin (LF) in Principal Cell. NU: Nucleus. [Cr(VI)-treated Monkey].

Figure 10.8: TEM: Lipofuscin (LF) in Basal Cell (BC). NU: Nucleus. [Cr(VI)-treated monkey].

Figure 10.9: TEM: Lipofuscin (LF) in Intraepithelial Macrophage (IM). [Cr(VI)-treated monkey].

observations, and provide scope for a directed study to investigate the management of sperm auto-antigens and germ cell debris by the epididymal epithelium, particularly, with a role by principal cells, basal cells, narrow cells and IE-Macs, through inducing a pathophysiological state of the epididymal luminal sperm and increased abundance of germ cell debris.

Unresolved Issues in Surveillance of Sperm Antigens by Epididymal Epithelium

We are currently investigating the following issues in this connection:

1. Do principal ells possess receptors for sperm auto-antigens and germ cell debris to bring about receptor-mediated endocytosis of sperm, or their portions/fragments enter the principal cells and are processed into lipofuscin material?
2. Is lipofuscin material in principal cells, under pathophysiological state of luminal sperm, really sperm auto-antigen mediated or only management of germ cell debris including dead sperm?
3. What is the mechanism of translocation of sperm- and germ cell debris-derived lipofuscin material from principal cells into basal cells?
4. Do basal cells and IE-Macs acquire lipofuscin material from the intercellular spaces around?
5. To what extent is lipofuscin material processed in the basal cells?
6. What is the fate of the basal cells ladenend with the lipofuscin material in them?
7. What is the relationship between basal cells and narrow cells in the management of sperm auto-antigens and/or germ cell debirs?
8. Do sperm, induced into a pathophysiological state, undergo change in the antigen status so as to provoke special immune defense mechanisms in the epididymal epithelium?

Acknowledgements

The transmission electron microscope facility of CMC and H, Vellore, is heartily acknowledged. We thank our students who generated the data presented in this article. The financial support to us from CSIR and DST, Government of India, is also acknowledged. The FIST facility of the DST, Government of India, awarded separately to our Departments, was extensively used in the investigations.

References

Adamali HI, and Hermo L. (1996). Apical and narrow cells are distinct cell types differing in their structure, distribution and functions in the adult rat epididymis. J Androl., 17, 208-222.

Agnes VF, and Akbarsha MA. (2001). Pale vacuolated epithelial cells in epididymis of aflatoxin-treated mice. Reproduction, 122, 629-641.

Akbarsha MA, and Averal, HI. (1999). Epididymis as a target for the toxic manifestation of vincristine: ultrastructural changes in the narrow cell. Biomedical Letters, 59, 113-120.

Akbarsha MA, and Sivasamy P. (1998). Reproductive toxicity of phosphamidon: histopathological changes in the epididymis. Indian J. Exp. Biol., 36, 34–38.

Akbarsha MA, Latha PNL, and Murugaian P. (2000a). Retention of cytoplasmic droplet by rat cauda epididymidal spermatozoa after treatment with cytotoxic and xenobiotic agents. J. Reprod. Fertil., 120, 385-390.

Akbarsha MA, Agnes VF, and Girija R. (2000b). Epididymal epithelial basal cell: Structure, development and role–A review. J. Endocrinol. Reprod., 4, 1-12.

Arrighi S, Romanello MG, and Gomeneghini C. (1993). Ultrastructure of epididymal epithelium in *Equs caballus*. Ann. Anat., 175, 1-9.

Aruldhas MM, Subramanian S, Sekhar P, Chandra Hasan G, Govindarajul P, and Akbarsha MA. (2004). Microcanalization in the epididymis to overcome ductal obstruction caused by chronic exposure to chromium–a study in the mature bonnet monkey (*Macaca radiata* Geoffroy). Reproduction, 128, 127-137.

Averal HI, Stanley A, Murugaian P, Palanisamy M, and Akbarsha MA. (1996). Specific effect of vincristine on epididymis. Indian J. Exp. Biol., 34, 53-56.

De Kerster DM, and Kerr JB. (1994). The cytology of testis. *In*: Knobil E, and Neill JD. (Eds.), *The Physiology of Reproduction*, Raven Press Ltd, New York. pp. 1177-1290.

Goyal HO, and Williams CS. (1991). Regional differences in the morphology of the goat epididymis: a light microscopic and ultrastructural study. Am. J. Anat., 190, 349-469.

Hermo L, Papp S, and Robaire B. (1994). Developmental expression of the Yf subunit of glutathione S-transferase P in epithelial cells of the testis, efferent ducts, and epididymis of the rat. Anat. Rec., 239, 421-440.

Holschbach C, and Cooper TG. (2002). A possible extratubular origin of epididymal basal cells n mice. Reproduction, 123, 517–525.

Nashan D, Malorny U, Sorg C, Cooper T, and Nieschlag E. (1989). Immuno-competent cells in the murine epididymis. Int. J. Androl., 12, 85-94.

Papp S, Robaire B, and Hermo L. (1995). Immunocytochemical localization of the Ya,Yc,Yb1 and Yb2 subunits of glutathione S-transferases in the testis and epididyis of adult rats. Microsc. Res. Tech., 30, 1-23.

Pollanen P, and Cooper TG. (1994). Immunology of testicular excurrent ducts. J. Reprod. Immunol., 26, 167-216.

Robaire B, and Hermo L. (1988). Efferent ducts, epididymis and vas deferens: structural functions and their regulation. *In*: Knobil E, and Neill JD. (Eds.), *The Physiology of Reproduction*, Raven Press Ltd, New York, pp 999-1080.

Seiler P, Wenzel I, Wagenfeld A, Yeung CH, Nieschlag E, and Cooper TG. (1998). The appearance of basal cells in the developing murine epididymis and their temporal expression of macrophage antigens. Int. J. Androl., 20, 217-226.

Seiler P, Cooper TG, Yeung CH, and Nieschlag E. (1999). Regional variation in macrophage antigen expression by murine epididymal basal cells and their regulation by testicular factors. J. Androl., 20, 738-746.

Seiler P, Cooper TG, and Nieschlag E. (2000). Sperm numbers and condition affect the number of basal cells and their expression of macrophages in the murine epididymis. Int. J. Androl., 23, 65-76.

Veri JP, Hermo L, and Robaire B. (1993). Immunocytochemical localization of the Yf subunit of glutathione S-transferae P shows regional variation in the staining of epithelial cells of the testis, efferent ducts and epididymis of the male rat. J. Androl., 14, 23-44.

Yeung CH, Nashan D, Cooper TG, Sorg C, Oberpenning F, Schulze H, and Nieschlag E. (1994). Basal cells of the human epididymis-antigenic and ultrastructural similarities to tissue-fixed macrophages. Biol. Reprod., 50, 917-926.

11

Estrogen Receptors: Signaling and Crosstalk

*Wright Jacob, Partha Roy and Ben M.J. Pereira**

Reproductive Biology Laboratory, Department of Biotechnology, Indian Institute of Technology Roorkee, Roorkee – 247 667, Uttaranchal, India

ABSTRACT

Estrogens belong to the family of steroids whose biological activities are regulated by binding with receptors. It is generally recognized that these receptors are members of nuclear receptor super family and depending upon their interaction with ligands, shuttle between the cytoplasm and the nucleus, to effect transcriptional activation. In addition, several other extracellular and intracellular signals also modulate the transcriptional activity of estrogens. This is primarily due to the fact that estrogen receptors are activated by a wide spectrum of ligands and factors that act through other signaling pathways. Cellular signaling pathways can be linear, but more complex patterns are rule than exception. Signaling pathways are extensively intertwined and crosstalk at many different levels. Steroid receptors lead a double life by coupling to extra signaling molecules that elicit very rapid non-genomic action. The signaling pathways of steroid-independent activation of steroid receptors and of non-genomic signaling by membrane receptors, display a remarkable reciprocal relationship suggesting that these two modes of signaling crosstalk may be two faces of the same coin. The relevance of the signaling and crosstalk is discussed.

***Keywords**: Estrogen, Receptor, Signaling, Crosstalk, Endocrine Disrupting Compound (EDC).*

**Corresponding Author*: E-mail: benmjfbs@iitr.ernet.in; Telephone: +91-1332-285790; Fax: +91-1332-273560

Introduction

The endocrine system is an impressive collection of hormones that carry out a startlingly complex set of metabolic functions. Each hormone is a chemical messenger that provokes a response in its target tissue. Estrogens are C-18 steroid hormones, produced from androgens by adrenals, placenta and ovaries. The principal estrogenic hormone is 17β-estradiol (E2), which is more potent, but other hormones such as estriol and estrone are also produced as products of estradiol metabolism. Just as most other hormones, the estrogens bring about their effects in target tissues by interacting with receptors. Estrogen receptors (ER) are generally classified under the nuclear receptor family. Nuclear receptors are a large family of structurally related ligand-inducible transcription factors.

Molecular Architecture of ER

Nuclear receptors for steroid hormones have four structurally distinct functional domains (Figure 11.1). The N-terminal A/B region contributes to transcriptional activation function and harbors the activation function 1 (AF-1). The mid-region also called the C region, or the DBD, harbors the DNA binding domain that mediates specific DNA binding. The DNA binding domain (DBD) of the two ER isoforms share the same response elements. The topology of ER-DBDs is characterized by a zinc finger-like motif with eight cysteines that constitute the tetrahedral coordination of two zinc ions. Residues participating in the 'D box' have been shown to be involved in the dimerization interface, whereas residues present in the 'P box' are implicated in specific interaction with DNA and are in contact with the central base pairs of the palindrome response element. DBDs are monomers in solution and form dimers when bound to their respective response elements.

Next to the DBD is the D domain or the hinge region, which is less characterized. At the C-terminus is the Ligand-Binding Domain (LBD or E/F domain), a pocket that not only binds ligands but also co-factors and facilitates transactivation, nuclear localization and interactions with heat

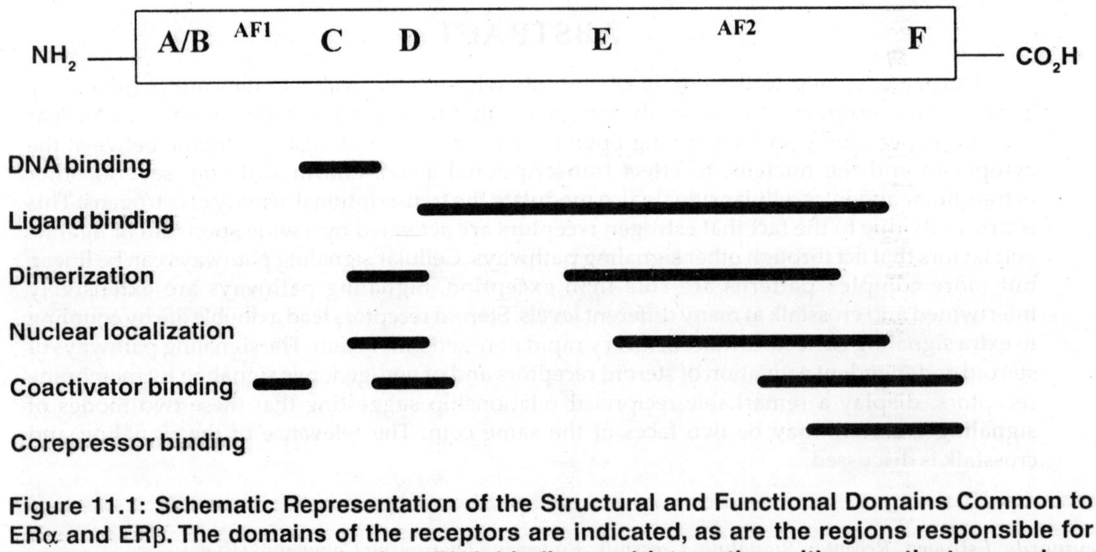

Figure 11.1: Schematic Representation of the Structural and Functional Domains Common to ERα and ERβ. The domains of the receptors are indicated, as are the regions responsible for nuclear translocation, dimerization, DNA binding and interaction with co-activators and co-repressors.

shock proteins. Ligand binding domain (LBD) is a globular domain that harbors a hormone binding site, a dimerization interface (homo- and hetero-dimerization), and a coactivator and corepressor interaction function. ER-LBDs are arranged in an antiparallel α-helical 'sandwich' fold. The liganded ER-LBD contains 11 α-helices (H1-H12) organized in a three-layered sandwich structure with H4, H5, H6, H8 and H9 flanked on one side by H1 and H3, and on the other side by H7, H10, and H11. The ligand pocket is closed on one side by an antiparallel β-sheet and on the other by H12, which is directly involved in the transactivation function AF-2. For this reason the receptor is able to distinguish between 'agonist' or 'antagonist' conformations (Moggs and Orphanides, 2001).

The ER-LBDs form dimers with both agonist and antagonist complexes and the overall homodimeric arrangement is the same whatever the class of the ligand or the ER isotypes.The dimerization interface involves residues from helix H8 up to helix H11, but the most important contact surface is located on H10 through a hydrophobic leucine zipper-like interaction zone and hydrophilic contacts (direct hydrogen bonds or via water molecules). The dimer contacts in both ER isoforms are constituted mainly by helices H10 and H11, which are also in contact with the ligand, providing the link between ligand binding and dimerization.

Agonist binding induces a conformational rearrangement in the LBD resulting in the formation of a specific binding site for the helical NR-box module of nuclear coactivators. This binding site is a hydrophobic groove formed by residues from helices H3, H4, H5 and H12, and the turn between helices H3 and H4.The coactivator LXXLL motif functions as a hydrophobic docking module that binds on the surface of the LBD. In ERs, both partial and pure antagonists induce conformations of the AF-2 region that are distinct from that observed in the presence of pure agonists. The binding of raloxifene and tamoxifen is accompanied by major structural reorganization in the ternary structure in both ER isotypes. The large piperidine extension of this ligand provokes steric clashes that prevent the transactivation helix H12 to adopt its characteristic conformation. Instead, H12 lies tightly in the coactivator recognition groove. H12 possesses a NR box-like sequence (LXXML versus LXXLL) that perfectly mimics the interactions made by NR-box peptides in the ERα complex, whereas there is a shift in the H12 position of between 1.9 and 3.2 Å in the ERβ complex. This displacement does not affect the final location of the two key leucine residues, which coincide with the first and third NR-box leucines (Shao and Brown, 2004).

The agonist and antagonist recognition by ER was obtained by studying the crystal structure of ERα LBD. It provided the molecular basis of the interaction of the receptor with its natural ligand 17β-estradiol (E2) (Brzozowski *et al.*, 1997). The E2 cavity is completely shielded from the external environment and buries the ligand in a highly hydrophobic environment mostly defined by 22 residues. Two polar regions located at opposite sides of the ligand-binding pocket are involved in the anchoring of the E2 hydroxyl moiety at positions 3 and 17. The phenolic hydroxyl group of the A-ring (3-OH) is hydrogen bonded to Glu353 from H3, and to Arg394 from H5 and a water molecule. The hydroxyl group of the D-ring (17β-OH) forms a single hydrogen bond with His524 (H11). The cavity delimited by the protein exhibits a probe accessible volume of 450 Å, which is much larger than the molecular volume of the natural ligand (250 Å) (Ruff *et al.*, 2000).

The first estrogen receptor, now named ERα, was cloned in 1986, more than 20 years after it was identified by its affinity for 17β-estradiol. In 1996, the second ER, ERβ, was reported. Since then the existence of these two types of nuclear receptors has been widely accepted. Although their basic attributes remain the same these are several points of differences, which are described below.

ERα maps to 6q25.1 of the human chromosome. It is a very large gene spanning more than 140 kb and containing 8 exons and 7 introns. The position of its introns has been highly conserved. It codes an mRNA of size 6.6 kb. ERα has 1,785 nucleotides of the cDNA correspond to a polypeptide of 595 amino acids (66.2 kDa). As described earlier, a highly conserved 66-aa region of the ERα, which corresponds to part of the receptor DNA-binding domain, determines the specificity of target gene recognition. ER-alpha is a ligand-activated transcription factor composed of several domains important for hormone binding, DNA binding, and activation of transcription.

The ERβ gene maps to 14q22-24 of the human chromosome and has a size of the about 40-kb comprising 8 exons and 7 introns. The first human cDNA sequence published reported a length of 487 amino acids (Mosselman *et al.*, 1996). Longer cDNA sequences were subsequently reported, predicting a primary translation product of 530 amino acid residues. The size of the mRNA and its splice variants range from ~1.0 kb to 10.0 kb which perhaps explain this anomaly. A comparison of ERβ and ERα reveals a 96 per cent amino acid homology between the two receptors in the DBD but in the LBD it is only 53 per cent. This suggests that ERβ would recognize and bind to similar EREs as ERα but that each receptor would have a distinct spectrum of ligands. There is less conservation between the two receptors in the N-terminal AF-1 and C-terminal AF-2. This would mean that different sets of proteins in the transcription complexes may interact with ERα and ERβ and direct them to specific targets. Such a mechanism could explain why at AP-1 sites, ERβ and ERα function oppositely (Muramatsu and Inoue, 2000).

Distribution of Estrogen Receptor

E2 target tissues can be divided into two groups, the classical and non-classical E2 target tissues, respectively. The classical targets are the uterus, mammary gland, placenta, liver, central nervous system (CNS), cardiovascular system and bone. These tissues have a high ERα content and respond to E2 challenge with increases in transcription of certain E2-responsive genes. The non-classical target tissues include prostate, testis, ovary, pineal gland, thyroid gland, parathyroids, adrenals, pancreas, gallbladder, skin, urinary tract, lymphoid and erythroid tissues. In these tissues, expression of ERα is either very low or not measurable but ERβ is highly expressed.

Mechanism of Estrogen Action

E2 circulates in blood either free or bound (95 per cent) to plasma carrier proteins. The unbound and loosely albumin-bound E2 is generally believed to diffuse freely through the capillary wall and the cell membrane bilayer. Once E2 is in the cytoplasm, it is not yet clear whether a transport protein is required for movement of the hydrophobic steroid molecule through the aqueous cytoplasm to the nucleus. The inactive ER is bound to various receptor-associated proteins like Hsp-90. The current model (Figure 11.2) is that E2 diffuses freely in the cytoplasm and enters the nucleus by two processes: 1) passive diffusion through the 'ever open' central channel of the nuclear pore or 2) active transport that is mediated by interaction of the NLSs (nuclear localization signals) on the ER with the receptor-Hsp90 complex. ER resides in the nucleus in an unliganded state (Guiochon–Mantel *et al.*, 1996).

ER is thought to be complexed with the Hsp90 chaperon complex. Hsp90 and its accompanying proteins are thought to stabilize the receptor until hormone binding occurs. (Pratt and Toft, 1997). Alternatively, unliganded ER may be bound by co-repressors SMRT or NCoR. Binding of E2 to the ER 'activates' the receptor by inducing alterations in the conformation of the receptor protein, triggering dimerization. ER forms homodimers, or heterodimers of ERα and ERβ and such dimers binds to EREs in the regulatory region of the target gene. Once bound to the ERE, the liganded ER induces a DNA bend and, in the presence of agonist ligand, *e.g.* E2, recruits co-activator proteins such as SRC-1, a component of a co-activator complex that includes p300 and PCAF (Mueller and Korach KS 2001). The histone acetyltransferase (HAT) activity of co-activators acetylates lysine residues in histones H3 and

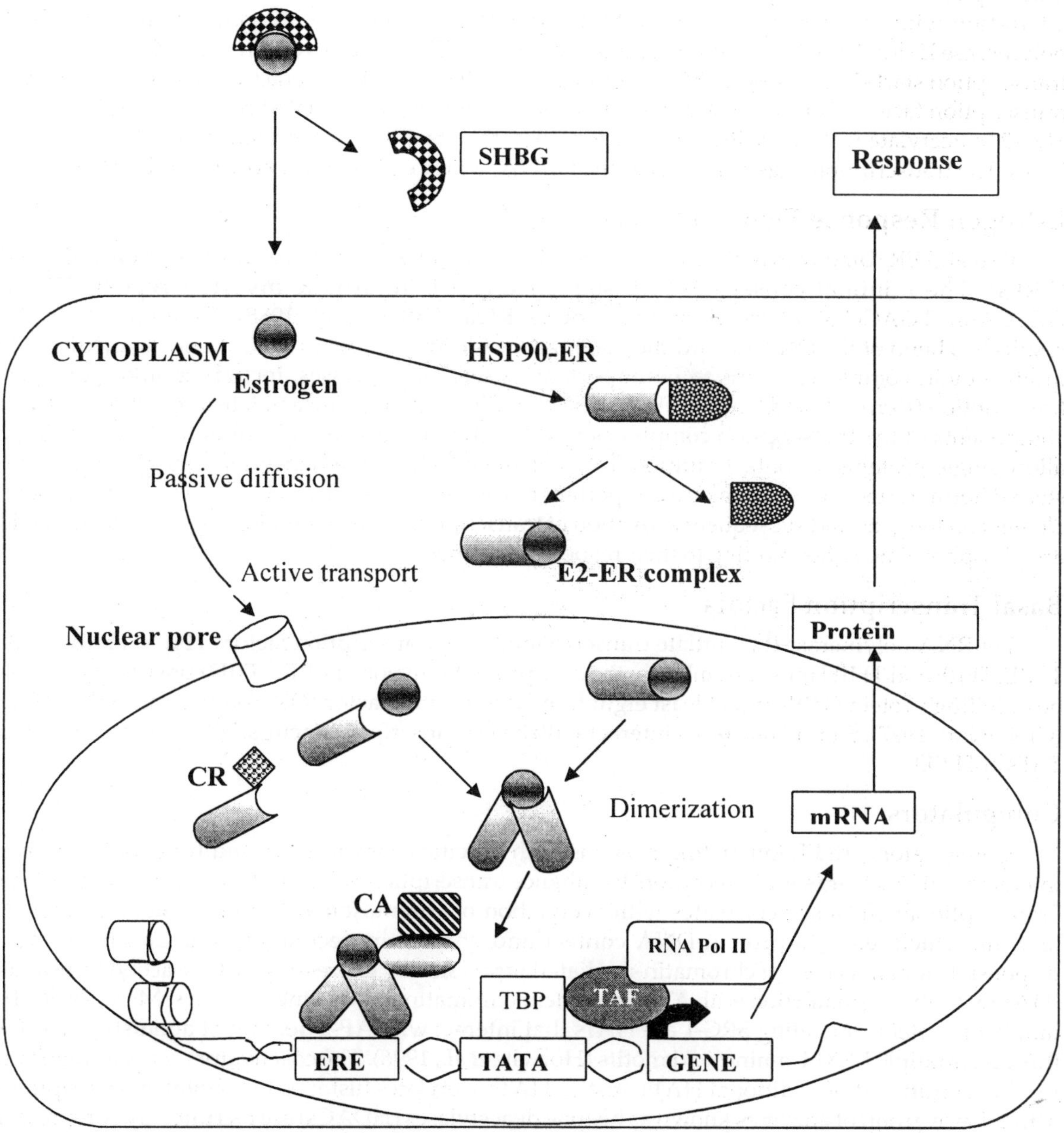

Figure 11.2: Schematic Diagram of the Classical Mechanism of Estrogen action

CR: Corepressor; CA: Coactivators: ERE-Estrogen Response Element; TBP: TATA box Binding Protien; TAF: TBP Associated Factors.

H4 within the nucleosome, relieving chromatin repression and facilitating the binding of general transcription factors, TBP (TATA box Binding Protein) and TAFs (TBP Associated Factors) and interaction with the promoter. Co-activators and ERα also interact with components of the RNA polymerase II (RNA Pol II) initiation complex including TBP and TAFs and recruit RNA Pol II to the transcription start site (Nilsson 2000). Ligand-activated ERα can also interact with other DNA-bound transcription factors. This protein–protein action may allow ERα-associated co-activators, SRC-1 and SRC-2 to acetylate histones within the nucleosome, thus facilitating template accessibility (Klinge, 2000). The transcription of gene produces the mRNA, which codes for the corresponding protein.

Estrogen Response Element (ERE)

ERα and ERβ bind with high affinity to specific DNA sequences called estrogen response elements (EREs). The minimal consensus ERE sequence is a palindromic inverted repeat (IR): 5′-GGTCAnnnTGACC-3′, where *n* is any nucleotide (Klein–Hitpass *et al.*, 1988). ER binding to an ERE results in a bend of the DNA toward the major groove (Kim *et al.*, 1997). DNA bending mediated by multi-protein complexes seems to be important in cellular processes, including eukaryotic gene transcription (Kerppola and Curran, 1997). DNA bending is thought to facilitate interactions between components of the transcription complex bound to different sites and to promote DNA looping to allow single proteins to contact multiple DNA elements. Hormone-dependent phosphorylation of steroid hormone receptors may play an important role in binding of the receptor to its specific response element on the gene and subsequent activation of transcription. All the steroid receptors, including ER are phosphorylated after binding to their respective ligands.

Basal Transcription Factors

For RNA polymerase II to initiate transcription, basal transcription factors TFIIA, TFIIB, TFIID, TFIIE, TFIIF and TFIIH must assemble on the core promoter region of ERE. TFIID consists of the TATA box binding protein (TBP) and at least eight tightly associated factors (TAFs) of 18 through 250 kDa (Greenblatt, 1997). Steroid receptors interact with basal transcription factors TFIIB, TBP, and various TAFs of TFIID.

Coregulators

Coregulators are ER interacting proteins with specific characteristics and functions. One type interacts with nuclear/steroid receptors to enhance transcription while the other represses the same. Transcriptional efficiency correlates with acetylation of chromosomal histone proteins at their N-termini, which leads to protein-DNA contact and chromatin decomposition. Coactivators are responsible for alleviation of chromatin-mediated repression by expressing histone acetyl transferase (HAT) activity in conjunction with ATP dependent chromatin remodeling enzymes. Member of p160 family of proteins including SRC-1 are HATs that interact with AF-2 domain of agonist-bound ERs through multiple LXXLL amino acid motifs (Horwitz *et al.*, 1996). The coactivators serve as platforms for the recruitment of additional HATs. Just as HATs acetylates histones to facilitate transcriptional activation, a group of enzymes known as histone deacetylases (HDACs) are recruited by corepressors to deacetylate histones and promote transcriptional repression (Robyr *et al.*, 2000). An alternate model for repression involves factors that compete with p160 family coactivators to agonist- bound ER. These include REA (Repressor of Estrogen Activity) and orphan receptor SHP and DAX1.

The current model to account for cell-specific regulation of estrogen target gene expression suggests that target cells express different levels of co-activators and co-repressors. This, along with different concentrations of ERα, ERβ, and ligand, allows fine-tuning of target gene transcription in response to estrogens. Northern blot analysis confirmed the idea that different rat tissues and cell lines express different amounts of the mRNA for the coactivators SRC-1, RIP-140, p300, and CBP, as well as the co-repressors SMRT, and NCoR. McKenna *et al.* (1999) reviewed in detail nuclear coregulators and their molecular biology. A few coregulators and their function are listed in Table 11. 1.

Table 11. 1: Coregulators of ER transcription.

Coregulators	*Functions*
BRG1	Belongs to SWI/SNF ATP dependent chromatin remodeling complex.
CARM-1	Coactivator with intrinsic protein methytransferases activity.
CIA	Coactivator independent of AF-2 function.
Cyclin D1	Cell cycle protein.
	Facilitates ligand-independent interactions between ER and protein acetyltransferase.
MAT-1	Mediates the recruitment of protein deacetylase-complex.
N-CoR, SMRT	Corepressor
	Antagonist dependent interaction with ER.
	Mediate the recruitment of protein deactylase-containingcomplex.
p300, CBP, pCAF	Coactivator with intrinsic protein methytransferases activity.
RIP140	Coregulator.
	Recruit protein deacetylase.
SRC-1, TIF-2 (also known as GRIP-1)	Belongs to p160 family of coactivators. Mediates recruitment of protein acetyl transferase and methytransferases activity.
TFIIH	General transcription factor.
	Ligand–regulated interaction of p62 and XPD subunits directs CDK7 kinase-mediated phosphorylation of ERα.
TRAP220	Component of coactivator complex.
SHP, DAX-1	Orphan nuclear receptor.
	Compete for binding of p160 coactivator to ligand–bound ER.
REA	Competes for binding of p160 coactivators to ligand-bound ER.

Ligand–independent Activation of Estrogen Receptor

Initially, steroid receptors were thought to act as transcription factors only when bound to their cognate ligands. Agonistic ligands induce a conformational change in the hormone-binding domain that allows the recruitment of transcriptional coregulators. In the late 1980s and early 1990s it was reported that steroid receptors could be activated by a host of extracellular signals in the absence of their cognate ligands. This is because most nuclear receptors are phosphoproteins, and their function can be altered by changes in their phosphorylation in the absence of a hormonal ligand. The activators of protein kinases, such as growth factors, can elicit estrogen-independent activation of the receptor molecule by inducing phosphorylation of the receptor at sites that differ according to the identity of the activator. Phosphorylation occurs predominantly at specific serine or tyrosine residues and is catalyzed by enzymes such as receptor tyrosine kinase and mitogen-activated protein kinases. Mitogen-activated protein kinases are composed of several serine–threonine kinases that are activated in response to various cell-growth signals and transduce extracellular signals to intracellular targets by way of membrane receptors. *In vitro* results have proved the existence of such crosstalk between signal pathways in the case of the activation of estrogen-independent receptors by dopamine, epidermal growth factor, transforming growth factor a, A insulin or insulin-like growth factor-1, heregulin, and cyclic AMP (Table 11. 2). Epidermal growth factor consistently induced markers of estrogenic action in vivo, such as the augmentation of progesterone messenger RNA transcripts, whereas it failed to have similar effects in estrogen-receptor–knockout mice. The activation by intracellular factors is of interest for two reasons: (i) It may mimic the stimulation by certain extracellular factors by plugging into their signaling pathways, and (ii) it may be physiologically relevant since intracellular signaling components can be overexpressed or constitutively activated in certain tumors. This phenomenon may also be attributed to the cell-specific partial agonism of many anti-hormones. In transfection experiments, a well-known ER antagonist tamoxifen is switched to an agonist in the presence of growth factors or their downstream signaling molecules. The activation of ER by non-steroids such as growth factors involves their kinase receptor and kinase enzymes. Constitutive activation of the MAPK pathway mimics activation of ERα by the extracellular factor, and its inhibition blocks it and GnRH activates ERα using protein kinase C and MAPK pathway. But, constitutive activation of the PI3K-Akt pathway is not only able to activate ERα , but this pathway is also required for activation of ERα by IGF-1. These results point out a possibility of peptide growth factors signaling to steroid receptors through multiple or branched pathways and existence of tissue specific signaling circuitry.

Table 11. 2: Steroid-independent activators of estrogen receptors.

Factors	*Receptors*
Dopamine	ERα
Epidermal growth factor (EGF)	ERα, ERβ
Heregulin	ERα
TGFα	ERα
Insulin and insulin-like growth factors (IGF)	ERα
Keratinocyte growth factor	ERα
Fibroblast growth factor 2 (FGF-2)	ERα
Gonadotropin release hormone (GnRH)	ERα
Sex hormone binding globulin	ERα
Interleukin-6	ERα
erbB2	ERα
PI3K	ERα
Protein kinase B	ERα
v-Src	ERα
Ras	ERα
MEKK1	ERα
MAPKK	ERα
Cyclin D1	ERα
Cyclin A-Cdk2	ERα
Ets-1	ERα
Activators of protein kinase A	ERα
Inhibitors of protein phosphatases 1 and 2A	ERα
Inhibitor of phosphotyrosine phosphatases	ERα
Activator of protein kinase C	ERα

There is ample evidence that steroid-independent activation is through direct and indirect phosphorylation of specific amino acid residues in ER. Peptide growth factor activates ER by phosphorylation of serine residues in the AF-1 domain of the ERs. Cyclin A-Cdk2 complex can hyperphosphorylate ER serine residues 104 and 106 in presence of estrogen thereby enhancing ERα activity.

There are alternative activation pathways that do not seem to involve the direct phosphorylation of ERα. One and the same signal can target different ERα domains in different cell types and can target the same ERα domain through different pathways in different cell types: MAPK and S118 phosphorylation are not involved in the activation of ERα by EGF in vascular cells even though the same domain, AF-1, is the target. Likewise, the stimulatory effect of v-Src on AF-1 activity is only in part due to the phosphorylation of S118 by MAPK; v-Src also stimulates AF-1 activity indirectly through the Rac-MEKK-JNKK-JNK pathway (Liu *et al.*, 1995). However, in all of these cases one cannot exclude a role for minor phosphorylation sites of ERα itself.

Dopamine as depicted in Figure 11.3 functions through a G-protein coupled membrane receptor. Its activation results in the stimulation of adenylate cyclase, elevation of cAMP, and activation of PKA. Despite the fact that both cAMP and PKA mimic the ability of dopamine to activate steroid receptors in the absence of ligand, there is no evidence that PKA elicits the steroid-independent activity by directly phosphorylating the steroid receptors. Rather, the phosphorylation of S236 in the DNA-binding domain of ERα by PKA is inhibitory for dimerization and DNA binding. The issue is complicated by the fact that cAMP signaling can either inhibit or activate the MAPK pathway depending on cell type (Endoh *et al.*, 1999). However, cAMP-mediated activation of ERα also works in cells where PKA inhibits the MAPK pathway. Moreover, EGF and cAMP appear to activate ERα through distinct pathways targeting different domains; cAMP may target AF-2, or possibly both AF-1 and AF-2. Further progress towards dissecting steroid-independent activation of steroid receptors by cAMP has been made in whole animal experiments with rodents which has been discussed elsewhere (Mani *et al.*, 1996).

The challenge is to explain how the transcriptional activity of aporeceptors is stimulated in the absence of steroid at the molecular level. As alluded above, the role of phosphorylation of the ERs for transcriptional coactivator recruitment has received particular attention. The EGF-induced phosphorylation of the ERβ AF-1 domain promotes its recruitment of the coactivator SRC1, and transcriptional activity is further increased in the presence of the CREB binding protein, CBP. The steroid-independent activation of ERα by cyclin D1 overexpression represents an interesting paradigm. Cyclin D1 functions independently of its kinase partner to provide a bridge between p160 coactivators and the ERα, which is unable to bind these coactivators directly in the absence of hormone. Remarkably, cAMP signaling promotes the interaction between ERα and cyclin D1, and this correlates with increased transcriptional activity of apo-ERα. It remains to be determined how cAMP stimulates this interaction and whether the activation of apo-ERα by cAMP or by other signals requires cyclin D1. Close to 100 ER interacting proteins or proteins interacting with the other nuclear receptors could serve as a platform or conduit for signaling crosstalk and many are. The induced release of a repressor would constitute an alternative mechanism for steroid-independent activation. In the absence of ligand, steroid receptors are associated with molecular chaperones, which are thought to keep them transcriptionally inactive. Other targets might be BRCA1, which can function as a repressor of apo-ERα, and the transcriptional corepressors NCoR and Smrt. NCoR appears to interact with ERα in the presence of the antagonist tamoxifen, and EGF stimulates the release of NCoR and concomitant activation of ERα (Lavinsky *et al.*, 1998).

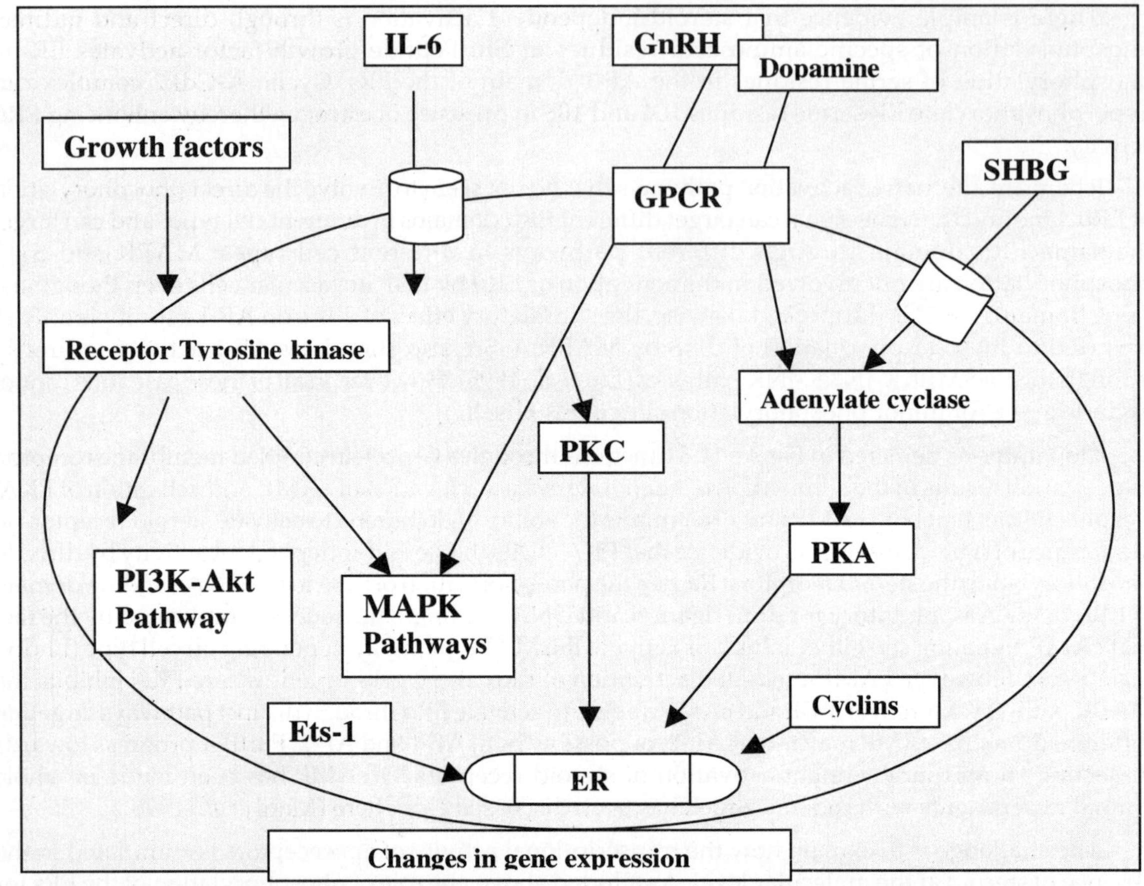

Figure 11.3: Schematic Diagram of Different Pathways Involved in Steroid Independent Activation of Estrogen Receptor. GPCR- G-protein coupled receptor; SHBG- Sex Hormone Binding Globulin, MAPK-Mitogen Activated Protein Kinase

Physiological Significance

The physiological significance of steroid-independent activation of the ERα is best documented for the growth regulation of the rodent uterus. Both estrogen and peptide growth factors stimulate uterine growth, and both types of signals depend on ERα function. In ERα knockout mice, IGF-1 signaling normally activates Akt and MAPK but fails to stimulate uterine growth; moreover, an ERE-luciferase transgene is activated by either IGF-1 or estrogen. Steroid-independent activation of progesterone receptor by dopamine may be involved in the lordosis response in female rodents. EGF can also elicit this behavioral response, but in this case it depends on ERα rather than PR. But despite the elegance of these *in vivo* experiments, there is no formal proof yet in this system that ERα and PR are transcriptionally activated by stimulation with EGF and dopamine, respectively. It is still possible that these nuclear receptors must be there as competence factors at some point preceding the stimulation. Essentially all of the other evidence comes from studies with tissue culture cell lines. For example, estrogen-dependent breast cancer cell lines can lose their steroid- but not their steroid receptor-dependence upon overexpression of the tyrosine kinase receptor HER-2 (Lui *et al.*, 1995). Considerable circumstantial evidence has led to the speculation that steroid-independent activation of ERα contributes to breast cancer progression, notably to estrogen-independent and tamoxifen resistance of ERα-positive tumors. A large number of growth factors are indeed expressed in breast tumors, and act in both autocrine and paracrine fashions (Ethier 1995). In addition to extracellular factors, HER-2 and cyclin D1 play a role and are often overexpressed in breast tumors. In transfection experiments, both are able to activate the ERα in the absence of estrogen and overexpression of HER-2 renders breast cancer cell lines less dependent on estrogen and reduces their tamoxifen sensitivity. Thus, these pathways may not only stimulate the proliferation of ERα-dependent breast tumor cells by themselves, but they must also be considered for their potential to pervert the action of estrogen antagonists by switching them to agonists. By stimulating the AF-1 function of ERα, growth factors reduce the requirement, for AF-2, and render ERα "resistant" to antagonists such as tamoxifen, which primarily block AF-2. Even worse, tamoxifen may then even facilitate the activation of ERα. Remarkably, this phenomenology in tissue culture seems to correlate with the observation that tamoxifen in a small percentage of breast cancer patients, ends up stimulating tumor progression. In any case, HER-2 overexpression seems to predict a poor response to endocrine therapy (Lui *et al.*, 1995), but the weakness of the correlation, emphasizes that signaling crosstalk is but one of several causes of tamoxifen resistance.

Nongenomic Signaling by Steroid Receptors

There have been several reports over the decades that steroid hormones also signal through nongenomic pathways with very rapid cellular effects. (Cato *et al.*, 2002). Within seconds or minutes, steroids can induce an increase in several second messengers such as inositol triphosphate, cAMP, Ca2+, and the activation of MAPK and PI3 kinase (Aronica *et al.*, 1994).

Biological responses attributed to short-term effects range from increased levels of vasodilator nitric oxide in endothelial cells, Ca+ accumulation in maturing oocytes and granulosa cells, neuro-protection, and bone protection to the stimulation of proliferation of carcinoma cells. The molecular mechanisms and notably the receptor have only recently begun to be unraveled. It now appears that at least some of the reported effects are mediated by the nucleus localized estrogen receptors. However, it is important to note that they clearly do not account for all nongenomic phenomena elicited by estrogens. Several unrelated membrane receptors contribute to a large diversity of rapid responses.

The conclusion that the known steroid receptors can themselves be at the top of a signaling cascade for rapid responses is primarily based on the following findings:

1. Rapid nongenomic responses to estrogen, can be reconstituted in unresponsive cell lines by transient expression of ER and the responses seem to be independent of the transcriptional competence of the steroid receptor.
2. A small fraction of the total cellular (and mostly nuclear) ER complement associates with the cell membrane, extra nuclear localization is necessary, and cytoplasmic sequestration by a variant of the metastatic tumor antigen 1 promotes nongenomic signaling; in the case of ER signaling to eNOS in endothelial cells, this nongenomic response can even be demonstrated with isolated caveolae, a specialized membrane compartment.
3. ER can be immunoprecipitated with a whole series of signaling molecules that are typical of signaling pathways triggered by membrane receptors.
4. Antibodies to " nuclear " ERα added to the culture medium can modulate estradiol responses. In line with their proposed membrane association, the ERs have been shown to couple to a large number of signaling molecules (Figure 11.4). Signaling of ERα to eNOS takes place in

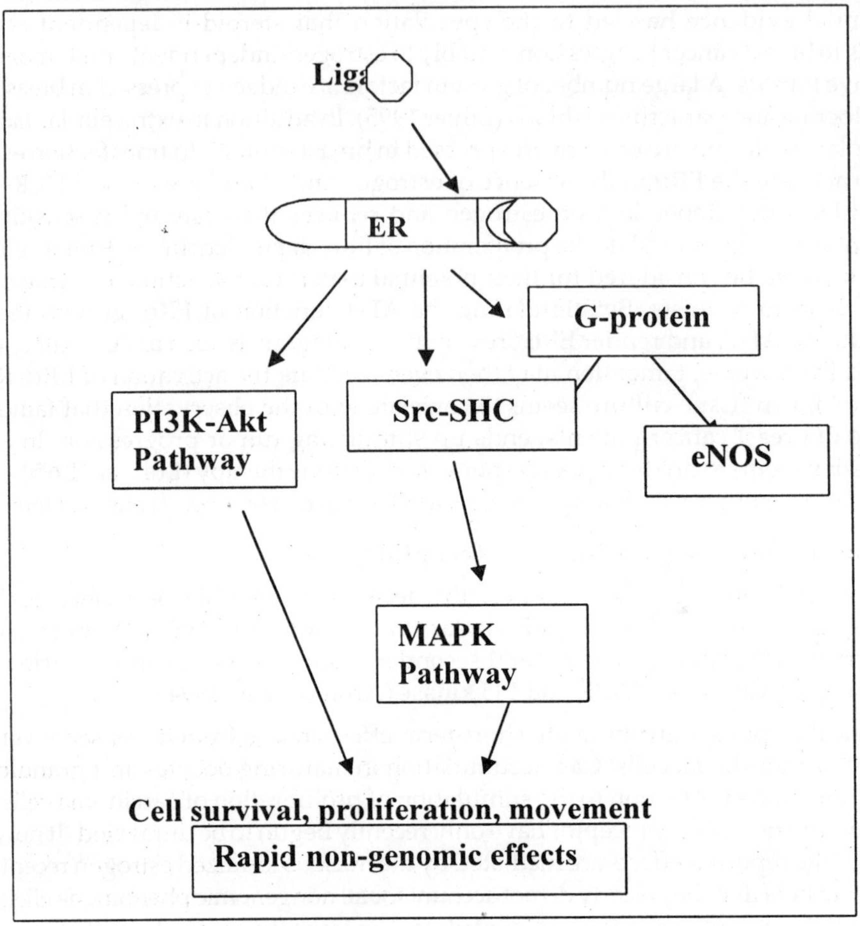

Figure 11.4: Schematic Diagram of Non-genomic Action of Estrogen

caveolae and involves the recruitment of a Gαi protein to ERα (Wyckoff *et al.*, 2001). ERs are also coupled to other types of G proteins. Many groups have reported that activation of estrogen receptors can activate the MAPK pathway.

ERα can elicit this rapid response by binding and activation of c-Src. Additional adaptor proteins may also be involved in MAPK activation by estrogen since several reports demonstrate a role for SHC and its adaptor proteins Grb2 and Sos, and the modulator protein MNAR promotes the formation and activity of ternary complexes between ERα and c-Src. The formation of ternary ERα-Src-PI3K complexes may explain why both PI3K and the Src-MAPK pathway have been placed downstream of ERα.

Surprisingly, the kinetics of activation of PI3K by estradiol has been shown to be slower than that of MAPK. As a result of the activation of PI3K by ERα, Akt is activated. In adipocytes, MAPK activation by estradiol requires a G protein, Src, PI3K as well as protein kinase C (PKC). Signaling pathways are likely to differ substantially between cell types and experimental systems. For example, the requirement for c-Src for estradiol signaling mentioned above has not always been observed. Moreover, the existence of negative and positive feedback mechanisms may often render a linear view of signaling inappropriate.

Indeed, the IGF-I receptor is both upstream and downstream of ERα in a pathway leading to the activation of MAPK. Likewise, Akt is both upstream and downstream of ERα. A prediction from this type of positive feedback is that the system should respond to lower levels of stimuli.

Signaling Crosstalk and Endocrine Disrupting Compounds (EDC)

Signaling cross talk modifies and enhances the response to extracellular signals. In this context, two types of interactions with EDC's (xenoestrogens, metalloestrogens and phytoestrogens) that are steroid receptor ligands must be considered. (1) Their agonistic or antagonistic genomic effects could be altered by crosstalk with other signals. (2) In addition to their well-documented genomic effects, EDC's could elicit nongenomic responses mediated by membrane-associated steroid receptors, a possibility that has not yet received much attention (Nadal *et al.*, 2000). By acting synergistically, extracellular factors might be expected to lower the EC50 and/or to increase the maximal transcriptional response of EDC's. Thus, signaling crosstalk must be considered as an additional risk factor because the disruptive potential of EDC's may otherwise be underestimated. Studies with estrogen-dependent breast cancer cells have shown an increased proliferative response to low levels of estrogens upon constitutive activation of the MAPK pathway. Augmented MAPK activity may also contribute to the estrogen hypersensitivity that ERα-dependent breast cancer cells acquire upon long-term estrogen deprivation (Nazareth and Weigel 1996). Moreover, as in the case of tamoxifen and RU486, signaling crosstalk could convert weakly antagonistic EDC's into weak agonists.

Estrogen Signal Integration

Action of estrogen through receptors both nuclear and membrane/cytoplasmic pools activates or represses gene transcription and subsequent protein expression. It is also reported that membrane ER signaling through kinase cascades, calcium, and other second messengers impacts transcription and expression. But many sex steroid effects reflect the integration of actions at several receptor pools. Work over the past decade has identified important mechanisms whereby membrane initiated signaling through growth factor receptors or membrane ER impacts the function of nuclear ER. Integrative signaling by E2 from several sites in the cell results both in rapid and prolonged actions, and provides plasticity for the cell response to the estrogen. An example of signal from the membrane ER to the nucleus potentially impacts various functions of the sex steroid. In breast cancer, the endogenous membrane ER crosstalks with the *trans* activation of the EGF, ErbB2 and IGF1 receptors (Fujimoto and Katzenellenbogen 1994). This results in downstream signaling to ERK MAP kinase and PI3/AKT kinases. Important non-receptor tyrosine kinases such as Src are also involved as intermediary steps. ERK and PI3 kinase phosphorylate discrete residues of the endogenous nuclear ER, upregulating its transcriptional activity or stability. In response to EGF or IGF1, the nuclear ER can be activated in steroid ligand independent fashion (Kushner *et al.*, 2000).

Knock Out Mice and ER

Several mouse models for the study of estrogen function are available today. These include the knockout of aromatase (Ar–/–), ERα (ERα–/–), ERβ (ERβ –/–), as well as the knockout of both ERα and ERβ (ERα–/–ERβ –/–). Since aromatase (CYP19) is the enzyme, which converts C19 steroids to estrogens, its inactivation means that these mice are completely free of E2 and all of its metabolites. Complete characterization of the abnormalities in all these four E2-related gene knockout models is still under extensive investigation but the phenotypes of these mice have provided us with an outline of the roles of E2 and its receptors. It is clear that, from the differences between the three single knockout mouse models, both ERα and ERβ have distinct non-redundant physiological roles; in addition, the relatively milder phenotype of Ar–/– mice suggests the presence in the body of estrogenic molecules other than and in addition to E2 (Korach *et al.*, 2003).

ER and Cancer Treatment

Although estrogen receptors have been implicated in human breast cancer, their exact role remains elusive. The choice of treatment of breast cancer depends on the presence or absence of ER in the tumor. Shao and Brown (2003) have reviewed the current strategies used in endocrine therapy against breast cancer. According to the authors the drugs that target ER fall into two major classes: these include Selective Estrogen Receptors Modulators (SERM) such as Tamoxifen and the pure antiestrogens such as Fulvestrant (Flasodex), both of which reduce estrogen concentrations. Tamoxifen exhibits antagonistic effect in the breast and agonistic effect in the bone, uterus, and serum lipid profile and cardio vascular system. This agonistic action of tamoxifen raised concerns on the overall risk: benefit ratio on the prophylactic uses of tamoxifen against breast cancer. There is an urgent need to design new SERMs with desired clinical profiles, which can be achieved by in-depth understanding of their mechanism of action.

Another fact that has generated considerable interest among pharmacists is the distinct patterns of tissue distribution of the two receptors in the body and has opened up new sites and means for pharmacological interventions in clinical conditions such as osteoporosis, stroke, hypertension, infertility, hyperplasia, dementia, depression and cancer (Gustafsson, 2003).

ER and Endocrine Disrupting Substance

There is increasing concern over the unknown effects of various Endocrine Disrupting Substance (EDS) released into the environment on the reproduction of humans and other species. Threats to the reproductive capabilities of birds, fish, and reptiles have become evident and similar effects in humans have been proposed (Nimrod and Benson 1996, Gimeno *et al.*, 1996). In the past 50 year, the incidence of testicular cancer and developmental abnormalities of the male reproductive tract appear to have increased in some developed countries (Sharpe and Skakkebaek 1993). Several reports have also provided evidence for a decline in semen quality and/or sperm count over the same period, although this change may not be universal. Male offspring born to mothers who were given diethylstilbestrol (DES), a very potent synthetic estrogen, to prevent miscarriages have an increased incidence of undescended testes, urogenital tract abnormalities, and reduced semen quality compared with those from mothers who did not take DES. Among steroids, Endocrine Disrupting Substance (EDS) could be classified into three groups estrogen mimics, anti-estrogen and anti-androgens. Steroid hormones and endocrine disruptors directly interact with intracellular steroid receptors, which leads to the recruitment of cofactors for specific interaction with hormone responsive elements (HREs) in hormone responsive genes. EDS could also interact with receptors for growth factors or neurotransmitters to up-regulate mitogen-activated protein (MAP) kinase signaling. This leads to the ligand independent activation of steroid receptors (Indarto and Izawa 2001). EDS could also bind with ERs and bring about change in intracellular signaling pathway. Apart from this, EDS could also inactivate or destroy endogenous hormones (Sonnenschein and Soto 1997). *In vitro* assessment techniques that are cheap and easy to handle to assess the potency of EDC's are the need of the hour (Mueller 2004). It is therefore vital that we have deep understanding of the signaling pathways and their crosstalk with EDC's

ER and Phytoestrogens

Human diet contains several plant-derived, nonsteroidal weakly estrogenic compounds (Korach *et al.*, 1994). They are found in plant products (phytoestrogens) or plant products infected by fungi (mycoestrogens). Phytoestrogens can be divided into three main classes: flavonoids (flavones, isoflavones, flavanones and chalcones) such as genistein, naringenin, and kaempferol; coumestans (such as coumestrol); and lignans (such as enterodiol and enterolactone). Mycoestrogens are mainly zearalenone (resorcylic acid lactone) or their derivatives. Messina *et al.* (1997) observed that incidence of breast and prostate cancers is low in countries where intake of phytoestrogen is significantly higher. This observation suggested their use as chemopreventive agents. Phytoestrogens are believed to exert their chemopreventive action by interacting with estrogen receptors, although alternative mechanisms, most notably inhibition of protein tyrosine kinase activity, have been proposed by Messina *et al.* (1997) and, Kurzer and Xu (1997). Kuiper *et al.* (1998) evaluated the potency of phytoestrogens on two types of ERs and listed them based on relative binding affinity (RBA) values. 17β-estradiol >> zearalenone = coumestrol > genistein > daidzein > apigenin = phloretin > (biochanin A) = kaempferol = naringenin >formononetin = ipriflavone = quercetin = chrysin for ERα and 17β-estradiol >>genistein = coumestrol > zearalenone > daidzein > (biochanin A) = apigenin = kaempferol = naringenin > phloretin = quercetin = ipriflavone = formononetin = chrysin for ERβ. The estrogenic potency of compounds is a complicated phenomenon that is being attributed to a number of factors, such as differential effects on the transactivation functionalities of the receptor, coactivators recruitment and is also cell- and target gene promoter-specific (Brzozowski *et al.*, 1997). The extent of interplay among these factors at the moment remains unknown because of the extensive crosstalk in the signaling pathways.

Conclusion

Estrogen receptors are localized to many sites within the cell, potentially contributing to overall estrogen action. In the nucleus, estrogen mainly modulates gene transcription, and the resulting protein products determine the cell biological actions of the sex steroid. In addition, a small pool of ER localize to the plasma membrane and signal mainly though coupling, directly or indirectly, through GTP activating (G) proteins. In response to steroid, signal transduction modulates both non-transcriptional and transcriptional events and impacts both the rapid and more prolonged actions of estrogen. Crosstalk from membrane-localized ER to nuclear ER can be mediated through growth factor receptor tyrosine kinases, such as EGFR and IGF1R. Growth factor receptors enact signal transduction to kinases such as ERK and PI3 kinase that phosphorylate and activate nuclear ER, and this can also occur in the absence of sex steroid. A complex relationship between the membrane and nuclear effects of estrogen also involves membrane-initiated phosphorylation of co-activators, recruiting these proteins to the nuclear transcriptosome. The integration of sex steroid effects at distinct cellular locations of its receptor leads to important cellular physiological outcomes, and are manifest in both reproductive and non-reproductive organs. This in-depth knowledge on ER biology was greatly made possible due to the development of ER knockout mice. Researchers worldwide are unraveling the intricate interplay between ERα, ERβ and their respective isoforms and are continuing to discover new modes of action in ER signaling networks. More improved design of ERα and ERβ-selective ligands and identification of subtype-specific signaling pathways will help in planning novel therapeutic strategies for the treatment of estrogen-dependent pathologies.

Acknowledgement

Wright Jacob thanks the ICMR and the AICTE for the award of fellowships for carrying out this work. The authors are also thankful to DST, Government of India for financial support.

References

Brzozowski AM, Pike ACW, Dauter Z, Hubbard RE, Bonn T, Engstrom O, Ohman L, Greene GL, Gustafsson J-Å, and Carlquist M. (1997). Molecular basis of agonism and antagonism in the oestrogen receptor. Nature, 389, 753–758.

Cato AC, Nestl A, and Mink S. (2002). Rapid Actions of Steroid Receptors in Cellular Signaling Pathways. Science's STKE RE9.

Endoh H, Maruyama K, Masuhiro Y, Kobayashi Y, Goto M, Tai H, Yanagisawa J, Metzger D, Hashimoto S, and Kato S. (1999). Purification and identification of p68 RNA helicase acting as a transcriptional coactivator specific for the activation function 1 of human estrogen receptor α Mol. Cell. Biol., 19, 5363–5372.

Ethier SP. (1995). Growth factor synthesis and human breast cancer progression J. Natl. Cancer Inst., 87, 964-973.

Fujimoto N, and Katzenellenbogen BS.(1994). Alternation in the agonist/antagonist balance of antiestrogens by activation of protein kinase A signaling pathways in breast cancer cells: antiestrogen selectivity and promoter dependence. Mol. Endocrinol., 8, 296-304.

Gimeno S, Gerritsen A, Bowmer T, and Komen H. (1996). Feminization of male carp. Nature, 384, 221–222.

Greenblatt J. (1997). RNA polymerase II holoenzyme and transcriptional regulation. Curr. Opin. Cell Biol., 9, 310–319.

Guiochon–Mantel A, Delabre K, Lescop P, and Milgrom E. (1996). Intracellular traffic of steroid hormone receptors. J. Steroid Biochem. Mol. Biol., 56, 1–6.

Gustafsson JA. (2003). What pharmacologists can learn from recent advances in estrogen signaling Trends Pharmacol. Sci., 24, 479-485.

Horwitz K, Jackson T, Bain D, Richer J, Takimoto G, and Tung L. (1996). Nuclear receptor coactivators and co-repressors. Mol. Endocrinol., 10, 1167–1177.

Indarto D, and Izawa M. (2001). Steroid Hormones and Endocrine Disruptors: Recent Advances in Receptor–Mediated Actions. Yonago Acta medica, 44, 1–6.

Kerppola TK, and Curran T. (1997). The transcription activation domains of Fos and Jun induce DNA bending through electrostatic interactions. EMBO J., 16, 2907–2916.

Kim J, DeHaan G, Nardulli AM, and Shapiro DJ. (1997). Prebending the estrogen response element destabilized binding of the estrogen receptor DNA binding domain. Mol. Cell. Biol., 17, 3173–3180.

Klein–Hitpass L, Ryffel GU, Heitlinger E, and Cato AC. (1988). A 13 bp palindrome is a functional estrogen responsive element and interacts specifically with estrogen receptor. Nucleic Acids Res., 16, 647–63.

Klinge CM. (2000). Estrogen receptor interaction with co-activators and co-repressors. Steroids, 65, 227–251.

Korach KS, Emmen JMA, Walker VR, Hewitt SC, Yates M, Hall JM, Swope DL, Harrell JC, and Couse JF. (2003). Update on animal models developed for analyses of estrogen receptor biological activity. J. Steroid Biochem. Mol. Biol., 86, 387–391.

Korach KS, Migliaccio S, and Davis VL. (1994). Estrogens. *In:* Munson PL. (Ed.), *Principles of Pharmacology-Basic Concepts and Clinical Applications.* Chapman and Hall, New York, pp 809–825.

Kuiper GG, Lemmen JG, Carlsson B, Corton JC, Safe SH, van der Saag PT, van der Burg B, and Gustafsson JA. (1998). Interaction of estrogenic chemicals and phytoestrogens with estrogen receptor beta. Endocrinol., 139, 4252-63.

Kurzer MS, and Xu X. (1997) Dietary phytoestrogens. Ann. Rev. Nutr., 17, 353–381.

Kushner PJ, Agard DA, Greene GL, Scanlan TS, Shiau AK, Uht RM, and Webb P. (2000). Estrogen receptor pathways to AP-1. J. Steroid Biochem. Mol. Biol., 74, 311-317.

Lavinsky RM, Jepsen K, Heinzel T, Torchia J,Mullen TM, Schiff R, Del Rio RL, Ricote M, Ngo S, Gemsch J, Hilsenbeck SG, Osborne GK, Glass GK, Rosenfeld MG, and Rose D.W. (1998). Diverse signaling pathways modulate nuclear receptor recruitment of N-CoR and SMRT complexes. Proc. Natl. Acad. Sci. (USA), 95, 2920–2925.

Liu Y, el-Ashry D, Chen D, Ding IY, and Kern FG. (1995). MCF-7 breast cancer cells overexpressing transfected c-erbB-2 have an in vitro growth advantage in estrogen-depleted conditions, reduced estrogen-dependence, tamoxifen-sensitivity in vivo. Breast Cancer Res. Treat., 34, 97-117.

Mani SK, Allen JM, Lydon JP, Mulac-Jericevic B, Blaustein JD, DeMayo FJ, Conneely O, and O'Malley BW. (1996). Dopamine requires the unoccupied progesterone receptor to induce sexual behaviour in mice. Mol. Endocrinol., 10, 1728-1737.

McKenna NJ, Lanz RB, and O'Malley BW. (1999). Nuclear receptor co-regulators: cellular and molecular biology. Endocr. Rev., 20, 321–344.

Messina MJ, Persky V, Setchell KDR, and Barnes S. (1994). Soy intake and cancer risk: a review of the *in vitro* and *in vivo* data. Nutr. Cancer, 21, 113–131.

Moggs JG and Orphanides G. (2001). Estrogen receptors:orchestrators of pleiotrophic cellular responses. EMBO reports, 2, 775-781.

Mosselman S, Polman J, and Dijkema R. (1996). ERβ: Identification and characterization of a novel human estrogen. FEBS Lett., 392, 49-53.

Mueller SO. (2004). Xenoestrogens: mechanisms of action and detection methods. Ann. Bioanal. Chem., 378, 582-587.

Mueller SO, and Korach KS. (2001). Mechanisms of estrogen receptor-mediated agonistic and antagonistic effects. *In:*. Metzler M (Ed), *Endocrine Disruptors*, Springer, Berlin Heidelberg New York, pp 125.

Muramatsu M, and Inoue S. (2000). Estrogen Receptors: How Do They Control Reproductive and Nonreproductive Functions? Biochem. Biophys. Res. Commun., 270, 1–10.

Nadal A, Ropero A.B, Laribi O,Maillet M, Fuentes E, and Soria B. (2000). Nongenomic actions of estrogens and xenoestrogens by binding at a plasma membrane receptor unrelated to estrogen receptor alpha and estrogen receptor beta. Proc Natl Acad Sci (USA), 97, 11603–11608.

Nilsson S, and Gustafsson JA. (2000). Estrogen receptor transcription and transactivation: Basic aspects of estrogen action. Breast Cancer Res., 2, 360–366.

Nimrod AC, and Benson WH. (1996). Environmental estrogenic effects of alkylphenol ethoxylates. Crit. Revi. Toxicol., 26, 335–363.

Pratt WB, and Toft DO. (1997). Steroid receptor interactions with heat shock protein and immunophilin chaperones. Endocr. Rev., 18, 306–60.

Robyr D, Wolffe AP, and Wahli W. (2000). Nuclear Hormone Receptor Coregulators In Action: Diversity For Shared Tasks. Mol. Endocrinol., 14, 329-347.

Ruff M, Gangloff M, Wurtz JM, and Moras D. (2000). Estrogen receptor transcription and transactivation structure–function relationship in DNA- and ligand-binding domains of estrogen receptors. Breast Cancer Res., 2, 353–359.

Shao W, and Brown M. (2004). Advances in estrogen receptor biology: prospects for improvements in targeted breast cancer therapy. Breast Cancer Res., 6, 39-52.

Sharpe RM, and Skakkebaek NE.(1993). Are oestrogens involved in falling sperm counts and disorders of the male reproductive tract? Lancet, 341, 1392–1395.

Sonnenschein C, and Soto AM. (1998). An updated review of environmental estrogen and androgen mimics and antagonists. J. Steroid Biochem. Mol. Biol., 65, 143–150.

Wyckoff MH, Chambliss KL. Mineo C, Yuhanna IS, Mendelsohn ME, Mumby S M, and Shaul PW. (2001). Plasma membrane estrogen receptors are coupled to endothelial nitric-oxide synthase through Gαé. J. Biol. Chem., 276, 27071–27076.

12

Ovarian Tissue Remodeling: The Role of Extra Cellular Matrix

Samir Kumar Saha[1], *Pamela Ghosh*[1], *Debanjali Mitra*[1], *Aditya Konar*[1], *Samir Bhattacharya*[2], *Satinath Mukherjee*[3] *and Sib Sankar Roy*[1]*

[1]*Molecular Endocrinology Laboratory, Indian Institute of Chemical Biology, 4 Raja S. C. Mullick Road, Kolkata – 700 032, India*

[2]*School of Life Science, Visva Bharati University, Santiniketan – 731 235, India*

[3]*Department of Endocrinology and Metabolism, IPGMER and SSKM Hospital, Kolkata, India*

ABSTRACT

Ovaries can be considered very dynamic organ in which follicles and corpora lutea continually grow and regress. Follicles have both epithelial and stromal layers in which cell migration or movement, cell division, specialization and differentiation, and death occur. A fluid-filled antrum develops and at ovulation the epithelial cells undergo an epithelial to mesenchymal transition into luteal cells. Although growth factors and hormones are very important in some of these processes, the extracellular matrix participates in all of them. On atresia the follicular basal lamina is not degraded as occurs at ovulation. The matrix component present, change structurally and biochemically during follicular development. Resent studies are directed at investigating the regulation of the matrix and its function in the ovary. MMP and TIMP isoforms play critical role in remodeling the components of ECM during the different developmental stages of ovarian tissue.

**Corresponding Author*: E-mail: sibsankar@iicb.res

Abbreviations

ECM: Extra Cellular Matrix, MMP: Matrix Metalloproteinase, TIMP: Tissue inhibitor of Metalloproteinase, LH: Luteinizing hormone, GF: Growth factor.

Introduction

The mammalian ovary produces mature oocytes capable of being fertilized and sustaining embryonic development. It also produces steroid hormones. These functions of the ovary are accomplished by follicles and corpora lutea. The ovary contains a reserve of inactive primordial follicles. Each contains a small non-growing oocyte and a layer of non-dividing pregranulosa cells encapsulated by the follicular basal lamina. Each day a number of primordial follicles become active, and the oocyte commences growing and the granulosa cells begin to divide. As the granulosa cells divide, the number of layers of granulosa cells (called the membrana granulosa or follicular epithelium) around the oocyte increases, and the basal lamina expands. Later in development a fluid-filled cavity or antrum forms in the follicles. Only follicles that reach the stage of having a large antrum ovulate their oocytes in response to the surge release of LH. On ovulation, the epithelioid granulosa cells redifferentiate into the parenchymal cells of the corpus luteum. Non-ovulating follicles undergo atresia and regression.

These temporary endocrine units within the adult ovary (follicles and corpora lutea) continually develop and regress, thus giving rise to the day-to-day variation in hormone secretion by the ovary. The formation of follicles and corpora lutea requires tissue remodeling, cellular multiplication and its specialization. Ovarian cells respond to external stimuli that signal changes in gene regulation, growth or differentiation. Considerable research effort has been directed at how hormones and growth factors regulate such processes (Luck, 1994; Rodgers RJ *et al.*, 1999, 2000). The ECM provides specialized microenvironment in ovarian tissue that is helpful for maintenance of normal physiological and endocrine function of ovary. The ECM is generally a diverse mixture of components, therefore the composition of the matrix is important. MMP and TIMP isoforms play critical role in stabilizing the ECM status, thus these enzymes are essential for maintaining the normal ovarian texture and function. The present article attempts to review the importance and relationship among ECM, MMPs and TIMPs to perform these functions.

Role of ECM in Follicle Development

The extracellular matrix has many different roles (Hay, 1991). These include effects on cell shape and behaviour, such as migration, division, differentiation, cell death and cell anchorage. All these behaviours occur in follicle development. The matrix can play a role in the fluid dynamics of a tissue, either providing osmotic forces which could be important for follicular fluid formation, or filtering soluble materials, which could occur across the follicular basal lamina. The matrix can provide rigid or elastic mechanical support for tissues. In addition, nutrients and hormones and other extracellular signals are often required to traverse the matrix in order to reach target cells. The matrix has the ability to bind growth factors including those found in follicles. In essence, the extracellular matrix defines or rather provides specialized microenvironments. It thus determines both cell and tissue function. It is not surprising then that the extracellular matrix is generally a diverse mixture of components, and that the composition of the matrix is important.

Granulosa Cells-ECM Interactions

A number of studies on the effects of the components of basal lamina and other ECM components have been undertaken by a number of workers in a variety of species. One model ideal for studying

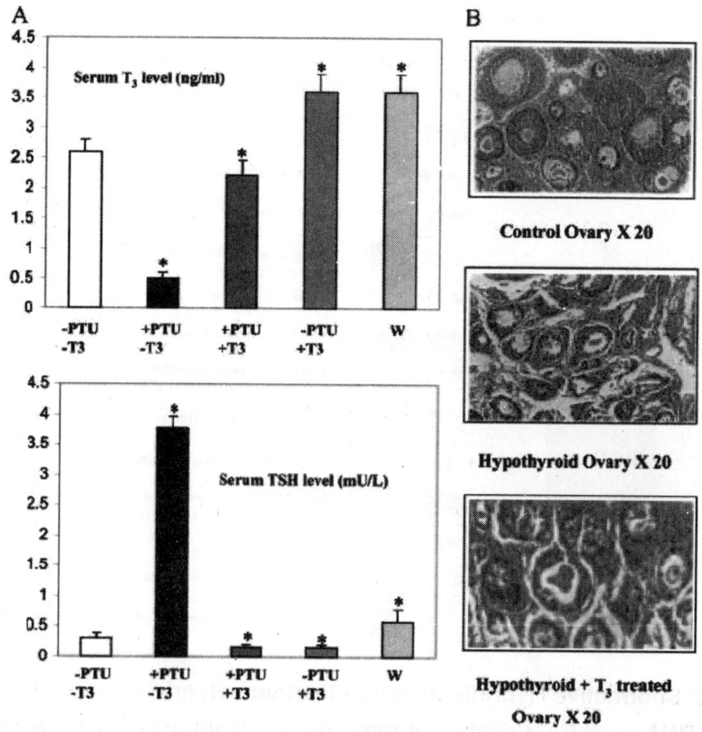

Figure 12.1: Effect of PTU on Serum T_3 and TSH Level and Histological Study of the Hypothyroid Rat Ovary

T_3 and TSH were measured in blood serum collected from control (-PTU-T_3), hypothyroid (+PTU-T_3), hypothyroid+ T_3 (+PTU +T_3), control + T_3 (-PTU +T_3). PTU was withdrawn from the drinking water after two weeks of treatment (W). Three experiments were performed in duplicate, the data represented as mean + SD. *, $P<0.05$ (Figure 1A). Figure 1B shows the histological staining of control and hypothyroid rat ovary. The histological sections of rat ovary were stained with haematoxylin and eosin. It shows that both number and size of the ovarian follicles in hypothyroid condition were drastically reduced compared to control. The hypothyroid +T_3 treated ovarian section shows indication of recovery of number and size of the follicles.

effects of the follicular basal lamina on granulosa cells is that of chicken follicles. Methods for isolating the follicular basal lamina of chickens have been available since the late 1970s (Gilbert *et al.*, 1977) but it is only recently that it has been used for this purpose (Asem *et al.*, 2002). In addition, other components of basal lamina are available commercially: laminin 1 ($\alpha 2\beta 1\gamma 1$) and laminin 2 ($\alpha 2\beta 1\gamma 1$), and even an assembled laminin 1 matrix (Matrigel). These have been used to determine effects of these components on follicular cells in culture. However, it is the components of laminin 3 ($\alpha 2\beta 1\gamma 1$) that are present in the follicular basal lamina (Rodgers *et al.*, 1999, 2000). Other matrix molecules (for example collagens to coat surfaces or three-dimensional matrices, vitronectin, fibronectin, peptides containing cellular attachment sequences) are available for culturing granulosa cells but none of these truly represent the appropriate matrix to mimic that found *in vivo*. Cells interact with the matrix through cell-surface adhesion receptors including the integrins. Integrins are heterodimeric glycoproteins composed of α and β subunits. Over seventeen α and eight β subunits are known, producing over 23 different

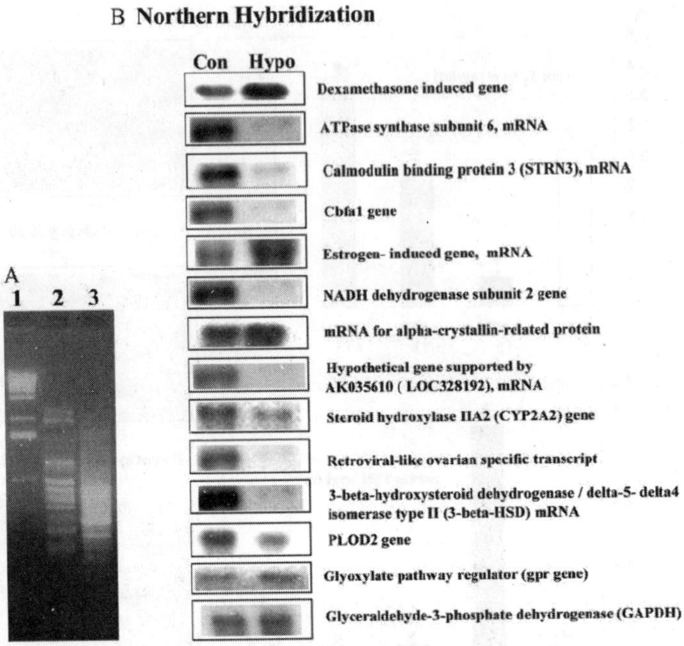

Figure 12.2: Subtractive Hybridisation and Northern Hybridisations of Selective Clones

3 mg poly (A+) RNA each from control or hypothyroid granulosa cells used as tester and driver and vice versa. Two rounds of PCRs were performed with these cDNAs and then subjected to agarose gel electrophoresis and loading on each lane as follows: DNA marker, *i.e.* 1 Kb DNA ladder parched from In Vitrogen, USA, (lane1), forward subtracted clones (lane2), reverse subtracted clones (lane 3) (Figure 2A). After secondary screening by dot-blot hybridisation the selective clones were again screened (tertiary screening) by Northern hybridisation. Figure 2B shows the northern hybridisation results of some of the differentially expressed genes that we have got by subtractive hybridisation.

heterodimeric combinations (Belkin, 2000). Additional molecules within the cell associate with integrins in order for a complex molecular cell-matrix junction to form. These junctions or 'focal adhesions' can transduce multiple intracellular signals as well as provide the cells with anchorage. However, it is not simply a ligand-receptor interaction that determines the effects of the matrix on cells. Only a few integrins have been localized to granulosa cells: $\alpha6\beta1$ in non-luteinized granulosa cells (Fujiwara *et al.*, 1997; Le Bellego *et al.*, 2002) and $\alpha2$ and $\alpha5$ in luteinizing cells. Tetraspanins also interact with the integrins as part of the process by which the matrix links with intracellular signaling molecules, and indeed one of the most highly expressed proteins in granulosa cells from ovulating follicles was identified long ago (Rapp *et al.*, 1990).

ECM: Its Turnover and Degradation

During follicle development, continual remodeling of the follicular wall occurs as the follicle enlarges. In particular, turnover of structural collagens of the theca externa and in nearby stroma and tunica allow for expansion of the follicle. The follicular basal lamina must also expand with follicle growth. The extent of vasculature within the theca layers must also increase to keep up with follicle

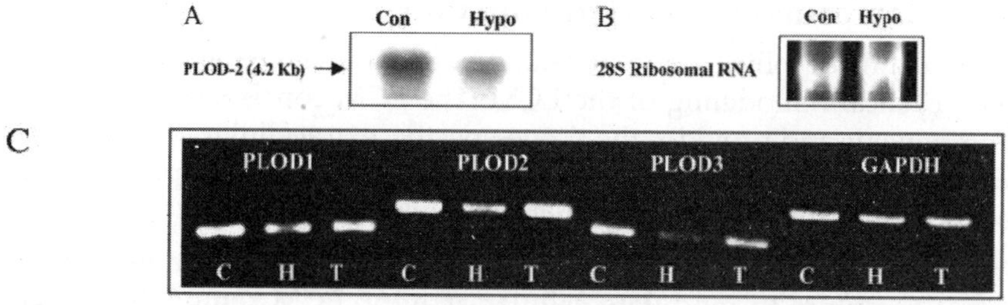

Figure 12.3: Plod Isoforms are Down-regulated in Hypothyroid Condition and the Effect of T3 in their Expression in Ovary

A, Northern hybridization of Plod-2 cDNA. This Northern blot was hybridized with radiolabeled Plod-2 cDNA obtained from subtractive hybridization. B, 28S rRNA from the same gel to show equal loading of RNAs in each lanes. C, Expression of Plod-1,–2, and–3 in the ovary of control, hypothyroid, and T3 injected-hypothyroid rats by semiquantitative RT-PCR. The gene-specific oligonucleotide primers of Plod-1,–2, and–3 and GAPDH primers (for loading control) were used for RTPCR and the amplified products were loaded in agarose gel as control (C), hypothyroid (H), and T3 injected-hypothyroid animals (T). All the experiments were performed three times in duplicate, and the mean + SD values have been shown. *, P =0.05.

enlargement, and this probably involves all three mechanisms of angiogenesis: sprouting, intussusception and elongation. Hence, the vasculature-associated matrix must also increase. These processes of matrix turnover require discrete control, as the final result is an expansion of the matrix, not a total degradation. The precise mechanism by which this occurs is not well understood, but clearly involves simultaneous degradation and synthesis of the matrix. Enzymes involved in matrix degradation, including matrix metalloproteinases 1, 2, 9, and 13; tissue inhibitor of metalloproteinases 1 and 2; tissue plasminogen activator; urokinase plasminogen activator; and plasminogen activator inhibitor 1, have been studied in the ovary and recently reviewed by Curry and Osteen (2001).

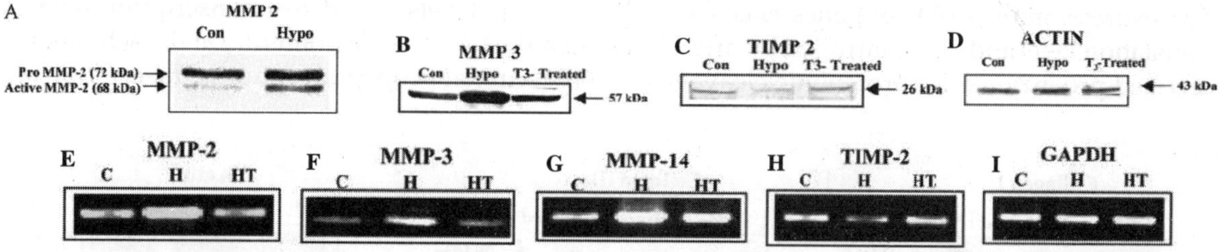

Figure 12.4: Effect of Hypothyroidism on the Expression of MMP-2, MMP-14, MMP-3, and TIMP-2 in Hypothyroid Rat Ovary

A–C, Thirty micrograms of protein from each set were electrophoresed on 10 per cent SDS-polyacrylamide gel, transferred onto PVDF membrane and subjected to Western blot using rabbit anti-MMP-2 antibody, goat anti-MMP-3 antibody, and rabbit anti-TIMP-2 antibody, respectively. Western immunoblotting with mouse antiactin antibody was performed to show equal loading on each lanes (D). E–H, Semiquantitative RT-PCR data of MMP-2, 3, and 14 and TIMP-2, respectively, which was performed with the respective gene-specific primers using ovarian RNA. RT-PCR with GAPDH primers was performed to show equal loading in each lane (I). The RNAs used for the RT-PCR were isolated from control (C), hypothyroid (H), and T3-injected hypothyroid ovaries (HT). All experiments were performed three times in duplicate and the mean + SD values have been shown. *, *P* = 0.05.

MMP and TIMP Isoforms in Ovarian Function

Growth and ovulation of follicles, and development, maintenance, and regression of the corpus luteum depend on cyclical remodeling of the ECM. The ECM consists of proteinaceous and non-proteinaceous components and provides the tissue-specific, extracellular architecture to which cells attach, and modulates the activities of cells through cell surface receptors. Specific components of the ECM are cleaved by MMPs, the activities of which are inhibited by TIMPs.

The cells are attached in the "scaffold" provided by the ECM and contains domains that bind to specific cell surface receptors to modulate cellular activity. ECM remodeling is very important phenomena for many physiological and pathological processes. Examples of ECM remodeling by MMPs during normal physiological processes within the reproductive system include follicular growth, ovulation, corpus luteum formation and regression, endometrial remodeling at menstruation, blastocyst outgrowth, trophoblast invasion, uterine involution, and mammary growth and involution. Examples of MMP involvement in pathological ECM remodeling include wound healing, osteoarthritis, breast carcinoma, and other forms of tumor cell metastasis. Two families of enzymes involved in remodeling of ECM are the MMP family and the plasmin family (plasminogen activator). The role of the plasmin in ovarian function has been reviewed by Ny *et al.* (1993). In this review we want to focus the role of MMPs and TIMPs in ovarian ECM remodeling and the subsequent effects on ovarian function.

MMP

MMPs are zinc- and calcium-dependent enzymes that are largely responsible for degrading ECM proteins (Mignatti and Rifkin, 1993). Individual family members differ in their specificity for ECM components. Twelve members of the MMP family of enzymes have been characterized to date. Although most MMPs appear to be secreted, two membranes bound MMPs, which preferentially activate MMP-2 (gelatinase A), are also known (Sato *et al.*, 1994; Takino *et al.*, 1995). MMPs are secreted as latent proenzymes, which are activated by proteolytic cleavage of an N-terminal peptide. ProMMPs are activated *in vivo* by serine proteases (plasminogen activator and kallikriens; Espey, 1992), mast cell proteases (Suzuki *et al.*, 1995), and other MMPs including plasma membrane MMPs (Sato *et al.*, 1994). The expression of proMMP genes is controlled at various levels, *i.e.* at the transcription level, at translation level and also controlled by mRNA stability. Further control points are at the activation of secreted proenzymes by proteolytic cleavage of an N-terminal propeptide, and the association and

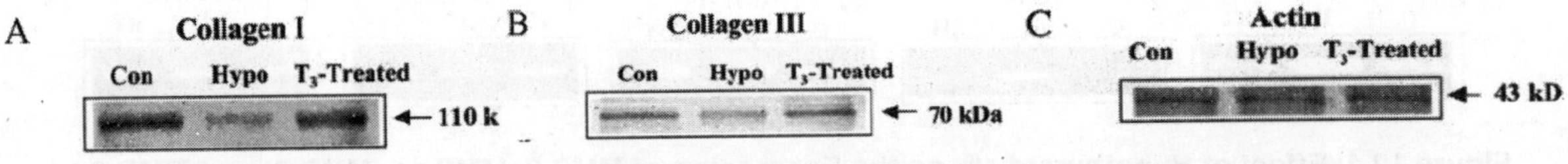

Figure 12.5: Western Immunoblot Shows the Collagen Status in Hypothyroid Condition and the Effect of T3 in their Expression in Ovary

Thirty micrograms of ovarian proteins from each set were fractionated on 10 per cent SDS-polyacrylamide gel, transferred onto polyvinyl difluoride membrane, and subjected to immunodetection with either rabbit anticollagen I (A), rabbit anticollagen III (B), or mouse anti-_-actin antibody as an internal control (C). The lanes indicated by Con, Hypo, and T3-treated represent the protein loaded were isolated from control, hypothyroid, and T3-injected hypothyroid animals, respectively. All experiments were performed three times in duplicate and the mean + SD values have been shown. *, *P*=0.05.

dissociation of MMPs with specific inhibitors. Another way in which ECM remodeling can be regulated spatially and temporally is illustrated by the observation that MMP-2 is localized to the plasma membrane of invasive cells by association with an integrin (Brooks *et al.*, 1996).

TIMPS

The activities of MMPs are regulated by a family of TIMPs, which includes TIMP-1,–2,–3 and–4. All TIMPs share 12 cysteine residues, which are believed to form six disulfide bonds that arrange the molecule into two distinct domains. Despite similar structural properties, TIMP-1,–2,–3, and–4 differs in size, glycosylation, and expression *in vitro* and *in vivo*. Although they are separate gene products, each TIMP inhibits most members of the MMP family by non-covalent binding with a 1:1 stoichiometry and high affinity (Willenbrock, *et al.*, 1994). The extent of ECM remodelling depends upon the ratio of active MMPs to TIMPs. A ratio in favour of MMP activity results in ECM degradation, whereas a ratio favouring TIMPs inhibits ECM degradation and is permissive for ECM deposition. The molecular interactions between TIMPs and MMPs have been studied extensively but much less is known about the post-secretary mechanisms regulating the activity of TIMPs. The mechanisms by which MMPs are liberated from complexes with TIMPs or by which TIMPs are rendered inactive have not been fully investigated. Factors regulating the function of TIMP-1 include: (1) endothelial-stimulating angiogenesis factor (ESAF), liberates active MMP-1 from MMP1-TIMP1 complexes (McLaughlin *et al.*, 1991); (2) proteinases (Itoh and Nagase, 1995); and (3) peroxynitrite, a product of superoxide radicals and hydrogen peroxide (Frears *et al.*, 1996). TIMPs are multifunctional molecules that, in addition to inhibiting MMP activity, stimulate proliferation of various cell types (Edwards *et al.*, 1996), and stimulate progesterone production by steroidogenic cells (Boujrad *et al.*, 1995). The mechanism by which TIMPs stimulate cell proliferation is unknown but may involve plasma membrane receptors (Edwards *et al.*, 1996).

ECM and Follicular Growth

Proliferation of theca and granulosa cells as well as expansion of the basement membrane is required for follicular growth within the confines of the ovarian stroma. The development of a bovine primordial follicle (d 0.1 mm) into a preovulatory follicle (d 19 mm) requires approximately a 36,000-fold increase in follicular surface area. MMPs and TIMPs are obvious candidates for modulating the expansion of the follicular basement membrane and ECM remodelling within the theca interna, also during angiogenesis. The collagenous components of the ovarian stroma include interstitial collagens (types I and III), whereas collagen type IV is the collagenous component of the basement membrane (Luck, 1994). Interstitial collagenase (MMP-1) occurs in theca and granulosa cells at all stages of follicular growth from primordial to Graafian follicles (Tadakuma *et al.*, 1993). Many growth factors, like FGF, TGFb, PDGF, HGF etc. participate in the proliferation and differentiation of ovarian follicular cells and are often constitutively expressed (Logan and Hill, 1992, Fowlkes *et al.*, 1994) and release active TNFa by processing its proform (Moss *et al.*, 1997). Interactions between components of the ECM and growth factors increase their stability, concentrate them near the surface of specific cells, protect them from proteolysis, and alter their biological activity (Flaumenhaft and Rifkin, 1992). So, the regulated proteolytic degradation of the ECM may provide an important mechanism for regulating growth factor availability and activity during follicular growth.

ECM and Ovulation

The preovulatory LH surge initiates a cascade of proteolytic activity within the preovulatory follicular wall that resembles an inflammatory response (Espey, 1992, 1994). The apex region of the preovulatory follicular wall consists of granulosa cells, basement membrane, theca interna, theca externa, tunica albuginea, and superficial epithelium with its underlying basement membrane. The tensile strength of the follicular wall is determined by its collagen content. After a preovulatory LH surge, ovarian collagenase activity increases (Curry *et al.*, 1985) and collagen degradation is greatest in the apical region of the preovulatory follicular wall (Murdoch et al, 1992). In the stigma region, proteolytic degradation of collagenous fibers progressively weakens the follicular wall and this, in conjunction with moderate and constant intrafollicular pressure, culminates in follicular rupture (Espey, 1994).

Hypothyroidism and Ovarian Extra Cellular Matrix

Hypothyroid induced reproductive malfunction in both the sexes is a common phenomenon of global concern. In an attempt to characterize the differentially expressed genes that might be responsible for these disorders, our laboratory has identified a number of differentially expressed clones from hypothyroid rat ovary by PCR-select cDNA subtractive hybridization technique. One of such clone was procollagen lysyl hydroxylase2 (Plod-2), the key enzyme for the first step of collagen biosynthetic pathway, which was down regulated in hypothyroid condition (Saha *et al.*, 2005, Figures 12.1 and 12.2). We have also demonstrated the reduced expression of other isoforms of Plods, namely Plod1 and 3 in hypothyroid rat ovary (Figure 12.3). This report showed that thyroid hormone regulates Plod isoforms in rat ovary. Our study also provides evidence of the differential expression of MMPs and TIMP-2 in hypothyroid ovary (up regulation of MMPs whereas down regulation of TIMPs in hypothyroid rat ovary; Figure 12.4). Therefore, in hypothyroid rat ovary, the enzymes that enhance the collagen biosynthesis are down regulated, concomitantly the enzymes degrade extracellular matrix are up regulated. It has been shown that in hypothyroid condition collagen biosynthesis in ovary seems to be disturbed with (Figure 12.5) concomitant enhancement in collagen degradation resulting in disintegration of overall ovarian structure (Saha *et al.*, 2005).

Conclusion

In conclusion, growth of an ovarian follicle culminating in ovulation or atresia and the development, maintenance and regression of a corpus luteum are cyclic events that depend upon extensive remodelling of the ECM. Components of the ECM modulate cellular processes (including gene expression; cellular proliferation, migration and differentiation; and apoptosis) via cell surface receptors and serve as a reservoir for a number of biologically active factors. Consequently, the controlled degradation of ECM proteins by MMPs may be essential for preserving a microenvironment conducive to follicular and luteal function. The role of TIMPs are also very much important maintaining the ovarian function.

Acknowledgements

We gratefully acknowledge the financial support provided by the Council of Scientific and Industrial Research (CSIR), Govt. of India for funding the project work that have been carried out in the authors' laboratories.

References

Asem EK, Qin W, and Rane SG. (2002). Effect of basal lamina of ovarian follicle on T-and L-type Ca (2^+) currents in differentiated granulosa cells. Am. J. Physiol. Endocr. Metabol., 282, E184-E196.

Belkin AM, and Stepp MA. (2000). Integrins as receptors for laminins Microscopy Res. Tech., 51, 280-301.

Boujrad N, Ogwuegbu SO, Garnier M, Lee C, Martin BM, and Papadopoulos V. (1995). Identification of a stimulator of steroid hormone synthesis isolated from testis. Science, 268, 1609-1612.

Brooks PC, Stromblad S, Sanders LC, von Schalscha TL, Aimes RT, Stetler-Stevenson WG, Quigley JP, and Cheresh DA. (1996). Localization of matrix metalloproteinase MMP-2 to the surface of invasive cells by interaction with integrin αV B3. Cell, 85: 683-693.

Curry TE, and Osteen KG. (2001). Cyclic changes in the matrix metalloproteinase system in the ovary and uterus. Biol. Reprod., 64, 1285-1296.

Curry TE, Jr, Dean DD, Woessner JF, and Le Maire WJ. (1985). The extraction of a tissue collagenase associated with ovulation in the rat. Biol. Reprod., 33, 981-991.

Edwards, D R, Beaudry PP, Laing TD, Kowal V, Leco KJ, Leco PA, and Lim SA. (1996). The roles of tissue inhibitors of metalloproteinases in tissue remodeling and cell growth. Int. J. Obesity, 20, 9-15.

Espey LL. (1992). Ovulation as an inflammatory process. In: Sjoberg N, Hamberger L, Janson P, Owman C, and Coelingh-Bennink H. (Eds.), Local Regulation of Ovarian Function, Parthenon Publishing Group, Park Ridge, pp 183–200.

Espey LL. (1994). Current status of the hypothesis that mammalian ovulation is comparable to an inflammatory reaction. Biol. Reprod., 50, 233-238.

Flaumenhaft R, and Rifkin DB. (1992). The extracellular regulation of growth factor action. Mol. Biol. Cell, 3, 1057-1065.

Fowlkes JL, Enghild JJ, Suzuki K, and Nagase H. (1994). Matrix metalloproteinases degrade insulin-like growth factor-binding protein 3 in dermal fibroblast cultures. J. Biol. Chem., 269, 25742-25746.

Frears ER, Zhang Z, Blake DR, O'Connell JP, and Winyard PG. (1996). Inactivation of tissue inhibitor of metalloproteinase-1 by peroxynitrite. FEBS Lett., 381, 21-24.

Fujiwara H, Honda T, Ueda M, Nakamura K, Yamada S, Maeda M, and Mori T. (1997). Laminin suppresses progesterone production by human luteinizing granulosa cells via interaction with integrin alpha 6 beta 1. J. Clin. Endocr. Metabol., 82, 2122-2128.

Gilbert AB, Evans AJ, Perry MM, and Davidson MH. (1977). A method for separating the granulosa cells, the basal lamina and the theca of the preovulatory ovarian follicle of the domestic fowl (*Gallus domesticus*). J. Reprod. Fertil. 50, 179-181.

Hay ED. (1991). *Cell Biology of Extracellular Matrix*, 2nd Ed. Plenum Press, New York.

Itoh Y, and Nagase H. (1995). Preferential inactivation of tissue inhibitor of metalloproteinases-1 that is bound to the precursor of matrix metalloproteinase 9 (progelatinase B) by human neutrophil elastase. J. Biol. Chem., 270, 16518-16521.

Le Bellego F, Pisselet C, Huet C,Monget P, and Monniaux D. (2002). Laminin alpha6 beta1 integrin interaction enhances survival and proliferation and modulates steroidogenesis of ovine granulosa cells. J. Endocr., 172, 45-59.

Logan A, and Hill DJ. (1992). Bioavailability: is this a key event in regulating the actions of peptide growth factors? J. Endocr., 134, 157-161.

Luck MR (1994) The gonadal extracellular matrix. Oxford Rev. Reprod. Biol., 16, 33-85.

McLaughlin B, Cawston T, and Weiss JB. (1991). Activation of the matrix metalloproteinase inhibitor complex by a low molecular weight angiogenic factor. Biochim. Biophysic. Acta, 1073, 295-298.

Mignatti P, and Rifkin DB. (1993). Biology and biochemistry of proteinases in tumor invasion. Physiol. Rev., 73, 161-195.

Moss ML, Jin SLC, Milla ME, Burkhardt W, Carter HL, Chen W, Clay WC, Didsbury JR, Hassler D Hoffman CR, Kost TA, Lambert MH, Leesnitzer MA, McCauley P, McGeehan G, Mitchell J, Moyer M, Pahel G, Rocque W, Overton LK, Schoenen F, Seaton T, Su J, Warner J, Murdoch WJ, and McCormick RJ. (1992). Enhanced degradation of collagen within apical versus basal wall of ovulatory ovine follicles. Am. J. Physiol., 385, E221-E225.

Ny T, Peng X, and Ohlsson M. (1993). Hormonal regulation of the fibrinolytic components in the ovary. Thrombosis Res., 71, 1-45.

Rapp G, Freudenstein J, Klaudiny J, Mucha J, Wempe F, Zimmer M, and Scheit KH. (1990). Characterization of three abundant mRNAs from human ovarian granulosa cells. DNA Cell Biol., 7, 479-485.

Rodgers RJ, Irving-Rodgers HF, and vanWezel IL. (2000). Extracellular matrix in ovarian follicles. Mol. Cell. Endocr., 163, 73-79.

Rodgers RJ, van Wezel IL, Rodgers HF, Lavranos TC, Irvine CM, and Krupa M. (1999). Roles of extracellular matrix in follicular development. J. Reprod. Fertil. Suppl., 54, 343-352.

Saha, SK, Ghosh P, Konar A, Bhattacharya S, and Roy SS. (2005). Differential Expression of Procollagen Lysine 2-Oxoglutarate 5-Deoxygenase and Matrix Metalloproteinase Isoforms in Hypothyroid Rat Ovary and Disintegration of Extracellular Matrix. Endocr., 146, 2963-2975.

Sato H, Takino T, Okada Y, Cao J, Shinagawa A, Yamamoto E, and Seiki M. (1994). A matrix metalloproteinase expressed on the surface of invasive tumor cells. Nature, 370, 61-65.

Suzuki K, Lees M, Newlands G, Nagase H, and Woolley DE. (1995). Activation of precursors for matrix metalloproteinases 1 (interestial collagenase) and 3 (stromelysin) by rat mast-cell proteinases I and II. Biochem. J., 305, 301-306.

Tadakuma H, Okamura H, Kitaoka M, Iyama K, and Usuku U. (1993). Association of immunolocalization of matrix metalloproteinase 1 with ovulation in hCG treated rabbit ovary. J. Reprod. Fertl., 98, 503-508.

Takino T, Sato H, Shinagawa A, and Seiki M. (1995). Identification of the second membrane-type matrix metalloproteinase (MT-MMP-2) gene from a human placenta cDNA library: MT-MMP form a unique membrane type sub-class in the MMP family. J. Biol. Chem., 270, 23013-23020.

Willard D, and Becherer JD. (1997). Cloning of a disintegrin metalloproteinase that processes precursor tumor-necrosis factor alpha. Nature, 385, 733-736.

Willenbrock F, and Murphy G. (1994). Structure–function relationships in the tissue inhibitors of metalloproteinases. Am. J. Respr. Critic. Med., 150, 165-170.

13

Melatonin and Immunomodulation in Seasonally Breeding Mammals

Chandana Haldar and Seema Rai*

Pineal Research Laboratory, Department of Zoology, Banaras Hindu University, Varanasi – 221 005, India

ABSTRACT

Immunity, a primeval feature of living organism is essential for reproduction and to face the challenges originating from seasonal variation and environmental threat as well. Seasonal breeding is a silent component of the life history of the most of the vertebrates and possesses a characteristic rhythm of various physiological activities, immunity being one of them. The immune system is not only driven by cytokines but also by hormone and neurohormones. Among these neurohormones popularity of melatonin in immunomodulation is a current topic of research. This is because of rhythmic secretory pattern of melatonin and its inverse correlation with breeding activity and direct relation with immune status. In the present review we wish to update the current knowledge of melatonin as immunomodulator specially in seasonally breeding mammals because of an adaptive significance of melatonin in balancing immune status. It acts during winter as immunostimulator to save the seasonally breeding mammals from the adverse condition of the environment (low temperature, scarcity of food and shelter etc) and free radical load. We proposed the mechanism of immunomodulation by melatonin and believe that it is the direct one in modulating the immunity according to the season. However, interdependence of melatonin with other endocrine gland (adrenal and thyroid) is well known and such a correlation of melatonin with endocrine system in regulation of immune function may throw some light to the clinical problems of endocrine dysfunction and immune deficiency.

**Corresponding Author.* E-mail: chaldar2001@yahoo.com

Introduction

Maintenance of health depends to a significant extent on the ability of the exposed host to respond appropriately and, eventually, to adapt with environmental stressors. It is now well established that inappropriate or maladaptive response to such stressors weakens the body's resistance to other stimuli from the environment such as pathogenic. Immunity, a primeval feature of living organisms is essential for reproduction and to face the challenges originating from seasonal variation and environmental threats as well social environment can be considered as a part of general environment, which has an impact on the body via redundant and reciprocal interaction between the body and the brain. These are linked by nervous endocrine and immune systems and utilize large array of chemical messengers including hormones, cytokines and neurotransmitters (Spector *et al.*, 1996). There is abundant evidences for the functional afferent nerve endings in the tissue of the immune and hematopoietic system arising from both the sympathetic and parasympathetic systems (Ader *et al.*, 1991, 1995). It is also clear that many neurotransmitters, neuroendocrine factors and hormone can drastically change immune function and that on the contrary, cytokines derived from immunocompetent cells can profoundly affect the central nervous system (Ader *et al.*, 1991, 1995). As a consequence, any environmental stimulus to the nervous system will affect the immune system and *vice versa* essentially *via* the endocrine system. In this conceptual basis it should not be surprising that the day/night or a photoperiod, which constitutes a basic environmental cue for any organism, can also influence the immune hematopoietic system. As for many other adaptive responses, a major mediator of this influence seems to be the neuroendocrine pineal gland, which transduces the light/dark rhythm into the circadian synthesis and release of melatonin (Yu and Reiter, 1993).

The study of immune system is quite common among clinicians and physicians. But, the interaction of endocrine organs along with immune function was mostly noted by the basic biologist. Further, we are still far from complete understanding of the mechanism underlying the immunological and hematopoietic action of many neurohormones, neurotransmitters including melatonin. Further, this field of study received very less attention among the clinicians. All the literature so far surveyed suggests that immunity response had been studied only in laboratory animal model, but, to note the variation in immune status of wild seasonally breeding species is equally important if we want to save some endangered fauna. We therefore, have tried to cover the general aspect of immunomodulation by neuroendocrine pineal gland and its hormone melatonin in general with a special attention to seasonally breeding mammals.

The Pineal Gland-Immune Network

During the last two decades the pineal gland became an organ of potential interest, since it was noticed that it plays an important role in the annual cyclicity of reproduction in vertebrates. Today melatonin effects are much wider than previously suspected and indeed involve every cell in the organism. It has been shown to influence nearly every biochemical, physiological, endocrinological and behavioural pattern (Armstrong and Redman, 1991).

Over the years, it has been recognized that the central nervous system, the neuroendocrine and the immune systems are reciprocally and functionally connected. The interaction of these system influences the nature of their ongoing functions in the maintenance of organismal homeostasis (Nir, 1995). The study of the two–way relationships between pineal gland, melatonin and the immune system has raised notably importance in recent years. Experiments have shown that, there is a functional (direct) relationship between the pineal gland and immune system in experimental mammals and avian model (Maestroni, 1993; Rodriguez and Lea, 1994; Skwarlo-Sonta, 1996; Rodriguez *et al.*, 1997,1999) while a similar relation exist in seasonal breeders as well which suggest that melatonin treatment increased percent lymphocyte count (per cent LC) of bone marrow and peripheral blood, total leukocyte count (TLC), percent stimulation ratio (per cent SR) of thymocytes and splenocytes in normal squirrels, which was noted low in the pinealectomized (Px) Indian palm squirrel, *Funambulus pennanti* squirrels. Melatonin supplementation, which enhanced all immune parameters in both *in vivo* and *in vitro*, suggesting that the immune system was sensitive during the reproductively active phase of the seasonally breeding squirrels to melatonin treatment and the importance of melatonin for maintenance of immune status (Rai, 2004).

In humans, circadian rhythms are present in most if not all, immune functions *e.g.* the peak level (acrophase) of circulating lymphocytes is studied close to that of melatonin in normal (24 hr) environment. However it has been reported that circadian rhythm in the serum levels of mononuclear cells, T and B lymphocytes, as well as the serum levels of melatonin, affecting notably the functionality of the natural killer cells in rats (McNaulty *et al.*, 1990). Both functional (permanent lighting) and pharmacological (evening administration of the β-adrenergic blocker, propanolol, p-chlorophenylalanine) inhibition of melatonin synthesis in mice resulted in depressed cellular and humoral immune functions (Maestroni and Pierpaoli, 1981; Maestroni *et al.*, 1986). Exogenous melatonin administration in the evening reverses this effect, suggesting that circadian melatonin synthesis is mandatory for optimal immune functions (Maestroni *et al.*, 1986).

The pineal gland and its principal neurohormone, melatonin can affect lymphatic tissue sizes. For example, exposure male and female hamster to short days or daily afternoon melatonin injections elevated splenic mass (Vaughan *et al.*, 1987). Elevated splenic mass could be prevented in short-days hamsters by pinealectomy (Vaughan *et al.*, 1987). The pineal gland seems to exert a stimulatory action on thymic growth, since the administration of pineal extracts may determine hyperplasia of peripubertally, around 60 days of age, suggesting an involvement of sex steroid hormones (Vermeulen *et al.*, 1993). Pinealectomy also ameliorates Collagen II-induced arthritis in mice (Hansson *et al.*, 1993), as well as inhibits humoral immune function and suppress bone marrow progenitors for granulocytes and macrophages (Kuci *et al.*, 1983; Haldar *et al.*, 1992a,b). Additionally natural killer (NK) cell activity and IL-2 production are reduced in mice after pinealectomy (del Gobbo *et al.*, 1989).

Melatonin and Immunity

Within a decade of its discovery, melatonin had been shown to be functionally related to neuroendocrine physiology, with the most obvious link being its control of reproductive physiology in photoperiodic rodents (Hoffman and Reiter, 1965). By 1972 a scheme defining the role of pineal gland, seasonally changing photoperiods and annual fluctuations in reproductive competence had been presented (Reiter, 1973). Because of the circadian nature of the melatonin rhythm, with elevated levels always being with the dark phase of the light/dark cycle (Arendt, 1988). The pineal hormone melatonin has also been implicated in the regulation of 24-h-cycle via an action on the suprachiasmatic nuclei (SCN), the so-called biological clock (Cassone, 1990). Melatonin interactions with cells in the SCN are now known to have important physiological implication for the entire organism (McArthur *et al.*, 1991). In the last decades, it has been suggested that the pineal gland may act as a neuroendocrine transducer, able to convey environmental information into hormonal signals, mainly through the synthesis and release of its hormone melatonin, which is involved in the modulations of several neuroendocrine activities (Reiter, 1982). The role played by melatonin is analogous to that played by light in the external milieu (Nir, 1995).

Evening administration of melatonin in mice enhanced the suppressed primary antibody production and thymus cellularity induced either by acute restraint and/or corticosteroid treatment. One of the significant observations made was that the naltrexone, a specific opioid antagonist abolished completely the immuno-enhancing effect of melatonin in normal and stressed mice. In addition it has also been noted that daily subcutaneous injection of melatonin (25μg/100g body mass.) at 17.30–18.00 hr. to squirrels during reproductively active phase (May–June) significantly increased the lymphocyte count of blood and bone marrow and blastogenic response/percent stimulation ratio of spleen and thymus. Histological observation showed densely packed thymocytes and splenocytes. During this period, peripheral testosterone level was high and melatonin was low establishing an inverse relationship as noted earlier for this squirrel. In pinealectomized squirrels decreased total leukocyte count, percent lymphocyte count in peripheral blood and bone marrow along with the decreased cell density of spleen and thymus was observed histologically (Rai and Haldar, 2003) Further, melatonin treatment of pinealectomized palm squirrels resulted in restoration of the immune parameters according normal control level. We suggest that during the reproductively active period of male Indian palm squirrels, the lymphoid organs were sensitive to melatonin hence; the exogenous melatonin treatment had immuno–enhancing effect.

The circadian synthesis and release of melatonin modulates antibody response and also alters tumorigenesis (Blask, 1985). At the normal cellular level, melatonin is believed to affect antimitotic processes as well as cytotoxic activity (Boucek and Alvarez, 1970; Winston *et al.*, 1974; Poffenbarger and Fuller, 1976). When the synthesis of endogenous melatonin is blocked, antibody production is depressed (Maestroni and Pierpaoli, 1981; Maestroni *et al.*, 1986). Pharmacological and surgical pinealectomy also modulates other immune parameters including plaque-forming cells and blastogenic responses of splenocytes and thymocytes to various mitogen (Kuci *et al.*, 1983; Becker *et al.*, 1988). Furthermore, elimination of melatonin synthesis by pinealectomy profoundly decreased mass of thymus gland (Csaba and Barath, 1975). However, pinealectomy reduces some immune reactions like arthus activity to bovine serum albumin in the rat (Jankovic *et al.*, 1970), while the administration of pineal extract showed enhanced antibody production in mice (Belokrylov *et al.*, 1976). Some other authors (Rella and Lapin, 1976) did not observe any significant change in immune reactions after pinealectomy. Importantly melatonin mediates immune function (Maestroni, 1993). In virtually all cases examined, melatonin enhanced humoral and cell mediated immunity (Guerrero and Reiter, 1992; Maestroni, 1993). Melatonin treatment of both normal and immunocompromised house mice elevated *in vitro* and *in vivo* antibody responses (Maestroni, 1993; Caroleo *et al.*, 1992). Impaired T-helper cell activity in immunocompromised mice are stored by melatonin treatment (Caroleo *et al.*, 1992). Antigen presentation by splenic macrophages to T-cells is also enhanced by melatonin; furthermore this enhancement is coincident with an increase in major histocompatibility (MHC) class II molecules, as well as Interleukin-1 (IL-1) and tumor necrosis factor (TNFa) production (Pioli *et al.*, 1993). Murine antibody-dependent cellular toxicity (ADCC) is reduced in adult mice that were pinealectomized prior to 7 days of age (Vermeulen *et al.*, 1993). ADCC is a lytic process that occurs when lymphocytes bind to specific antibody coated-target cells through receptors for the Fc portion of the Ig molecule expressed on their membrane. The impairment of the ADCC appears the proliferation of bone marrow progenitors for granulocytes and macrophages (CFU-GM); disruption of the nighttime peak of melatonin completely abolished CFU-GM proliferation (Kuci *et al.*, 1983; Haldar *et al.*, 1992a,b). Melatonin is important in many disease processes especially cancer (Blask, 1985; Caroleo *et al.*, 1992; Guerrero and Reiter, 1992; Giordano *et al.*, 1993; Maestroni, 1993; Pioli *et al.*, 1993; Poon *et al.*, 1994; Nelson and Demas, 1996, 1997). The overwhelming majority of studies indicate that melatonin is an oncostatic hormone. A number of treatments for cancer now incorporate melatonin as a part of the immunomodulatory therapy (Neri *et al.*, 1994; Lissoni *et al.*, 1994,a,b,c, 1995; Barni *et al.*, 1995). As predicted (Guerrero and Reiter, 1992; Maestroni, 1993), melatonin receptors have been isolated on circulating lymphocytes (Pang and Pang, 1992; Poon and Pang, 1992; Pang *et al.*, 1993; Liu and Pang 1993; Calvo *et al.*, 1995) as well as on thymocytes and splenocytes (Lopez-Gonzales *et al.*, 1993,a,b; Martin-Cacao *et al.*, 1993; Rafii-EL-Idrissi *et al.*, 1995) suggesting a direct effect of melatonin on the regulation of the immune system.

Mechanism of Immunomodulation by Melatonin

The mechanism by which melatonin modulates the immune function is still unresolved (Guerrero and Reiter, 1992), but the following hypothesis have been suggested:

The Pineal Gland-Immune System-Opioid Network

Opioids have been suggested to be the mediators of melatonin action on the immune system (Maestroni and Conti, 1992; Maestroni, 1993). The immunoenhancing and antistress effects of melatonin were abolished by the opioid antagonist, naltrexone (Maestroni *et al.*, 1987a, b, 1988, 1999). Dynorphin 1-b and β-endorphin mimicked the immunological effects of melatonin (Maestroni and Conti, 1992, 1989). Melatonin also stimulated the release of opioid peptides by activated T-helper thymocytes (Maestroni and Conti, 1990). These melatonin–induced immuno-opioids mediated the immunoenhancing and antistress effects of melatonin and cross-reacted immunologically with anti β-endorphin and anti-met-enkaphalin anticera (Maestroni and Conti, 1990, 1991).

Opioid peptides particularly β-endorphin and Met-enkaphalin, have also been implicated as immunomodulators since, they can affect several immune mechanisms. However, it is difficult to conclude whether, as a whole these peptides are immunosuppressive or immunostimulant since, the effects reported differ depending on the type of immune process studied, on the cell source and type, on the species, on the concentrations of the opioid peptide used, and on experimental conditions *i.e.* in vivo or in vitro (Plotnikoff *et al.*, 1986; Homo-Delarche and Durant, 1994) (Figure 13.1).

Lymphokines as Mediators of Melatonin Action on the Immune System

The production of Interleukin-2 and γ-interferon was stimulated by melatonin administration (Caroleo *et al.*, 1992; Muscettola *et al.*, 1992). As interleukins and interferons were known stimulators of natural killers cell activity and/other natural immune responses (Saksela and Timonen, 1980; Kuribayashi *et al.*, 1981), it is possible that these lymphokines mediate the observed effects of melatonin on the mammalian immune responses (Maestroni, 1993).

Indirect Action of Melatonin Through Other Endocrine System

Photoperiod and melatonin treatment have been shown to affect reproductive competence, sex hormone secretion together with selected parameters of immune function (Nelson *et al.*, 1995). This suggests that the hypothalamo-pituitary-gonadal axis may be involved in the mediation of melatonin action on the immune system (Brainard *et al.*, 1987; 1988; Vaughan *et al.*, 1987; Champney and Mc Murray, 1991). In addition, pineal melatonin and adrenal steroids may interact to modulate the immune functions. Melatonin antagonized the immunosuppressive effect of corticosteroids (Maestroni *et al.*,

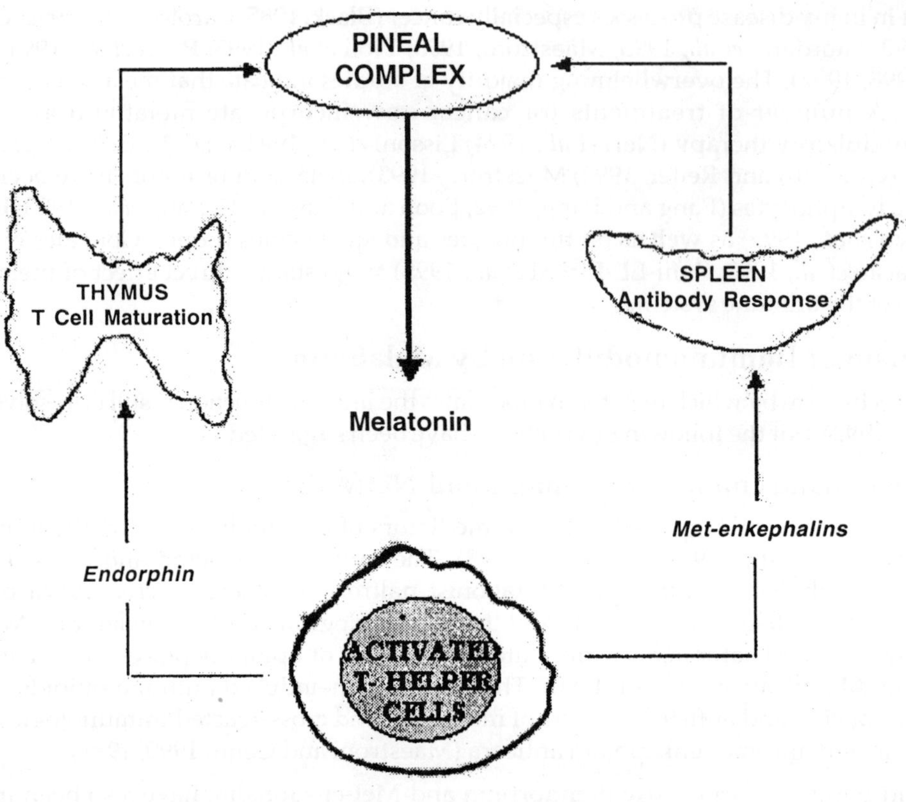

Figure 13.1: The Melatonin-immune Opioid Network

1986; Aoyama *et al.*, 1986, 1987). The immunostimulatory action of melatonin also varied with the steroid *milieu* (Persengiev *et al.*, 1991). With the demonstration of a steroid-dependent modulation of the rat thymic glucocorticoids receptors by melatonin (Persengiev *et al.*, 1991) and modulation of the duck thymic 2[^{125}I] iodomelatonin binding sites by changes in the immune status of the animal (Poon *et al.*, 1994b), the thymus has been suggested to be a target site for the immunomodulatory interactions between pineal melatonin and adrenal steroids (Persengiev *et al.*, 1991; Poon *et al.*, 1994a). Moreover, melatonin stimulated the release of hypothalamic thyrotropin-releasing hormone (TRH), which possesses immuno-reconstituting and antiviral activity. Suggesting that TRH may also be important in the mediation of melatonin action in the immune system (Pierpaoli and Yi, 1990).

Direct Action of Melatonin on Lymphoid Tissue

It is hypothesized that melatonin acts directly on the immune system (Pang *et al.*, 1991; Yu *et al.*, 1991; Guerrero and Reiter, 1992; Poon *et al.*, 1994a). This hypothesis is based on the fact that melatonin, being a lipid soluble molecule, should be readily available to all lymphoid tissues. Moreover, the melatonin induced stimulation of lymphokine and opioid productions (Maestroni and Conti, 1990, 1991; Caroleo *et al.*, 1992; Muscettola *et al.*, 1992) suggested a possible existence of melatonin receptors on the lymphoid tissues. Evidence for such a hypothesis is provided by the presence of 2[^{125}I] iodomelatonin binding sites in the tissue homogenates of primary and secondary lymphoid organs of various birds and mammals (Niles, 1989; Pang *et al.*, 1991; Yu *et al.*, 1991; Poon and Pang, 1992; Liu and Pang, 1992, b; 1993; Poon *et al.*, 1993; Lopez–Gonzalez *et al.*, 1993 a,b; Poon *et al.*, 1994b), rat splenocytes (Rafii-El-Idrissi *et al.*, 1995) and on human peripheral lymphocytes (Lopez-Gonzalez *et al.*, 1992 a,b) as well as T-helper lymphocyte on bone marrow (Maestroni, 1995).

Bone Marrow and Melatonin

Bone marrow is an important tissue in generation of immunocompetent and peripheral blood cells. The progenitors of hematopoietic cells in bone marrow exhibit continuous proliferation and differentiation and they are highly vulnerable to environmental oxidative insults, such as irradiation, chemotherapeutic compounds, oxidative stresses and other environmental pollutants. Tan and his coworkers (1999) identified a highly elevated level of pineal melatonin by using immunocytochemistry, radioimmunoassay, high performance liquid chromatography with electrochemical detection and mass spectroscopy. In their finding the night time melatonin concentrations in the bone marrow of rats were roughly two times of magnitude higher that those in peripheral blood. Important point is that bone marrow cells have a measurable N-acetyltransferase (NAT) activity but they have very low level of hydroxyindole-O-methyltransferase (HIOMT) activity. Moreover, melatonin level in the bone marrow followed the same pattern as in the serum *i.e.* bone marrow melatonin concentration showed a reasonably good correlation with circulating levels. This potential finding about the identification melatonin synthesis in the bone marrow was a very interesting one, which turned us towards the direction to find about the relationship between the bone marrow lymphocytes with that of the pineal melatonin. Consistent to this result high expression for Mel 1a R was observed on bone marrow lymphocyte of the *Funambulus pennanti* (Rai, 2004). It is now well recognized that a main actor in the continuous interaction between the nervous and immune system is the pineal hormone melatonin. Hematopoiesis is apparently influenced by the action of the melatonin induced immuno-opioid (MIIO), which is recognized as novel opioid cytokines on kappa-opioid receptors present on stromal bone marrow macrophages. Most interestingly, gamma–interferon (γ-INF) and colony stimulating factors (CSFs) may modulate the production of melatonin in the pineal gland. A hypothetical pineal-immune-hematopoietic network is therefore, taking a shape. From the immuno-pharmacological and ethical point of view, clinical studies on the effect of melatonin in combination with IL-1, IL-2 or other cytokines in viral diseases including human immunodeficiency virus infected patients and cancer patients are needed. Melatonin seems to play a crucial role in the homeostatic interactions between the brain and the immune hematopoietic system and deserve to be further studied to identify its therapeutic indications and its adverse effects.

Melatonin and Haemopoiesis

It has been reported that melatonin may rescue haemopoiesis in mice transplanted with Lewis Lung carcinoma (LLC) and treated with cancer chemotherapeutic compounds (Maestroni *et al.*, 1994a). This effect apparently involved in the endogenous release of granulocyte/macrophage colony stimulating factor and melatonin induced opioid cytokines (Maestroni *et al.*, 1994a,b; 1999). These opioid cytokines comprise two peptides of 15 and 67 kDa with the common opioid sequence (Tyr-Gly-Gly-Phe) at their amino terminal and a carboxy terminal extension, which resembles both IL-4 and dynorphin (Maestroni *et al.*, 1996). Activated lymph nodes, Th cells and bone marrow Th cells release these opioids cytokines. Recently, the protective effect of melatonin in rats and mice against the hemopoietic drugs was confirmed (Anwar *et al.*, 1998; Rapozzi *et al.*, 1998). Administration of melatonin resulted in an increased survival time and reduced chemotherapeutic toxicity (Lissoni *et al.*, 1997).

Exogenous melatonin administration enhanced the production of natural killer cells and monocytes in the bone marrow of mice (Currier *et al.*, 2000). Melatonin seems also to promote the survival of precursor b-cells in mouse bone marrow (Yu *et al.*, 2000) and had been reported to restore the human B tonsillar lymphocytes in recurrent acute tonsillitis (Lopez-Gonzales *et al.*, 1998). Interestingly, the presence of high level of melatonin was found to correlate with tonsillar hypertrophy in the pediatric patients (Lopez-Gonzales *et al.*, 1998) (Figure 13.2).

The survey of literature points to an important and clinically useful haemopoietic effect of melatonin. The relative mechanism might depend on the ability of melatonin to enhance the production of haemopoietic cytokines, as well as on its reported antioxidant and antiapoptotic action (Maestroni *et al.*, 1994a; Reiter and Maestroni, 1999).

Melatonin Receptors in Histocompatible Tissues

Melatonin binds specific high affinity plasma membrane receptors, which are coupled to GTP binding protein. Two different G proteins coupled with melatonin receptors have been described

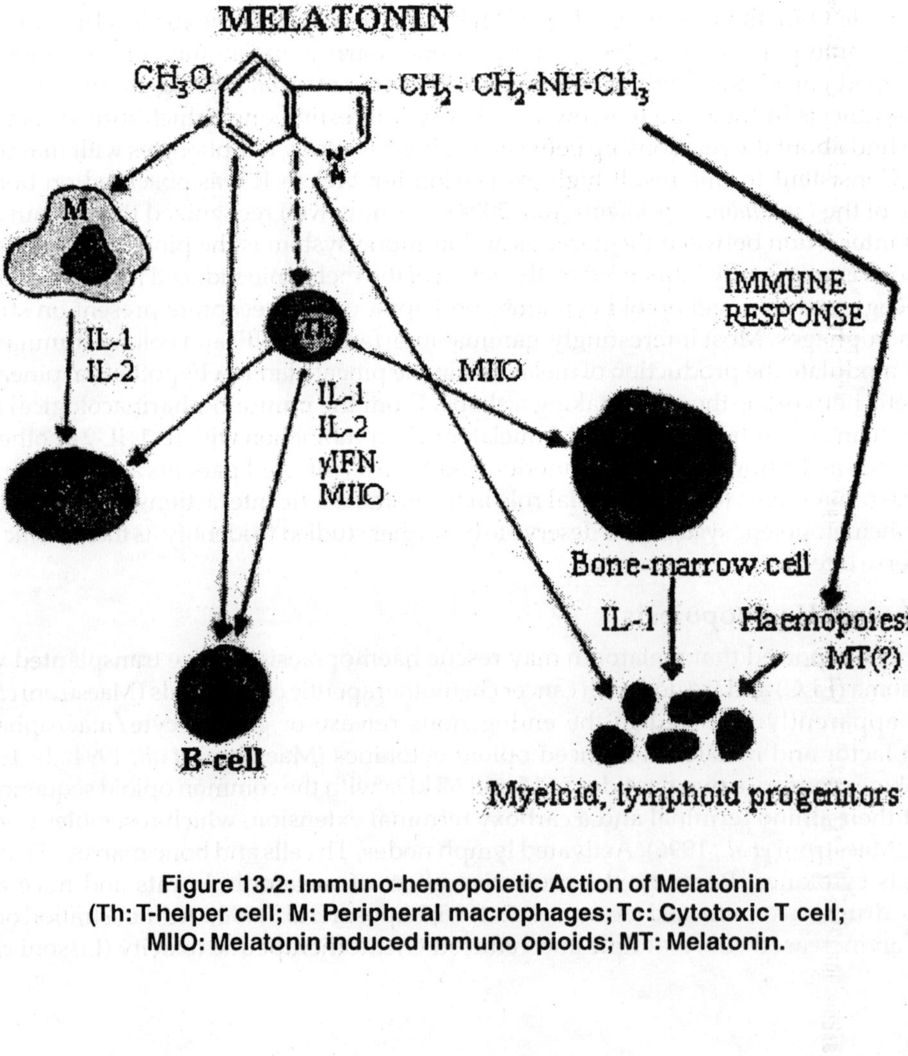

Figure 13.2: Immuno-hemopoietic Action of Melatonin (Th: T-helper cell; M: Peripheral macrophages; Tc: Cytotoxic T cell; MIIO: Melatonin induced immuno opioids; MT: Melatonin.

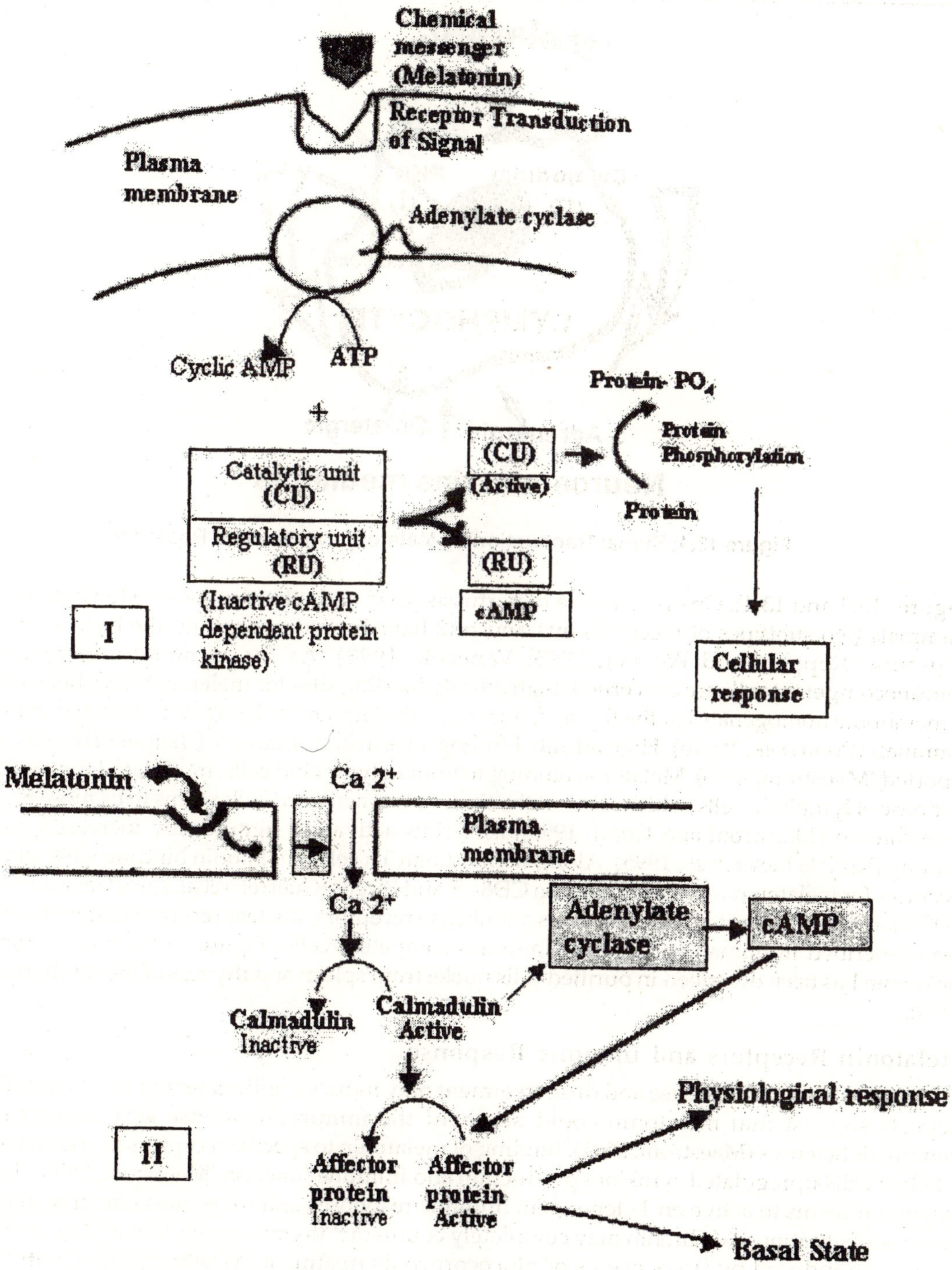

Figure 13.3: Signal Transduction by Melatonin via cAMP (I) and Calmudlin (II)

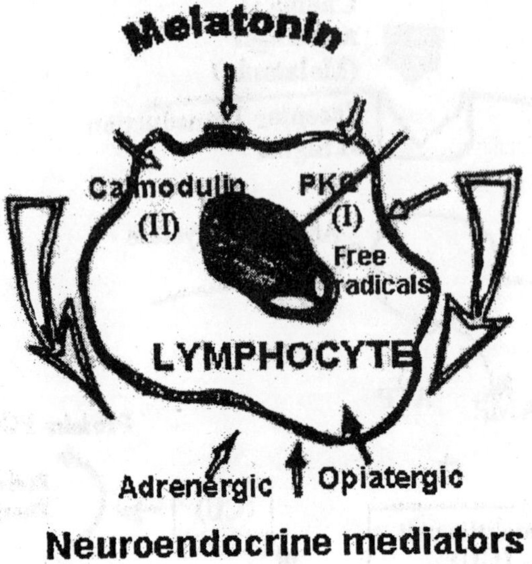

Figure 13.4: Signal Transuction by Melatonin via Nuclear Receptor

(Figures 13.3 and 13.4). One is sensitive to pertussis toxin and other is sensitive to cholera toxin. In mammals two subtypes of receptors mt1 and mt2 have been described on the basis of molecular structure (Reppert and Weaver, 1995; Vanecek, 1998). As for as melatonin receptors in immunocompetent cells are concerned, high affinity binding sites for melatonin have been described in membrane homogenates of the thymus, bursa of Fabricius and spleen from a number of birds and mammals (Poon *et al.*, 1994b). High affinity binding sites in bone marrow T-helper (Th) cells was also reported (Maestroni, 1995). Melatonin binding to human lymphoid cells, modulated the proliferative response of lymphoid cells, which support the finding that activated cells seems to be more responsive to melatonin (Maestroni and Conti, 1990), as T-cells activation significantly increased melatonin binding (Konakchieva *et al.*, 1995). Along with human Th cells, melatonin binding sites and mRNA receptors for melatonin was also present on CD8+ T and B-cells (Guerrero et al, 1996; Garcia-Preganeda, 1997; Garcia-Maurino et al, 1997). Besides membrane receptors, nuclear receptors for melatonin have been described in human and murine immuno competent cells (Figure 13.4). Specific binding of melatonin has been described in purified cells nuclei from spleen and thymus of the rat (Barjavel *et al.*, 1998).

Melatonin Receptors and Immune Response

Acute stress, viral disease and drug treatment may induce life-threatening immunodeficiences. Reports suggest that melatonin could augment the immune response and counteract these immunodeficiences (Maestroni, 2001). Binding of melatonin to specific receptors in antigen activated T-helper cells upregulated cytokines production and immune function (Maestroni, 1998). In general melatonin seems to active on T-dependent immune responses and to be most effective in immuno-depressed situations. Melatonin may completely counteract thymus involution and immunological depression induced by stress events or glucocorticoids treatment (Maestroni and Conti, 1990), or restored depressed immunological functions after soft tissue trauma and hemorrhagic shock (Wichmann *et al.*, 1996).

In our present investigation the down regulation of percent expression of receptor for melatonin (Mel 1a R) in melatonin treated lymphoid organs were noted suggesting a direct action of melatonin on lymphatic cells of this squirrel. The melatonin receptor Mel 1a R is a member of the super family of guanine nucleotide-binding regulatory protein *i.e.* (G-protein)–coupled receptor (Brydon *et al.*, 1999; Roka *et al.*, 1999). Signaling through melatonin receptors inhibits adenylate cyclase and stimulates phospholipase Cβ upon activation of pertusis toxin (PTX)–sensitive G-proteins (Brydon *et al.*, 1999; Roka *et al.*, 1999). It is expressed as 37 kd protein localized on to splenocytes, thymocytes and bone marrow lymphocytes and implicated in controlling cellular growth in response to melatonin (Shiu *et al.*, 2000; Rai, 2004). Therefore, it can be suggested that melatonin may act directly and indirectly to affect immune function. Melatonin can act directly on high affinity receptors located on lymphoid tissue to increase immune function (Pang and Pang, 1992; Calvo *et al.*, 1995). The result of the present study support the hypothesis that melatonin is acting, at least in part, directly on lymphocyte proliferation. It is possible that melatonin binds to receptors localized to the splenocyte and thymocytes and increased the general proliferative ability of these cells, leading to enhanced cell mediated immunity.

Melatonin is active only when injected in afternoon or in the evening *i.e.* with a schedule constant with its physiological rhythm (Maestroni, 1998; Rafii-El-Idirissi *et al.*, 1998). The immunoenhancing properties of melatonin has been confirmed in birds as well as in mammals (Yellon *et al.*, 1999; Kliger *et al.*, 2000; Moore and Siopes, 2000), and showed species-specific differences. Melatonin stimulated the production of IL-2, γIFN and IL-6 in peripheral blood mononuclear cells (Garcia-Maurino *et al.*, 1997). Melatonin rhythm positively correlated with the rhythmicity of Th1/Th2 ratio in the circulation which reflect physiologically the nocturnal high level of γ–IFN and IL-10 in circulation with nocturnal peak of melatonin (Petrovsky and Harrison, 1996). In mammals (human and rats), the nocturnal increase in serum concentration of thymosin α and thymulin seems to depend on melatonin (Molinero *et al.*, 2000).

Melatonin: Receptors Genes in the Immune System

The melatonin receptors gene was first cloned from an immortalized cell line of *Xenopus laevis* melanophores (Ebisawa *et al.*, 1994), and since then several receptors and receptors fragments have been cloned (Kokkola and Laitinen, 1998). In non-mammalian vertebrates, melatonin receptors can be divided into three subtypes, Mel_{1a}, Mel_{1b} and Mel_{1c}, based on their DNA and amino acid sequences. Within a subtype, these receptors have 69-96 per cent homology, across species. Mel_{1a} and Mel_{1b} mRNA sequences, detectable by reverse-transcription PCR (RT-PCR) are present in all vertebrates, although Mel_{1c} mRNA is only present in non-mammalian vertebrates (Kokkola and Laitinen, 1998). According to the present terminology, in mammals the designation of MT1 and MT2 is accepted for the pharmacologically distinct products of Mel_{1a} and Mel_{1b}, respectively (Dubocovich *et al.*, 1998).

Within the immune system, Mel_{1a} mRNA expression has been demonstrated in the rat thymus and spleen, as well as in isolated T and B cells (Pozo *et al.*, 1997). In contrast, the presence of mRNA was demonstrated for Mel_{1a} and Mel_{1b} in the chicken spleen, thymus, bursa of Fabricius, and in isolated lymphocytes in the absence of Mel1a mRNA using the RT-PCR technique (Skwarlo-Sonta *et al.*, 2003). Mel1a mRNA is, however, present in the chick central nervous system (CNS), indicating a tissue-specific pattern of melatonin receptors gene expression. This pattern would also appear to be age related, since Mel_{1c} transcripts are present from the first week of the life in the chick thymus but from the second week in the spleen. In the bursa of Fabricius, transcripts of Mel_{1c} mRNA have been found in one week old chicks, and Mel_{1b} transcripts in two week old chicks (Skwarlo-Sonta *et al.*, 2003).

As immune system in the chick becomes functionally mature between 2 and 4 weeks post hatch (Chen *et al.*, 1994), these changes in the expression of melatonin receptors mRNA may be related to developmental changes in the immune function.

Melatonin: Signal Transduction in Immune Cells

The signal transduction pathway activated by melatonin is largely unknown. Although a decrease in cAMP level elevated by previous forskolin treatment, has been established in mammalian immune cells (Garcia-Preganeda *et al.*, 1997, 1999). This effect is prevented by preincubation with pertussis toxin, indicated that it is mediated via a G_i protein. The inhibition of cAMP accumulation is also likely to be via a subunit of the G_i protein and is accompanied by an increase in diacylglycerol production, most probably as a result of the activation of phospholipase C by β and γ subunits (Garcia-Preganeda *et al.*, 1997). At least some membrane bound melatonin receptors in the immune system therefore appear to be coupled with $G_{i/o}$ proteins. The increased concentration of cGMP in human monocytes following melatonin treatment does, however, suggest the involvement of other second messengers in the signal transduction of melatonin in mammalian immune cells (Sandler *et al.*, 1975).

Melatonin, Steroids and Immunity

Interactions between glucocorticoids and immune function have been reported in relation to environmental stress. Adrenocortical hormones, especially glucocorticoids, suppress immune function in both humans and non-humans (Black, 1994) and were found necessary for growth of cultured lymphocyte (Ambrose, 1970). Reports exist to demonstrate that environmental stresses elevate blood glucocorticoid levels and that high glucocorticoid levels suppress cellular and humoral immune function (Besedovsky and Del Rey, 1991; Ader and Cohen, 1993) and increases susceptibility to both infections and neoplastic processes (Hall and Goldstein, 1983). Corticosterone has been reported to suppress antibody production; nucleated spleen cells and causes thymic atrophy (Maestroni *et al.*, 1986).

Further, melatonin, a principal pineal neurohormone has been reported to ameliorate the immunocompromising effects of glucocorticoids (Caroleo et al, 1992; Maestroni, 1993). In contrast, melatonin has been reported to cause immune suppression in human (Rogers *et al.*, 1997). However, nothing is known about how circulating melatonin can help the seasonal breeders to protect them from environmental stress. It is important to note that during reproductively inactive phase (Oct.–Dec.) of male Indian palm squirrels, environmental temperature is low with scarcity of food grains and shelter in nature and is very crucial for tropical rodent since they has to cope with environmental stress (low temperature/less food availability/lack of shelter) while gonadal steroid level is low. This phase also us to establish the adaptive significance of melatonin in preventing animal from winter stress (Haldar *et al.*, 2004).

Melatonin may counteract the immunossuppressive effects of glucocorticoids that are induced during seasons via infections, by inducing T helper cells to release opiod peptides (Maestroni and Conti, 1993; Conti and Maestroni, 1994). The immunostimulating properties of melatonin seem to depend on activation of CD4+, T cells, which upon melatonin stimulation shows an enhanced synthesis and or release of opioid peptides, IL-2 and γ-interferon (Conti and Maestroni, 1994; Maestroni *et al.*, 1995).

Further, melatonin-binding sites have been described on the circulating lymphocytes (Calvo *et al.*, 1995) as well as thymocytes and splenocytes (Martin- Cacao *et al.*, 1993; Rafii-el-Idrissi *et al.*, 1995) suggesting a direct effect of melatonin on the regulation of the immune system (Guerrero and Reiter,

1992). However, modulatory effects of dexamethasone and melatonin on immune function may be explained in terms of receptor density variation (Poon *et al.*, 1994a). Similarly, chronic melatonin treatment has been reported to decrease the density of thymic glucocorticoid receptors in rats (Persengiev *et al.*, 1991). Our study with seasonal breeder *F. pennanti* clearly suggests an inverse relation and feedback mechanism between pineal and adrenal gland (Figure 13.5).

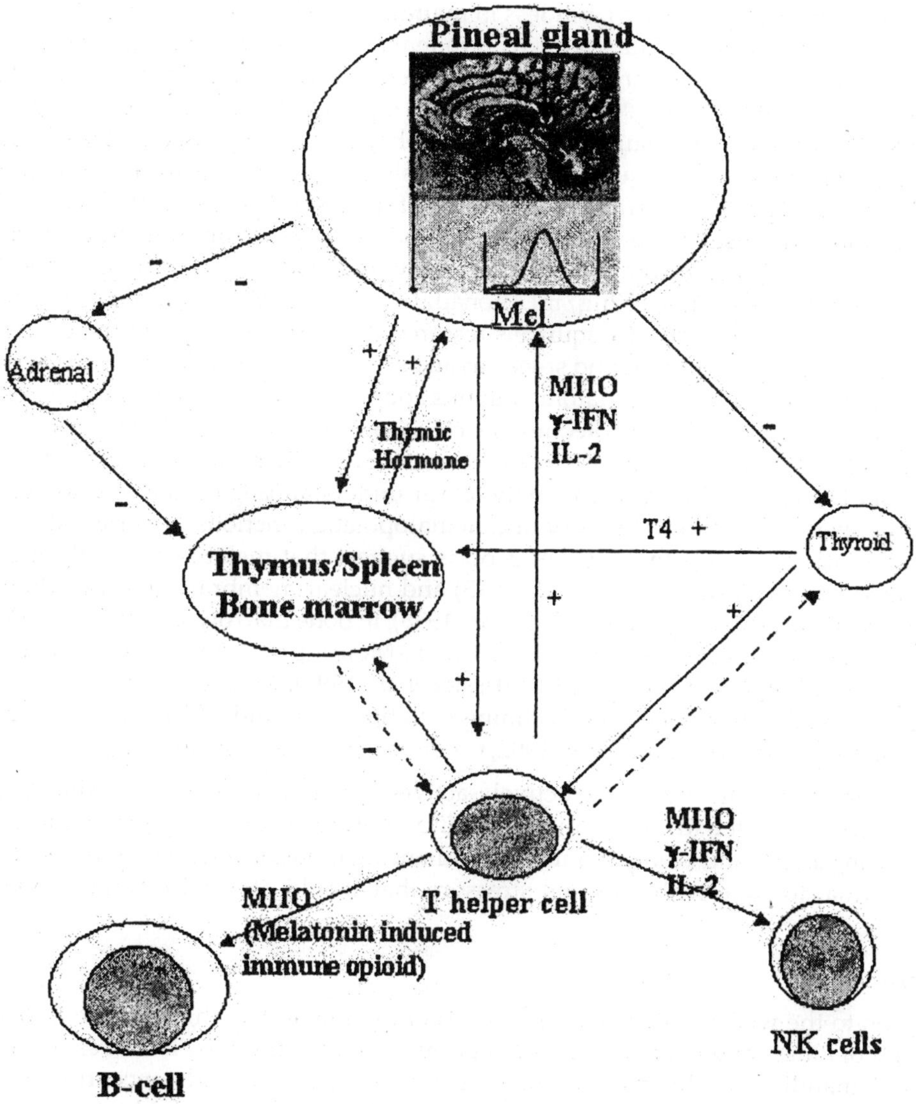

Figure 13.5: Hypothetical Diagram Showing Interactions Between Melatonin, Steroids, Thyroid and Immune System

In addition to seasonal changes in reproductive function, melatonin also mediates seasonal changes in immune function. Variation in circulatory melatonin (high) which is also known as anti-stress hormone might have played an important role in adaptation of the immune status of this seasonally breeding rodent Indian palm squirrels during reproductively inactive phase (Haldar and Singh, 2001a).

Seasonal changes in lymphatic tissue have been thoroughly investigated in birds (Skwarlo-Sonta *et al.*, 1994; Haldar and Singh, 2001a,b), but less in mammals. Therefore, we concentrated upon the immune status of a seasonally breeding semi-wild rodent during reproductive active phase with an attention that high gonadal steroids are associated with reduced lymphatic organ mass and function. Some reports are available on seasonal change of lymphatic tissue of hibernating mammals (Shivatcheva and Hadjioloff, 1987) suggesting that lymphatic tissue regressed in the autumn in both hibernating and non hibernating squirrels, but regression was reported to be more complete in hibernating animals and proliferation and hypertrophy of spleen and gut associated lymphoid tissues were observed prior to arousal in the spring. Photoperiodic changes in immune function are known for mammals; however, pinealectomy is not commonly included in such studies. Our Px results suggested slightly increased (~1.5-1-52ng/mL) gonadal steroid thereby suppressing immune status. However, injection of melatonin to Px squirrels restored the immune status to the controls. Other studies (*in vivo and in vitro*) on thymus and spleen as an immune organ in relation with the exogenous melatonin treatment suggested that melatonin inhibits apoptosis of thymocytes and splenocytes (Rai, 2004; Sainz *et al.*, 1995) and thereby protecting the cells from DNA damage. The literature suggests that melatonin enhances immune function in most normal cases where the basal level of melatonin is low (Nelson and Demas, 1997.) The complexity to our understanding on the physiological role of melatonin towards its immunomodulatory and hematopoietic functions in vertebrates, including humans is still a matter of debate, however reports suggest that melatonin acts through specific receptors located on the plasma (Calvo *et al.*, 1995) and nuclear membrane (Garcia-Mourino *et al.*, 1998) of lymphocytes and monocytes (Barjavel *et al.*, 1998), in bone marrow T- helper cells (Maestroni, 1995), in $CD8^+$–T and B-cells (Guerrero et al.,1996; Garcia-Mourino *et al.*, 1997) and purified cell nuclei from the spleen and thymus of rat (Rafii-El-Idrissiet *et al.*, 1998). Melatonin may interact with the immune system via other hormones like declining secretion of steroids, PRL, and opiatergic *i.e.* via opioid receptor mediated (Poon and Pang, 1992; Calvo *et al.*, 1995 Maestroni *et al.*, 1988).

We also suggest that during reproductively active period of male Indian palm squirrels, the lymphoid organs were sensitive to injection; and therefore, the exogenous melatonin has immunoenhancing effects and is only required for enhancing immune function of seasonal breeder to help the individual to cope with seasonal stressors that would otherwise compromise immune functions.

Conclusion

In addition to the well–studied seasonal cycles of mating and birth, there are also significant seasonal cycles of illness and death among field population of mammals. The stress of coping with energetically demanding conditions may increase adrenocortical steroid concentration that could indirectly cause illness and death by compromising immune function. Individuals in tropics would enjoy a survival advantage if seasonal recurring stressors like winter could be anticipated and countered by bolstering immune function. Immune function is compromised during the winter in field studies of mammals. Hence, we hypothesize that winter stressors present in field studies counteract short-day induced melatonin and thus enhancement of immune function.

In summary, melatonin may be part of an integrative system to coordinate reproductive, immunologic and other physiological processes. From the survey of literature along with our own finding it is evident that regulation of immune system is a complex phenomenon (Figure 13.6). On one side it is regulated by cytokines, immuno-opioids, while on the other hand it is under strong influence of some highly important metabolic endocrine glands like adrenal, thyroid and gonad. We have tried to summarize the aspects of (positive and negative) these factors in immune regulation with particular reference to melatonin secretion in seasonally breeding mammals.

References

Ader R, Cohen N, and Felten D. (1995). Psychoneuroimmunology: interaction between the nervous system and immune system. Lancet, 345, 99-103.

Ader R, and Cohen N. (1993). Psychoneuroimmunology: Conditioning and stress. Ann. Rev. Psychol., 44, 53-85.

Ader R, Felten D, and Cohen N. (1991). Psychoneuroimmunology II, Academic Press, San Diego.

Ambrose CT. (1970). The essential role of the corticosteroid in the induction of the immune response *in vitro. In:* Wolstenholme GEW, and Knight J. (Eds.), *Hormone and Immune Response*, Churchill, London, pp100–125.

Anwar MM, Mahfoug HA, and Sayed AS. (1998). Potential protective effects of melatonin on bone marrow of rats exposed to cytotoxic drugs. Comp. Biochem. Physiol., 119, 493-591.

Aoyama H, Mori W, Mori N. (1986). Anti–glucocorticoid effects of melatonin in young rats. Acta Pathol. Jpn., 36, 423–428.

Aoyama H, Mori W, and Mori N. (1987). Anti–glucocorticoid effects of melatonin in adult rats. Acta.Pathol. Jpn., 37, 1143–1148.

Arendt J. (1988). Melatonin. Clin. Endocrinol. (Oxford). 29, 205-229.

Armstrong SM, and Redman JR. (1991). Melatonin: A chronobiotic with anti-aging properties. Med. Hypothesis., 34, 300- 310.

Barjavel MJ, Mamdouh Z, and Raghbate N. (1998). Differential expression of the melatonin receptors in human monocytes. J. Immunol., 160, 1191-1197.

Becker J, Veit G, Handgretinger R, Allantio A, Bouchett G, Treuner, I, Niethammer, D, and Das Gupta TK.(1988). Circadian variations in the immunomodulatory role of the pineal gland. Neuroendocrinol. Lett.,10, 65-72.

Belokrylov GA, Morozov VG, Khavinson VKH, and Sofronov BN. (1976). Effect of low molecular extracts of heterologous thymus and pineal gland and hypothalamus on the immune response in mice. Bull. Exp. Biol. Med., 81, 218-222.

Besedovsky HO, and Del Rey A. (1991). Feed back interaction between immunological cells and the hypothalamo-pituitary-adrenal axis. Netherlands J. Med., 39, 274-280.

Black PH. (1994). Central nervous system–immune system interactions: Psychoneuroendocrinology of stress and its immune consequences. Antimicrob. Agents Chemother., 38, 1- 6.

Blalock JE, and Smith E M. (1985). A complete regulatory loop between the immune and neuroendocrine systems. Fed. Proc., 44, 108-111.

Blask DE. (1985). The pineal: an oncostatic gland? *In:* Reiter RJ. (Ed.), *The Pineal Gland*, Raven Press, New York, pp 253-284.

Boucek RJ, and Alvarej, TR. (1970). 5-hydroxytryptamine: a cytospecific growth mediator of cultured fibroblasts. Science,167, 898-899.

Brainard GC, Watson-Whimeyer M, Knobler RL, and Lubin FD. (1988). Neuroendocrine regulation of immune parameters: photoperiod control of the spleen in Syrian hamster. Ann. NY Acad. Sci., 540, 704-706.

Brainard GC, Knobler RL, Podolin PL, Lavasa M, and Lubin FD. (1987). Neuroimmunology: Modulation of the hamster immune system by photoperiod. Life Sci., 40, 1319–1326.

Brydon L, Roka F, Petit L, Copper P, Tissot M, Barrett P, Morgan PJ, Nanoff C, Strosberg AD, and Jockers R. (1999). Dual signaling of human Mel1 a melatonin receptor via G (12), G (r3) and G (G/11) protein. Mol. Endocrinol.,13, 2025-2038.

Calvo JR, Rafii-El-Idrissi M, Pozo D, and Guerrero G M. (1995). Immunomodulatory role of melatonin: specific binding sites in human and rodent lymphoid cells. J. Pineal Res., 18, 119-126.

Caroleo MC, Frasca D, Nistico G, and Doria, G. (1992). Melatonin as immunomodulator in immunodeficient mice. Immunopharmacol., 23, 81-89.

Cassone V. (1990). Effect of melatonin on vertebrate circadian systems. Trend. Neurosci., 13, 475-465.

Champney TH, and Mc Murray DN. (1991). Spleen morphology and lymphoproliferative activity in short photoperiod exposed hamsters. *In:* Fraschini F, and Reiter RJ. (Eds.), *Role of Melatonin and Pineal Peptides in Neuroimmunomodulation*, Plenum Press, London, pp. 219-225.

Chen Chen-Lo H, Pickel JM, Lathi J M, and Cooper MD. (1991). Surface markers on avian immune cells. *In:* Sharma JM. (Eds.), *Avian Cellular Immunology*, CRC Press, Boca Raton, pp. 1-22.

Chen LD, Kumar P, Reiter RJ, Tan D-X, Manchester LC, Chambers JP, Poeggeler B, and Saarela S. (1994). Melatonin prevents the suppression of cardiac Ca^{+2} stimulated ATPase activity induced by alloxan. Am. J. Physiol., 267, E57-E62.

Conti A, Maestroni GJM. (1994). Melatonin induced immuno-opioids: Role in lymphoproliferative and autoimmune diseases. Adv. Pineal Res., 7, 83-100.

Csaba C, and Barath P, (1975). Morphological changes of thymus and thyroid gland after postnatal extirpation of pineal body. Endocrinol. Exp. (Bristol), 9, 59-65.

Currier NL, Sun LZ, and Miller SC. (2000). Exogenous melatonin: quantitative enhancement in, *in vitro* of cells mediating non-specific immunity. J. Neuroimmunol., 104, 101-108.

del Gobbo V, Libri V, Villani N, Calio R, and Nistico G. (1989). Pinealectomy inhibits interleukin-2 production and natural killer cell activity in mice. Int. J. Immunopharmacol., 11, 567-573.

Dubocovich MS, Yun K, Al-Ghoul WM, Benloucif S, and Masana MI. (1998). Selective MT2 melatonin receptor antagonists block melatonin mediated phase advances of circadian rhythms. FASEB J., 12, 121-1220.

Ebisawa T, Karne S, Lerner MR, and Reppert SM. (1994). Expression cloning of high affinity melatonin receptor from *Xenopus* dermal melanophores, Proc. Natl. Acad. Sci. USA., 91, 6133-6137.

Garcia-Maurino S, Gonzales-Haba MG, Calvo JR. (1998). Involvement of the immune binding sites for melatonin in the regulation of IL-2 and IL-6 production in human blood monolayer cells. J. Neuroimmunol., 92, 76-84.

Garcia-Maurino S, Gonzalez-Haba MG, Calvo JR, Rafii-El-Idrissi M, Sanchez-Margalet V, Goberna R, and Guerrero JM. (1997a). Melatonin enhances IL-2, IL-6, and IFN-gamma production by human circulating CD4+ cells: a possible nuclear receptor–mediated mechanism involving T helper type I lymphocytes and monocytes. J. Immunol., 159, 574-581.

Garcia-Perganeda A, Pozo D, and Guerrero JM. (1997b). Signal transduction for melatonin in human lymphocytes. Involvement of a pertussis toxin sensitive G-protein. J. Immunol., 159, 3774-3781.

Garcia-Preganeda A, Guerrero JM, Rafii-El-Idrissi M, Paz Romero M, Pozo D, and Calvo JR. (1999). Characterization of membrane melatonin receptor in mouse peritoneal macrophages: inhibition of adenyl cyclase by a pertusis toxin-sensitive G-protein. J. Neuroimmunol., 95, 85-94.

Guerrero JM, and Reiter RJ. (1992). A brief survey of pineal gland-immune system interrelationships. Endocrinol. Res., 18: 91-113.

Guerrero JM, Garcia-Mourino S, Gil-Haba M, Rafii-El-Idrissi M, Pozo D, Garcia-Perganeda A, and Calvo, JR. (1996). Mechanism of action on the human immune system: Membrane receptors versus nuclear receptors. *In*: Maestroni GJM, Conti A, and Reiter RJ. (Eds.), *Therapeutic Potential of the Pineal Hormone Melatonin.* Karger, Basel, pp. 43-52.

Haldar C, Haussler D, and Gupta D. (1992a). Response of colony forming units for granulocyte macrophages (CFU-GM) of intact and pinealectomized rat bone marrow cells to murine recombinant interleukin-3 (rIL-3), recombinant granulocyte macrophage colony stimulating factor (rGM-CSF) and human recombinant erythropoietin (rEPO). Prog. Brain Res., 91, 323-325.

Haldar C, Haussler D, Gupta D. (1992b). Effect of the pineal gland on circadian rhythmicity of colony forming units for granulocytes and macrophages (CFU-GM) from rat bone marrow cell cultures. J. Pineal Res., 12, 79-83.

Haldar C, and Singh SS. (2001a). Seasonal changes in melatonin and immunological adaptation in birds. J. Endocrinol. Reprod., 5, 13–24.

Haldar C, and Singh SS. (2001b). Melatonin and immunological functions/expression by the bursa of Fabricius in Indian jungle bush quail, *Perdicula asiatica*. In: Dawson A, and Chaturvedi CM. (Eds.), *Avian Endocrinology*, Narosa Publishing House, New Delhi, pp. 427-435.

Hall NR, and Goldstein AL. (1983). Role of thymosin and the neuroendocrine system in the regulation of immunity. *In*: Fabris N, Garaci E, Hadden J, and Mitchison N. (Eds.), *Immunoregulation*, Plenum Press, New York, pp. 141- 146.

Hansson I, Holmdahl R, and Mattsson R. (1993). Pinealecto my ameliorates collagen II-induced arthritis in mice. Clin. Exp. Immunol., 92, 432-436.

Hoffman RA, and Reiter RJ. (1965). Pineal gland: Influence on gonads of male hamsters. Science, 148, 1609-1611.

Homo-Delarche F, and Durant S. (1994). Hormones, neurotransmitters and neuropeptides as modulators of lymphocyte functions. *In*: Rola–Pleszczynski M. (Ed.), *Immunopharmacology of lymphocytes*, Academic Press Ltd., London 169–240.

Jancovich BD, Isakovic C, and Petrovic S. (1970). Effect of pinealectomy on immune reactions in rat. Immunol., 18, 1-6.

Kliger CA, Gehad AE, Hulet RM, Roush WB, Lillehoj HS, and Mashaly MM. (2000). Effect of photoperiod and melatonin on lymphocyte activities in male broiler chicken. Poul. Sci., 79, 18-25.

Kokkola T, and Laitinen A. (1998). Melatonin receptors genes. Ann. Med., 30, 88-94.

Konakchieva R, Kyurkchiev S, and Kehayou I. (1995). Selective effects of methoxyindoles on the lymphocyte proliferation and melatonin binding to activated human lymphoid cells. J. Neuroimmunol., 63, 125-132.

Kuci S, Becker J, and Veit G. (1983). Circadian variations of the immunomodulatory role of the pineal gland. Neuroendocrinol. Lett., 10, 65-79.

Kuribayashi K, Grillis S, Kern DE, and Henny CS. (1981). Murine NK cell culture: Effect of interleukin–2 and interferon on cell growth and cytotoxic reactivity. J. Immunol., 126, 2321–2327.

Lissoni P, Paolorossi F, and Ardizzoia A. (1997). A randomized study of chemotherapy with cisplatin, etoposide and the pineal hormone melatonin as first line treatment of advanced non-small cell lung cancer patients in a poor clinical state. J. Pineal Res., 23, 15-19.

Liu ZM, and Pang SF. (1993). [^{125}I] iodomelatonin binding sites in the bursa of fabricius of birds: Binding characteristics, sub-cellular distribution, diurnal variations and age studies. J. Endocrinol., 138, 51-57.

Lopez–Gonzalez MA, Calvo JR, Osuna C, Rubio A, and Gurrero JM. (1992a). Intercation of melatonin with human lymphocytes: evidence for binding sites coupled to potentiation of cAMP stimulated by vasoactive intestinal peptide and activation of cAMP. J. Pineal Res., 12, 87–104.

Lopez–Gonzalez MA, Calvo JR, Osuna C, Rubio A, and Gurrero JM. (1992b). Synergestic action of melatonin and vasoactive intestinal peptide in stimulating cAMP production in human lymphocytes. J. Pineal Res., 12, 174–180.

Lopez–Gonzalez MA, Calvo JR, Osuna C, Sequra JJ, and Gurrero JM. (1993a). Characterization of melatonin binding sites in human peripheral blood neutrophils. Biotechnol. Ther., 4, 253–262.

Lopez–Gonzalez MA, Mortin Cacao A, Calvo JR, Reiter RJ, Osuna C, Rubio A, and Gurrero JM. (1993b). Specific binding of 2- [^{125} I] iodomelatonin by partially purified membranes of rat thymus. J. Neuroimmunol., 45, 121–126.

Lopez-Gonzales MA, Guerrero JM, and Ceballo-Pedraja JM. (1998). Melatonin in palate tonsils with recurrent acute tonsillitis and tonsillar hypertrophy. Acta Otorrinolaringol. Exp., 49, 625-528.

Maestroni GJM. (1993). The immunoneuroendocrine role of melatonin. J. Pineal Res., 14, 1-10.

Maestroni, GJM. (1995). T helper 2 lymphocytes as a peripheral target of melatonin. J. Pineal Res., 18, 84–89.

Maestroni GJM. (1998). The photoperiod transducer melatonin and the immune-hematopoietic system. J. Photoch. Photobiol., 43, 186-192.

Maestroni GJM. (2001). The immunotherapeutic potential of melatonin. Expert Opin Investig. Drugs., 10, 467-76.

Maestroni GJM, and Conti, A, (1989). Beta–endorphin and dynorphin mimic the circadian immunoenhancing and anti stress effect of melatonin. Int. J. Immunopharmacol., 11, 333–340.

Maestroni GJM, and Conti A. (1990). The pineal neurohormone melatonin stimulate activated CD4+, Thy–1+ cells to release opioid agonist(s) with immunoenhancing and anti stress properties. J. Neuroimmunol., 28, 167–176.

Maestroni GJM, and Conti A. (1991). Anti-stress role of the melatonin-immuno-opioid network. Evidence for a physiological mechanism involving T cell cell derived immunoreactive beta-endorphin and met-enkaphalin binding to thymic opioid receptors. Int. J. Neurosci., 61, 289-298.

Maestroni GJM, and Conti A. (1992). Melatonin and the immune system. *In:* Touitou Y, Arendt J, and Pevet P. (Eds.), *Melatonin and Pineal Gland. From Basic Science to Clinical Application,* Excerpta Medica, Amsterdam, pp. 295-302.

Maestroni GJM, and Pierpaoli W. (1981). Pharmacological control of hormonally mediated immune response. *In:* Ader, R. (Ed.), *Psychoneuroimmunology,* Academic Press, New York, pp 405-425.

Maestroni GJM, Conti A, and Pierpaoli W. (1986). Role of the pineal gland in immunity: Circadian synthesis and release of melatonin modulates the antibody response and antagonizes the immunosuppressive effect of corticosterone. J. Neuroimmunol., 13, 19-30.

Maestroni GJM, Conti, A, and Pierpaoli W. (1987a). The pineal gland and the circadian opiatergic immunoregulatory role of melatonin. Ann. N Y Acad. Sci., 521,140-148.

Maestroni GJM, Conti A, and Pierpaoli W. (1987b). Role of pineal gland in immunity. II. Melatonin enhances the antibody response via an opiatergic mechanism. Clin. Exp. Immunol., 68, 384-391.

Maestroni GJM, Conti A, and Pierpaoli W. (1988). Role of the pineal gland in immunity. III, Melatonin antagonizes the immunosuppressive effect of acute stress via an opiatergic mechanism. Immunology, 63, 465-469.

Maestroni GJM, Conti, A, and Lissoni A. (1994a). Hematopoietic rescue via T- cell dependent endogenous GM–CSF by the pineal neurohormone melatonin in tumor bearing mice. Cancer Res., 54, 2429-2432.

Maestroni GJM, Conti A, and Lissoni P. (1994b). Colony stimulating activity and hematopoietic rescue from cancer chemotherapy compounds are induced by melatonin via endogenous interleukin-4. Cancer Res., 54, 4740-4743.

Maestroni GJM, Flamigni LF, Hertens H, and Conti A. (1995). Biochemical and functional characterization of melatonin- induced-opioids in spleen and bone marrow T- helper cells. Neuroendocrinol. Lett., 17, 145-152.

Maestroni GJM, Hertens E, and Galli P. (1996). Melatonin induced T-helper cell hematopoietic cytokines resembling both interleukin-4 and dynorphin. J. Pineal Res., 21, 131-139.

Maestroni GJM, Zammaretti F, and Pedrinis E. (1999). Hematopoietic effect of melatonin. Involvement of k-1 opioid receptor on bone marrow macrophages and interleukin-1. J. Pineal Res., 27, 145-153.

Martin-Cacao A, Lopez-Gonzalez NA, Reiter RJ, Calvo JR, and Guerrero JM. (1993). Binding of 2[^{125}I] melatonin by rat thymus membranes during postnatal development. Immunol. Lett., 36, 59-64.

Mc Arthur AJ, Gillette MV, and Prosser RA. (1991).Melatonin directly resets the rat suprachiasmatic circadian clock *in vitro*. Brain Res., 566, 158-161.

McNaulty JA, Relfson LM., Fox LM, Kus L, Handa RJ, and Schneider GB. (1990). Circadian analysis of mononuclear cells in the rats following pinealectomy and superior cervical ganglionectomy. Brain. Behav. Immunol., 4, 292-307.

Molinero P, Soutto M, and Benot S, (2000). Melatonin is responsible for the nocturnal increased observed in serum and thymus of thymosin α1 and thymulin concentrations: observations in rats and humans. J. Neuroimmunol., 103, 180-188.

Moore CB, and Siopes TD. (2000). Effect of lighting conditions and melatonin supplementation of the cellular and humoral immune responses in Japanese quail, *Coturnix coturnix japonica*. Gen. Comp. Endocrinol., 119, 95-104.

Muscettola M, Grasso G, Fanetti G, Tanganalli C, and Borghesi–Nicoletti C. (1992). Melatonin and modulation of interferon–γ production. Neuropeptides, 22, 46- 51.

Nelson RJ, Demas GE, Klein SL, and Kriegsfeld LJ. (1995). The influence of season, photoperiod and pineal melatonin on immune function. J. Pineal Res., 19, 149-165.

Nelson RJ, and Demas GE. (1997). Role of melatonin in mediating seasonal energetic and immunologic adaptations. Brain Res. Bull., 44, 423-431.

Nir I. (1995). Biorhythms and the biological clock involvement of melatonin and the pineal gland in life and disease. Biomed. Env. Sci., 8, 968-975.

Pang CS, and Pang SF. (1992). High affinity specific binding of 2[^{125}I] iodomelatonin by spleen membrane preparations of chicken. J. Pineal Res., 12, 167-173.

Pang SF, Yuan H, Yu ZH, Wong XL, Tang PL, and Yau M. (1991). Melatonin binding sites in the nervous system and immune system. *In*: Reiter RJ, and Fraschini F. (Eds.), *Role of Melatonin and Pineal Peptides in Neuroimmunemodulation*. Plenum Press, New York, pp. 107–117.

Persengiev S, Patchev V, and Velev B. (1991a). Steroid hormone receptors in the thymus: A site of immunomodulatory action of melatonin. Int. J. Biochem., 23, 1483-1485.

Persengiev S, Patchev V, and Velev B. (1991b). Melatonin effects on thymus steroid receptors in the course of primary antibody responses: significance of circulating glucocorticoid level. Int. J. Biochem., 23, 1487-1489.

Petrovsky N, and Harrison L. (1996). Diurnal rhythmicity of human cytokine production. A dynamic disequilibrium in T helper cell type1 / T helper cell type 2 balance. J. Immunol., 158, 5163-5168.

Pierpaoli W, and Yi C. (1990). The involvement of pineal gland and melatonin in immunity and aging, thymus mediated immuno-reconstituting and antiviral activity of thyrotropin releasing hormone. J. Neuroimmunol., 27, 99–109.

Pioli C, Caroleo MC, Nistico G, and Doria G. (1993). Melatonin increases antigen presentation and amplifies specific and non-specific signals for T-cell proliferation. Int. J. Immunopharmacol., 15, 463-468.

Plotnikoff NP, Faith RE, Murgo AS, and Good A. (Eds.). (1986). *Enkephalins and endorphins: stress and the immune system*. Plenum Press, New York.

Poffebbarger M, and Fuller GM. (1976). Is melatonin a micro tubular inhibitor? Exp. Cell. Res., 14, 151-165.

Poon AMS, and Pang SF. (1992). 2[125I] iodomelatonin binding sites in spleens of guinea pigs. Life Sci., 50, 1719-1726.

Poon AMS, Wang XL, and Pang, SF. (1993). Characteristics of 2- [^{125} I] iodomelatonin binding sites in the pigeon spleen and modulation by guanine nucleotides. J. Pineal Res., 14, 169–177.

Poon AMS, Liu ZM, Tang F, and Pang SF. (1994a). Cortisol decreases 2[^{125}I] iodomelatonin binding sites in the duck thymus. Eur. J. Endocrinol., 130, 320-324.

Poon AMS, Liu ZM, Tang F, and Pang SF. (1994b). Evidence for a direct action of melatonin on the immune system. Biol. Signals, 3, 107-117.

Pozo D, Delgado M, Calvo JR, Garcia-Perganeda A, and Guerrero JM. (1997). Expression of membrane melatonin receptor mRNA in rat thymus and spleen. *In*: Maestroni GJM, Conti A, and Reiter RJ. (Eds.), *Therapeutic Potential of Melatonin*, Karger, Basel, pp. 36-42.

Rafii-el-Idrissi M, Calvo JR, Pozo D, Harmouch A, and Guerrero JGM. (1995). Specific binding of 2[^{125}I] iodomelatonin by rat splenocytes: Characterization and its role on regulation of cyclic AMP production. J Neuroimmunol., 57, 171-178.

Rafii-El-Idrissi M, Calvo JR, Harmouch A. (1998). Specific binding of melatonin by purified cell nuclei from spleen and thymus of the rat. J. Neuroimmunol., 86, 190-197.

Rai S. (2004). Immunomodulatory role of melatonin and thyroxin in a seasonally breeding vertebrate, *Funambulus pennanti*. Ph.D. thesis, Department of Zoology. Banaras Hindu Uiniversity, Varanasi, India.

Rai S, and Haldar C. (2003). Pineal Control of immune status and hematological changes in blood and bone marrow of male squirrels (*Funambulus pennanti*) during their reproductively inactive phase. Comp. Bioch. Physiol. (C), 136, 319-328.

Rapozzi V, Zorzet S, and Comelli M. (1998). Melatonin decreases bone marrow and lymphatic toxicity of adryamycin in mice bearing T L X5 lymphoma. Life Sci., 63, 171-173.

Reiter RJ. (1982). The mammalian pineal gland. Structure and functions. Am. J. Anat., 162, 287-313.

Reiter RJ, Maestroni GJM. (1999). Melatonin in relation to the antioxidative defense and immune systems: possible implications for cell and organ transplantation. J. Mol. Med., 77, 36-39.

Rella W, and Lapin V. (1976). Immunocompetence of pinealectomized and simultaneously pinealectomized and thymectomized rats. Oncol., 33, 3-8.

Reppert SM, and Weaver DR. (1995). Melatonin madness. Cell, 3, 1059-1062.

Rodriguez AB, and Lea RW. (1994). Effect of pinealectomy upon the nonspecific immune response of the ring-dove (*Streptopelia risorcia*). J. Pineal Res., 16, 159-166.

Rodriguez AB, Marchena JM, Nogales G, Duran G, and Barriga C. (1999). Correlation between the circadian rhythm of melatonin, phagocytosis and superoxide anion levels in ring-dove heterophils. J. Pineal Res., 26, 35-42.

Rodriguez AB, Ortega E, Lea RW, and Barriga C. (1997). Melatonin and phagocytotic process of heterophils from ring-dove (*Streptopelia risorcia*). Mol. Cell. Biochem., 168, 185-190.

Rogers N, Vanden Heuve lC, and Dawson D. (1997). Effect of melatonin and corticosteroid on in vitro cellular immune function in humans. J. Pineal Res., 22, 75-80.

Sainz RM, Mao JC, Uria H, Kotler M, Antolin F, Roadrignez C, and Menadez-Palacz AZ. (1995). The pineal neurohormone melatonin prevents *in vivo* and *in vitro* apoptosis of thymocytes. J. Pineal Res., 19, 178-188.

Saksela E, and Timonen T. (1980). Cellular interaction in the augmentation of human NK activity by interferon. Ann. N. Y. Acad. Sci., 350, 102–111.

Shiu SY, Li L, Siu SW, Xi SC, Fong SW, and Pang SF. (2000). Biological basis and possible physiological implication of melatonin receptor mediated signaling in rat epididymis. Biol. Signal Receptors, 2, 172-187.

Shivatcheva TM, and Hadjioloff AI. (1987). Seasonal involution of the splenic lymphoid tissue of the European ground squirrel. Arkhiv. Anat. Gist. Embriol., 92, 48–53.

Skwarlo-Sonta K. (1996). Functional connection between the pineal gland and immune system. Acta Neurobiol. Exp., 56, 341-357.

Skwarlo-Sonta K, Majewski P, Markowska M, Olap R, and Olazanska B. (2003). Bidirectional communication between the pineal galnd and immune system. Can. J. Physiol. Pharmacol., 81, 342-349.

Skwarlo-Sonta K, Wolinska-Witort E, Dziwinski T, Snochowski M, and Sotowska-Brochocka J. (1994). Melatonin binding sites in chicken lymphoid glands, gonads and central nervous system. Acta Neurobiol. Exp., 54,132–133.

Spector NH, Dolina S, Cornelessen G, Halberg F, Markovic BM, Yankovic BD. (1996). Neuroimmunomodulation: Neuroimmuno interactions with the environment. *In:* Melvin J, Fregly and Blatteis CM. (Eds.), Handbook of Physiology, Amer. Physiol. Soc. Book, pp. 1437-1549.

Vanecek J. (1998). Cellular mechanism of melatonin action. Physiol. Rev. 78, 687-721.

Vaughan MK, Hubbard GB, Champney GM., Vaughan GM, Little JC, and Reiter RJ. (1987). Splenic hypertrophy and extramedullary hematopoiesis induced in male Syrian hamsters by short photoperiod or melatonin injection and reversed by melatonin pillete or pinealectomy. Am. J. Anat., 179, 131-136.

Vermeulen M, Palerno M, Giordano M. (1993). Neonatal pinealectomy impairs murine antibody dependent cellular toxicity. J. Neuroimmunol., 43, 97-101.

Wichmann M, Zellweger R, Demaso, A. (1996). Melatonin administration attenuates depressed immune functions after trauma-hemorrhage. J. Surg. Res., 63, 256-262.

Winston M, Johnson E, Kelleher JK, Benerjee S, Margulis L. (1974). Melatonin: Cellular effects on line stensor correlated with the inhibition of choline binding on microtubule protein. Cytobios, 9, 237-243.

Yellon SM, Teasley LA, and Fagoaga OR. (1999). Role of photoperiod and pineal gland in T cell dependent humoral immune reactivity in the Siberian hamsters. J. Pineal Res., 27, 243-248.

Yu HS, and Reiter RJ. (1993). *Melatonin: Biosynthesis, Physiological Effects and Clinical Applications*. CRC Press, Boca Raton.

Yu Q, Miller SC, and Osmond DG. (2000). Melatonin inhibits apoptosis during early B-cell development in mouse bone marrow. J. Pineal Res., 29, 86-93.

Yu ZH, Yuan H, Lu Y, and Pang SF. (1991). [^{125}I] Iodomelatonin binding sites in spleen of birds and mammals. Neurosci. Lett., 125, 175-178.

— Section C —
Comparative Endocrinology and Biotechnology

14

Phylogenetic Diversity and Evolution of Pituitary Gonadotropin Beta Subunits

John Yuh-Lin Yu and San-Tai Shen*

Endocrinology Laboratory, Institute of Zoology, Academia Sinica, Taipei, Taiwan R.O.C.

ABSTRACT

All vertebrates have a pituitary gland, which synthesizes and secretes gonadotropic hormone (gonadotropin, GTH). There are two distinct types of GTHs in gnathostomatous vertebrates: luteinizing hormone (LH) and follicle-stimulating hormone (FSH). All GTHs are glycoproteins consisting of α and β subunits. Within a given species, α subunits are common, while β subunits are hormone specific. For better understanding of their phylogenetic diversity and evolution, we have cloned cDNAs of β-subunits of the pituitary LHs and FSHs from species representing various vertebrate classes. Based on their deduced amino acid sequence of our observations, together with finding from the literature, the following conclusions are derived: (1) Twelve cysteine residues, forming six disulfide bonds, have been conserved for β subunits of LHs during evolution of vertebrates. (2) The positions of the 12 cysteine residues in FSH-βs are all conserved in tetrapods. In certain fish, the 3rd cysteine residues are positionally shifted to the N-terminus. (3) One asparagine (N)-linked glycosylation site (Asn-X-Thr/Ser) is found between Cys 1 and Cys 2 of LH-βs in all vertebrates. However, two N-linked glycosylation sites (between Cys 1 and Cys 2, and between Cys 3 and Cys 4) are conserved in FSH-βs of all tetrapods and most of fish; there is only one N-linked glycosylation site is present in FSH-β of certain fish. (4) The inter-animal class homologies of LH-βs and FSH-βs are as low as 32 per cent and 25 per

**Corresponding Author*: E-mail: johnylyu@gate.sinica.edu.tw

cent, respectively. (5) The intra-order animal class homology is greater than inter-animal class homology in both LH-βs and FSH-βs. And, 6) FSH-βs are more conserved than LH-βs in tetrapods; however, LH-βs are more conserved than FSH-βs in teleosts.

Introduction

All vertebrate species are characteristics of possession of a pituitary gland which secretes glycoprotein hormones: gonadotropins (GTH) and thyrotropin (TSH) that is synthesized, respectively by gonadotrope and thyrotrope. Gonadotropins control the function of gonad including spermatogenesis, folliculogenesis, spermiation, ovulation, steroidogenesis and others; while thyrotropin regulates the thyroidal function, stimulating formation and release of thyroid hormones. There are two distinct types of GTH in gnasthomas: LH and FSH. In fish, FSH and LH were originally names as GTH-I and GTH-II, respectively (Suzuki *et al.*, 1988). All these pituitary glycoprotein hormones are composed of two structurally different subunits, α and β; α-subunits are common within a given species, while β-subunits are specific and determine the hormonal activity and species specificity (Pierce and Parsons, 1981). The subunits are synthesized initially by different genes as separate peptides from different messenger RNAs and then are associated with non-covalent linkage to form hormonal molecules (Gharib *et al.*, 1990).

In 1970's, pituitary glycoprotein hormones were isolated and separated to two subunits before chemical analysis of amino acids. Since the development and advancement of cDNA cloning techniques in 1980's, investigators have cloned the cDNAs of these molecules, and their amino acids were then deduced from the nucleotide codons.

For better understanding of the phylogenetic diversity and evolution of the pituitary glycoprotein hormones, we have cloned cDNAs of α and β subunits of pituitary glycoprotein hormones from teleosts (Chatterjee *et al.*, 2001; Han *et al.*, 2003; Han *et al.*, 2004; Chatterjee *et al.*, 2005), birds (Hsieh *et al.*, 2000; Hsieh *et al.*, 2001; Shen and Yu, 2002), and reptiles (Chien *et al.*, 2005). In this article, we present phylogenetic comparison of the deduced amino acids of β-subunits of GTH of species representing various vertebrate classes. Our investigations together with studies by others demonstrated that where and which amino acids of GTH β-subunits have altered, yet retaining their original biological activities, during the evolution of vertebrates.

Pylogenetic Diversity of LH-β and FSH-β Subunits

The LH-βs have been cloned for more than 55 species (18 mammals, 4 birds, 2 reptiles, 3 amphibians, and 28 fish). The approximate identities of amino acids in LH-β mature proteins of respective animal class are: 75 per cent among mammals; 84 per cent among birds; 59 per cent between 2 reptiles (Aizawa and Ishii, 2003); 83 per cent among amphibians (Komoike and Ishii, 2003); 71 per cent among fish. The approximate identities between the two animal classes are: 42 per cent between mammals and birds; 44 per cent between mammals and reptiles; 41 per cent between mammals and amphibian; 41 per cent between mammals and fish; 49 per cent between birds and reptiles; 39 per cent between birds and amphibians; 43 per cent between birds and fish; 50 per cent between reptiles and amphibians; 48 per cent between reptiles and fish; and 49 per cent between amphibians and fish.

The FSH-βs have been cloned for more than 55 species (17 mammals, 4 birds, 2 reptiles, 4 amphibians and 28 fish). The approximate identities of amino acids in FSH-β mature proteins of respective animal classes are: 86 per cent among mammals; 95 per cent among birds; 91 per cent amphibians; 97 per cent among reptiles; 54 per cent among fish. The approximate similarities between the two animal classes are: 69 per cent between mammals and birds; 68 per cent between mammals and reptiles; 55 per cent between mammals and amphibians; 41 per cent between mammals and fish; 85 per cent between birds and reptiles; 62 per cent between birds and amphibians; 40 per cent between birds and fish; 61 per cent between reptiles and amphibians; and 40 per cent between amphibian and fish.

Conservation of Disulfide Bonds and Asparagine-Linked Glycosylation Sites

The multiple amino acid sequence alignment of LH-β and FSH-β subunits of species representing different classes and orders of vertebrates are shown in Figures 14.1 and 14.2, respectively. As indicated

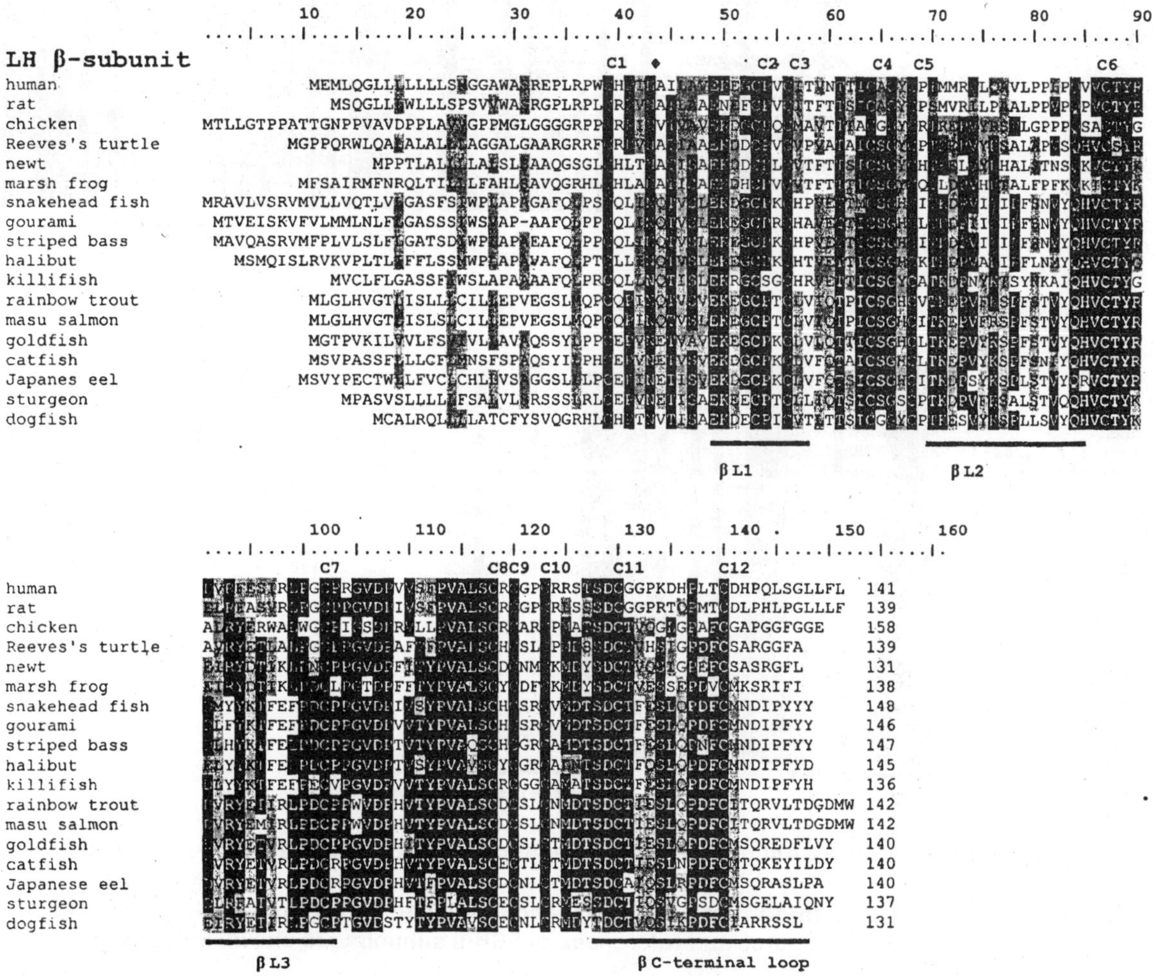

Figure 14.1: Multiple Amino Acid Sequence Alignments of Vertebrate LH-β subunits

The sequences include both signal peptides and mature proteins. Amino acid positions are indicated on the top. Twelve conserved cysteine residues are numbered C1 through C12. βL1, βL2, and βL3 denote the three *b*-hairpin loops. Conserved amino acids are shaded in black boxes. Putative Asp-linked glycosylation sites are indicated with (◆). The source and references of LH-β subunits are shown in Table 14.1.

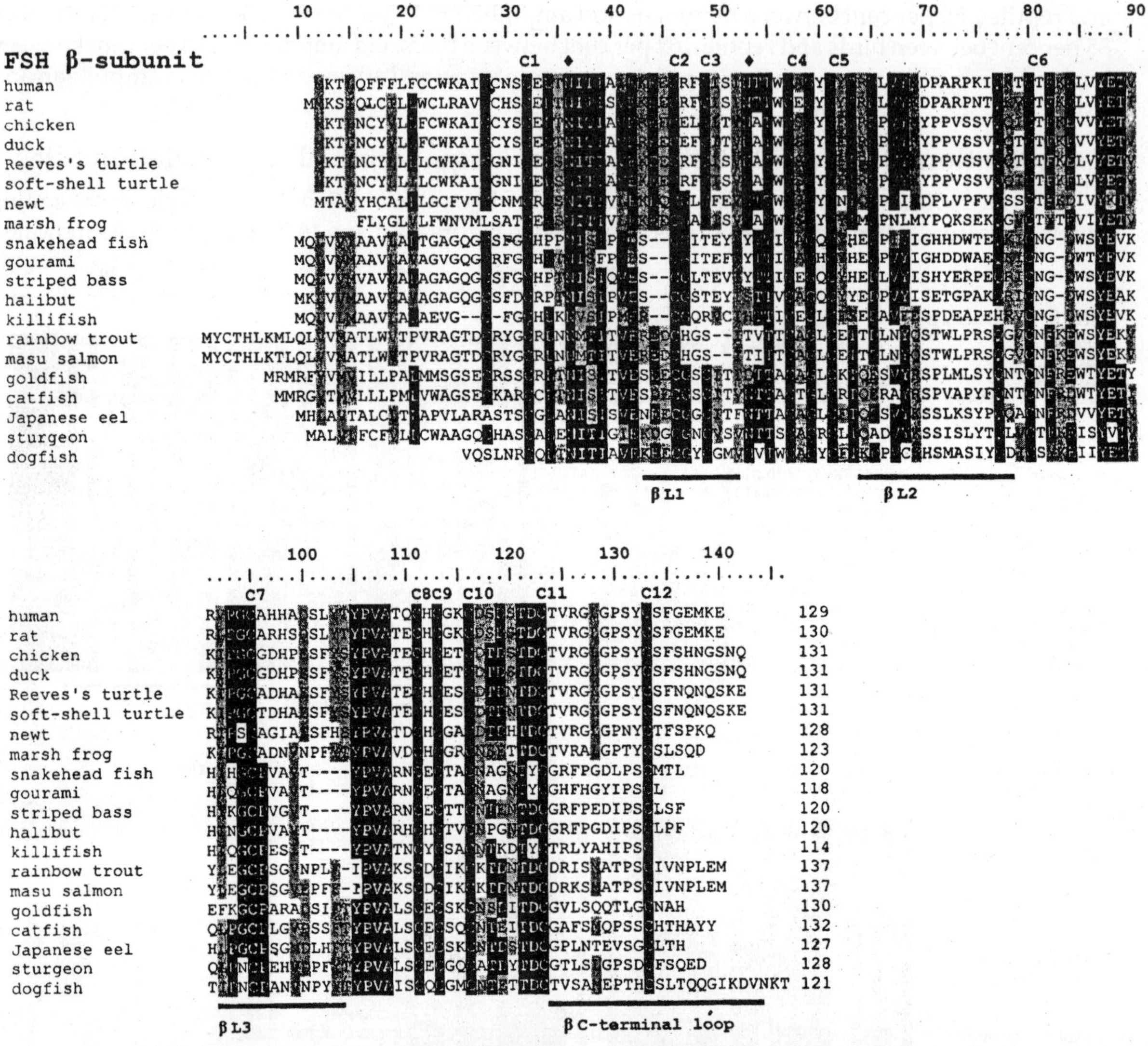

Figure 14.2: Multiple Amino Acid Sequence Alignments of Vertebrate FSH-β subunits.

See Figure 14.1 legend for further details. Hyphen (-) indicates the gap used for obtaining the maximal homology. The source and references of FSH-β subunits are shown in Table 14.1.

in Figure 14.1, twelve cysteine residues, forming six disulfide bonds, are conserved in LH-βs of all vertebrate species so far examined. From the crystal structure of hCG (Lapthorn *et al.*, 1994) and human FSH (Fox *et al.*, 2001), the six disulfide bonds are: C1-C6, C2-C9, C3-C12, C4-C8, C5-C9, and C10-C11. One asparagine- linked (N-linked) glycosylation site (between C1 and C2) is conserved in LH-βs of all vertebrate species.

Table 14.1: The Sources (GeneBank Accession Numbers and References) of Vertebrate FSH-β and LH-β Used for Protein Sequence Alignment and Phylogenetic Analysis

Class/Order	*Species*		*Accession Number*		*References*
			FSH-β	LH-β	
MAMMALIA					
Primates	human	*Homo sapiens*	P01225	P01229	Saxena and Rathnam, 1976; Talmadge *et al.*, 1984
Rodentia	rat	*Rattus norvegicus*	A32893	P01230	Maurer, 1987; Chin *et al.*, 1983
AVES					
Galliformes	chicken	*Gallus gallus*	AAK31580	A61091	Shen and Yu, 2002; Noce *et al.*, 1989
REPTILIA					
Testudines	Reeves's turtle	*Geoclemys reevesi*	BAB92948	BAB92949	Aizawa and Ishii, 2003
	C. soft-shell turtle[a]	*Pelodiscus sinensis*	b	—	Chien *et al.*, 2005
AMPHIBIA					
Anura	marsh frog	*Rana ridibunda*	CAC39253	CAC39252	Unpuslibhed.
Caudata	newt	*Cynops pyrrhogaster*	BAB92958	BAB92959	Saito *et al.*, 2002
OSTEICHTHYES					
Perciformes	snakehead fish	*Channa maculata*	AY447038	AY447037	Chatterjee, *et al.*, 2005
	gourami	*Trichogaster trichopterus*	AAD51934	AAD51935	Jackson *et al.*, 1999
	striped bass	*Morone saxatilis*	AAC38035	AAC38019	Hassin *et al.*, 1995
Pleuronectiformes	J. flounder[a]	*Paralichthys olivaceus*	BAB47387	BAB47388	Kajimura *et al.*, 2001
Cyprinodontiformes	At. killifish[a]	*Fundulus heteroclitus*	AAB59962	AAB59963	Lin *et al.*, 1992
Salmoniformes	rainbow trout	*Oncorhynchus mykiss*	BAB17686	BAB17687	Weil *et al.*, 1995
	masu salmon	*Oncorhynchus masou*	AAB30423	AAB30424	Kato *et al.*, 1993
Cypriniformes	goldfish	*Carassius auratus*	BAA13530	BAA13531	Yoshiura *et al.*, 1997
Siluriformes	catfish	*Ictalurus punctatus*	AAG32155	AAG32156	Liu *et al.*, 1997
Anguilliformes	J. eel[a]	*Anguilla japonica*	Q9YGK3	AAL93618.	Han *et al.*, 2003; Yoshiura *et al.*, 1999
Acipenseriformes	sturgeon	*Acipenser baeri*	CAB93504	CAB93502	Querat *et al.*, 2000
CHONDRICHTHYES					
Lamniformes	Dogfish	*Scyliorhinus canicula*	CAC43236	CAC43235	Querat *et al.*, 2001

a: C. Chinese; J.: Japanese; At.: Atlantic; b: In submission.

As shown in Figure 14.2, twelve cysteine residues of FSH-βs are conserved in tetrapods (mammals, birds, reptiles, and amphibian), chondrichthyes (ex. dogfish), and certain osteichthyes (ex. goldfish, catfish, eels, sturgeon), similar to those of LH-βs. In certain fish including Perciformes (snakehead fish, gourami, and striped bass), Salmoniformes (trout and salmon), Pleuronectiformes (halibut, flounder), Cyprinodontiformes (killifish), eleven out of twelve cysteine residues are conserved; but the 3rd cysteine residues are positionally shifted to the N-terminus. We have proposed such positionally shifted cysteine can form disulfide bond with the 12th cysteine as indicated by hypothetical structural model of snakehead fish FSH (Chatterjee *et al.*, 2005) based on crystal structure of human FSH (Fox *et al.*, 2001). The disulfide bonds are important for the correct folding and structural integrity of the molecules. In particular, the disulfide bond formed by the 3rd and 12th cysteine residues of the β-subunit ties the carboxyl-terminal loop with the βL1 hairpin loop; such organization serves as a "seat-belt" to wrap one end of the elongated α-subunit.

Studies revealed that the mutation of the second glycosylation site from Asn to Gln of human FSH-β led to reducing its biological activity (Flack *et al.*, 1994). Whether or not the absence of the second N-linked glycosylation of the FSH-βs in certain fish affects the biological activity requires further elucidation.

Phylogenetic Tree of Vertebrate GTH-β Subunits

No GTH has been isolated from agnathous vertebrates so far. GTH-β was cloned very recently; there is only type of GTH-β existing in sea lamprey (Kawauchi, 2005). Thus, it appears that a single type of GTH existing in agnatha; while in gnasthomas vertebrates GTH evolved to two distinct types of molecules. *i.e.* LH and FSH.

A phylogenetic tree of vertebrate LH-β and FSH-β based on their homology in amino acid sequences is presented in Figure 14.3. As indicated, LH-β and FSH-β are clearly separated into two groups. In general, they are grouped in accordance with vertebrate class. In both LH-βs and FSH-βs, the species in the same animal classes are grouped together, with exception of sturgeon and dogfish. These two fish are separated from fish subgroup, but are grouped closer to tetrapods. In tetrapodian FSH-βs, amniotes are separated from amphibians. However, in LH-β, mammals and bird (chicken) are separated from reptile and amphibian.

The amino acid homology of FSH-βs between birds (including chicken, quail and ostrich) and mammals (N=8) is 69 per cent in average (in the range of 63-76 per cent); while the amino acid homology of their LH-β subunits counterpart is 42 per cent in average (in the range of 34-51 per cent). This fact indicates that FSH-β subunit is more conserved than LH-β subunit during evolution in higher vertebrates. These observations agree well with the LH and FSH bioassay results from our laboratory: LHs from various tetrapodian species show remarkable variation of potency (>10,000 fold) in the LH bioassays employing the stimulation of testicular androgen formation systems *in vitro* (Yu and Wang, 1987; Yu *et al.*, 1995; Yu, 1997; Yu, 2000); by contrast, FSHs from various tetrapodian species show less variation of potencies (~1,000 fold) in a FSH bioassay in stimulating estradiol-17ß formation of immature rat Sertoli cells *in vitro* (Yu *et al.*, 1996).

Summary and Conclusions

For the past two decades, cDNAs of pituitary GTH-β subunits have been cloned for species representing different animal classes and/or orders of vertebrates from various laboratories. Their deduced peptide sequences were thus available for investigation of phylogenetic diversity and evolution of GTH-β in vertebrates. We have cloned cDNAs of pituitary GTH-β subunits from birds (chicken and duck), reptile (Chinese soft-shell turtle), and fish (snakehead fish and Japanese eel) for better understanding of evolution of these hormonal molecules. The following conclusions may be derived from our results together with the findings from literature:

1. The numbers of amino acids in both signal peptide and mature protein of GTH-β subunits vary with species.

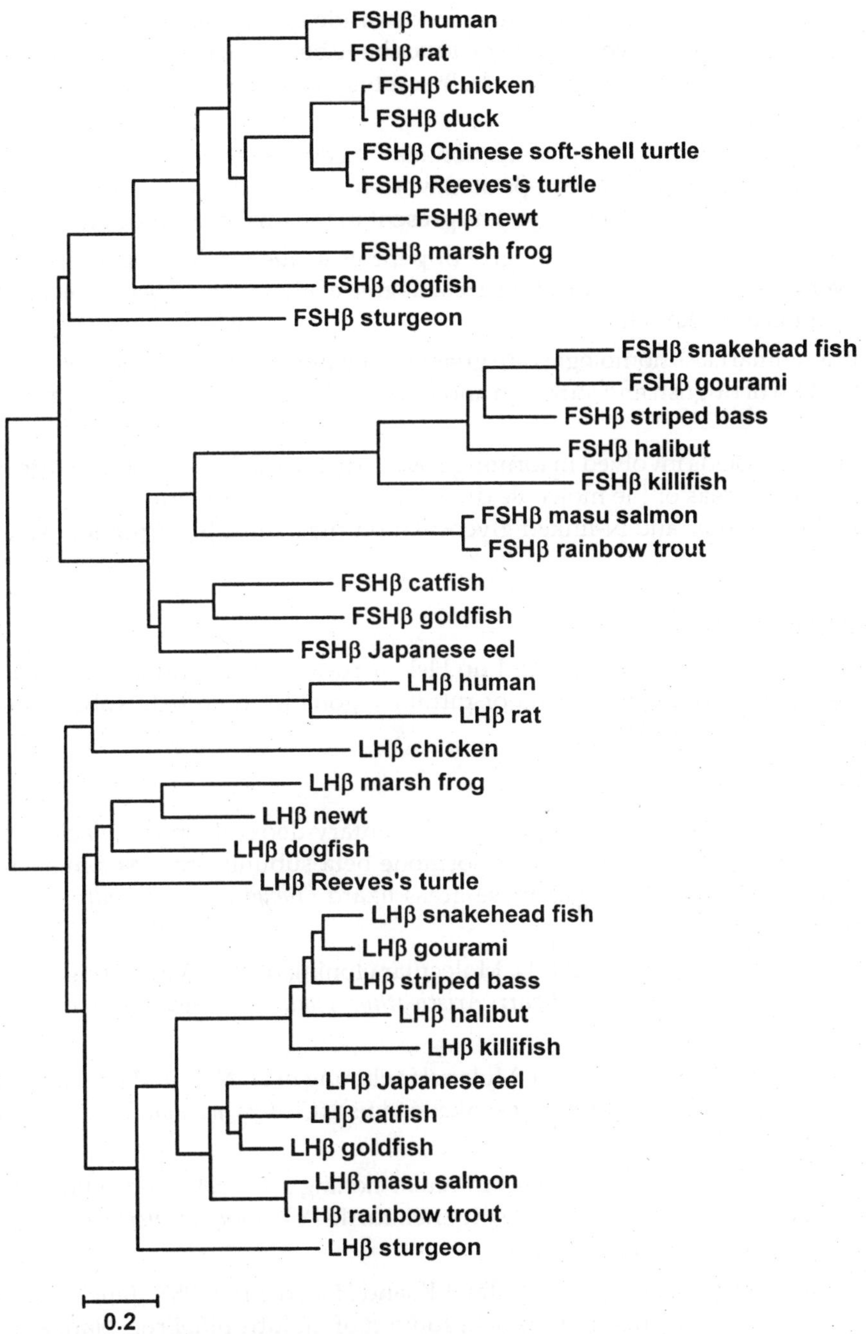

Figure 14.3: A Phylogenetic Tree of Vertebrate LH-βs and FSH-βs Based on their Amino Acid Homologies

Data were calculated according to the amino acid sequence alignments in Figures 14.1 and 14.2, with Blosum-62 amino acid substitution matrix, and the tree was constructed by neighbor-joining method.

2. Twelve cysteine residues, forming six disulfide bonds, have been conserved in LH-βs of vertebrates. The twelve cysteine residues have been also conserved in FSH-βs of tetrapods and most of fish; in certain fish the 3rd cysteine residues are positionally shifted to the N-terminus.
3. One asparagine-linked glycosylation site, located between Cys 1 and Cys 2, have been conserved in LH-βs and FSH-βs of all vertebrates. Additional N-linked glycosylation site, located between Cys 3 and Cys 4, is present in FSH-βs of tetrapods and most fish.
4. The inter-animal class homologies of peptide sequence of LH-β and FSH-β subunits are lower than 30 per cent, indicating great degree of alteration of these molecules occurs during evolution of vertebrates.
5. Intra-animal class homologies are greater than inter-animal class homologies with exception of FSH-β of dogfish and sturgeon and LH-β of dogfish. More studies in elasmobrahches are needed.
6. The amino acids involved in forming basic structure of GTH-β subunits are more conserved than other areas of the molecule during evolution of vertebrates, those cysteines forming disulfide bonds, and N-linked glycosylation sites, three β hairpin loops, and βC-terminal loop.

Acknowledgements

We thank Drs. Abhijit Chatterjee, Ya-Lun Hsieh, Yu-Shan Han, Intrajee Chowdhury, and Jung-Tsun Chien for their cloning of cDNAs of pituitary gonadotropin beta subunits of various non-mammalian species in our laboratory.

References

Aizawa Y, and Ishii S. (2003). Cloning of complimentary deoxyribonucleic acid encoding follicle-stimulating hormone and luteinizing hormone beta subunit precursor molecules in Reeves's turtle (*Geoclemys reevesii*) and Japanese grass lizard (*Takydromus tachydromoides*). Gen. Comp. Endocrinol., 132, 465-473.

Chatterjee A, Hsieh YL, and Yu JY. (2001). Molecular cloning of cDNA encoding thyroid stimulating hormone beta subunit of bighead carp *Aristichthys nobilis* and regulation of its gene expression. Mol. Cell Endocrinol., 174, 1-9.

Chatterjee A, Shen ST, and Yu JY. (2005). Molecular cloning of cDNAs and structural model analysis of two gonadotropin beta-subunits of snakehead fish (*Channa maculata*). Gen. Comp. Endocrinol., 143, 278-286.

Chien JT, Shen ST, Lin YS, and Yu JY. (2005). Molecular cloning of the cDNA encoding follicle-stimulating hormone beta subunit of the Chinese soft-shell turtle *Pelodiscus sinensis*, and its gene expression. Gen. Comp. Endocrinol., 141, 190-200.

Chin WW, Godine JE, Klein DR, Chang AS, Tan LK, and Habener JF. (1983). Nucleotide sequence of the cDNA encoding the precursor of the beta subunit of rat lutropin. Proc. Natl. Acad. Sci. USA, 80, 4649-4653.

Flack MR, Froehlich J, Bennet AP, Anasti J, and Nisula BC. (1994). Site-directed mutagenesis defines the individual roles of the glycosylation sites on follicle-stimulating hormone. J. Biol. Chem., 269, 14015-14020.

Fox KM, Dias JA, and Van Roey P. (2001). Three-dimensional structure of human follicle-stimulating hormone. Mol. Endocrinol., 15, 378-389.

Gharib SD, Wierman ME, Shupnik MA, and Chin WW. (1990). Molecular biology of the pituitary gonadotropins. Endocr. Rev., 11, 177-199.

Han YS, Liao IC, Huang YS, Tzeng WN, Yu JYL. (2003). Profiles of PGH-alpha, GTH I-beta, and GTH II-beta mRNA transcript levels at different ovarian stages in the wild female Japanese eel *Anguilla japonica*. Gen. Comp. Endocrinol., 133, 8-16.

Han YS, Liao IC, Tzeng WN, and Yu JY. (2004). Cloning of the cDNA for thyroid stimulating hormone beta subunit and changes in activity of the pituitary-thyroid axis during silvering of the Japanese eel, *Anguilla japonica*. J. Mol. Endocrinol., 32, 179-194.

Hassin S, Elizur A, and Zohar Y. (1995). Molecular cloning and sequence analysis of striped bass (*Morone saxatilis*) gonadotrophin-I and–II subunits. J. Mol. Endocrinol., 15, 23-35.

Hsieh YL, Chatterjee A, Lee G, and Yu JYL. (2000). Molecular cloning and sequence analysis of the cDNA for thyroid-stimulating hormone beta subunit of Muscovy duck. Gen. Comp. Endocrinol., 120, 336-344.

Hsieh YL, Chatterjee A, Chien JT, and Yu JYL. (2001). Molecular cloning of the cDNAs for pituitary glycoprotein hormone alpha subunits of two species of duck and their gene regulation. J. Mol. Endocrinol., 27, 339-347.

Jackson K, Goldberg D, Ofir M, Abraham M, and Degani G. (1999). Blue gourami (*Trichogaster trichopterus*) gonadotropic beta subunits (I and II) cDNA sequences and expression during oogenesis. J. Mol. Endocrinol., 23, 177-187.

Kajimura S, Yoshiura Y, Suzuki M, and Aida K. (2001). cDNA cloning of two gonadotropin beta subunits (GTH-Ibeta and–IIbeta) and their expression profiles during gametogenesis in the Japanese flounder (*Paralichthys olivaceus*). Gen. Comp. Endocrinol., 122, 117-129.

Kato Y, Gen K, Maruyama O, Tomizawa K, and Kato T. (1993). Molecular cloning of cDNAs encoding two gonadotrophin beta subunits (GTH-I beta and–II beta) from the masu salmon, *Oncorhynchus masou*: rapid divergence of the GTH-I beta gene. J. Mol. Endocrinol., 11, 275-282.

Kawauchi H. (2005). The dawn and evolution of hormones in the adnohypohysis. Proc. 15th Int. Cong. Comp. Endocrinol., Boston, Mass, USA. May 22-27, p. 82.

Komoike Y, and Ishii S. (2003). Cloning of cDNAs encoding the three pituitary glycoprotein hormone beta subunit precursor molecules in the Japanese toad, *Bufo japonicus*. Gen. Comp. Endocrinol., 132, 333-347.

Lapthorn AJ, Harris DC, Littlejohn A, Lustbader JW, Canfield RE, Machin KJ, Morgan FJ, and Isaacs NW. (1994). Crystal structure of human chorionic gonadotropin. Nature, 369, 455-461.

Lin YW, Rupnow BA, Price DA, Greenberg RM, and Wallace RA. (1992). *Fundulus heteroclitus* gonadotropins. 3. Cloning and sequencing of gonadotropic hormone (GTH) I and II beta-subunits using the polymerase chain reaction. Mol. Cell Endocrinol., 85, 127-139.

Liu Z, Li P, Argue BJ, Dunham RA. (1997). Gonadotropin alpha-subunit glycoprotein from channel catfish (*Ictalurus punctatus*) and its expression during hormone-induced ovulation. Mol. Mar. Biol. Biotechnol., 6, 217-227.

Maurer RA. (1987). Molecular cloning and nucleotide sequence analysis of complementary deoxyribonucleic acid for the beta-subunit of rat follicle stimulating hormone. Mol. Endocrinol., 1, 717-723.

Noce T, Ando H, Ueda T, Kubokawa K, Higashinakagawa T, and Ishii S. (1989). Molecular cloning and nucleotide sequence analysis of the putative cDNA for the precursor molecule of the chicken LH-beta subunit. J. Mol. Endocrinol., 3, 129-137.

Pierce JG, and Parsons TF. (1981). Glycoprotein hormones: structure and function. Annu. Rev. Biochem., 50, 465-495.

Querat B, Sellouk A, Salmon C. (2000). Phylogenetic analysis of the vertebrate glycoprotein hormone family including new sequences of sturgeon (*Acipenser baeri*) beta subunits of the two gonadotropins and the thyroid-stimulating hormone. Biol. Reprod., 63, 222-228.

Querat B, Tonnerre-Doncarli C, Genies F, and Salmon C. (2001). Duality of gonadotropins in gnathostomes. Gen. Comp. Endocrinol., 124, 308-314.

Saito A, Kano Y, Suzuki M, Tomura H, Takeda J, and Tanaka S. (2002). Sequence analysis and expressional regulation of messenger RNAs encoding beta subunits of follicle-stimulating hormone and luteinizing hormone in the red-bellied newt, *Cynops pyrrhogaster*. Biol. Reprod., 66, 1299-1309.

Saxena BB, and Rathnam P. (1976). Amino acid sequence of the beta subunit of follicle-stimulating hormone from human pituitary glands. J. Biol. Chem., 251, 993-1005.

Shen ST, Yu JYL. (2002). Cloning and gene expression of a cDNA for the chicken follicle-stimulating hormone (FSH)-beta-subunit. Gen. Comp. Endocrinol., 125, 375-386.

Suzuki K, Kawauchi H, and Nagahama Y. (1988). Isolation and characterization of two distinct gonadotropins from chum salmon pituitary glands. Gen. Comp. Endocrinol., 71, 292-301.

Talmadge K, Vamvakopoulos NC, and Fiddes JC. (1984). Evolution of the genes for the beta subunits of human chorionic gonadotropin and luteinizing hormone. Nature, 307, 37-40.

Weil C, Bougoussa-Houadec M, Gallais C, Itoh S, Sekine S, Valotaire Y. (1995). Preliminary evidence suggesting variations of GtH 1 and GtH2 mRNA levels at different stages of gonadal development in rainbow trout, *Oncorhynchus mykiss*. Gen. Comp. Endocrinol., 100, 327-333.

Yoshiura Y, Kobayashi M, Kato Y, and Aida K. (1997). Molecular cloning of the cDNAs encoding two gonadotropin beta subunits (GTH-I beta and–II beta) from the goldfish, *Carassius auratus*. Gen. Comp. Endocrinol., 105, 379-389.

Yoshiura Y, Suetake H, and Aida K. (1999). Duality of gonadotropin in a primitive teleost, Japanese eel (*Anguilla japonica*). Gen. Comp. Endocrinol., 114, 121-131.

Yu JYL. (1997). Functional evolution of pituitary glycoprotein hormones and receptors in vertebrates: hormone-receptor interaction. *In*: Kawashima, S, and Kikuyama S. (Eds.), *Advances in Comparative Endocrinology, Proceedings of the XIIIth International Congress of Comparative Endocrinology, Yokohama, Japan*, November 17-21, 1977, Monduzzi Editore, International Proceedings Division, Bologna, Italy, Vol. 2, pp.811-816.

Yu JYL. (2000). Phylogenic diversity and evolution of pituitary glycoprotein hormones in vertebrates. Proceeding of 4th Congress of Asia and Oceanina Society for Comparative Endocrinology. *In*: Yu JYL. (Ed.), *Advances in Comparative Endocrinology*, May 12-18, 2000, Taipei, Taiwan R.O.C.

Yu JYL, and Wang LM. (1987). Comparative effects of diverse vertebrate gonadotropins on androgen formation in vitro from testes of roosters and mice. Biol. Reprod., 36, 816-824.

Yu JYL, Shen ST, Liu CT, Weng CF, Peng HK, and Liu FK. (1995). Comparative effects of avian and picine gonadotropins on gonadal steroidogenesis, and of avian and picine pituitaries on induction of spermiation and ovulation in the loach and white silver carp. Aquaculture,135, 59-72.

Yu JYL, Shen ST, Yang WH, Papkoff H, and Ishii S. (1996). Comparative effects of diverse vertebrate gonadotropins on estradiol-17 beta formation in vitro in an immature rat Sertoli cell bioassay. Gen. Comp. Endocrinol., 104, 253-261.

15

Molecular Aspects of Reproductive Neuroendocrinology in Salmon

Akihisa Urano[1*] and Hironori Ando[2]

[1]*Field Science Center for Northern Biosphere, Hokkaido University, Sapporo, Hokkaido 060-0810, Japan*

[2]*Laboratory of Advanced Animal. and Marine Bioresources, Grad. School of Bioresource and Bioenvironmental Sciences, Kyushu University, Fukuoka 812-8581, Japan*

ABSTRACT

Pacific salmons are anadromous fish, which return to their natal river for spawning after oceanic migration. Accumulation of genetic information on hormone precursors and their receptors enabled us to examine various neuroendocrine events that regulate salmon reproduction including spawning migration at a molecular level. We mainly used two salmonid species: masu salmon (*Oncorhynchus masou*) as model fish and chum salmon (*O. keta*) as wild fish, and assessed changes in expression of particular genes in relation to physiological events using a real-time PCR method. Regulatory mechanisms of hormonal gene expression were also examined to some extent, particularly of subunits of gonadotropins.

Cloning and sequencing studies indicated the presence of two salmon (s) GnRH and five GnRH receptor (GnRH-R) genes in masu salmon. Seasonal variation of gene expression for sGnRH and GnRH-R were then estimated through determination of the absolute amounts of mRNAs by the real-time PCR in the brains and the pituitaries of growing and maturing fish, in association with changes in expression of pituitary hormone genes. The expression of two sGnRH genes in the ventral telencephalon and the preoptic area elevated twice in late spring to

**Corresponding Author*: E-mail: aurano@sci.hokudai.ac.jp; Phone: +81-11-706-3525, Fax: +81-11-706-4448

early summer and in breeding season in autumn, whereas patterns of changes in expression of GnRH-R genes in the pituitaries differed among different types. Elevation of gene expression for LH subunits in the pre-spawning season coincided well with that for two sGnRH precursors and some types of GnRH-R. These coincident molecular events may be important for control of final maturation accompanied by upstream migration.

Introduction

The essential requirements for reproductive success are that the partners of different sexes are at the same place at the same time in the same state of reproductive readiness. Spawning migration, which is seen in varieties of animal species including many fishes, is one of evolutionally successful strategies that fulfill the above-mentioned requirements. In vertebrates, it is principally the important part of life history in individual migratory animals. Spawning migration of anadromous salmon is a part of large scale oceanic migration, and considered as a typical stereotyped instinctive behavior which is believed to be genetically programmed.

Migration of salmon is composed of several phases: downstream migration of juveniles following par-smolt transformation; seawater adaptation at the entrance of estuarine habitat; offshore migration for feeding and growth; and then spawning migration to the natal river. The spawning migration includes homing, fresh-water adaptation and upstream migration in maturing adults. Each phase of migration is controlled by the particular neuroendocrine system (Figure 15.1). Par-smolt transformation and downstream migration are under the control of the brain-pituitary-thyroid axis, although roles of thyrotropin-releasing hormone (TRH) are ambiguous in the control of secretion of thyroid-stimulating hormone (TSH). The initial step of seawater adaptation is promoted by cortisol, and then growth hormone (GH) functions in both seawater adaptation and somatic growth through production of insulin-like growth factor I (IGF-I).

Following feeding migration the duration of which is largely different among salmonid species, salmon initiate sexual maturation and spawning migration in late spring probably under the control of the brain-pituitary-gonadal axis (see Urano *et al.*, 1999). When fish arrive at the coastal area or their natal river, prolactin (PRL) and vasotocin serve to adapt fresh-water environment, and sex steroids may stimulate upstream migration. Meantime, final maturation is prompted by luteinizing hormone (LH), which is previously termed as gonadotropin II (GTH II). The neuroendocrine systems are thus involved in the controls of various phases of migratory behavior in salmon, so that we presumed that hypothalamic neurosecretory neurons have crucial roles in the control of salmonid migration.

It is well known that hypothalamic neurosecretory neurons have important roles in the control of instinctive behavior in vertebrates, probably because they have appropriate neuroanatomical features to coordinately control functions of both the central nervous system and the endocrine system (Urano, 1984 and 1988). In terms of the control of spawning migration, projections of gonadotropin-releasing hormone (GnRH) neurons and vasotocin neurons possess the most favorable features (Figure 15.2). They send their axons to many brain loci including the optic tectum, which is crucial for the control of orientation behavior in all vertebrate species, and to the neurohypophysis where GnRH and vasotocin are released into hypophysial vasculatures. Our recent result showed interaction of GnRH neurons and classical neurosecretory neurons (Saito *et al.*, 2003). However, since the aim of present article is an overview of molecular aspects of reproductive neuroendocrinology in salmon, we focus on the GnRH-GTH system and spawning migration.

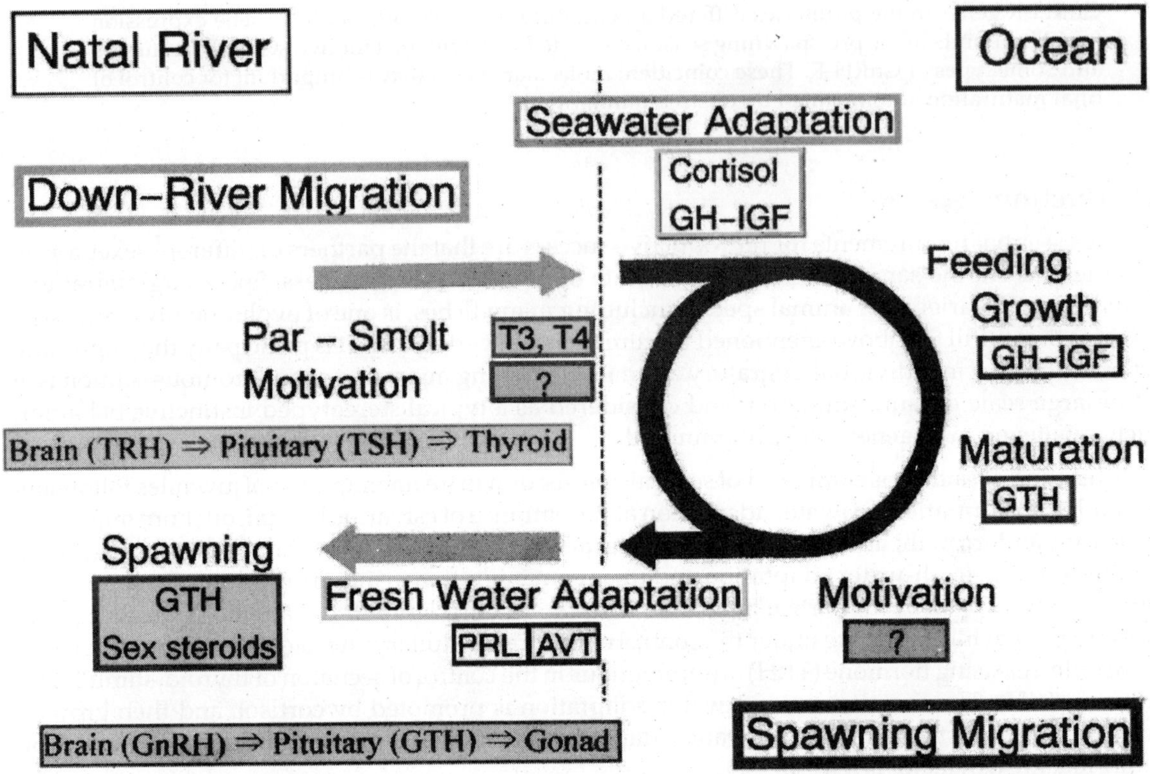

Figure 15.1: The Neuroendocrine Systems Involved in the Control of Migration in Salmon. See the text for explanation. (modified from Urano *et al.*, 1999)

Experimental Fishes: Model Fish and Wild Fish

So-called pacific salmons that belong to genus *Oncorhynchus* include seven species. Their life histories, in particular the duration of oceanic life, differ among species (Groot and Margolis, 1991). Recently diverged species like chum salmon (*O. keta*) spend several years in the North Pacific Ocean and the Bering Sea until they initiate spawning migration, whereas ancestral species like masu salmon (*O. masou*) return their natal river after one year of oceanic life. Salmonids in the former group can not sexually mature when they are aquacultured in fresh water, while species in the latter group can mature even in a freshwater tank. Nonetheless, neuroendocrine features of sexual maturation seemed to be almost the same among different species. We therefore selected masu salmon as a model salmon species to clarify molecular events during sexual maturation, and chum salmon as a representative wild salmon to study molecular mechanisms of spawning migration on the basis of experimental facts shown in the model system.

In our research, we used masu salmon of Mori hatchery strain, which was established for propagation of high return population in the Mori branch of Hokkaido Fish Hatchery. Fish of this strain retain ability of homing migration, nonetheless can grow, mature and spawn in a FW environment. Meanwhile, as wild fish, pre-spawning chum salmon of Ishikari stock, which were obtained from the

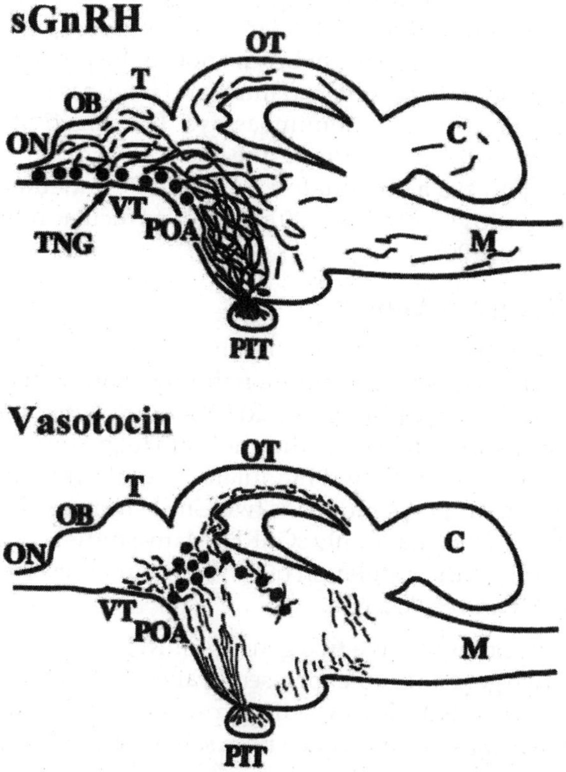

Figure 15.2: Schematic Diagrams Showing Projections of sGnRH Neurons (A) and vasotocin neurons (B).

Note that both sGnRH neurons and vasotocin neurons send their axons to various brain loci and the pituitary (PIT), favorable features for dual controls of the central and neuroendocrine systems. C, cerebellum; M, medulla oblongata; OB, olfactory bulb; ON, olfactory nerve; OT, optic tectum; POA, preoptic area; TNG, terminal nerve ganglion.

Ishikari River-Ishikari bay water system in Hokkaido, Japan, were mainly used in our study. In this vast water system, about 30 million juveniles were released every year from the Chitose Salmon Hatchery, which is located at the riverside of the Chitose River, a tributary of the Ishikari River, so that almost all fish used in the present study were regarded as returnees to the Chitose Salmon Hatchery. The distance from the mouth of the Ishikari River to the Chitose Salmon Hatchery is about 70 km, and chum salmon usually required two to three weeks to reach the hatchery from the mouth of the Ishikari River.

Quantitative Analysis of mRNA Amounts for Estimation of Hormonal Gene Expression

Description of molecular events in the neuroendocrine system in relation to reproductive function, in particular in seasonal breeders, requires information on changes of the absolute amounts of particular mRNAs for hormone precursors. We therefore developed an assay method to determine magnitudes of changes in the amount of particular mRNA, as indices of expression of hormone genes in connection with salmon reproduction. First, we developed a quantitative dot blot analysis in which single-stranded sense cDNA was used as the standard for particular mRNA encoding a precursor of certain pituitary hormone. Validation tests showed that cross hybridization of the probe to mRNAs for other pituitary hormone precursors was less than 1 per cent, and intra- and inter-assay coefficients of variations were comparable to or less than those of radio immunoassays. The sensitivity was between 1 to 10 amol. However, this value is not sufficient to determine the levels of neurohormone mRNAs in the discrete brain loci. Hence, we incorporated a quantitative real-time PCR method, and established a reliable protocol which yields the sensitivity of about 100 copies of particular mRNA with satisfactory ranges of variations (Urano and Ando, 2002).

GnRH and GnRH Receptor Genes

Single vertebrate species usually have two or three GnRH forms. One form of GnRH, so-called hypothalamic GnRH, regulates gonadal maturation through stimulation of pituitary gonadotropes. The second form of GnRH is chicken GnRH-II (cGnRH-II), which is highly conserved from fish through mammals. cGnRH-II neurons are localized in the midbrain tegmentum and send their axons widely throughout the central nervous system, and modulates sexual behavior in some vertebrate species (Millar, 2003). The brain of salmonid fish contains two GnRH forms, salmon (s) GnRH and cGnRH-II, whereas the pituitary was shown to have only sGnRH. Many studies conducted in salmonids showed that sGnRH has a central role in the regulation of synthesis and release of GTHs, and thus promotes the gonadal maturation (Amano *et al.*, 1997).

Because of tetraploidy, salmons have two genes (sGnRH-I and–II) encoding sGnRH precursors (Higa *et al.*, 1997). These two genes are co-expressed in almost all sGnRH neurons in the forebrain of sockeye salmon (Amano *et al.*, 1998). However, the expression level of sGnRH-II gene is much higher than that of sGnRH-I gene in masu salmon and chum salmon (Ando *et al.*, 2001; Kitahashi *et al.*, 2004; Onuma *et al.*, 2005).

Multiple genes which encode GnRH receptors (GnRH-Rs) are present in single species (Lethimonier *et al.*, 2004; Millar *et al.*, 2004). Several GnRH-R subtypes with different structural characteristics are expressed in the brain, pituitary and various peripheral organs. We recently showed that, in masu salmon, five different GnRH-R genes, termed as msGnRH-R1, R2, R3, R4, and R5, are expressed in the brain, the pituitary and other peripheral tissues (Jodo *et al.*, 2003). Binding of GnRH to its receptors may activate multiple signal transduction pathways in target cells, and thus express multiple functions of GnRH in reproduction.

Seasonal Changes in Expression of Genes for sGnRH and GTHs in Masu Salmon

In the forebrains of masu salmon of both sexes, the amounts of sGnRH mRNAs were high during winter through spring in the prepubertal stage, declined toward summer, and then increased again in the spawning period (Figure 15.3) (Ando *et al.*, 2001; Kitahashi *et al.*, 2004). These changes correspond well with the changes in the content of sGnRH in the forebrain, whereas the amount of sGnRH in the pituitary gradually increased along with sexual maturation and reached its maximum in the spawning period (Amano *et al.*, 1992, 1993). Such difference in the contents of GnRH between the forebrain and the pituitary may indicate that the central roles of GnRH are independent of endocrine roles. The prepubertal elevation of sGnRH gene expression in the forebrain suggests a neuromodulatory action of sGnRH that is involved in homing migration, because masu salmon initiate homing migration at this stage.

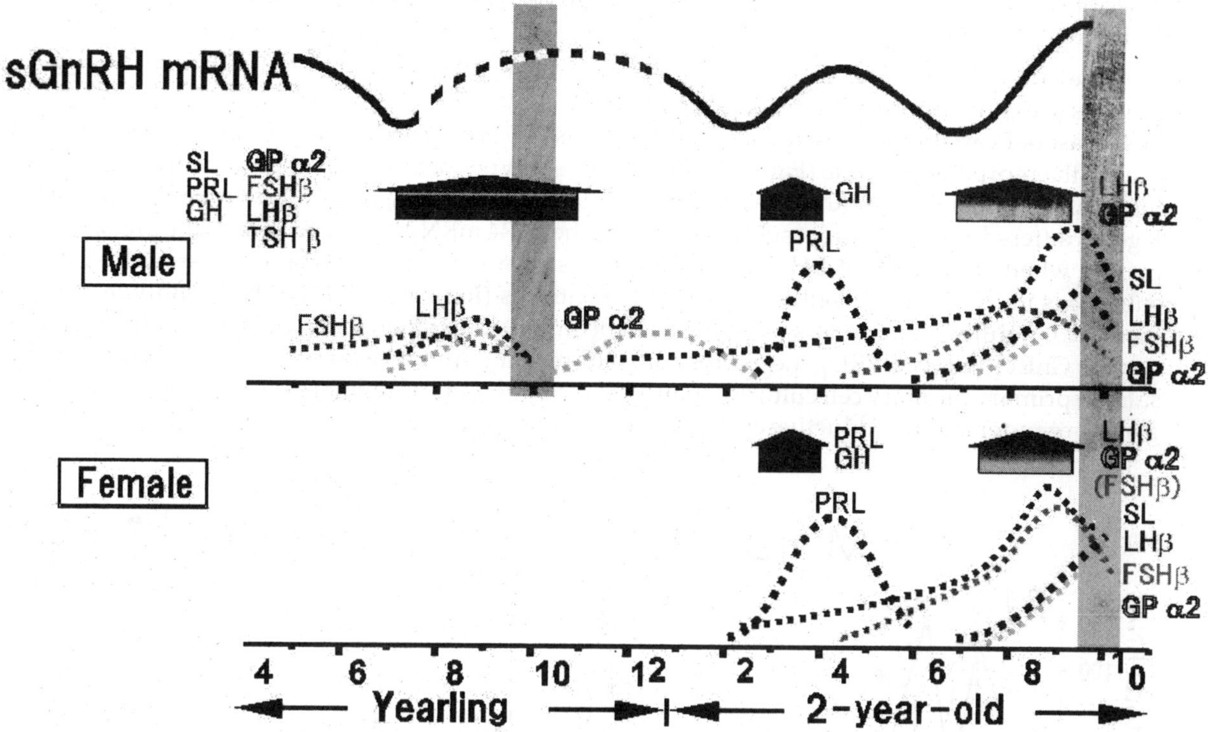

Figure 15.3: Seasonal Changes of Gene Expression for Precursors of sGnRH and Pituitary Hormones in Growing and Maturing Masu salmon in Aquaculture as the Model System. Yellow vertical bars show the breeding season. One-year old males, but not females, showed precocious maturation. See the text for further explanation. (from Kitahashi *et al.*, 2004).

The amounts of mRNAs encoding follicle-stimulating hormone (FSH) β, which was previously called as GTH I, and LHβ increased along with sexual maturation, and reached to considerably high magnitudes in the spawning period. Nevertheless, the increase in the content of FSHβ mRNA was initiated at early stages of gametogenensis, and the elevation of that of LHβ mRNA was initiated at late stages of gametogenensis (Gomez *et al.*, 1999; Kitahashi *et al.*, 2004). In the primary pituitary cells of masu salmon, the amounts of α2, FSHβ, and LHβ mRNAs increased from March through May and reached their maximum at the pre-spawning stage in July (Ando *et al.*, 2004). The amounts of α2 and FSHβ mRNAs then declined in September, while that of LHβ mRNA remained at high levels.

We recently examined GTH release activity of masu salmon primary pituitary cell cultures at four reproductive stages in March (initiation of sexual maturation), May (early maturation), July (pre-spawning), and September (spawning period) (Ando *et al.*, 2004). FSH levels in the culture medium increased with sexual maturation and peaked in September, whereas LH release remained low until July and considerably increased in September. These results indicate that the differential regulation of releases of two GTHs can be achieved in part *in vitro* condition without hypophysiotropic stimuli and neuronal inputs such as dopaminergic and γ-aminobutyric acid (GABA)-ergic (Yaron *et al.*, 2003). It thus appears that two different types of gonadotropes (FSH and LH cells) release GTH autonomously in different manners.

The different effects of GnRH on synthesis and release of two GTHs may be attributed in part to changes in expression of GnRH-Rs in FSH and LH cells. We therefore determined by use of real-time PCR seasonal variations in expression of the five msGnRH-R genes in the pituitary of masu salmon during the reproductive cycle (Figure 15.4). All five subtypes of msGnRH-R genes were expressed in the pituitary, although R4 mRNA was dominant. Interestingly, the expression patterns of five msGnRH-R genes differed among different subtypes. Among them, R4 mRNA increased only in the pre-spawning period, when expression of LH subunit genes was stimulated by GnRHa, suggesting that the R4 subtype is involved in the GnRH-induced LH synthesis (Jodo *et al.*, 2004). Other subtype mRNAs increased in different periods, so that they can be involved in other action of GnRH in the pituitary, such as GnRH-induced PRL gene expression. Our recent unpublished results indicate that, in masu salmon primary pituitary cell cultures, sGnRH and sex steroids upregulate expression of msGnRH-R genes in gender and reproductive cycle dependent manners.

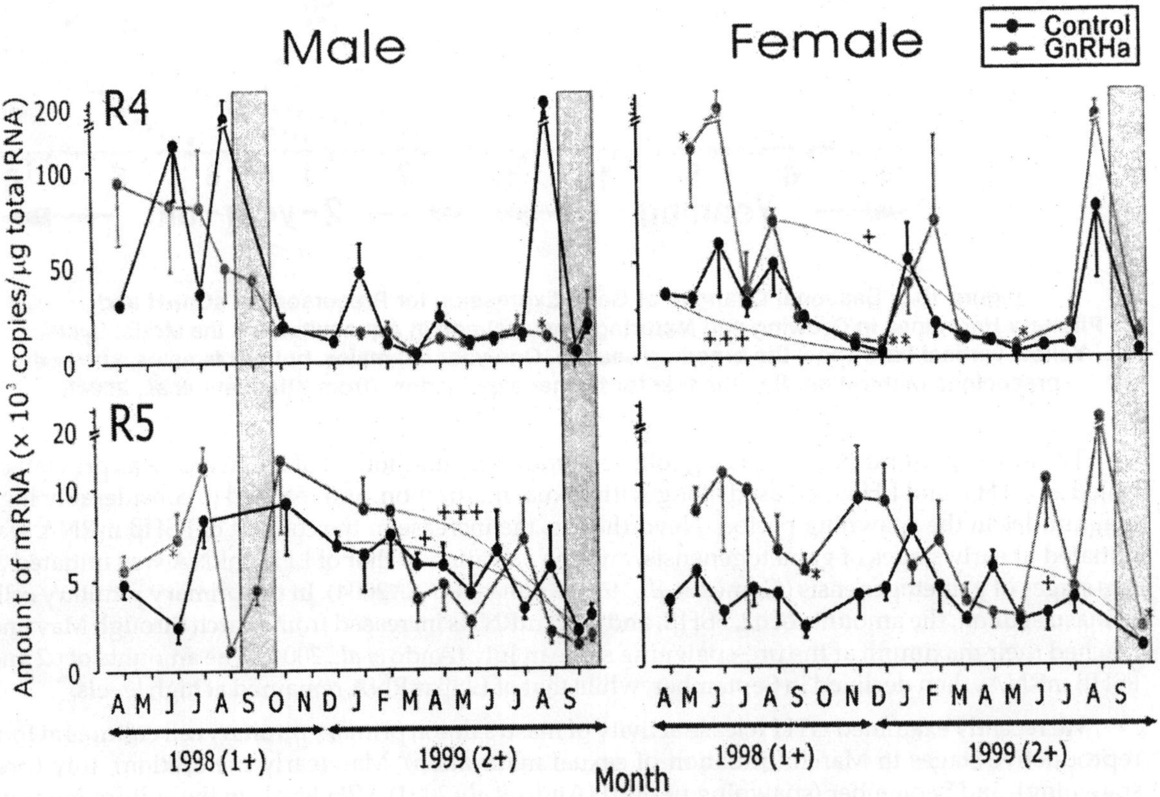

Figure 15.4: Seasonal Changes in Expression of GnRH-R Genes in the Pituitaries of Growing and Maturing Masu Salmon whose Gene Expression for sGnRH and Pituitary Hormones is Shown in Figure 15.3. See the text for explanation. (from Jodo *et al.*, 2004).

Gene Expression for sGnRH and GTHs in Homing Chum Salmon

Our previous studies suggested that sGnRH neurons regulate both final maturation and migratory behavior in homing salmonids (Urano *et al.*, 1999, Kitahashi *et al.*, 2004). As was seen in the model system of masu salmon, sGnRH neurons in the forebrain of chum salmon are considered to be active during pre-spawning to spawning periods when they show upstream migration. We therefore examined expression of two genes encoding precursors of sGnRH in discrete forebrain loci of wild pre-spawning chum salmon, which were captured along their migratory pathway in the Ishikari River-Ishikari bay water system. As was mentioned previously, salmon has two genes for sGnRH precursors, which are referred to as sGnRH-I and sGnRH-II, so that we determined the amounts of two mRNAs. Since the many previous reports showed that the salmon pituitary does not contain cGnRH II (see Ando and Urano, 2005), we did not determine amounts of cGnRH II mRNA.

Patterns of changes in activity of the GnRH-GTH system in wild homing chum salmon are subjects of year to year variations, since we found that there were year-to-year differences in plasma

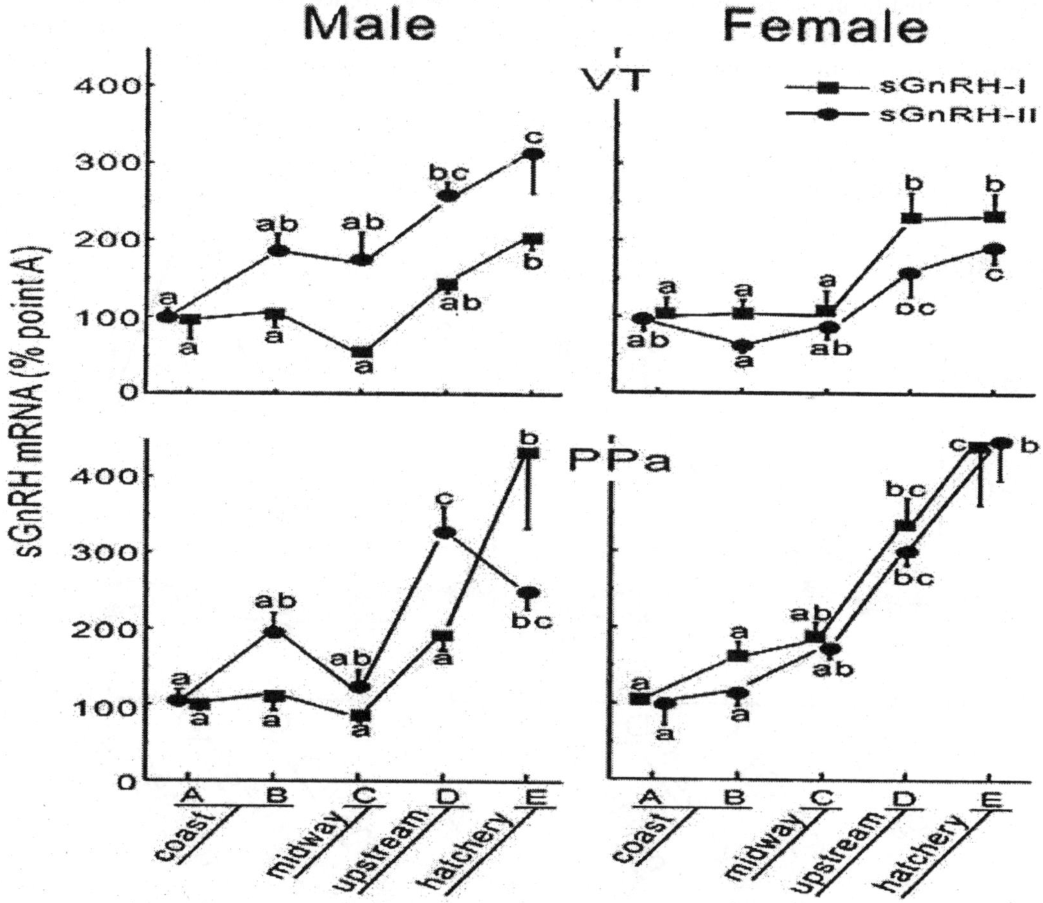

Figure 15.5: Elevation of sGnRH Gene Expression in the Ventral Telencephalon (VT) and the Nucleus Preopticus Parvocellularis Anterioris (PPa) of Homing Chum Salmon During Upstream Migration in 1999. Fish were captured at several points along their homing pathway in the Ishikari water system. (based on Onuma *et al.*, 2005)

levels of steroid hormones and sexual maturation during our six-year survey (Onuma *et al.*, 2003). By this reason, we used fish that were captured from 1997 through 1999 in our research in which changes in gene expression for sGnRH and GTHs were determined in homing chum salmon (Onuma *et al.*, 2005). They were fished along their homing pathway: coastal areas, a midway of the river, 4 km downstream of the natal hatchery, and the hatchery, and the brains and pituitaries were taken out to be frozen in liquid nitrogen. Amounts of sGnRH mRNAs were determined by real-time PCR in fresh frozen sections that were grouped into ten particular brain loci: the olfactory bulb (OB), terminal nerve (TN), ventral telencephalon (VT), nucleus preopticus parvocellularis anterioris (PPa) and nucleus preopticus magnocellularis (PM). The amount of sGnRH-II mRNA was higher than that of sGnRH-I mRNA, while they showed similar changes during upstream migration. In the OB and TN, the amounts of sGnRH mRNAs elevated from the coast to the natal hatchery. In the VT and PPa, they elevated along with the progress of final maturation (Figure 15.5). Such elevation was also observed in the rostroventral, middle and dorsocaudal parts of the PM. The amounts of gonadotropin IIβ (Figure 15.6) and somatolactin mRNAs in the pituitary also increased in consistent with the elevation of gene expression for sGnRH. These results in combination with lines of previous evidence (see Ando and Urano, 2005) indicate that sGnRH neurons are activated in almost all the forebrain loci during the last phases of spawning migration, resulting in coordination of final gonadal maturation and migratory behavior to the spawning ground.

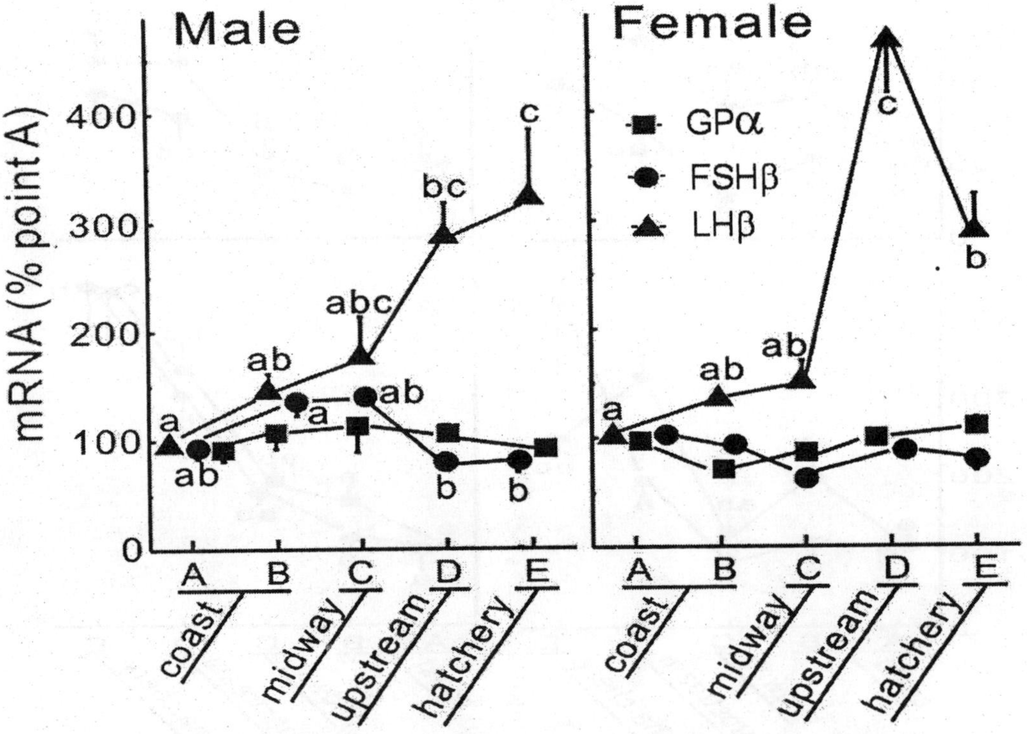

Figure 15.6: Changes in Expression of Genes for Precursors of GTH Subunits in the Pituitaries of the Same Homing Chum Salmon Shown in Figure 15.5. Note that expression of LHβ gene was highly elevated during upstream migration. This elevation is coincident with that of GnRH gene expression. (based on Onuma *et al.*, 2005)

Conclusion

In masu salmon as the model system and chum salmon as wild fish, seasonal variation of gene expression for sGnRH and GnRH-R were estimated through determination of the absolute amounts of mRNAs by the real-time PCR in the brains and the pituitaries of growing and maturing fish, in association with changes in expression of pituitary hormone genes. The expression of two sGnRH genes in the ventral telencephalon and the preoptic area of masu salmon elevated twice in late spring to early summer and in breeding season in autumn, whereas patterns of changes in expression of GnRH-R genes in the pituitaries differed among different types. Nonetheless, in masu salmon, elevation of gene expression for LH subunits in the pre-spawning season coincided well with that for two sGnRH precursors and some types of GnRH-R. Homing chum salmon of the Ishikari stock showed similar patterns of changes in expression of genes for sGnRH precursors and LH subunits during upstream migration. These coincident molecular events may be important at least for control of final maturation accompanied by upstream migration. Further, a similarity between masu salmon and chum salmon seen in pre-spawning and spawning periods promises that a use of masu salmon as the model system should yield fruitful results to clarify cellular and molecular bases of neuroendocrine mechanisms in reproduction of salmons and also other teleost fishes.

Acknowledgements

We would like to express our sincere thanks to Dr. M. Ban (National Salmon Resources Center), Professor H. Ueda (Hokkaido University), Dr. H. Okada, Mr. N. Koide and Mr. Y. Sasaki (Hokkaido Fish Hatchery) for their collaboration. Our thanks are extended to all of former and present laboratory members for their help and discussion. Parts of our research were supported by Grants-in-Aid from the Fisheries Agency and the Ministry of Education, Culture, Sports, Science and Technology, Japan.

References

Amano M, Aida K, Okumoto N, and Hasegawa Y. (1992). Changes in salmon GnRH and chicken GnRH-II contents in the brain and pituitary, and GTH contents in the pituitary in female masu salmon, *Oncorhynchus masou*, from hatching through ovulation. Zool. Sci., 9, 375-386.

Amano M, Aida K, Okumoto N, and Hasegawa Y. (1993). Changes in levels of GnRH in the brain and pituitary and GTH in the pituitary in male masu salmon, *Oncorhynchus masou*, from hatching to maturation. Fish Physiol. Biochem., 11, 233-240.

Amano M, Urano A, and Aida K. (1997). Distribution and function of gonadotropin-releasing hormone (GnRH) in the teleost brain. Zool. Sci., 14: 1-11.

Amano M, Ashihara M, Yoshiura Y, Kitamura S, Ikuta K, and Aida K. (1998). Two differing salmon GnRH precursor mRNAs are co-expressed in the brain of sockeye salmon. Cell Tissue Res., 292, 267-273.

Ando H, Sasaki Y, Okada H, and Urano A. (2001). Prepubertal increases in the levels of two salmon gonadotropin-releasing hormone mRNAs in the ventral telencephalon and preoptic area of masu salmon. Neurosci. Letters, 307, 93-96.

Ando H, Swanson P, Kitani T, Koide N, Okada H, Ueda H, and Urano A. (2204). Synergistic effects of salmon gonadotropin-releasing hormone and estradiol-17β on gonadotropin subunit gene expression and release in masu salmon pituitary cells *in vitro*. Gen. Comp. Endocrinol., 137, 109-121.

Ando H, and Urano A. (2005). Molecular regulation of gonadotropin secretion by gonadotropin-releasing hormone in salmonid fishes. Zool. Sci., 22(4), in press.

Gomez JM, Weil C, Ollitrault M, Le Bail P-Y, Breton B, and Le Gac FL (1999). Growth hormone (GH) and gonadotropin subunit gene expression and pituitary and plasma changes during spermatogenesis and oogenesis in rainbow trout (*Oncorhynchus mykiss*). Gen. Comp. Endocrinol., 113, 413-428.

Groot C, and Margolis L. (2001). Pacific Salmon Life Histories. UBC Press/Vancouver.

Higa M, Kitahashi T, Sasaki Y, Okada H, and Ando H. (1997). Distinct promoter sequences of two precursor genes for salmon gonadotropin-releasing hormone in masu salmon. J. Mol. Endocrinol., 19, 149-161.

Jodo A, Ando H, and Urano A. (2003). Five different types of putative GnRH receptor gene are expressed in the brain of masu salmon (*Oncorhynchus masou*). Zool. Sci., 20, 1117-1125.

Jodo A, Kitahashi T, Taniyama S, Ueda H, Urano A, and Ando H. (2004). Changes in expression of GnRH receptor genes and responsiveness of GnRH analog during growth and sexual maturation in masu salmon. *In*: Oishi T, Tsutsui K, Tanaka S, and Kikuyama S. (Eds.), *Trends in Comparative Endocrinology, Proceedings of the Fifth Congress of the Asia and Oceania Society for Comparative Endocrinology*, AOSCE, pp 408-410.

Kitahashi T, Bhandari RK, Taniyama S, Ando H, Ueda H, and Urano A. (2004). Changes in expression of salmon GnRH genes and responsiveness of pituitary hormone genes to GnRH analog during growth and sexual maturation in masu salmon. *In*: Oishi T, Tsutsui K, Tanaka S, and Kikuyama S. (Eds.), *Trends in Comparative Endocrinology, Proceedings of the Fifth Congress of the Asia and Oceania Society for Comparative Endocrinology*, AOSCE, pp 92-94.

Lethimonier C, Madigou T, Muñoz-Cueto JA, Lareyre JJ, and Kah O. (2004). Evolutionary aspects of GnRHs, GnRH neuronal systems and GnRH receptors in teleost fish. Gen. Comp. Endocrinol., 135, 1-16.

Millar RP. (2003). GnRH II and type II GnRH receptors. Trends Endocrinol. Metab., 14, 35-43.

Millar RP, Lu ZL, Pawson AJ, Flanagan CA, Morgan K, and Maudsley SR. (2004). Gonadotropin-releasing hormone receptors. Endocrine Rev., 25, 235-275.

Onuma T, Higashi Y, Ando H, Ban M, Ueda H, and Urano A. (2003). Year-to-year differences in plasma levels of steroid hormones in prespawning chum salmon: influence of sea surface temperature. Gen. Comp. Endocrinol., 133, 199-215.

Onuma T, Higa M, Ando H, Ban M, and Urano A. (2005). Elevation of gene expression for salmon gonadotropin-releasing hormone in discrete brain loci of prespawning chum salmon during upstream migration. J. Neurobiol., in press.

Saito D, Hasegawa Y, and Urano A. (2003). Gonadotropin-releasing hormones modulate electrical activity of vasotocin and isotocin neurons in the brain of rainbow trout. Neurosci. Letters, 351, 107-110.

Urano A. (1984). A Golgi-electron microscopic study of anterior preoptic neurons in the bullfrog and the toad. Zool. Sci., 1, 89-99.

Urano A. (1988). Neuroendocrine control of anuran anterior preoptic neurons and initia-tion of mating behavior. Zool. Sci., 5, 925-937.

Urano A, Ando H, and Ueda H. (1999). Molecular neuroendocrine basis of spawning migration in salmon. *In*: Kwon HB, Joss JMP, and Ishii S. (Eds.), *Recent Progress in Molecular and Comparative Endocrinology*, HRC, Chonnam National Univ./Kwangju. pp 46-56.

Urano A, and Ando H. (2002). Quantitative analyses of the levels of hormonal mRNAs in the salmon neuroendocrine system. *In*: Shimizu N, Aoki T, Hirono I, and Takashima F. (Eds.), *Aquatic Genomics*, Springer Ferlag/Tokyo, pp 225-235.

Yaron Z, Gur G, Melamed P, Rosenfeld H, Elizur A, and Levavi-Sivan B. (2003). Regulation of fish gonadotropins. Int. Rev. Cytol., 225, 131-185.

16

Could Autonomic Sympathetic Nerves of the Pineal Organ Mediate Light-Specific Information?

B. Vigh*, L.C. Frank, C. David, G. Racz and A. Szel

Department of Human Morphology and Developmental Biology, Semmelweis University, Tuzolto 58, 1094, Budapest, Hungary

ABSTRACT

Since the mammalian pineal is generally considered to be light insensitive, it was supposed, that light information for the periodic melatonin secretion reaches the organ from the retina via autonomic nerves innervating pinealocytes. In earlier electron microscopic investigations we have found that a part of these autonomic perivasal nerves form endings on smooth muscle cells of pineal vessels, consequently, they represent vasomotor fibers.

In the present work we investigated in various vertebrates the light- and electron microscopy and tyrosine hydroxylase (TH)–immunoreactivity of the autonomic nerves reaching the organ by the conarian nerve and by perivasal plexuses. We found that autonomic nerves form terminals around microvessels similar to those present around brain vessels and central artey of the retina. Some of the perivasal pineal fibers show TH-immunoreaction indicating their noradrenergic character. Noradrenergic nerves are known to originate from the superior cervical

**Corresponding Author.* E-mail: vigh@ana2.sote.hu; Fax: 36-1-215-3064.

ganglia, and were postulated to mediate light information to the pineal. We found that most of the TH-immunoreactive autonomic nerves remain in the perivasal meningeal septa, and do not enter the pineal tissue proper. Intrapineal TH-immunoreactive fibers in some species are processes of intrapineal neuronal perikarya and were followed in serial section to the pineal stalk. The fibers leaving the pineal by its stalk show the existance of a monoaminergic neuronal efferentation of the organ. Some of the TH-immunoreactive fibers of the stalk may be pinealopetal as well entering the organ from the brain stem.

The presence of TH-immunoreactive perivasal nerve terminals on vascular smooth muscle cells show that these pineal autonomic fibers are vasomotor and perhaps vasoconstrictor in nature. We suppose that the vasomotor regulation of the pineal microvessels by autonomic nerves may serve the vascular support for circadian and circannual periodic changes in the metabolic activity of the pineal tissue. Since in general there is no neuroanatomical evidence for the mediation of light-specific information by autonomic nerves, a supposed direct pineal light sensitivity in mammals or a direct retinal afferentation by its stalk can explain the light-dependent melatonin secretion of the pineal.

Keywords: *Pineal vessels, Central artery of the retina, Vasomotor nerves, Fine structure, TH-immunoreactivity, Various vertebrates.*

Introduction

Called "glands" in mammals and human, the pineal organs of vertebrates (pineal-, parapineal organs, frontal- and parietal eyes) represent nonvisual photoreceptors of the environmental illumination. Some of them, like the parietal eye also have a semivisual function directionally detecting light sources (insolation). Most of these organs possess synaptic neural efferentation as well as a non-synaptic neurohormonal efferentation.

Generally, the significance of pineal sympathetic nerves was studied in connection to the innervation of pinealocytes and experimental works demonstrated the importance of sympathetic fibers in the melatonin secretion of the organ. Noradrenergic pineal fibers were thought to mediate light information to the pineal organ from the retina in mammals (Kappers, 1968; Axelrod, 1974; Huang and Lin, 1984; Klein, 1985; Ling *et al.*, 1990; Reiter, 1991; Korf, 1996).

Like the retina, the pineal organs are parts of the brain, histologically composed of nervous tissue containing photoreceptor-like pinealocytes, neurons, glial cells, neve fibers and synapses. Determining its histophysiology, the pineal nervous tissue (like the brain tissue) is separated from intrapineal perivascular spaces by glial limiting membrane and basal lamina. The pineal "capsule" is formed by the meninges (Vígh and Vígh-Teichmann, 1988; Vígh-Teichmann and Vígh, 1994; Vígh *et al.*, 2002, 2003).

Autonomic nerve fibers of the organ contain several bioactive substances: tyrosine hydroxylase containing noradrenergic and NPY-immunoreactive fibers originate from the superior cervical ganglia. VIP-containing nerves take their origin in the pterygopalatine ganglia, while substance P and CGRP immunoreactive peripheral fibers originate from the trigeminal ganglia (Shiotani *et al.*, 1986; Moller and Baeres, 2002).

Since well developed photoreceptor outer segments were found in cyclostomes, fishes and amphibians and several granular vesicles in the pinealocytes of reptiles and birds, the rudimentation of pineal photoreceptor outer segments in mammals was explained by a phylogenetic transformation of photoreceptor pinealocytes to glandular cells. The central connection of the mammalian pineal with the brain was denied by this theory, and the periodic melatonin secretion of the organ was thought to be the effect of circadian retinal light information reaching the organ via peripheral nerves (Wurtman *et al.*, 1968; Kappers, 1968; Karasek and Reiter, 1992; Kramm *et al.*, 1993).

In earlier investigations we demonstrated that photoreceptor outer segments and immunoreactive opsins are present not only in lower vertebrates, but also in reptiles, birds and mammals, first of all in young animals (Vígh and Vígh-Teichmann, 1981; Vígh *et al.*, 1982; Vígh-Teichmann and Vígh, 1986). Similar results were obtained later by several authors (Korf *et al.*, 1985; Zhao *et al.*, 1997; Blackshaw and Snyder, 1997; Tosini *et al.*, 2000). Concerning autonomic nerves, we have found that some of them form endings on smooth muscle cells of pineal vessels, consequently, they represent vasomotor fibers (Vígh and Vígh-Teichmann, 1992; Vígh *et al.*, 2001; Frank *et al.*, 2003).

In the present work we compared the microscopic structure of vascular innervation of the pineal organ including retina and brain tissue of mammals with those of submammalians. Using antisera against tyrosine hydroxylase an enzyme taking part in the biosynthesis of catecholamines (dopamine, norepinephrine, and epinephrine) we also investigated the immunoreaction of perivasal and intrapineal fibers.

Materials and Methods

Both sexes of the following species–3 to 5 from each–were investigated: mammals (laboratory rat, rabbit, Mongolian gerbil, ferret), birds (*Gallus domesticus, Coturnix coturnix, Buteo buteo*), reptiles (*Lacerta muralis, Natrix natrix*), amphibians (*Rana esculenta, Triturus cristatus* and *Triturus vulgaris*), fish (*Raja clavata, Carassius auratus*) and the cyclostome Petromyson fluviatilis. The animals were kept under normal laboratory conditions and anesthetised prior fixation with phenobarbital or MS222. For light- and electron microscopy 1 per cent glutaraldehyde were perfused through the aorta, some pineals were dissected and immersed in the fixative. The pineal organs were further fixed in 1 per cent OsO4 for 1,5 h. The material was dehydrated and embedded in Poly Bed (Polysciences, St. Goar, FRG). Semithin serial-sections and ultrathin sections were cut on a Reichert Ultracut S ultramicrotome. The semithin sections were stained by toluidine blue-azur II and the ultrathin sections with uranyl acetate and led citrate and were examined and photographed in a Hitachi H7600 electron microscope.

TH-immunoreaction was performed by polyclonal and monoclonal antisera (Sigma Bio Sciences, St. Louis, MO., U.S.A). The material was fixed in 4 per cent paraformaldehyde and cut with a Leica EM FCS cryotome. The primary antibody was diluted in 1:1000 to 1:2000, and the secondary antibody (antirabbit or antimouse) diluted 1:100 to 1:400. For visualisation avidin biotin-HRP complex, diaminobenzidine reaction and chloronaphthol staining was used.

Results

In all species investigated, autonomic nerves accompany vessels entering the arachnoidal capsule and interfollicular meningeal septa of the pineal organ. More fibers are present in reptiles, birds and mammals than in cyclostomes, fishes and amphibians. The perivasal autonomic nerve bundles contain some myelinated and numerous unmyelinated fibers encapsulated by Schwann-cells and thin endo- and perineural connective tissue.

Several axons of the perivasal nerve bundles form terminal enlargements on or near smooth muscle cells of pineal arterioles [Figures 16.1(a–c)]. Attached by half-desmosomes to the basal lamina of the nerves, the terminal part of the axons are free from covering Schwann-cells. The terminals

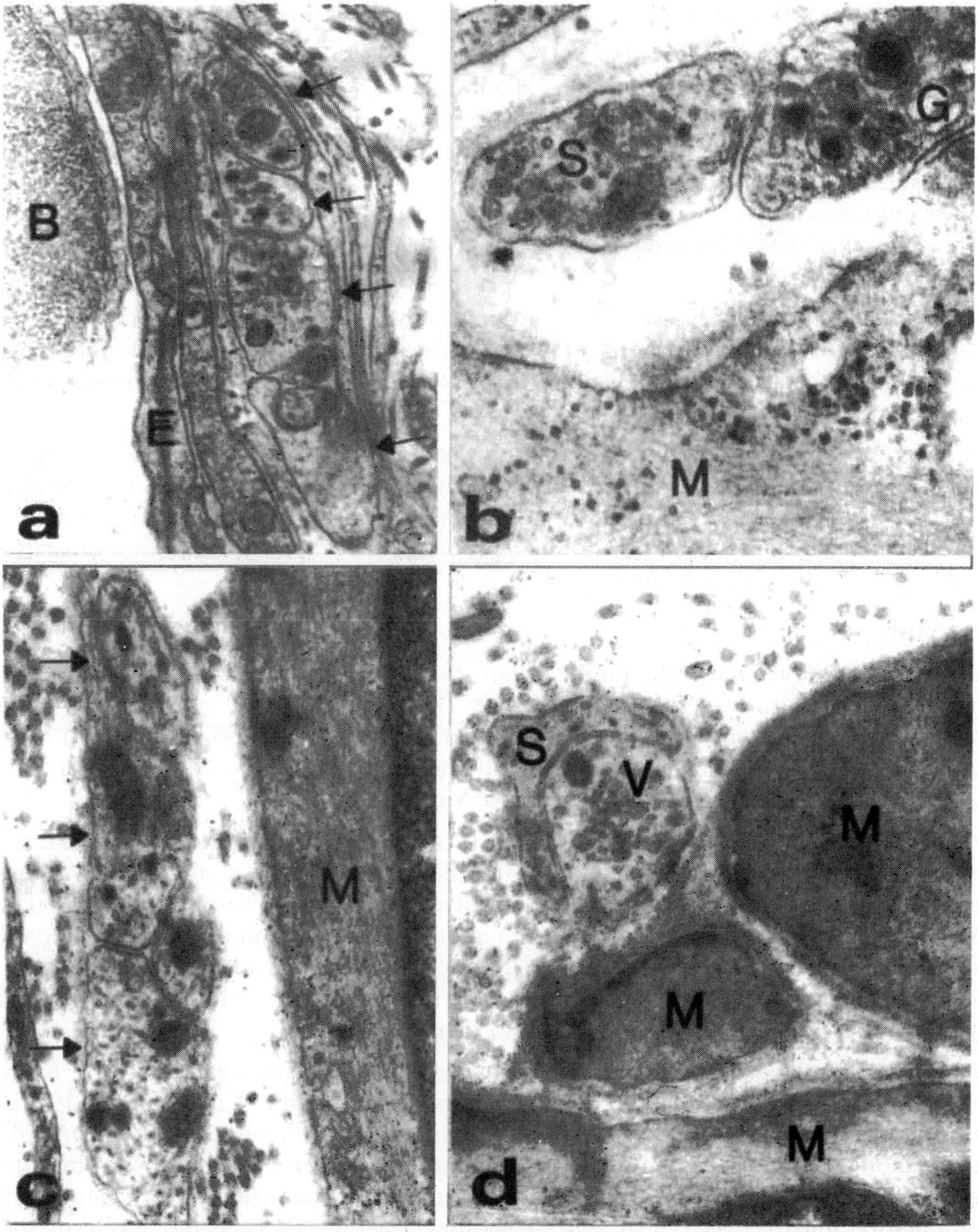

Figure 16.1: Autonomic Nerves Supplying the Vessels of Pineal Organs and Retina

(a) Nerve terminals (arrows) containing granular and synaptic vesicles on a pineal vessel in the lizard (*Lacerta muralis*). E: endothelial cell, B: blood cell, x 13000; (b) Nerve terminals containing synaptic (S) and granular vesicles (G) on an arteriolar smooth muscle cell of a pineal arterole of the buzzard *Buteo buteo*, M: smooth muscle cell, x 5300; (c) Nerve terminals (arrows) in the wall of a pineal arteriole of the rabbit, M: myofilaments of the smooth muscle cell, x 40.800; (d) Nerve terminal containing granular and synaptic vesicles (V) on the smooth muscle cells (M) of the central artery of the retina in the rat, S: Schwann-cell, x 40500.

contain several synaptic and some granular vesicles. Axon terminals are present not only in the wall of arterioles but also around pineal veins.

In serial sections, we found that the so-called conarian autonomic nerves reach the pineal organ alongside pineal veins draining into the great internal cerebral vein. Similar nerves present near arteries of the arachnoid enter the pineal meningeal capsule and septa by arterioles, both perivenous and periarterial nerves form terminals of vasomotor type. Nerves forming terminals around pineal arterioles and veins may have different origin: arteriolar nerves seem to be the continuation of the periarterial nerve plexus of the arteries supplying the brain and perivenous nerves from the conarian nerves following veins.

In all species studied perivasal nerves around meningeal vessels and arterioles of epithalamic and various brain areas are morphological similar to those of the pineal autonomic nerves. In the optic nerve, autonomic fibers following the central artery of the retina terminate close to the smooths muscle cells of arterial branches [Figure 16.1(d)].

Some of the pineal perivasal nerves and terminals are TH-immunoreactive, they are present in mammals as well as in submammalians studied [Figure 16.2(a–d)]. There are TH-immunoreactive fibers in the pineal nervous tissue as well, in the Mongolian gerbil they originate from intrapineal TH-immunoreactive neuronal perikarya. Axon-like processes of intrapineal TH-immunoreactive neurons run to the pineal stalk [Figure 16.2(e)].

Discussion

Sympathetic nerves reach the pineal organ by the so-called conarian nerves and periarterial plexuses (Le Gros Clark, 1940; Romijn, 1975; Ueck, 1979; Vollrath, 1981, Schröder and Vollrath, 1986; Shiotani et al,1989; Reuss and Schröder, 1988; Zang *et al.*, 1991; Moller, 1992; Korf, 1996). Most of these nerves run around vessels in the menigeal capsule and septa of the organ (Romijn, 1975; Bowers *et al.*, 1984; Matsuura and Sano, 1986; Mikkelsen, 1989; Moller, 1992).

A considerable number of the autonomic fibers are catecholaminergic, detected by monoamine fluorescence and/or immunohistochemistry (Pellegrino de Iraldi *et al.*, 1965; Wartenberg and Baumgarten, 1963; Hess, 1981; Schon *et al.*, 1985; Matsuura and Sano, 1986; Cozzi *et al.*, 1990; Mikkelsen and Moller, 1991; Moller *et al.*, 1999; Matsushima *et al.*, 1999). TH-immunoreactive, dopamine beta-hydroxilase-negative cells have been reported in the pineal organ of hamsters. Bilateral removal of the superior cervical ganglion strikingly diminished the number of TH-positive nerve fibers, but did not affect the number of TH-positive cells (Jin *et al.*, 1988).

Abundant tyrosine hydroxylase and neuropeptide-Y immunoreactive fibers form a dense meshwork in the superficial ad deep pineals of hamster and gerbil (Li and Welsh, 1991). TH- and NPY-immunoreacting nerve fibers were also found in the pineal arachnoid, the perivascular spaces, and throughout the superficial and deep pineal. A small number of TH- and NPY-immunoreacting nerve fibers were described in the posterior and habenular commissures as well, fibers penetrate from the commissures into the deep pineal. Two weeks after superior cervical ganglionectomy, the TH- and NPY-immunoreacting fibers in the superficial pineal had almost completely disappeared, but in the deep pineal and pineal stalk were still present (Zhang *et al.*, 1992).

In the present work the terminals of autonomic nerves found on smooth muscle cells of pineal microvessels in several vertebrate species strengthens the view on the vasomotoric role of these fibers. Vasomotoric nature of pineal autonomic fibers was already suggested by earlier authors (Holmgren,

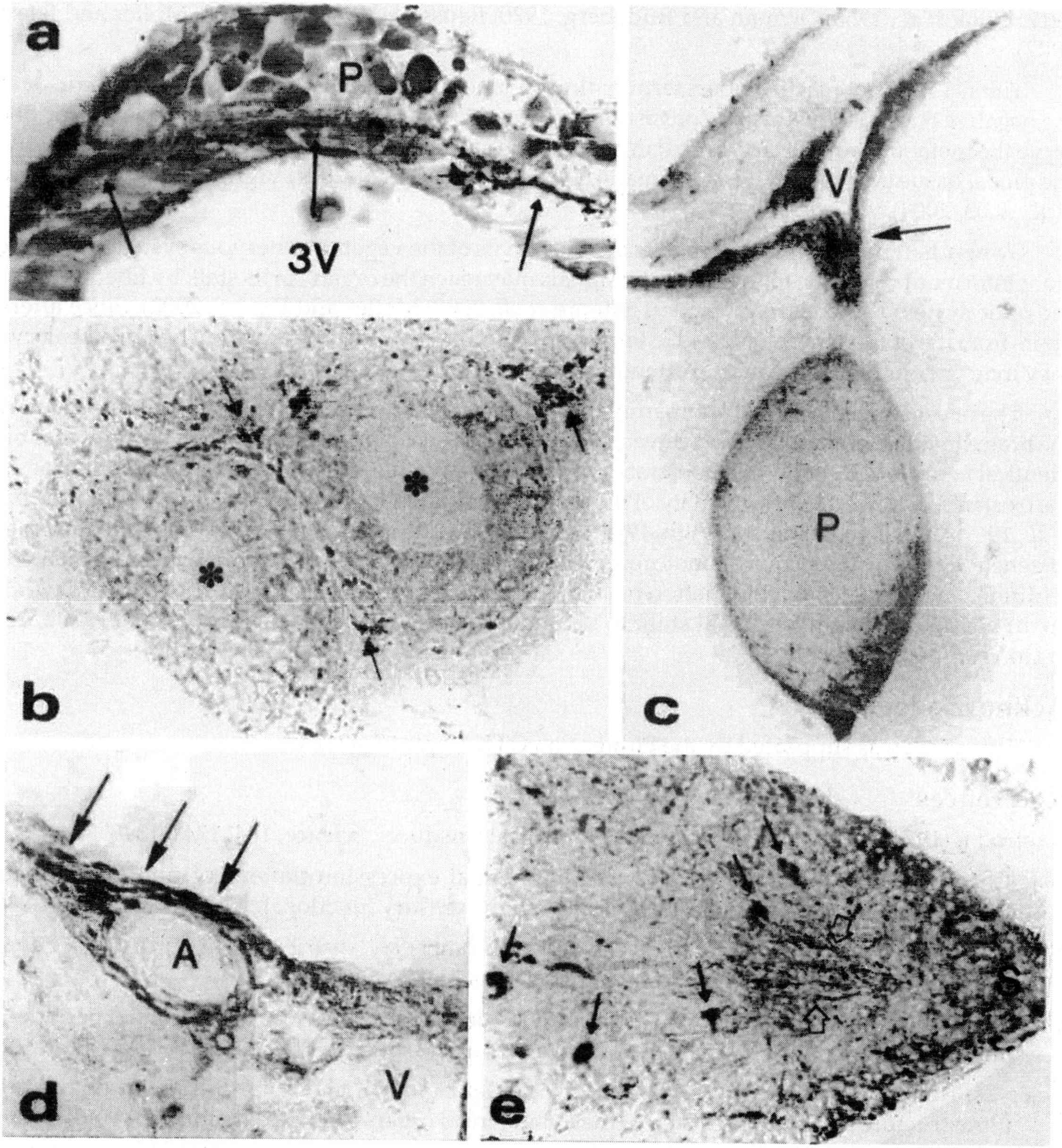

Figure 16.2: TH-immunoreactivity in the Autonomic Nerves of the Pineal Organ of Various Vertebrates

(a) TH-immunoreacting nerve fibers (arrows) in the epithalamic and pineal meninges of the newt *Triturus vulgaris*, 3V: third ventricle, P: pineal organ, x 560; (b) TH-immunoreactive periarterial nerve fibers (arrows) of the pineal organ of the quail, *Coturnix coturnix*. Asterisks: lumen of pineal follicles, x 420; (c) TH-immunoreactive fibers in the conarian nerve (arrow) of the Siberian hamster. P: pineal organ, V: great cerebral vein, x 360; (d) Perivasal TH-immunoreactive terminal and fibers (arrows) in the rat pineal, A: arteriole, V: vein, x 560; (e) TH-immunoreacting intrapineal neuronal perikarya (arrows) and fibers (hollow asterisks) of the hamster. S: pineal stalk, x 420.

1917; Quay *et al.*, 1968; Owman and Rüdeberg, 1970; Reuss and Schröder, 1988; Moller and Baeres, 2002).

The TH-immunoreactive fibers terminating on pineal vessels may represent vasoconstrictor and the negative ones vasodilator and/or vasosensor fibers. We can suppose, that these vasomotoric fibers serve the metabolic support for circadian and circannual periodic changes in the metabolic activity of the pineal tissue (Vígh and Vígh-Teichmann, 1988, 1999; Fejér et al, 2001; Vígh *et al.*, 2002, 2003; Frank *et al.*, 2003, 2004).

Sympathetic nerves are known to transmit signals of the vegetative nervous system, that differs from information specific for light. Photic signals may reach the organ via its stalk by fibers related to the optic system (Mikkelsen *et al.*, 1991; Moore *et al.*, 2000; Moller and Baeres, 2002). Since afferent fibers from the brain stem were found to terminate on secondary pineal neurons but not on pinealocytes, they may only influence the neural efferentation of the pineal (Vígh *et al.*, 2002).

The neural efferentation of submammalian photoreceptor pineal organs transmit information to the brain stem nuclei about light perceived by pineal photoreceptor outer segments. The presence of an identical neural efferentation demonstrated in various mammals and human, strengthens the view on the existance of a direct light sensitivity of the mammalian pineal, predominently in newborn (Machado *et al.*, 1969; Vígh-Teichmann and Vígh, 1994; Tosini *et al.*, 2000; Frank *et al.*, 2004). The synchronization of genetic clocks by direct pineal photoreception in newborns may be differentially complemented by additional nonvisual light information from the retina (*e.g.* eyes of hermelin are opened 45 days after birth) passing through the pineal stalk to secondary pineal neurons (Vígh *et al.*, 2001, 2002, 2003; Frank *et al.*, 2004).

Acknowledgement

This work was supported by the Hungarian OTKA grant T 42524.

References

Axelrod J. (1974). The pineal gland: a neurochemical trnsducer. Science, 184, 1341-1348.

Blackshaw S, and Snyder SH. (1997). Developmental expression pattern of phototransduction components in mammalian pineal implies a light-sensory function. J. Neurosci., 17, 8074-8082.

Bowers CW, Dahm LM, and Zigmond RE. (1984). The number and distribution of sympathetic neurons that innervate the rat pineal gland. Neuroscience, 13, 87-96.

Cozzi B, Mikkelsen JD, Merati D, Capsoni S, and Moller M. (1990). Vasoactive intestinal peptide-like immunoreactive nerve fibers in the pineal gland of the sheep. J. Pineal Res., 8, 41-47.

Fejér Zs, Röhlich P, Szél Á, Dávid C, Zádori A, Manzano MJ, and Vígh B. (2001). Comparative ultrastructure and cytochemistry of the avian pineal organ. Micr. Res. Techn., 53, 12-24.

Frank CL, Dávid C, Czirok S, Vincze C, Manzano MJ, and Vígh B. (2003). Autonomic nerves terminating on smooth muscle cells of vessels in the pineal organ of various mammals. Acta Biol. Hung, 54, 233-240.

Frank CL, Czirok S, Vincze C, Rácz G, Szél A, and Vígh B. (2005). Autonomic nerves terminating on microvessels in the pineal organ of various submammalian vertebrates. Acta Biol. Hung., 56, 35-41.

Hess A. (1981). Localization of noradrenaline and serotonin in nerves in the pineal gland of rats and guinea-pigs studied by glyoxylic acid histofluorescence and electron microscopy. Histochem. J., 13, 425-434.

Holmgren N. (1917). Zur Frage der Epiphysen-Innervation bei Teleostiern. Folia Neuro-biol. (Lpz.), 10, 1-15.

Huang HT, and Lin HS. (1984). Synaptic junction between adrenergic axon varicosity and the pinealocyte in the rat. J. Pineal Res., 1, 281-291.

Jin KL, Shiotani Y, Kawai Y, and Kiyama H. (1988). Immunohistochemical demonstration of tyrosine hydroxylase (TH)-positive but dopamine β¢-hydroxilase (DBH)-negative neuron-like cells in the pineal gland of golden hamsters. Neurosci. Lett., 93, 28-32.

Kappers JA. (1968). The morphological and functional evolution of the pineal organ during its phylogenetic development. Excerpta Med. Int. Congr. Ser., 184, 619-626.

Karasek M, and Reiter RJ. (1992). Morphofunctional aspects of the mammalian pineal gland. Micr. Res. Techn., 21, 136-157.

Klein DC. (1985). Photoneural regulation of the mammalian pineal gland. Ciba Found. Symp., 117, 38-51.

Korf HW, Foster RG, Ekström P, and Schanklen JJ. (1985). Opsin-like immunoreaction in the retinae and pineal organs of four mammalian species. Cell Tissue Res., 242, 645-648.

Korf HW. (1996). Innervation of the pineal gland. *In*: Unsicker K. (Ed.), *The Autonomic Nervous System*. Vol.10. (Burnstock, G. Ed), *Autonomic–Endocrine Interactions*, Harwood Academic Publishers, Amsterdam, pp. 129-180.

Kramm CM, De Drip WJ, and Korf HW. (1993). Rod-opsin immunoreaction in the pineal organ of the pigmented mouse does not indicate the presence of functional photopigment. Cell Tissue Res., 274, 71-78.

Le Gros Clark WE. (1940). The nervous and vascular relations of the pineal gland. J. Anat., 74, 470-494.

Li K, and Welsh M. (1991). Tyrosine dydroxylase and neuropeptide-Y immunoreactivity in the pineal glands developing in situ and in pineal grafts. Cell Tissue Res., 264, 515-527.

Ling EA, Tan SH, and Wong WC. (1990). Synaptic junctions between sympathetic axon terminals and pinealocytes in the monkey, *Macaca fascicularis*. Anat. Embryol., 182, 21-27.

Machado CRS, Machado AB-M, and Wragg LE. (1969). Circadian serotonin rhythm control: sympathetic and nonsympathetic pathways in rat pineals of different ages. Endocrinol., 85, 846-849.

Matsuura T, and Sano Y. (1986). Characteristic pattern of monoaminergic nerve fibers in the pineal organ of the monkey. Cell Tissue Res., 245, 453-456.

Matsushima S, Sakai Y, and Hira Y. (1999). Peptidergig peripheral nervous systems in the mammalian pineal gland. Micr. Res. Techn., 46, 265-280.

Mikkelsen JD, Cozzi B, and Moller M. (1991). Efferent projections from the lateral geniculate nucleus to the pineal complex of the Mongolian gerbil (*Meriones unguiculatus*). Cell Tissue Res., 264, 95-102.

Mikkelsen JD, and Moller M. (1991). Tyrosine hydroxylase- and neuropeptide Y- immunoreactive nerve fibers in the pineal complex of untreated rats and rats following removal of the superior cervical ganglia. Cell Tiss. Res., 265, 63-71.

Moller M. (1992). Fine structure of the pinealopetal innervation of the mammalian pineal gland. Micr. Res. Techn., 21, 188-204.

Moller M, and Baeres FMM. (2002). The anatomy and innervation of the mammalian pineal gland. Cell Tissue Res., 309, 139-150.

Moore RY, Weis R, and Moga MM. (2000). Efferent projections of the intergeniculate leaflet and the ventral lateral geniculate nucleus in the rat. J. Comp. Neurol., 420, 398-418.

Owman C, and Rüdeberg C. (1970). Light, fluorescence, and electron microscopic studies on the pineal organ of the pike, Esox lucius L., with special regard to 5-hydroxytryptamine. Z. Zellforsch., 107, 522-550.

Pellegrino de Iraldi A, Zieher LM, and De Robertis E. (1965). Ultrastructure and pharmacological studies of nerve endings in the pineal organ. Progr. Brain. Res., 10, 389-422.

QuayWB, Kappers JA, and Jogkind JF. (1968). Innervation and fluorescence histochemistry of monoamines in the pineal organ of a snake (*Natrix natrix*). J. Neurovisc. Rel., 31, 11-25.

Reiter RJ. (1991). Pineal melatonin: cell biology of its synthesis and its physiological interactions. Endocrinol. Reviews., 12, 151-180.

Reuss S, and Schröder H. (1988). Principal neurons projecting to the pineal gland in close association with small intensely fluorescent cells in the superior cervical ganglion of rats. Cell Tissue. Res., 254, 97-100.

Romijn HJ. (1975). Structure and innervation of the pineal gland of the rabbit, Oryctolagus cuniculus (L.). III. An electron microscopic investigation of the innervation. Z. Zellforsch., 139, 473-485.

Schon F, Allen JM, Yeats JC, Allen YS, Ballesta J, Polak JM, Kelly JS, and Bloom SR. (1985). Neuropeptide Y innervation of the rodent pineal gland and cerebral blood vessels. Neurosci. Lett., 57, 65-71.

Schröder H, and Vollrath L. (1986). Neuropeptide Y (NPY)-like immunoreactivity in the guinea pig pineal organ. Neurosci. Lett., 63, 285-289.

Shiotani Y, Jin KL, Kawai Y, and Kiyama H. (1989). Immunohistochemical studies on innervation of the mammalian pineal gland. *In*: Reiter RJ, and Pand SF. (Eds.), *Advances in Pineal Research*, Vol. 3, Libbey, London. pp. 49-54.

Tosini G, Doyle S, Geusz M, and Menaker M. (2000). Induction of photosensitivity in neonatal rat pineal gland. Proc. Natl. Acad, Sci. USA., 97, 11540-11544.

Ueck M. (1979). Innervation of the vertebrate pineal. Progr. Brain. Res., 52, 45-87.

Vígh B, and Vígh-Teichmann I. (1981). Light- and electron-microscopic demonstration of immunoreacting opsin in the pinealocytes of various vertebrates. Cell Tissue Res., 221, 451-463.

Vígh B, and Vígh-Teichmann I. (1988). Comparative neurohistology and immunocytochemistry of the pineal complex with special reference to CSF-contacting neuronal structures. Pineal Res. Rev., 6, 1-65.

Vígh B, and Vígh-Teichmann I. (1992). Two components of the pineal organ of the mink (Mustela vison): their structural similarity to submammalian pineal complexes and calcification. Arch. Histol. Cytol., 55, 477-489.

Vígh B, and Vígh-Teichmann I. (1999). Comparative morphophysiology of the pineal organs of vertebrates. *In*: Joy KP, Krishna A, and Haldar C. (Eds.), *Comparative Endocrinology and Reproduction*, Narosa Publishing House, New Delhi. pp. 479-506.

Vígh B,Vígh-Teichmann I, Röhlich P, and Aros B. (1982). Immunoreactive opsin on the pineal organs of reptiles and birds. Z. Mikrosk. Anat. Forsch., 96, 113-129.

Vígh B, Manzano MJ, Röhlich P, and Szél Á. (2001). The role of the pineal organ in extraretinal photoreception. Comparative fine structural organization and histochemistry. *In*: Haldar C, Singaravel M, and Maitra SK. (Eds.), *Treatise on Pineal Gland and Melatonin*, Oxford and IBH Publishing Co., New Delhi, India, pp. 17-50.

Vígh B, Manzano MJ, Zádori A, Frank CL, Lukáts A, Röhlich P, Szél A, and Dávid C. (2002). Nonvisual photoreceptors of the deep brain, pineal organs, and retina. Histol. Histopathol., 17, 555-590.

Vígh B, Manzano MJ, Frank CL, Dávid C, Lukáts Á, and Szél Á. (2003). Change in the control of the biological circadian rhythms during evolution. The role of the deep brain photoreceptors, pineal organs and retina. *In*: Csernus V, and Mess B. (Eds.), *Rhytmic Biological Processes. The Role of the Biological Clocks*, Dialóg Campus, Budapest, Pécs., pp. 43-92.

Vígh-Teichmann I, and Vígh B. (1986). The pinealocyte: its ultrastructure and opsin immunocytochemistry. Adv. Pineal Res., 1, 31-41.

Vígh-Teichmann I, and Vígh B. (1994). Postembedding light- and electron microscopic immunocytochemistry of the mammalian pineal photoneuroendocrinology. *In*: Gu J, and Hacker GW. (Eds.), *Modern Methods in Analytical Morphology*, Plenum Publ. Corp., New York, pp. 253-270.

Vollrath L. (1981). The Pineal Organ. In: Oksche A, and Vollrath H. (Eds.), *Handbuch der Mikroskopischen Anatomie des Menschen*. Vol. VI/7. Springer, Berlin, Heidelberg, New York, pp.1-665.

Wartenberg H, and Baumgarten HG. (1969). Untersuchungen zur fluoreszens- und elektronenmikroskopischen Darstellung von 5-Hydroxytryptamin (5-HT) im Pinealorgan von Lacerta viridis und Lacerta muralis. Z. Anat. Entw.-gesch., 128, 185-210.

Wurtmann RJ, Axelrod J, and Kelly DE. (1968). The Pineal. Academic Press, New York

Zhang ET, Mikkelsen JD, Moller M. (1991). Tyrosine hydroxylase and neuropeptide Y-immunoreactive nerve fibers in the pineal complex of untreated rats and rats following removal of the superior cervical ganglia. Cell Tissue Res., 265, 62-72.

Zhao X, Haeseleer F, Fariss RN, Huang J, Baehr W, Milam A, and Palczewski K. (1997). Molecular cloning and localization of rhodopsin kinase in the mammalian pineal. Visual Neurosci., 14, 225-232.

17

Control of Thyroid-Stimulating Hormone Secretion in Amphibians

*Reiko Okada, Yoichi Ito, Miyoko Kaneko, Kazutoshi Yamamoto, Sakae Kikuyama**

Department of Biology, School of Education, Waseda University, Nishi-Waseda 1-6-1, Tokyo 169-8050, Japan

ABSTRACT

Until recently, no direct approach has been made to elucidate the hypothalamic factors controlling TSH secretion, mainly because specific radioimmunoassay (RIA) system for amphibian TSH was lacking. We have developed RIA system for bullfrog (*Rana catesbeiana*) TSH employing a C-terminal peptide synthesized on the basis of the amino acid sequence deduced from bullfrog TSH β cDNA and an antiserum against the C-terminal peptide. Using the RIA system, we demonstrated that corticotropin-releasing factor (CRF) is a potent candidate for TSH-releasing factor. Bullfrog CRF that was synthesized on the basis of amino acid sequence predicted from a cDNA encoding bullfrog CRF precursor molecule stimulated the release of TSH from the pituitary markedly. Culture of pituitary cells in the presence of the bullfrog hypothalamic extract (HE) and α-helical CRF_{9-41}, a CRF-receptor (CRF-R) antagonist, revealed that it suppressed the TSH-releasing activity of the HE by 50 per cent in maximum. Furthermore, it was revealed that the action of CRF is mediated through type 2 CRF-R (CRF-R2). The CRF-induced TSH release was completely blocked by the non-specific CRF-R antagonist astressin and the CRF-R2-specific antagonist anitisauvagine-30, whereas the type 1 CRF-R-specific antagonist antalarmin showed

**Corresponding Author*: E-mail: kikuyama@waseda.jp

no effect on the release of TSH. On the other hand, thyrotropin-releasing hormone (TRH) and gonadotropin-releasing hormone (GnRH) exhibited a moderate effect on the adult pituitary but not on the larval one to cause the release of TSH. In addition to the effect of hypothalamic peptides, the effect of thyroid hormone on the release of TSH from the bullfrog pituitary was studied *in vitro*. Both triiodothyronine and thyroxine suppressed the CRF-induced release of TSH, but not the basal release of TSH, from the pituitary of both larval and adult, suggesting that negative feedback control of TSH release by thyroid hormone is functioning both in larvae and adults.

Introduction

Thyroid-stimulating hormone (TSH) is known as the major pituitary hormone responsible for the release of thyroid hormone. Until recently, however, a homologous radioimmunoassay (RIA) system for amphibian TSH has never been developed due to the lack of an adequate amount of TSH. This has made it difficult to quantify the TSH concentrations in the pituitary, plasma, or pituitary culture medium and thus to characterize the hypothalamic factors controlling TSH secretion. Under these circumstances, attempts have been made to assess indirectly the TSH-releasing activity of several hypothalamic peptides such as ovine corticotropin-releasing factor (CRF), thyrotropin-releasing hormone (TRH), and mammalian gonadotropin-releasing hormone (GnRH). All these peptides have been reported to induce the release of a pituitary factor that stimulates the release of thyroid hormone from the thyroid glands of adult frogs (Darras and Kühn, 1982; Denver, 1988; Jacobs *et al.*, 1988; Jacobs and Kühn, 1992). In the case of TRH, it stimulated the release of a factor that enhanced the release of thyroxine (T_4) from adult bullfrog (*Rana catesbeiana*) pituitaries but not from larval ones (Denver and Licht, 1989a). However, it is not clear yet whether the pituitary thyrotropic factor released in response to these peptides is the authentic TSH, because it was not TSH but thyroid hormone that was confirmed to be elevated in the plasma after injections of the releasing hormones or in the media in which thyroid glands were cultured in the presence of substances derived from the pituitary that had been subjected to the releasing hormones.

In the present article, we describe molecular cloning of a bullfrog TSH β-subunit (TSHβ) cDNA, production of antiserum against a partial sequence of bullfrog TSHβ deduced from the nucleotide sequence of the TSHβ cDNA and development of an RIA for bullfrog TSH by employing the antiserum and antigen. Information obtained by using this RIA system about the control of the release of TSH by the hypothalamic neuropeptides and thyroid hormones will be provided.

Molecular Cloning of Bullfrog TSHβ

Thyroid function in amphibian larvae has been considered to be under the control of TSH, for metamorphic stasis by hypophysectomy is prevented by treatment with mammalian TSH (see Dodd and Dodd, 1976) and immunoneutralization of endogenous TSH with antiserum against ovine TSH blocks spontaneous metamorphosis (Eddy and Lipner, 1976). Several years ago we obtained a thyrotropic glycoprotein fraction, which was virtually free of luteinizing hormone (LH) and follicle-stimulating hormone (FSH), from bullfrog pituitaries and demonstrated that it stimulated the release of T_4 from the larval thyroid gland several times as potently as bovine TSH (Sakai *et al.*, 1991). Up to the present, however, no data on the circulating levels and pituitary contents of TSH in metamorphosing tadpoles are available, since a homologous RIA for amphibian TSH has never been developed. Under these circumstances, information about the mRNA levels of TSH, especially of the TSH β-subunit, in the pituitaries of metamorphosing tadpoles would be suitable to fill this gap. Firstly we conducted molecular cloning of bullfrog TSHβ cDNA. Using the cDNA as a probe, expression of TSHβ mRNA during metamorphosis of bullfrog larvae was assessed (Okada *et al.*, 2000).

As shown in Figure 17.1, a TSHβ cDNA encoding both signal peptide and mature TSH molecule was cloned from a cDNA library constructed from total RNA of the bullfrog adenohypophysis. The bullfrog TSHβ mRNA was estimated by Northern blot analysis to be approximately 1 kb. The sequence homology between the bullfrog TSHβ and the bovine (Maurer *et al.*, 1984), rat (Chin *et al.*, 1985), mouse (Gurr *et al.*, 1983), chicken (Gregory and Porter, 1997), *Xenopus* (Buckbinder and Brown, 1993), European eel (Salmon *et al.*, 1993), and rainbow trout (Ito *et al.*, 1993) TSH β-subunits was 51.9, 50.4, 51.2, 50.0, 61.4, 39.5, and 43.5 per cent, respectively (Figure 17.2).

Using the cDNA as a probe, we measured changes in mRNA expression in metamorphosing tadpoles of *R. catesbeiana.* The TSHβ mRNA level increased progressively throughout prometamorphic stages, reaching its maximum at the end of prometamorphosis. The maximum level was maintained throughout early climax and mid-climax, declining at late climax (Figure 17.3).

Previously, Buckbinder and Brown (1993) cloned a *Xenopus* TSHβ cDNA. Using this cDNA as a probe, they measured TSHβ mRNA levels in the pituitaries of metamorphosing *Xenopus* larvae. In *Xenopus,* TSHβ mRNA levels increased throughout prometamorphosis, reaching maximum around the end of prometamorphosis and at early climax; thereafter, the levels declined considerably. Thyroid hormone levels in the plasma of *Xenopus* larvae had earlier been shown to rise during early climax stages and to decline thereafter (Leloup and Buscaglia, 1977).

The TSHβ mRNA expression profiles in the bullfrog larvae as revealed by us are somewhat different from those in *Xenopus* larvae. In the bullfrog larvae, TSHβ mRNA levels increased during prometamorphosis, reaching maximum at the end of prometamorphosis (stage XIX; Taylor and Korllos, 1946); and the maximum value was maintained throughout early and mid climax (stages XX–XXII). A considerable decline in TSHβ mRNA levels occurred when the larvae reached the late climax stage (XXIV). In parallel with the TSHβ mRNA levels, thyroid hormone levels also remained relatively high until the end of mid-climax (Miyauchi *et al.*, 1977; Mondou and Kaltenbach, 1979; Regard *et al.*, 1978; Suzuki and Suzuki, 1981). The discrepancy in the pattern of changes in TSHβ mRNA levels in *Xenopus* and *R. catesbeiana* larvae seems to be attributable to the difference in the time required for the completion of metamorphosis. In the bullfrog, tail resorption occurs more gradually than in *Xenopus.* Prolactin (PRL) is known to antagonize thyroid hormone to suppress the tail resorption (Bern et al.,1967; Etkin and Gona, 1967; Derby and Etkin, 1968).

More recently, over expression of PRL in transgenic *Xenopus* larvae was shown to produce tailed frogs (Huang and Brown, 2000). By immunoneutralization experiments using antisera against mammalian (Eddy and Lipner, 1975) and bullfrog (Clemons and Nicoll, 1977; Yamamoto and Kikuyama, 1982) PRLs, endogenous PRL was demonstrated to cause a retardation of tail resorption in bullfrog tadpoles. In this species plasma PRL levels start to rise throughout early climax and mid-climax, reaching maximum at the beginning of late climax and showing a moderate decline at the end of metamorphosis. In parallel with the change in plasma PRL levels, PRL mRNA levels in the pituitary exhibited a similar pattern of changes that started one stage in advance and ended one stage earlier (Takahashi *et al.*, 1990). In *Xenopus,* PRL mRNA levels abruptly rise around the end of climax (Buckbinder and Brown, 1993). Judging from the expression profiles of PRL mRNA, plasma PRL levels would probably be high during the late climax period when metamorphic changes are almost

```
  1  CCAGGATGACATCTATATTCATGGTGTCCTTTCTCCTTTGCTTTGCTTATGGGCATGCTA   60
    1  M  T  S  I  F  M  V  S  F  L  L  C  F  A  Y  G  H  A       18
 61  CTTTTCTATGTATGCTGACTGAATATACCATGTACGTGGAGATGGAAGAGTGTAGTCACT  120
 19  T  F  L  C  M  L  T  E  Y  T  M  Y  V  E  M  E  E  C  S  H   38
121  GCATTGCCATCAACACAACAATATGTTCTGGATACTGCTCAACTAAGGATCCTAATATGA  180
 39  C  I  A  I  N  T  T  I  C  S  G  Y  C  S  T  K  D  P  N  M   58
181  AGGGAAATTTACCGGAGGCCAAACTGAATCAGAACATTTGTACGTACAATGACTACATTC  240
 59  K  G  N  L  P  E  A  K  L  N  Q  N  I  C  T  Y  N  D  Y  I   78
241  TCAAGACGGTTTCAATTCCCAGTTGCCCAGTGCATGTCAATCCACACTATACCTATCCTG  300
 79  L  K  T  V  S  I  P  S  C  P  V  H  V  N  P  H  Y  T  Y  P   98
301  TAGCACTGAGTTGTAGGTGTGACAAATGTAACACAGGCTATATTGACTGTGTACAGGATA  360
 99  V  A  L  S  C  R  C  D  K  C  N  T  G  Y  I  D  C  V  Q  D  118
361  GTATTGAGTCTAATTACTGTACAAAGCCAAGGAAGCCGAAGGATTTCTTCTACAATTATG  420
119  S  I  E  S  N  Y  C  T  K  P  R  K  P  K  D  F  F  Y  N  Y  138
421  CCAAAAAATTCATTGGCCATAAATTTAAGTAAACCTGTTGTAATGGAACACTGCATGATC  480
139  A  K  K  F  I  G  H  K  F  K  *  148
481  CAGCATGTCTCTGCAACGCAGATGCCATGAAGTGCCTAACATACCCTCTGCTTGCCAAAT  540
541  GTAGCTCCAGGAAGTCCCCACATCTGGGCCATATACAGAGACCCATCTTTACAAGCTTGT  600
601  TGTTCCAGATCAACTCGCAGCAATTTTCAGCATCTGCATTTTTTGGAGCACCCTAGGTTC  660
661  AAGAACTTCAGGATATGTGTCAACAGCTGATGGTTCCTGCCGTTGCATATTCTTGCTGGT  720
721  TTATACAGGTTTACTTTAACAGATTTCAAAAAAATATATCTAGGTTTGCATTTGTTTAGA  780
781  GAAGCAGATAAGTTTGAAATATTATTTTAATTAGTTATACAGTTTCACTGTTTAAAAACA  840
841  TTAGATAAAATATACTGTTTTAACGTTCCTTTTACTGCAGAAAATTGAACTAAAATCAAT  900
901  ATCTATAATCAAAAAAAAAAAAAAAAAAAA  930
```

Figure 17.1: The Nucleotide Sequence of the Bullfrog TSH β cDNA

The predicted amino acids are shown below the nucleotide sequence. An asterisk indicates the termination codon. The signal peptide region (boxed) was estimated from the data by Sakai (1992), who determined the first 10 amino acid residues of its N-terminal sequence as FLCMLTEYTM. This sequence was identical with the amino acid sequence 20-29 deduced from the nucleotide sequence of TSHβ cDNA.

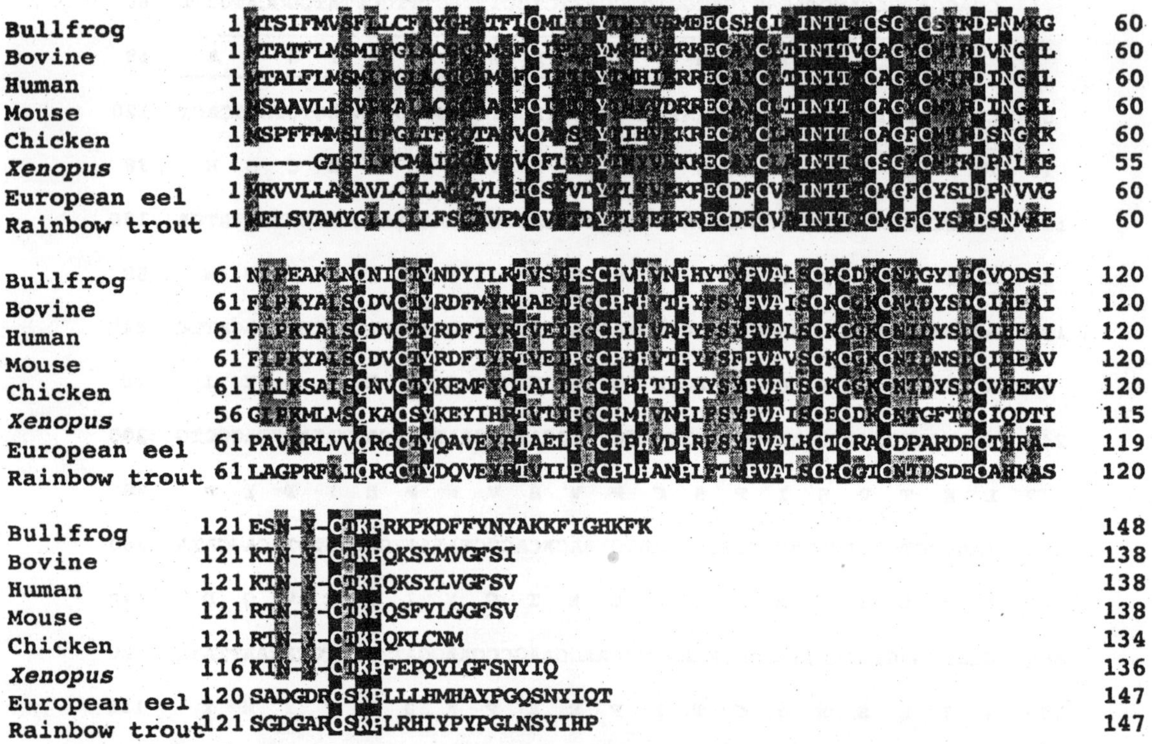

Figure 17.2: Amino Acid Sequences of Vertebrate TSH β-subunits

The amino acid residues which match those of bullfrog TSHβ are shown in white letters. Gaps as indicated by dashed lines have been introduced to obtain maximum homology.

complete. In *Xenopus*, therefore, metamorphosis may proceed rapidly with little interaction between thyroid hormone and PRL. In the bullfrog larvae, on the other hand, PRL may antagonize thyroid hormone to some extent, resulting in relatively slow metamorphic changes. This may account for the existence of thyroid hormone in relatively high concentrations and the elevated expression of TSH mRNA levels throughout early climax and mid-climax periods to ensure the completion of metamorphosis.

Development of RIA for Bullfrog TSH

A bullfrog TSHβ antiserum was produced by employing a C-terminal peptide (106-129) synthesized on the basis of the amino acid sequence deduced from bullfrog TSHβ cDNA (Okada *et al.*, 2000). Immunohistochemical studies revealed that the bullfrog adenohypophyseal cells that immunologically reacted with the anti-bullfrog TSHβ corresponded to those positively stained with an antiserum against human TSHβ. The antiserum was used for the development of a specific and sensitive RIA for the measurement of bullfrog TSH. The sensitivity of the RIA was 0.75 ± 0.07 ng TSH/ 100 *ml* assay buffer. The interassay and intraassay coefficients of variation were 7.6 and 5.3 per cent, respectively. Several dilutions of pituitary homogenates of larval and adult bullfrogs, or medium in which bullfrog pituitary cells were cultured, yielded dose–response curves that were parallel to the standard curve. Bullfrog PRL, growth hormone (GH), LH, FSH, and α-subunit derived from glycoprotein

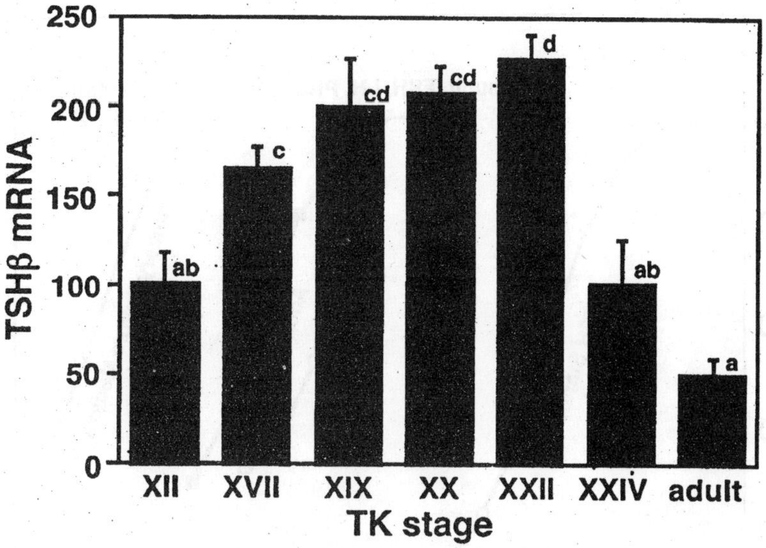

Figure 17.3: Developmental Changes in TSHβ mRNA in the Pituitary Gland of Metamorphosing Tadpoles

Total RNA from pituitaries of stages XII-XXIV larvae and of adult frogs was subjected to Northern blot analysis for TSHβ mRNA. Each sample was prepared from 20 pituitaries except for the case of adult frogs, for which each sample was prepared from a single gland. Total RNA was quantified, and equal amounts of RNA were electrophorsed. Each densitometry datum for TSHβ mRNA was normalized by that for bullfrog β-actin mRNA and expressed relative to that of the stage XII group. Each column and vertical bar represents the mean ± SEM (n=7). Values with same superscripts do not differ from each other at the 5 per cent level (ANOVA and Scheffe's test).

hormones did not react in this assay (Figure 17.4). Parallelism of inhibition curves between the C-terminal peptide of TSH β-subunit and the medium in which pituitary cells were cultured suggested that with this RIA, the intact form of bullfrog TSH is measurable, since the pituitary is expected to release TSH consisting of α- and β-subunits. In fact, immunoassayable TSH in the pituitary culture medium was completely precipitated with the antiserum against bullfrog α-subunit of glycoprotein hormones, indicating that the immunoassayable TSH in the culture medium was not the TSHβ alone but TSHβ coupled with the α-subunit. Moreover, the pituitary culture medium was confirmed to possess a potent bioactivity in promoting the release of T_4 from the thyroid gland of larval bullfrogs *in vitro*. Consequently, we concluded that with this new RIA system it would be possible to measure bullfrog TSH.

Using this RIA system, we measured pituitary TSH concentrations in the pituitaries from the larvae at various stages of development. As shown in Figure 17.5, TSH concentration in the pituitary is relatively low in early prometamorphosis (stage XIII) and increases as metamorphosis progresses, reaching maximum at late climax (stage XXIV), with a little decline at the end of metamorphosis (stage XXV).

Effect of Hypothalamic Peptides on the Release of TSH

The effects of hypothalamic hormones such as mammalian GnRH, ovine CRF, and TRH on the release of TSH by dispersed anterior pituitary cells of the bullfrog larvae and adults were studied. As

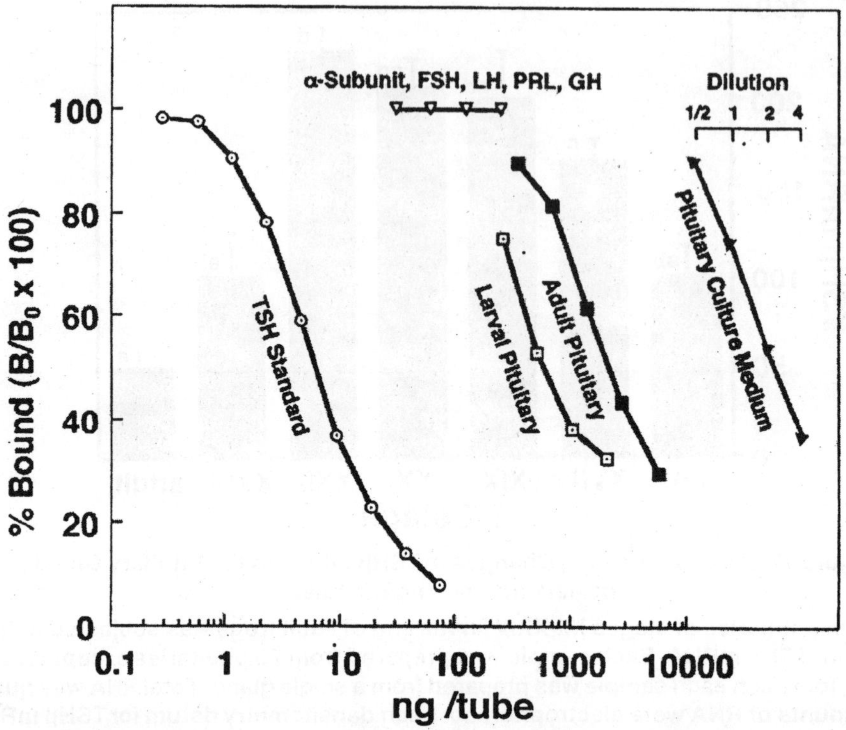

Figure 17.4: Displacement of ^{125}I-TSHβ$_{106-129}$ with homogenates of larval and adult bullfrog anterior pituitaries, 200 μl medium in which 70,000 pituitary cells of adult bullfrogs had been cultured for 24 h, α-subunit of pituitary glycoprotein hormones, LH, FSH, PRL, and GH. TSH standard (C-terminal peptide of TSHβ) is expressed as ng TSH (full length of TSHβ plus α-subunit).

shown in Figures 17.6 and 17.7, CRF markedly stimulated the release of TSH from both larval and adult pituitary cells. Both TRH and GnRH moderately stimulated the release of TSH from adult pituitary cells but not from the larval cells (Okada *et al.*, 2004).

Studies on the identification of the hypothalamic hormone(s) that stimulates the release of TSH from amphibian pituitary glands have been continued for the past 3 decades (see Denver, 1996; Dodd and Dodd, 1976; Kaltenbach, 1996; Kikuyama *et al.*, 1993; Shi, 1999). Due to the lack of a suitable RIA for amphibian TSH, indirect criteria such as the acceleration of metamorphosis, increase in the plasma thyroid hormone levels after administration of a hypothalamic hormone, and *in vitro* release of thyroid hormone from thyroid glands placed in medium in which a pituitary had been cultured in the presence of a hypothalamic hormone, have been adopted in most of these investigations. Initially TRH was considered to be the TSH-releasing hormone in amphibians. However, most experiments conducted to demonstrate the acceleration of metamorphosis by injecting TRH into larvae of various species of amphibians have not shown a definite positive effect of the peptide (see Norris and Dent, 1983). TRH was also inactive in inducing the release from larval pituitaries of a factor that stimulates the release of T_4 from the thyroid *in vitro* (Denver and Licht, 1989a). On the other hand, a possible TSH-releasing activity of TRH in amphibians was demonstrated by using adult *R. ridibunda* (Darras and Kühn, 1982)

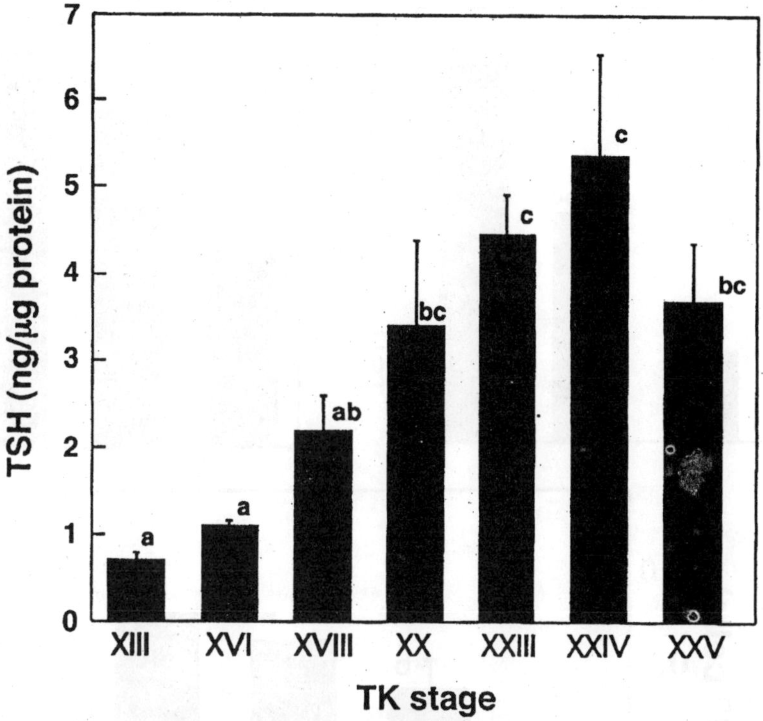

Figure 17.5: Changes of Immunoassayable TSH Concentration in Bullfrog Pituitary During and After Metamorphosis

The values with the same superscript do not differ from each other at the 5 per cent level of significance (ANOVA and Scheffe's test).

or the pituitary glands from adult *R. pipiens*, *Hyla regilla*, or *Xenopus laevis* (Denver, 1988). Ultrastructural morphometry of the pituitary TSH cells in adult *R. perezi* subjected to TRH stimulation revealed that the tripeptide enhanced the secretory function, *i.e.*, a decrease of secretory granules and development of endoplasmic reticulum and Golgi complex in the TSH cells (Castaño *et al.*, 1992).

In the mean time, GnRH and CRF have emerged as possible TSH-releasing factors. Intravenous injection of mammalian GnRH into ranid frogs (*R. ridibunda*, *R. esculenta*, and *R. temporaria*) was demonstrated to elevate plasma triiodothyronine (T_3) and T_4 levels (Jacobs *et al.*, 1988). The release of T_4 *in vitro* from the thyroid gland of *R. pipiens* was greatly enhanced when incubation was carried out in medium in which a pituitary gland from the same animal had been cultured in the presence of GnRH (Denver, 1988). Furthermore, it was demonstrated that culture medium in which larval pituitary gland of *R. catesbeiana* had been incubated in the presence of ovine CRF exhibited a potent stimulatory effect on the release of T_4 from the thyroid of adult *R. pipiens* (Denver and Licht, 1989a; Boorse and Denver, 2002). Furthermore, acceleration of metamorphosis (Gancedo *et al.*, 1992; Miranda *et al.*, 2000) and an increase in plasma thyroid hormone levels (Denver, 1993; Gancedo *et al.*, 1992) by injecting ovine CRF into larvae of *R. perezi*, *Bufo arenarum*, and *Ambystom a tigrinum*, or deceleration of metamorphosis by injecting an antiserum against CRF-like peptides into bullfrog larvae (Denver,

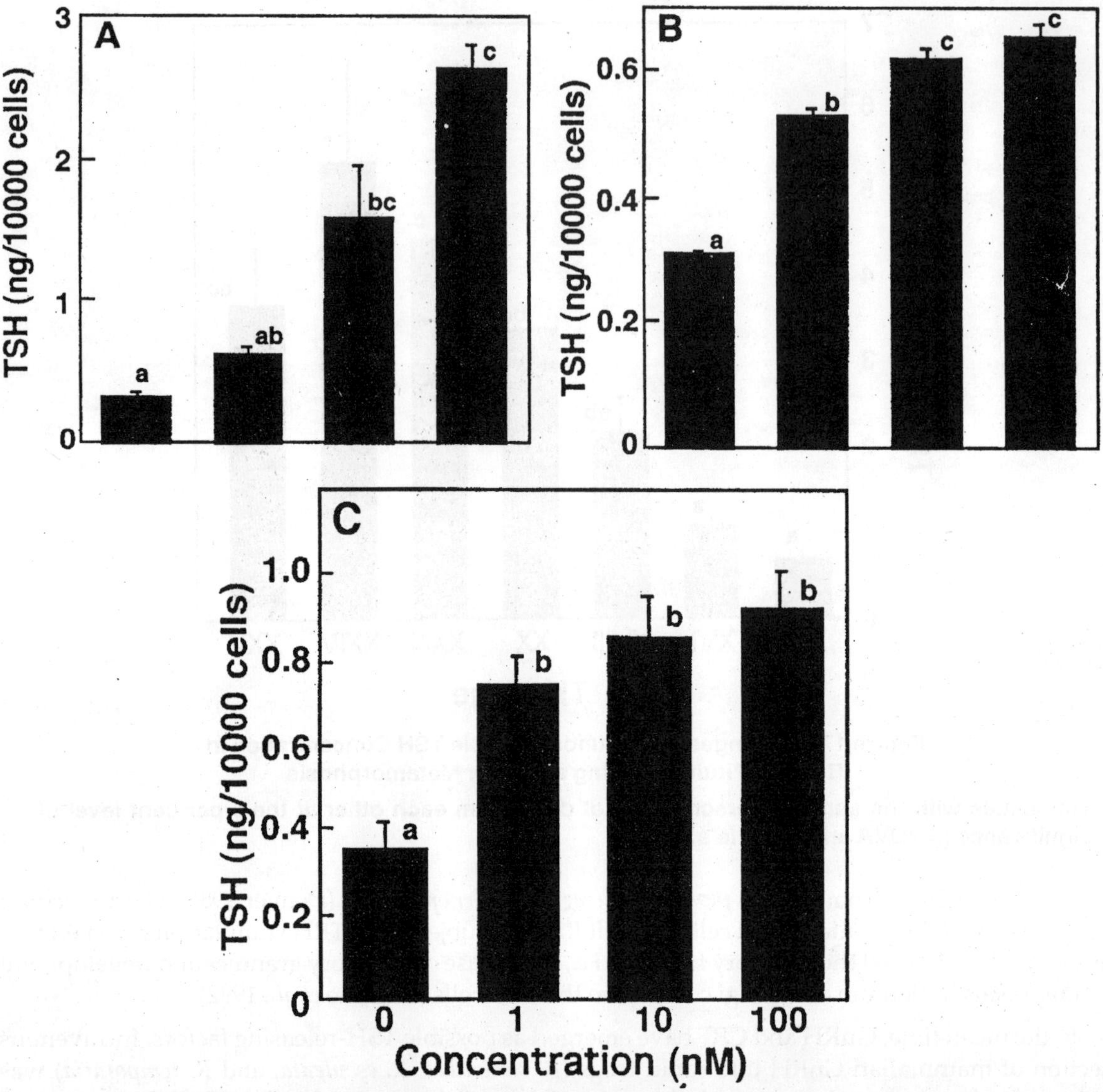

Figure 17.6. Effects of Various Concentrations of Ovine CRF (A), TRH (B), and Mammalian GnRH (C) on the Release of TSH from Dispersed Pituitary Cells of Adult Bullfrogs

The cells were incubated for 24 h. Data represent the means ± SEM (n=7). The values with the same superscript do not differ from each other at the 5 per cent level of significance (ANOVA and Scheffe's test).

1993) or CRF receptor (CRF-R) antagonist α-helical CRF_{9-41} into the western spadefoot toad (*Scaphiopus hammondii*) larvae (Denver, 1997) have been reported. Morphometric evidence of increased TSH release after *in vivo* administration of ovine CRF to adult *R. perezi* has also been reported (Malagon *et al.*, 1991). All these results indicate that CRF is a potent candidate for TSH-releasing hormone in amphibians.

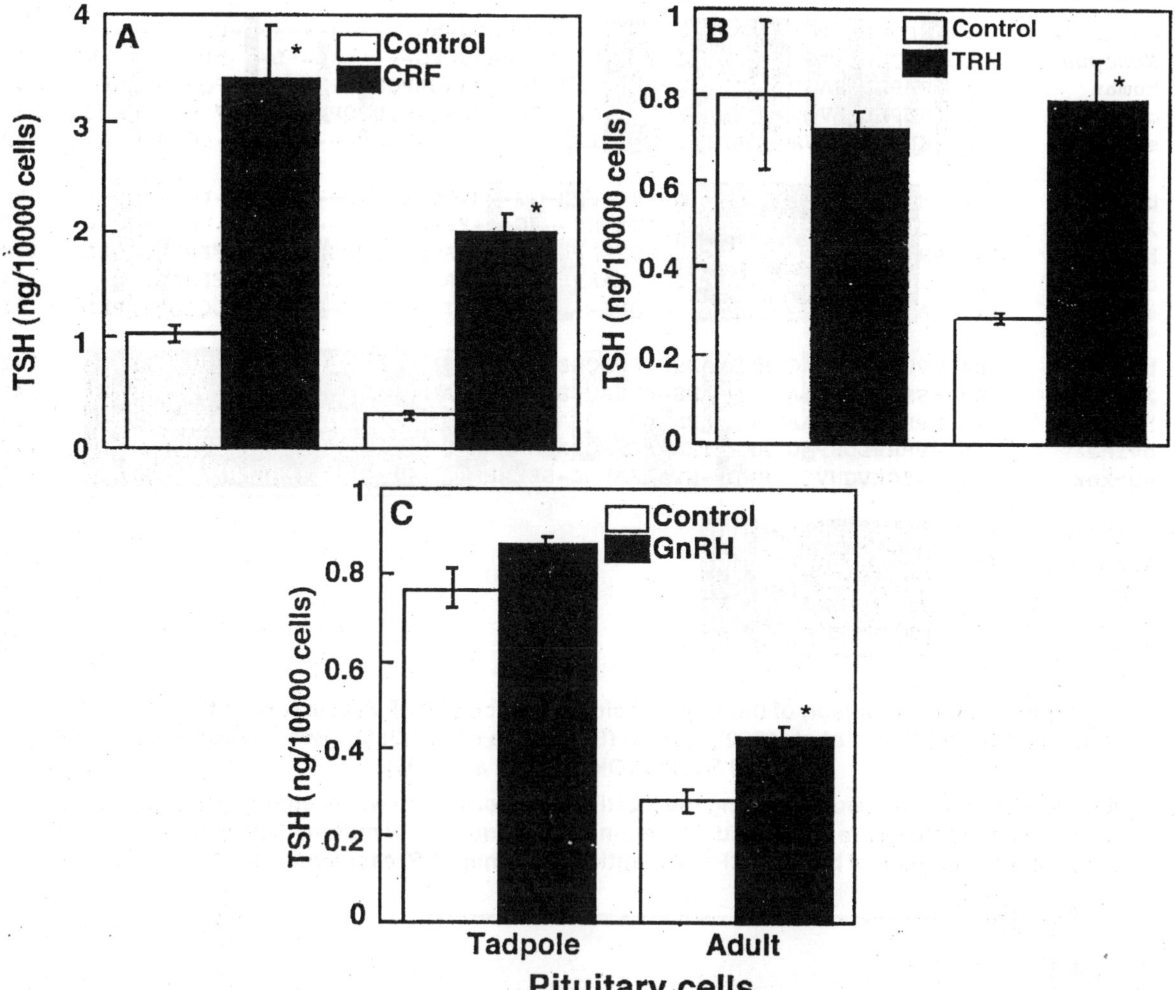

Figure 17.7: Effects of Ovine CRF, TRH, and Mammalian GnRH on the Release of TSH from the Dispersed Pituitary Cells of Larval or Adult Bullfrogs

After preincubation for 24 h, ovine CRF (A), TRH (B) or mammalian GnRH (C) was added to each well at a concentration of 100 nM, and the cells were incubated for 24 h. Data represent the means ± SEM (n=8).

***Significantly different from the control values at $p < 0.05$ (*F* test and *t* test).**

Effect of Homologous CRF on the Release of TSH from the Bullfrog Pituitary

As mentioned in the previous section, we have used ovine CRF but not the bullfrog counterpart. Recently, however, we obtained from a bullfrog hypothalamus cDNA library a cDNA encoding CRF with 83 per cent amino acid sequence identity with ovine CRF (Figure 17.8) (Ito *et al.*, 2004). It is notable that the amino acid sequences of human CRF and bullfrog CRF differed by only two amino acid residues. On the other hand, frog CRF differed in seven amino acids out of 41 amino acids as compared with the sequence of ovine CRF, which was the main CRF used in previous studies on non-mammalian

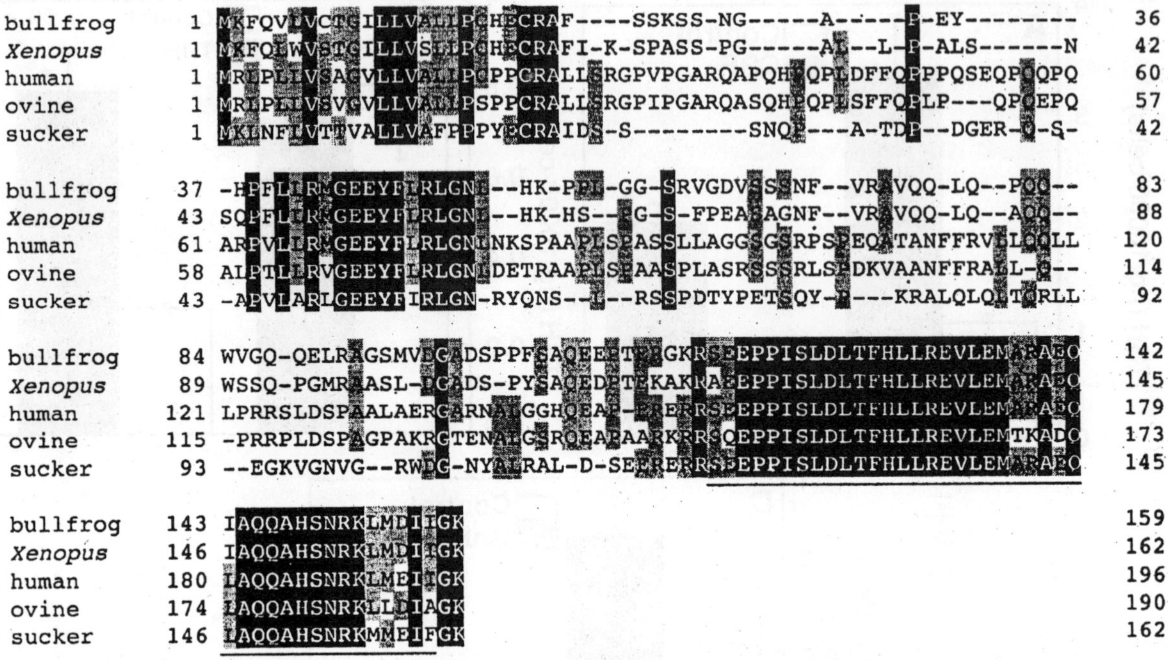

Figure 17.8: Comparison of the Amino Acid Sequence of CRF Precursors of the Bullfrog, *Xenopus* (Stenzel-Poore *et al.*, 1992), Human (Shibahara *et al.*, 1983), Ovine (Furutani *et al.*, 1983) and Sucker (Okawara *et al.*, 1998)

Identical amino acid residues among the 5 CRF precursors are in white letters. Identical residues among 3 or 4 of CRFs are shadowed. The amino acid sequence of mature CRF is underlined. The amino acid sequence of bullfrog CRF is identical with that of *R. esculenta* (Okada *et al.*, 2005).

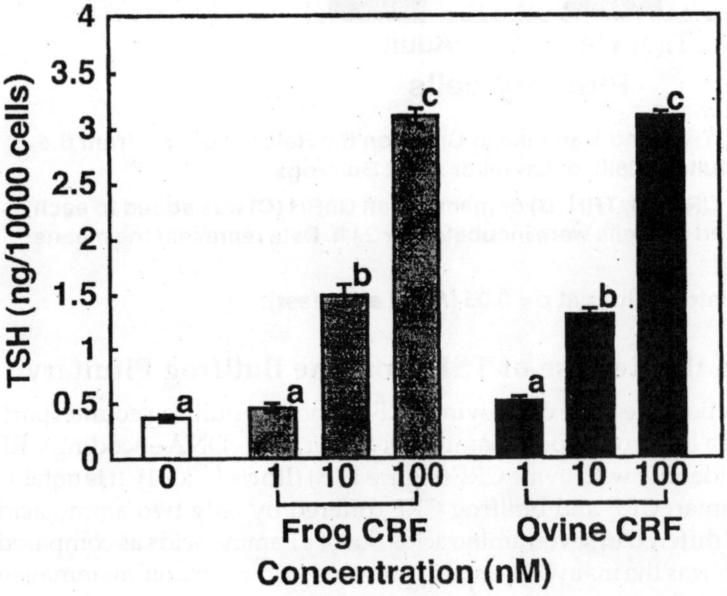

Figure 17.9: Effects of Frog CRF and Ovine CRF on the Release of TSH from Dispersed Pituitary Cells from Adult Bullfrogs

The cells were incubated for 24 h. Data represent means ± SEM (*n*=7). The values with the same superscript do not differ from each other at the 5 per cent level of significance (ANOVA and Scheffe's test).

vertebrates. (Denver, 1988; Denver and Licht, 1989a,b; Geris *et al.*, 1996; Larsen *et al.*, 1998). Nevertheless, both CRFs exhibited an equipotent activity in enhancing the release of TSH from the bullfrog pituitary cells (Figure 17.9).

As regards the biological activity of CRF family peptides on thyrotropes in non-mammalian vertebrates, sauvagine and urotensin I have been reported to stimulate the release of TSH from the pituitary of coho salmon *in vitro*. In amphibians an indirect evidence that sauvagine may induce the release of TSH from the pituitary of spadefoot toad tadpoles has been provided (Denver, 1993). He observed that injections of sauvagine as well as ovine CRF caused a precocious metamorphosis.

In the chicken, De Groef *et al.* (2003) have demonstrated that urocortin 3 evokes the release of TSH and that a type 2 CRF receptor (CRF-R2)-specific antagonist antisauvagine-30 blocks CRF-induced TSH secretion. Among the CRF family peptides, human urocortin 3 is known as a specific agonist for CRF-R2 in mammals, whereas it shows no affinity for type 1 CRF-R (CRF-R1) (Lewis *et al.*, 2001). De Groef *et al.* (2003) have also shown that CRF-R2 mRNA signals are present on thyrotropes but hardly on corticotropes. They concluded that the CRF-induced TSH release is mediated though CRF-R2.

In the bullfrog, human urocortin 3 was also effective in enhancing the release of TSH from the pituitary (Figure 17.10). This suggests the possible involvement of CRF-R2 in the release of TSH by CRF in amphibians as in the case of chicken. In fact, according to our unpublished data, mouse urocortin 2, a CRF-R2 selective agonist and human urocortin and sauvagine, non-specific CRF-R agonists were effective in enhancing the release of TSH from the dispersed bullfrog pituitary cells. CRF-R2 mediation of CRF-induced TSH release was further confirmed using CRF-R2 specific antagonist antisauvagine-30 and a CRF-R non-selective antagonist astressin. Both of them attenuated the TSH-releasing activity of CRF, whereas CRF-R1-selective antagonist antalarmin did not affect the release of TSH caused by CRF. The results clearly indicate the involvement of CRF-R2 in the CRF-induced release of TSH from the bullfrog thyrotropes.

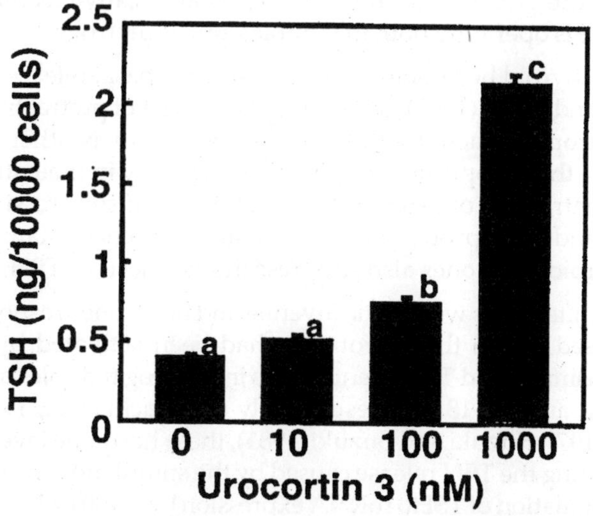

Figure 17.10: Effects of Human Urocortin 3 on the Release of TSH from Dispersed Pituitary Cells from Adult Bullfrogs

The cells were incubated for 24 h. Data represent means ± SEM ($n = 7$). The values with the same superscript do not differ from each other at the 5 per cent level of significance (ANOVA and Scheffe's test).

To ascertain whether the TSH-releasing activity of larval and adult hypothalamic extract (HE)s was due to CRF in them, we tested larval and adult HEs (one piece equivalent) for their TSH-releasing activity in the presence of a CRF-R antagonist, α-helical $CRF_{9\text{-}41}$. The addition of α-helical $CRF_{9\text{-}41}$ to the culture medium suppressed both larval and adult HE-induced release of TSH from the dispersed pituitary cells in a concentration-dependent manner. Synthetic bullfrog CRF (100 nM) used as a positive control exerted more potent TSH-releasing activity than adult HE or larval HE. The activity of the frog CRF was almost completely (more than 95 per cent) nullified by 100 ***m***M α-helical $CRF_{9\text{-}41}$, whereas approximately 56 and 43 per cent of one piece of adult and larval HE-induced TSH release, respectively, were suppressed by the same concentration of α-helical $CRF_{9\text{-}41}$ (Ito *et al.*, 2004). This indicates that HE of both larval and adult origin contained TSH-releasing substances other than CRF, although CRF is considered as one of the main TSH-releasing factors. In fact, it has been confirmed that a specific antiserum against the bullfrog CRF stained the axons that terminate in the median eminence as well as some of the cell bodies in the preoptic area, infundibularis dorsalis and nucleus hypothalamicus.

Effect of Thyroid Hormones on the Release of TSH from the Bullfrog Pituitary

Enzymatically dispersed pituitary cells of larval, juvenile, and adult bullfrogs were cultured in the absence or presence of 100 nM CRF of bullfrog origin, which is known to be a potent stimulator of the release of TSH. The amount of spontaneously released TSH was higher in late prometamorphic and climactic tadpoles than in early prometamorphic larvae and juvenile and adult frogs. Pituitary cells from tadpoles at metamorphic climax responded to CRF to release much more TSH than those from prometamorphic larvae and juvenile and adult frogs. In all cases, the CRF-induced release of TSH, but not the basal release of TSH, was suppressed significantly by 100 nM T_3. The suppressive effect of T_3 was revealed to be dependent on its concentrations (1-1000 nM). T_4 was also effective in suppressing the CRF-induced release of TSH, although the minimum effective concentration for T_4 is higher than that for T_3 (Figure 17.11). Thus, it can be said that negative feedback regulation of TSH release by thyroid hormone is operating both in tadpoles and adult frogs.

The ineffectiveness of thyroid hormones in suppressing the basal release of TSH agrees with the results obtained by Jacobs and Kühn (1992), who studied the effect of pretreatment of pituitary glands with T_3 or T_4 on the thyrotropic bioactivity of the pituitary *in vitro*. Neither T_3 nor T_4 pretreatment affected the basal release of thyrotropic factor(s) from the pituitary in their study. In *Xenopus* larvae, both T_3 and T_4 at the concentrations of 1-100 nM were effective in suppressing basal levels of gene expression of both TSHβ and glycoprotein hormone α-subunit genes (Manzon and Denver, 2004). This indicates that the thyroid hormones also suppress the synthesis of TSH.

The larval bullfrog pituitaries as well as the juvenile and adult ones responded to T_3 by reducing the amount of TSH released, when the thyrotropes had been activated by the CRF-stimulation. Considering that mean plasma T_3 and T_4 concentrations in bullfrog tadpoles at prometamorphic and climax stages are 0.4-1 nM and 1.3-13 nM, respectively (Miyauchi *et al.*, 1977; Regard *et al.*, 1978; Mondou and Kaltenbach, 1979; Suzuki and Suzuki, 1981), these hormone levels would be close to the range effective for suppressing the TSH release caused by the stimulation through the hypothalamus. In *Xenopus* larvae, down regulation of TSHβ mRNA expression by 1-100 nM T_3 or T_4 has been observed, although the responsiveness varies according to the metamorphic stages (Manzon and Denver, 2004). Since peak T_3 and T_4 concentrations in *Xenopus* larvae range from 8-12 nM and 7-10 nM, respectively (Leloup and Buscaglia, 1977; Krain and Denver, 2004), the synthesis of TSH in the pituitary can be considered to be controlled physiologically by thyroid hormones.

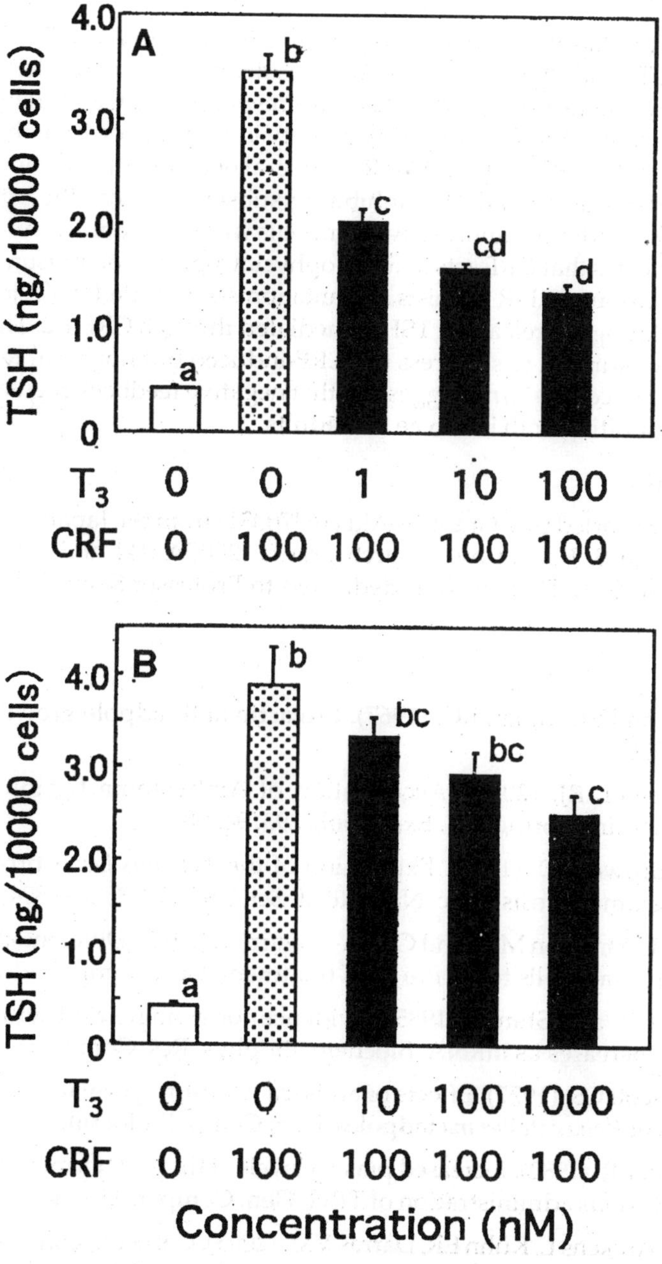

Figure 17.11: Effect of Various Concentrations of T_3 (A) or T_4 (B) on Bullfrog CRF-induced TSH Release from the Dispersed Pituitary Cells

The cells were incubated for 24 h. Each column and vertical bar represent the mean of 9 determinations and SEM, respectively. Values with the same superscript do not differ from each other at the 5 per cent level of significance (ANOVA and Scheffe's test).

Conclusion

RIA system for TSH has been developed for the first time in amphibians. This enabled us to measure TSH directly. The adult bullfrog pituitary revealed to respond to CRF markedly, and to TRH and GnRH moderately. Interestingly, larval pituitary responded to CRF but not to TRH and GnRH. Approximately 50 per cent of the TSH-releasing activity in the bullfrog hypothalamus was estimated to derive from CRF. Besides CRF, TRH and GnRH, several neuropeptides seem to affect the release of TSH. Isolation of TSH-releasing and TSH-inhibiting substances from the bullfrog hypothalami is under way. The fact that CRF-immunoreactive terminals were abundant around the capillaries in the median eminence indicates that CRF acts as a hypophysiotropic factor, presumably upon ACTH and TSH cells. The use of various CRF-R agonists and antagonists provided the evidence that the action of CRF in terms of stimulating the release of TSH is mediated through CRF-R2. Thyroid hormones (both T_3 and T_4) have been confirmed to suppress the CRF-induced but not basal release of TSH from the larval and adult pituitary cells *in vitro*, suggesting that negative feedback regulation of TSH release by thyroid hormone is operating both in larvae and adults.

Acknowledgments

This study was supported by a Grant-in-Aid (16370031) from the Japanese Ministry of Education, Science, and Culture to S. K. and by research grants (2005A-044 and 2005A-031) from Waseda University to R. O. and S. K. This article is dedicated to Professor Samir Bhattacharya, a long-time friend of S. K.

References

Bern HA, Nicoll CS, and Strohman RC. (1967). Prolactin and tadpole growth. Proc. Soc. Exp. Biol. Med., 126, 518–520.

Boorse GC, and Denver, RJ. (2002). Acceleration of Ambystoma tigrinum metamorphosis by corticotropin-releasing hormone. J. Exp. Zool., 293, 94–98.

Buckbinder L, and Brown DD. (1993). Expression of the *Xenopus laevis* prolactin and thyrotropin genes during metamorphosis. Proc. Natl. Acad. Sci. USA, 90, 3820–3824.

Castaño JP, Ramirez JL, Malagon MM, and Gracia-Navarro F. (1992). Differential response of amphibian PRL and TSH pituitary cells to *in vitro* TRH treatment. Gen. Comp. Endocrinol., 88, 178–187.

Chin WW, Muccini JAJ, and Shin L. (1985). Evidence for a single rat thyrotropin-β-subunit gene: Thyroidectomy increases its mRNA. Biochem. Biophys. Res. Commun., 128, 1152–1158.

Clemons GK, and Nicoll CS. (1977). Effects of antisera to bullfrog prolactin and growth hormone on metamorphosis of *Rana catesbeiana* tadpoles. Gen. Comp. Endocrinol., 31, 495–497.

Darras VM, and Kühn ER. (1982). Increased plasma levels of thyroid hormones in a frog *Rana ridibunda* following intravenous administration of TRH. Gen. Comp. Endocrinol., 48, 469–475.

De Groef B, Goris N, Arckens L, Kühn ER, Darras VM. (2003). Corticotropin-releasing hormone (CRH)-induced thyrotropin release in directly mediated through CRH receptor type 2 on thyrotropes. Endocrinol., 144, 5537-5544.

Denver RJ. (1988). Several hypothalamic peptides stimulate *in vitro* thyrotropin secretion by pituitaries of anuran amphibians. Gen. Comp. Endocrinol., 72, 383–393.

Denver RJ, and Licht P. (1989a). Neuropeptide stimulation of thyrotropin secretion in the larval bullfrog: evidence for a common neuroregulator of thyroid and interrenal activity in metamorphosis. J. Exp. Zool., 252, 101–104.

Denver RJ, and Licht P. (1989b). Neuropeptides influencing pituitary hormone secretion in hatchling turtles. J. Exp. Zool., 251, 306–315.

Denver RJ. (1993). Acceleration of anuran amphibian metamorphosis by corticotropin-releasing hormone-like peptides. Gen. Comp. Endocrinol., 91, 38–51.

Denver RJ. (1996). Neuroendocrine control of amphibian metamorphosis. *In*: Gilbert LI., Tata JR, and Atkinson BG (Eds.), *Metamorphosis: Post Embryonic Reprogramming of Gene Expression in Amphibian and Insect Cells*. Academic Press, New York, pp. 433–464.

Denver RJ. (1997). Environmental stress as a developmental cue: corticotropin-releasing hormone is a proximate mediator of adaptive phenotypic plasticity in amphibian metamorphosis. Hormones Behav., 31, 169–179.

Derby A, and Etkin, W. (1968). Thyroxine-induced tail resorption *in vitro* as affected by anterior pituitary hormones. J. Exp. Zool., 169, 1–8.

Dodd MHI, and Dodd JM. (1976). The biology of metamorphosis. In: Lofts, B. (Ed.), *Physiology of the Amphibia*, Vol. 3. Academic Press, New York, pp. 467–599.

Eddy L, and Lipner H. (1975). Accelation of thyroxine-induced metamorphosis by prolactin antiserum. Gen. Comp. Endocrinol., 25, 462–466.

Eddy L, Lipner H. (1976). Amphibian metamorphosis: The role of thyrotropin-like hormone. Gen. Comp. Endocrinol., 29, 333–336.

Etkin W, and Gona AG. (1967). Antagonism between prolactin and thyroid hormone in amphibian development. J. Exp. Zool., 165, 149–258.

Furutani Y, Morimoto Y, Shibahara S, Noda M, Takahashi H, Hirose T, Asai M, Inayama S, Hayashida H, Miyata T, and Numa S. (1983). Cloning and sequence analysis of cDNA for ovine corticotropin-releasing factor precursor. Nature, 301, 537-540.

Gancedo B, Corpas I, Alonso-Gomez AL, Delgado MJ, Morreale de Escobar G, and Alonso-Bedate M. (1992). Corticotropin-releasing factor stimulates metamorphosis and increases thyroid hormone concentration in prometamorphic *Rana perezi* larvae. Gen. Comp. Endocrinol., 87, 6–13.

Geris KL, Kotanen SP, Berghman LR, Kühn ER, and Darrs VM. (1996). Evidence of a thyrotropin-releasing activity of ovine corticotropin-releasing factor in the domestic fowl (*Gallus domesticus*). Gen. Comp. Endocrinol., 104, 139–146.

Gregory CC, and Porter TE. (1997). Cloning and sequence analysis of a cDNA for the β subunit of chicken thyroid-stimulating hormone. Gen. Comp. Endocrinol., 107, 182–190.

Gurr JA, Catterall JF, and Kourides IA. (1983). Cloning of cDNA encoding the pre-β subunit of mouse thyrotropin. Proc. Natl. Acad. Sci. USA, 80, 2122–2126.

Huang H, and Brown DD. (2000). Prolactin is not a juvenile hormone in *Xenopus laevis* metamorphosis. Proc. Natl. Acad. Sci. USA 97, 195–199.

Ito M, Koide Y, Takamatsu N, Kawauchi H, and Shiba T. (1993). cDNA cloning of the β subunit of teleost thyrotropin. Proc. Natl. Acad. Sci. USA, 90, 6052–6055.

Ito Y, Okada R, Mochida H, Hayashi H, Yamamoto K, and Kikuyama S. (2004). Molecular cloning of bullfrog corticotropin-releasing factor (CRF): effect of homologous CRF on the release of TSH from pituitary cells *in vitro*. Gen. Comp. Endocrinol., 138: 218-227.

Jacobs GFM, Goyvaerts MP, Vandorpe G, Quaghebeur AML, and Kühn ER. (1988). Luteinizing hormone-releasing hormone as a potent stimulator of the thyroidal axis in ranid frogs. Gen. Comp. Endocrinol., 70, 274–283.

Jacobs GFM, and Kühn ER. (1992). Thyroid hormone feedback regulation of the secretion of bioactive thyrotropin in frog. Gen. Comp. Endocrinol., 88, 415–423.

Kaltenbach JA. (1996). Endocrinology of amphibian metamorphosis. *In*: Gilbert LI, Tata JR, Atkinson BG (Eds.), *Metamorphosis: Postembryonic Reprogramming of Gene Expression in Amphibian and Insect Cells*. Academic Press, New York, pp. 403–431.

Kikuyama S, Kawamura K, Tanaka S, and Yamamoto K. (1993). Aspects of amphibian metamorphosis: hormonal control. *In:* Jeon KW, and Jarvik J. (Eds.), *International Review of Cytology*, Vol. 145. Academic Press, New York, pp. 105–148.

Krain LP, Denver RJ. (2004). Developmental expression and hormonal regulation of glucocorticoid and thyroid hormone receptors during metamorphosis in *Xenopus laevis*. J. Endocrinol., 181, 91-104.

Leloup J, and Buscaglia M. (1977). La triiodothyronine, hormone de la metamorphose des amphibiens. C. R. Acad. Sci., 284D, 2261-2263.

Larsen DA, Swanson P, Dickey JT, Rivier J, and Dickhoff WW. (1998). *In vitro* thyrotropin-releasing activity of corticotropin-releasing hormone-family peptides in coho salmon, *Oncorhynchus kisutch*. Gen. Comp. Endocrinol., 109, 276–285.

Lewis K, Li C, Perrin MH, Blount A, Kunitake K, Donaldson C, Vaughan J, Reyes TM, Gulyas J, Fischer W, Bilezikjian L, Rivier J, Sawchenko PE, and Vale WW. (2001). Identification of urocortin III, an additional member of the corticotropin-releasing factor (CRF) family with high affinity for the CRF2 receptor. Proc. Natl. Acad. Sci. USA, 98, 7570-7575.

Malagon MM, Ruiz-Navarro A, Torronteras R, and Gracia-Navarro F. (1991). Effects of ovine CRF on amphibian pituitary ACTH and TSH cells *in vivo*: a quantitative ultrastructural study. Gen. Comp. Endocrinol., 83, 487–497.

Manzon RG, and Denver RJ. (2004). Regulation of pituitary thyrotropin gene expression during *Xenopus* metamorphosis: negative feedback is functional throughout metamorphosis. J. Endocrinol., 182, 273-285.

Maurer RA, Croyle ML, and Donelson JE. (1984). The sequence of a cloned cDNA for the β subunit of bovine thyrotropin predicts a protein containing both NH_2- and COOH-terminal extensions. J. Biol. Chem., 259, 5024–5027.

Miranda LA, Affanni JM, and Paz DA. (2000). Corticotropin-releasing factor accelerates metamorphosis in *Bufo arenarum*: effect on pituitary ACTH and TSH cells. J. Exp. Zool., 286, 473–480.

Miyauchi H, LaRochell JFT, Suzuki M, Freeman M, and Frieden E. (1977). Studies on thyroid hormones and their binding in bullfrog tadpole plasma during metamorphosis. Gen. Comp. Endocrinol., 33, 254–266.

Mondou PM, and Kaltenbach JC. (1979). Thyroxine concentrations in blood serum and pericardial fluid of metamorphosing tadpoles and adult frogs. Gen. Comp. Endocrinol., 39, 343–349.

Norris DO, and Dent JN. (1983). Evolution of endocrine regulation of metamorphosis in lower vertebrates. Am. Zool., 23, 709–718.

Okada R, Iwata T, Kato T, Kikuchi M, Yamamoto K, and Kikuyama S. (2000). Cloning of bullfrog thyroid-stimulating hormone (TSH) β subunit cDNA: expression of TSHβ mRNA during metamorphosis. Gen. Comp. Endocrinol., 119, 224–231.

Okada R, Yamamoto K, Koda A, Ito Y, Hayashi H, Tanaka S, Hanaoka Y, and Kikuyama S. (2004). Development of radioimmunoassay for bullfrog thyroid-stimulating hormone (TSH): effects of hypothalamic releasing hormones on the release of TSH from the pituitary *in vitro*. Gen. Comp. Endocrinol., 135, 42-50.

Okada R, Ito Y, Kaneko M, Yamamoto K, Chartrel N, Conlon JM, Vaudry H, and Kikuyama S. (2005). Frog corticotropin-releasing factor (CRF); isolation, molecular cloning and biological activity. *In*: Vaudry H, Roubos E, Schoofs L, Flik G, and Larhammar D. (Eds.), *Trends in Comparative Endocrinology and Neurobiology*. New York Academy of Sciences, New York. in press.

Okawara Y, Morley SD, Burzio LO, Zwiers H, Lederis K, Richter D. (1998). Cloning and sequence analysis of cDNA for corticotropin-releasing factor precursor from the teleost fish *Catostomus commersoni*. Proc. Natl. Acad. Sci. USA, 85, 8439-8443.

Regard E, Taurog A, and Nakashima T. (1978). Plasma thyroxine and triiodothyronine levels in spontaneously metamorphosing *Rana catesbeiana* tadpoles and in adult anuran amphibia. Endocrinology, 102, 674–684.

Sakai M, Hanaoka Y, Tanaka S, Hayashi H, and Kikuyama S. (1991). Thyrotropic activity of various adenohypophyseal hormones of the bullfrog. Zool. Sci., 8, 929–934.

Sakai M. (1992). Isolation and characterization of amphibian thyrotropin. Thesis., Waseda University.

Salmon C, Marchelidon J, Fontaine Y-A, Huet J-G, and Querat B. (1993). Cloning and sequence of tyrotropin β-subunit of a teleost fish, the eel (*Anguilla anguilla* L.). C. R. Acad. Sci., 316,749–753.

Shi Y-B. (1999). *Amphibian Metamorphosis from Morphology to Molecular Biology*. Wiley-Liss, New York.

Shibahara S, Morimoto Y, Furutani Y, Notake M, Takahashi H, Shimizu S, Horikawa S, Numa S. (1983). Isolation and sequence analysis of the human corticotropin-releasing factor precursor gene. EMBO J., 2, 775-779.

Stenzel-Poore MP, Heldwein KA, Stenzel P, Lee S, and Vale WW. (1992). Characterization of the genomic corticotropin-releasing factor (CRF) gene from *Xenopus laevis*: two members of the CRF family exist in amphibians. Mol. Endocrinol., 6, 1716-1724.

Suzuki S, and Suzuki M. (1981). Changes in thyroidal and plasma iodine compounds during and after metamorphosis of the bullfrog, *Rana catesbeiana*. Gen. Comp. Endocrinol., 45, 74–81.

Takahashi N, Yoshihama K, Kikuyama S, Yamamoto K, Wakabayashi K, and Kato Y. (1990). Molecular cloning and nucleotide sequence analysis of complimentary DNA from bullfrog prolactin. J. Mol. Endocrinol. 5, 281-287.

Taylor AC, and Kollros JJ. (1946). Stages in the normal development of *Rana pipiens* larvae. Anat. Rec., 94, 7-23.

Yamamoto K, and Kikuyama S. (1982). Effect of prolactin antiserum on growth and resorption of tadpole tail. Endocrinol. Jpn., 29, 81-85.

18

Gonadotropin-Inhibitory Hormone (GnIH) in the Avian Brain

Kazuyoshi Tsutsui*

Laboratory of Brain Science, Faculty of Integrated Arts and Sciences, Hiroshima University Higashi-Hiroshima 739-8521, Japan and Core Research for Evolutional Science and Technology (CREST), Japan Science and Technology Corporation, Tokyo 150-0002, Japan

ABSTRACT

Probing undiscovered neuropeptides that play important roles in the regulation of pituitary function in birds is essential for the progress of avian endocrinology. Neuropeptide control of gonadotropin secretion at the level of the anterior pituitary gland is primarily through the stimulatory action of the hypothalamic decapeptide, gonadotropin-releasing hormone (GnRH). However, an inhibitory hypothalamic neuropeptide for gonadotropin release has, until recently, remained unknown in any vertebrate. In 2000, we discovered a novel hypothalamic dodecapeptide inhibiting gonadotropin release in quail and termed it gonadotropin-inhibitory hormone (GnIH). A gonadotropin inhibitory system is an intriguing concept and provides us with an unprecedented opportunity to study the regulation of avian reproduction from an entirely novel standpoint. To elucidate the mode of action of GnIH, we further identified the receptor for GnIH and characterized its expression and binding activity in quail. The identified GnIH receptor possessed seven transmembrane domains and specifically bound to GnIH in a concentration-dependent manner. The expression of GnIH receptor was found in the pituitary and several brain regions including the hypothalamus. These results indicate that GnIH acts directly on the pituitary via GnIH receptor to inhibit gonadotropin release. GnIH may also act on the hypothalamus to inhibit GnRH release. This paper summarizes the structure, function and mode of action of GnIH, a newly discovered hypothalamic neuropeptide, in birds.

*E-mail: tsutsui@hiroshima-u.ac.jp; Phone: +81 82-424-6571; Fax: +81 82-424-0759

Introduction

In vertebrates, the neuropeptide control of gonadotropin secretion is primarily through the stimulatory action of the hypothalamic decapeptide, gonadotropin-releasing hormone (GnRH), which was originally isolated from mammals and subsequently from nonmammals including birds. Until recently an inhibitory neuropeptide of gonadotropin secretion has remained unknown in vertebrates, although gonadal sex steroids and inhibin can modulate gonadotropin secretion. In 2000, we identified a novel hypothalamic dodecapeptide which directly inhibits gonadotropin release in quail and termed it gonadotropin-inhibitory hormone (GnIH; Tsutsui *et al.*, 2000). This was the first demonstration of a hypothalamic neuropeptide inhibiting gonadotropin release in any vertebrate. A gonadotropin inhibitory system is an intriguing concept and provides us with an unprecedented opportunity to study the regulation of avian reproduction from an entirely novel standpoint.

Discovery of GnIH

Since the molluscan cardioexcitatory neuropeptide Phe-Met-Arg-Phe-NH_2 (FMRFamide) was found in the ganglia of the venus clam (Price and Greenberg, 1977), neuropeptides that possess the RFamide motif at their C-termini (*i.e.*, RFamide peptides) have been characterized in various invertebrates. Subsequently, many immunohistochemical studies that used the antiserum against FMRFamide suggested that vertebrate nervous systems possess some unknown neuropeptides similar to FMRFamide (Raffa, 1988; Rastogi *et al.*, 2001). Immunohistochemical findings indicated that some of the FMRFamide-like immunoreactive neurons project to the hypothalamic region close to the pituitary gland, and thus were predicted to play an important role in the regulation of pituitary function. We therefore looked for a novel RFamide peptide in the avian brain.

To isolate the RFamide peptide from the brain, Japanese quails (*Coturnix japonica*) were used and the peptidergic molecule was probed with a competitive enzyme-linked immunosorbent assay (ELISA), employing the antibody against the dipeptide, Arg-Phe-NH_2 (Tsutsui *et al.*, 2000). Acetic acid extracts of quail brain were passed through C-18 reversed-phase cartridges, and the retained material was subjected to reversed-phase and cation-exchange high performance liquid chromatography (HPLC). Amino acid sequence analysis of the isolated substance by automated Edman degradation with a gas-phase sequencer revealed the following sequence: Ser(62)-Ile(252)-Lys(233)-Pro(226)-Ser(38)-Ala(194)-Tyr(173)-Leu(148)-Pro(104)-Leu(108)-Arg(45)-Phe(52) with the detected amount (pmol) of each amino acid indicated in parentheses. A protonated molecule ion $(M+H)^+$ peak in the fast atom bombardment-MS (FAB-MS) of this peptide at *m/z* 1389.4 indicated that the peptide is amidated at the C-terminus. Synthetic and native peptides showed identical retention times on a C-18 reversed-phase column and a cation-exchange column. The mixture of the synthetic and native peptides eluted as a single peak from each column. Thus the isolated native peptide was confirmed as a 12 amino acid sequence (SIKPSAYLPLRFamide) with RFamide at the C-terminus (Tsutsui *et al.*, 2000). This neuropeptide had not been previously reported in vertebrates, although the C-terminal LPLRFamide was identical to chicken pentapeptide LPLRFamide peptide (Dockray *et al.*, 1983). The chicken peptide may be a degraded fragment of the dodecapeptide, as suggested by Dockray *et al.* (1985).

Subsequently, the isolated novel peptide was shown to be located in the quail hypothalamo-hypophysial system and to decrease gonadotropin release from cultured anterior pituitary in a dose-dependent manner (Tsutsui *et al.*, 2000). We therefore designated this novel RFamide peptide as gonadotropin-inhibitory hormone (GnIH) (Tsutsui *et al.*, 2000; see Figure 18.1).

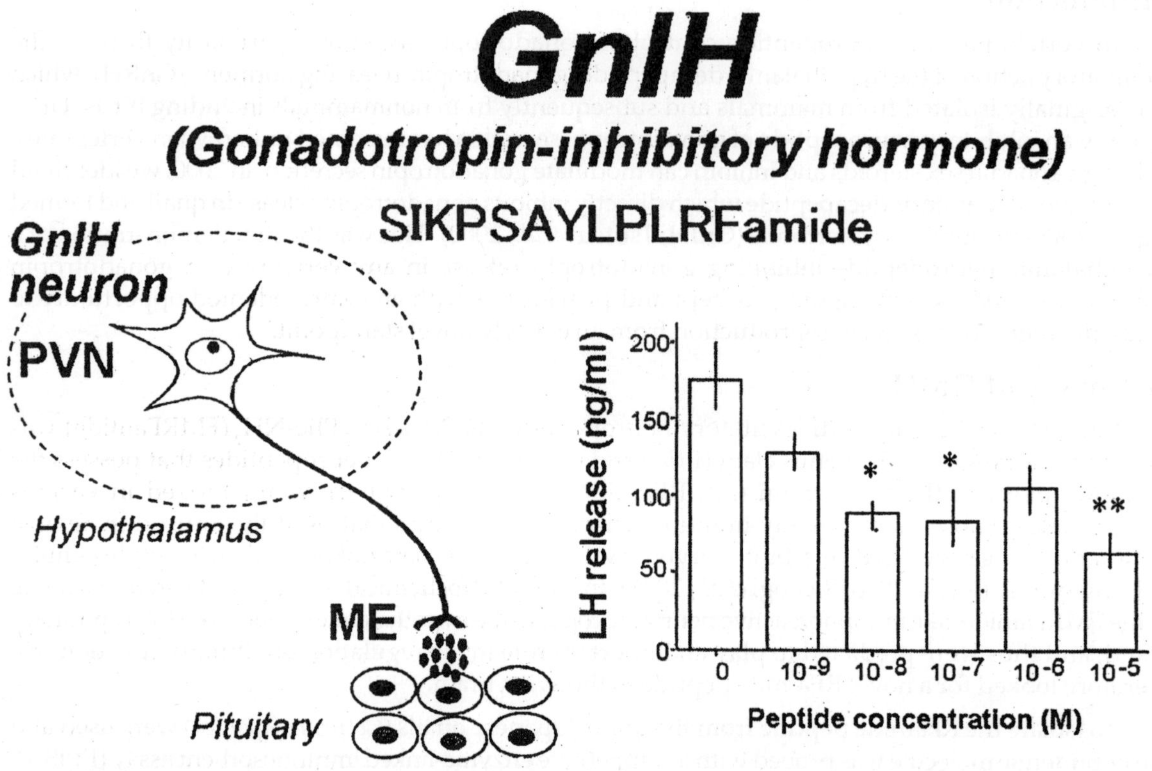

Figure 18.1: GnIH, A Newly Discovered Hypothalamic Neuropeptide, in the Avian Brain

We isolated a novel hypothalamic dodecapeptide (SIKPSAYLPLRFamide) inhibiting gonadotropin release in quail (Tsutsui *et al.*, 2000). Cell bodies and terminals containing the isolated novel neuropeptide were localized in the paraventricular nucleus (PVN) and median eminence (ME), respectively (Tsutsui *et al.*, 2000). The isolated novel neuropeptide was shown to decrease gonadotropin release from cultured anterior pituitary in a dose-dependent manner (Tsutsui *et al.*, 2000). We therefore designated this novel hypothalamic neuropeptide as GnIH (Tsutsui *et al.*, 2000).

Localization of GnIH in the Brain

The localization of GnIH in the brain is essential to understand its functions. As a first step to examine the localization of GnIH, we dissected out the quail brain into several regions, and quantified the concentration of GnIH by ELISA using a rabbit polyclonal antibody raised against GnIH (Tsutsui *et al.*, 2000). The concentration of GnIH in the diencephalon was much higher than that in the mesencephalon. In contrast, GnIH concentrations in the cerebrum and cerebellum were below the level of detectability. Subsequently, we investigated the precise localization of GnIH in the quail brain by immunohistochemistry (Tsutsui *et al.*, 2000; Ukena *et al.*, 2003; Ubuka *et al.*, 2003). Clusters of distinct GnIH-immunoreactive neurons were found in the paraventricular nucleus (PVN) in the hypothalamus. In addition to the PVN, some scattered small cells were immunoreactive in the septal area. In contrast to the highly-localized clusters of cell bodies, GnIH-containing fibers were widely distributed in the diencephalic and mesencephalic regions particularly in the ventral paleostriatum, septal area, preoptic area, hypothalamus, and optic tectum. The most prominent fibers were seen in the median eminence of the hypothalamus, and in the dorsal motor nucleus of the vagus in the medulla oblongata.

We further investigated GnIH localization in the brain of sparrows, seasonally-breeding avian species (Bentley *et al.*, 2003; Osugi *et al.*, 2004). Dense populations of GnIH-immunoreactive neurons were also found in the PVN of these birds. The PVN was the only location where immunoreactive neurons were located (Bentley *et al.*, 2003; Osugi *et al.*, 2004). Thus the presence of GnIH in the PVN appears to be a conserved property among several avian species. In addition, a widespread distribution of GnIH-containing fibers was also found in sparrows.

Interestingly, GnIH-containing fibers were further observed in extremely close proximity to GnRH neurons in the preoptic area (POA) in birds (Bentley *et al.*, 2003; Ukena *et al.*, 2003). It is therefore plausible that GnIH may act at the level of the hypothalamus to regulate gonadotropin release as well as at the pituitary.

Identification of GnIH precursor cDNA

We further examined the precursor polypeptide for GnIH and localization of its transcript, which would provide us with key information as to the regulation of the mature GnIH peptide, along with confirmation of brain area(s) that synthesize this novel peptide. A cDNA that encoded the GnIH precursor polypeptide was identified in the quail brain using a combination of 3' and 5' rapid amplification of cDNA ends (3'/5' RACE) (Satake *et al.*, 2001). The deduced GnIH precursor consisted of 173 amino acid residues that encoded one GnIH and two putative GnIH-related peptide (GnIH-RP-1 and–RP-2) sequences that included–LPXRF (X = L or Q) at their C-termini (Satake *et al.*, 2001). All these peptide sequences were flanked by a glycine C-terminal amidation signal and a single basic amino acid on each end as an endoproteolytic site.

We also cloned a cDNA that encoded GnIH in the brain of Gambel's white-crowned sparrow (Osugi *et al.*, 2004). The deduced sparrow GnIH precursor also consisted of 173 amino acid residues, encoding one sparrow GnIH and two sparrow GnIH-related peptides (sparrow GnIH-RP-1 and GnIH-RP-2) that included–LPXRFamide (X = L or Q) at their C-termini (Osugi *et al.*, 2004). Although the homology of sparrow and quail GnIH precursors was approximately 66 per cent, the C-terminal structures of GnIH, GnIH-RP-1 and GnIH-RP-2 were all identical in two species (Satake *et al.*, 2001; Osugi *et al.*, 2004). Subsequently, a cDNA encoding GnIH and GnIH-RPs was also reported in the chicken from a gene database (see Osugi *et al.*, 2004).

In situ hybridization further revealed the cellular localization of GnIH mRNA solely in the PVN of quail and sparrow hypothalami (Ukena *et al.*, 2003; Osugi *et al.*, 2004). As already described, immunohistochemical analysis using the quail and sparrow also showed that quail and sparrow GnIH-immunoreactive cell bodies and terminals were localized in the PVN and median eminence, respectively. Thus only the PVN expresses GnIH and, in birds, the immunoreactive peptide found in fibers in multiple brain areas including the median eminence appears to originate from the PVN only (Ukena *et al.*, 2003; Osugi *et al.*, 2004).

Functions of GnIH on Gonadotropin Release and Synthesis

In view of the immunohistochemical finding indicating that GnIH-immunoreactive neurons project to the median eminence close to the pituitary, we analyzed the effect of the isolated SIKPSAYLPLRFamide, GnIH, on the release of luteinizing hormone (LH), follicle-stimulating hormone

(FSH) and prolactin (PRL) using cultured quail anterior pituitaries (Tsutsui *et al.*, 2000). GnIH significantly inhibited LH release, after 100-min incubation. The inhibitory effect on LH release was dose-dependent and its threshold concentration ranged between 10^{-9} and 10^{-8} M. A possible similar tendency for GnIH to inhibit FSH release was also detected. However, there was no effect of GnIH on PRL release. Based on these results, we therefore named this novel RFamide peptide isolated from the quail brain GnIH (Tsutsui *et al.*, 2000).

We further showed that GnIH was effective in inhibiting circulating LH *in vivo*. When administered intraperitoneally to quail via osmotic pumps, GnIH significantly reduced plasma LH (Ubuka *et al.*, unpublished observation). GnIH also inhibited rapidly LH release *in vivo* in sparrows (Osugi *et al.*, 2004). Interestingly, *in vivo* treatment with GnIH rapidly inhibited GnRH-elicited LH release in sparrows (Osugi *et al.*, 2004).

In addition to the inhibitory effects of GnIH on gonadotropin release, there is evidence that GnIH inhibits gonadotropin biosynthesis *in vitro* (Ciccone *et al.*, 2005). The suppressive effect of GnIH on gonadotropin mRNA was associated with an inhibition of both LH and FSH release in the chicken (Ciccone *et al.*, 2005) and in the quail (Ubuka *et al.*, unpublished observation).

Mode of Action of GnIH

Identification of the receptor for GnIH is crucial to elucidate the mode of action of GnIH. We therefore identified the receptor for GnIH in the quail diencephalon and characterized its expression and binding activity (Yin *et al.*, 2005). We first cloned a cDNA encoding a putative GnIH receptor by a combination of 3′ and 5′ rapid amplification of cDNA ends (RACE) using PCR primers designed from the sequence for the receptor for rat RFamide-related peptide (RFRP), an orthologous peptide of GnIH. Hydrophobic analysis revealed that the putative GnIH receptor possessed seven transmembrane domains, indicating a new member of the G protein-coupled receptor (GPCR) superfamily (Yin *et al.*, 2005). The crude membrane fraction of COS-7 cells transfected with the putative GnIH receptor cDNA specifically bound to GnIH and GnIH-RPs in a concentration-dependent manner (Yin *et al.*, 2005). Scatchard plot analysis of the binding showed that the identified GnIH receptor possessed a single class of high affinity binding sites ($K_d = 0.752$ nM). Southern blotting analysis of reverse-transcriptase-mediated PCR products revealed the expression of GnIH receptor mRNA in the pituitary and several brain regions including diencephalon in the quail (Yin *et al.*, 2005). These results indicate that GnIH acts directly on the pituitary via GnIH receptor to inhibit gonadotropin release (see Figure 18.2). GnIH may also act on the hypothalamus to inhibit GnRH release.

Conclusions

Probing undiscovered neuropeptides that play important roles in the regulation of pituitary function in birds is essential for the progress of avian endocrinology. The discovery of GnIH has opened avenues for a new research field in avian endocrinology.

A gonadotropin inhibitory system by GnIH is an intriguing new concept and provides us with an unprecedented opportunity to study the regulation of avian reproduction from an entirely novel standpoint. In addition, the widespread distribution of GnIH in the avian brain indicates that GnIH may participate not only in neuroendocrine functions but also in behavioral, sensory, and autonomic functions. Future studies will bring to light previously unknown physiological functions of GnIH. Our recent study suggests that melatonin induces the expression of GnIH in the avian hypothalamus (Ubuka et al, 2005). We also need to clarify the regulatory mechanism governing the expression of GnIH.

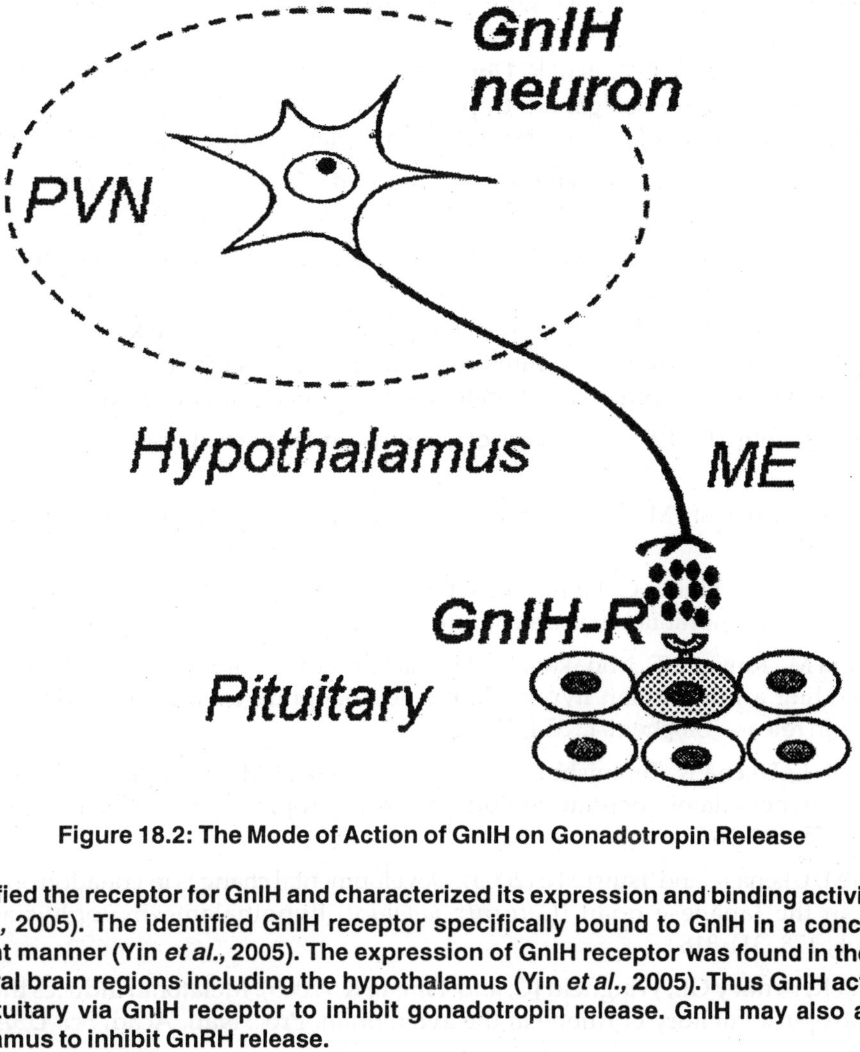

Figure 18.2: The Mode of Action of GnIH on Gonadotropin Release

We identified the receptor for GnIH and characterized its expression and binding activity in quail (Yin *et al.*, 2005). The identified GnIH receptor specifically bound to GnIH in a concentration-dependent manner (Yin *et al.*, 2005). The expression of GnIH receptor was found in the pituitary and several brain regions including the hypothalamus (Yin *et al.*, 2005). Thus GnIH acts directly on the pituitary via GnIH receptor to inhibit gonadotropin release. GnIH may also act on the hypothalamus to inhibit GnRH release.

Acknowledgements

I am grateful to all our collaborators, particularly Drs. K. Ukena, T. Ubuka, T. Osugi, E. Saigoh and H. Yin (Hiroshima University, Japan), G. E. Bentley, S O'Brien, IT Moore, IC Dunn, T Boswell and J. C. Wingfield (University of Washington, USA), H. Satake and H. Minakata (Suntory institute for Bioorganic Research, Japan), S. Ishii (Waseda University, Japan) and N. Ciccone and P. J. Sharp (Roslin Institute, UK). This work was supported in part by Grants-in-Aid for Scientific Research from the Ministry of Education, Science and Culture, Japan (13210101, 15207007 and 16086206) and NIH, USA (RO1 MH065974).

References

Bentley GE, Perfito N, Ukena K, Tsutsui K, and Wingfield JC. (2003). Gonadotropin-inhibitory peptide in song sparrows (*Melospiza melodia*) in different reproductive conditions, and in house sparrows

(*Passer domesticus*) relative to chicken-gonadotropin-releasing hormone. J. Neuroendocrinol., 15, 794-802.

Ciccone NA, Dunn IC, Boswell T, Tsutsui K, Ubuka T, Ukena K, and Sharp PJ. (2005). Gonadotrophin inhibitory hormone depresses gonadotrophin α and follicle-stimulating hormone β subunit expression in the pituitary of the domestic chicken. J. Neuroendocrinol., 16, 999-1006.

Dockray GJ, Reeve JR Jr., Shively J, Gayton RJ, and Barnard CS. (1983). A novel active pentapeptide from chicken brain identified by antibodies to FMRFamide. Nature, 305, 38-330.

Dockray GJ, and Dimaline R. (1985). FMRFamide- and gastrin/CCK-like peptides in birds. Peptides, 3, 333-337.

Osugi T, Ukena K, Bentley GE, O'Brien S, Moore IT, Wingfield JC, and Tsutsui K. (2004). Gonadotropin-inhibitory hormone in Gambel's white-crowned sparrows: cDNA identification, transcript localization and functional effects in laboratory and field experiments. J. Endocrinol., 182, 33-42.

Price DA, and Greenberg MJ. (1977). Structure of a molluscan cardioexcitatory neuropeptide. Science, 197, 670-671.

Raffa RB. (1988). The action of FMRFamide (Phe-Met-Arg-Phe-NH_2) and related peptides on mammals. Peptides, 9, 915-922.

Rastogi RK, D'Aniello B, Pinelli C, Fiorentino M, Di Fiore M.M., Di Meglio M. and Iela L. (2001). FMRFamide in the amphibian brain: a comprehensive survey. Micros. Res. Tech., 54, 158-172.

Satake H., Hisada M., Kawada T., Minakata H., Ukena K, and Tsutsui K. (2001). Characterization of a cDNA encoding a novel avian hypothalamic neuropeptide exerting an inhibitory effect on gonadotropin release. Biochem. J., 354, 379-385.

Tsutsui K, Saigoh E, Ukena K, Teranishi H, Fujisawa Y, Kikuchi M, Ishii S, and Sharp PJ. (2000). A novel avian hypothalamic peptide inhibiting gonadotropin release. Biochem. Biophys. Res. Commun., 275, 661-667.

Ubuka T, Ueno M, Ukena K, and Tsutsui K. (2003). Developmental changes in gonadotropin-inhibitory hormone in the Japanese quail (*Coturnix japonica*) hypothalamo-hypophysial system. J. Endocrinol., 178, 311-318.

Ubuka T, Ukena K, Bentley GE, Wingfield JC, and Tsutsui K. (2005). Melatonin induces the expression of gonadotropin-inhibitory hormone in the avian brain. Proc. Natl. Acad. Sci. USA, 102, 3052-3057.

Ukena K, Ubuka T, and Tsutsui K. (2003). Distribution of a novel avian gonadotropin-inhibitory hormone in the quail brain. Cell. Tissue Res., 312, 73-79.

Yin H., Ukena K., Ubuka T. and Tsutsui K. (2005) A novel G protein-coupled receptor for gonadotropin-inhibitory hormone in the Japanese quail (*Coturnix japonica*): identification, expression and binding activity. J. Endocrinol., 184, 257-266.

19

Aquaculture Biotechnology: An Indian Perspective

Arun K. Ray[1*] *and Basant K. Tiwary*[2]

[1]*Department of Animal Physiology, Bose Institute, Kolkata – 700 054, India*
[2]*Department of Life Science, Assam University, Silchar – 788 011, India*

ABSTRACT

Biotechnology can play a major role in development of aquaculture. Both genetic and endocrine biotechnologies have tremendous potential to contribute significantly to fish culture. Chromosome manipulation leads to production of polyploids especially triploids having better growth potential. Transgenesis is the transfer of foreign genes into a fish for better growth, disease resistance or to function as a bio-reactor. Endocrine biotechnology is equally effective for a considerable change in fish farming. Sex reversal may be utilized for production of desirable monosex population of a species through treatment with sex steroids. Reproductive manipulation through hormonal means gives us a unique opportunity for out-of-season egg production in a fish. These technologies may contribute significantly to generation of gainful employment to fish farmers and ongoing "aquaplosion" in India.

Introduction

Aquaculture, which is growing more rapidly than all other animal food-producing sectors, has produced 45.7 million metric tons in 2000 (FAO, 2002). India is the second largest producer in freshwater

**Corresponding Author.* E-mail: akray@bic.boseinst.ernet.in

aquaculture in the world having its total production increased from 0.75 million tons in 1951 to 6.1 million metric tons in 2003. Thus aquaculture has a significant contribution in increasing fish production (Dolandson, 1997). The vast potential of aquaculture in India is still under-exploited in water bodies deficient in oxygen such as paddy fields, artificially created ponds in lowland and saline–alkali wasteland. Therefore, a sustainable approach to aquaculture development is need of the hour in developing countries, especially in India.

Biotechnology has been used by man for thousands of years for producing wine by microorganisms and cross-breeding animals to improve their growth. The advent of new biotechnological tools have facilitated examining and manipulating genes, chromosomes, cells, individuals and populations of living organisms. Aquatic biotechnology has developed many technological innovations for modern aquaculture. The fish can be genetically modified through chromosome set manipulation, gene transfer, inter-specific hybridization, etc. Thus, biotechnology has a great potential for creation of brood stock with superior genetic merit. However, biotechnological means not only improve the genetic merit of a genetically modified organism (GMO) but also the endocrine functions of the body. Many hormones are known to have genomic action, and change the expression pattern of certain genes in an organism. As a result endocrine biotechnology can be used for expression of some target genes (*e.g.* related to growth and reproduction), rather than changing the whole genome as such. Thus endocrine biotechnology has a great potential for sex reversal and reproductive manipulation of fish. A synergy between genetic and endocrine biotechnology may perhaps boost-up productivity for a "blue revolution" in India.

In this chapter, two aspects of genetic biotechnology–chromosome manipulation and transgenesis and two aspects of endocrine biotechnology–sex reversal and reproductive manipulation, have been discussed with special reference to their feasibility in the prevalent socio-economic scenario in India (Table 19.1).

Table 19.1: A Comparative Account of Methods Applied in Aquaculture Biotechnology

Methods	*Effect*	*Economic Cost*	*Environmental Impact*
Chromosome manipulation	Polyploidy	low	low
Transgenesis	Expression of desirable gene	high	high
Sex reversal	Monosex	low	low
Reproductive manipulation	Out-of-season spawning	low	low

Chromosome Manipulation

The process of manipulating whole set of chromosomes for production of a polyploid fish is known as chromosome manipulation or chromosome engineering (Thorgaard, 1983). The technique of induced polyploidy has been actively investigated during the last three decades. The polyploid fish may be triploid (3n), tetraploid (4n), pentaploid (5n), hexaploid (6n) and heptaploid (7n) (Pandian and Koteeswaran, 1998). Triploid fish having three sets of con-specific chromosomes, are more useful to aquaculture owing to better survival and growth in comparison to rest of polyploids. Triploids are sterile due to irregular meiotic division during gametogenesis in their impaired gonads. Sterility of triploid fish provides them with some advantages. They can be used to control aquatic vegetation without any possible risk to environment (Cassani and Caton, 1985). The energy that is normally used for spawning may be redirected to growth. The taste of sterile fish may be superior due to higher levels of glycogen (Liao and Chao, 1997). Triploidy has been induced in many species of fish till date by various treatments designed to block the second meiotic division or to prevent extrusion of second polar body of the freshly fertilized egg. The treatments used are temperature, pressure or chemical shocks to the egg (Tiwary *et al.*, 2005). Triploids are identified by various direct methods such as chromosome counting, DNA measurement; and indirect methods like measurement of nuclear and cellular size, counting of nucleoli, microsatellite DNA, electrophoresis of proteins and examination of morphology (Tiwary *et al.*, 2005). Triploids are expected to have higher growth efficiency owing to sterility and impaired gonadal development. However, the growth performance has always been a variable factor between different species. Triploids of mud loach (*Misgurnus mizolepis*), Atlantic salmon (*Salmo salar*) and tilapia (*Oreochromis niloticus* and *O. aureus*) showed no significant changes in growth when compared with their diploid full-sibs (Chang *et al.*, 1993; Kim *et al.*, 1994; Hussain *et al.*, 1995; McGeachy *et al.*, 1995). Triploids of rainbow trout (*Oncorhynchus mykiss*), sunshine bass (*Lepomis gibbosus*), coho salmon (*Oncorhynchus kisutch*) and Atlantic salmon (*Salmo salar*) performed poorly than their diploid counterparts (Solar *et al.*, 1984; Galbreath *et al.*, 1995; Kerby *et al.*, 1995; Withler et. al., 1995; McCarthy *et al.*, 1996). The triploids of tilapia (*Oreochromis aureus*) and European catfish (*Silurus glanis*) exhibited early better growth potential than their normal diploids (Valenti, 1976: Krasznai and Marian, 1986) (Table II). Triploid Channel catfish (*Ictalurus punctatus*) and Nile tilapia (*Oreochromis niloticus*) have been proven to have late better growth potential at the time of gonadal maturation (Wolters *et al.*, 1982; Bramick *et al.*, 1995). Conversely, even long term treatment with anabolic steroid, Estradiol-17β could not improve weight gain in triploid brook trout, *Salvelinus fontinalis* (Schafhauser-Smith and Benfey, 2003).

Table 19.2: Positive Growth in Genetically Modified Fish (GMF)

Species	Method	Authors
Ictalurus punctatus	Triploidy	Wolters *et al.*, 1982
Oreochromis niloticus	Triploidy	Bramick *et al.*, 1995
Heteropneustes fossilis	Triploidy	Tiwary *et al.*, 1997
Oreochromis aureus	Triploidy	Valenti, 1976
Silurus glanis	Triploidy	Krasznai and Marian, 1986
Oncorhynchus kisutch	Transgenesis	Devlin *et al.*, 1995
Cyprinus carpio	Transgenesis	Chen *et al.*, 1993
Salmo salar	Transgenesis	Du *et al.*, 1992
Oncorhynchus mykiss	Transgenesis	Devlin *et al.*, 1994
Esox lucinus	Transgenesis	Gross *et al.*,1992
Oncorhynchus clarki lewisi	Transgenesis	Devlin *et al.*, 1994
Labeo rohita	Transgenesis	Venugopal *et al.*, 2004
Oreochromis niloticus	Transgenesis	Rahman *et al.*, 1998

Indian catfish (*Heteropneustes fossilis*) is a freshwater culture species highly regarded for its food value (22.8 per cent protein, 0.6 per cent fat and 226 mg per cent iron content; Anonymous, 1982). It thrives well in accumulated water in paddy fields and derelict waterbeds. As a result, culture of this species provides gainful employment to rural masses constituting majority of Indian populations. Despite its potential advantages, the culture of catfish is less lucrative to farmers due to its slower growth rate than Indian major carps and non-availability of quality seeds due to very brief spawning season (July-August). An attempt has been made to produce triploids of this species for overcoming the slower growth in normal diploids. Triploidy has been induced in this species by subjecting freshly fertilized eggs to cold shock (4° C) for 30 minutes (Figure 19.1; Tiwary *et al.*, 1997). Triploid catfish was detected by two different methods like assessment of Chromosome number and diameter of erythrocytes. Triploids showed three sets of chromosomes (3n=84) in comparison to two sets (2n=56) in normal diploids. The dimension of diameter of erythrocytes was also significantly increased in triploids (10.2±0.37) in comparison to diploids (6.5±0.31). Moreover, triploids showed better growth potential in comparison to their diploid siblings under controlled laboratory conditions (Figure 19.3). The growth performance was measured by three parameters: standard length, body weight, and condition factor. The standard length showed no significant changes between triploid catfish and its diploid full sib throughout the year. In contrast, body weight exhibited significant increment throughout the year except one period (August-October) in triploid catfish. The body weight gain was much impressive in females than males of triploid catfish (Figure 19.3B). Moreover, the condition factor of triploid catfish showed significant gain in comparison to their diploid siblings throughout the year. Thus, a field trial of this laboratory finding is underway to evaluate the growth of triploids under various environmental variables. There has also been a search for an easy and effective method to screen triploids from diploids in this species. Tiwary *et al.* (1999) have observed significant difference in ratio of nine morphometric characters between diploid and triploid catfish. Moreover, the ratio between standard length and body depth (SL/BD) was effective in separating triploids from diploids. Some

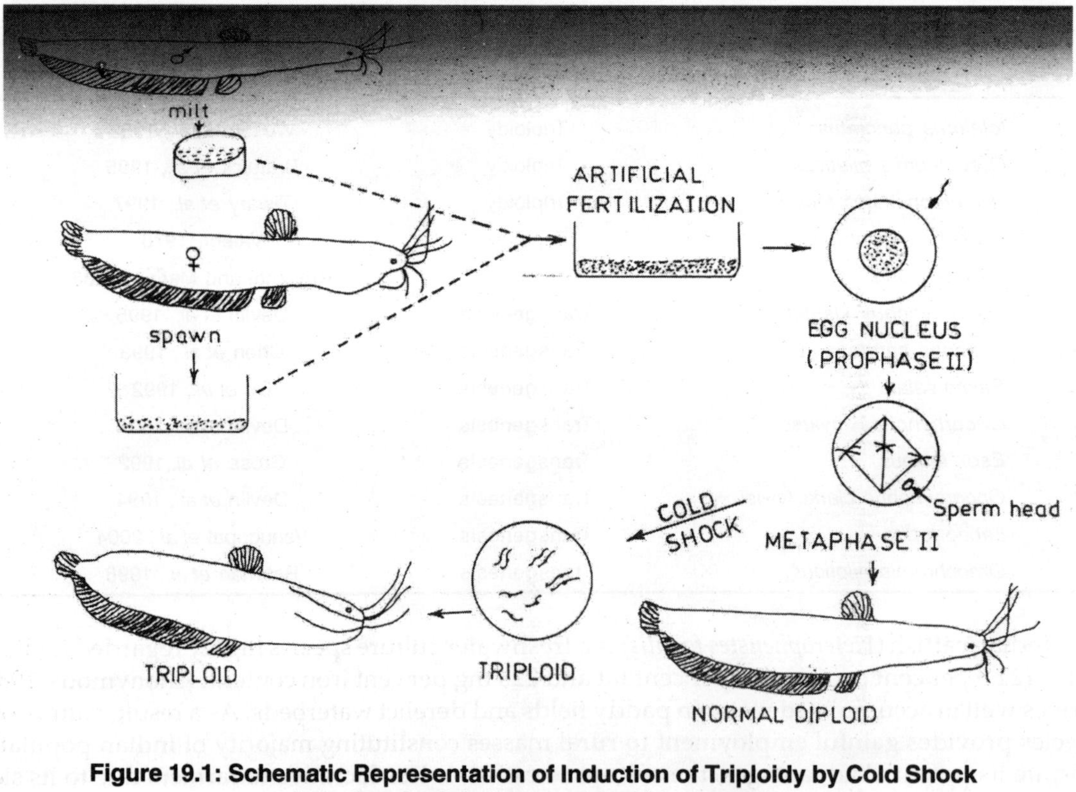

Figure 19.1: Schematic Representation of Induction of Triploidy by Cold Shock in Indian Catfish (*Heteropneustes fossilis*)

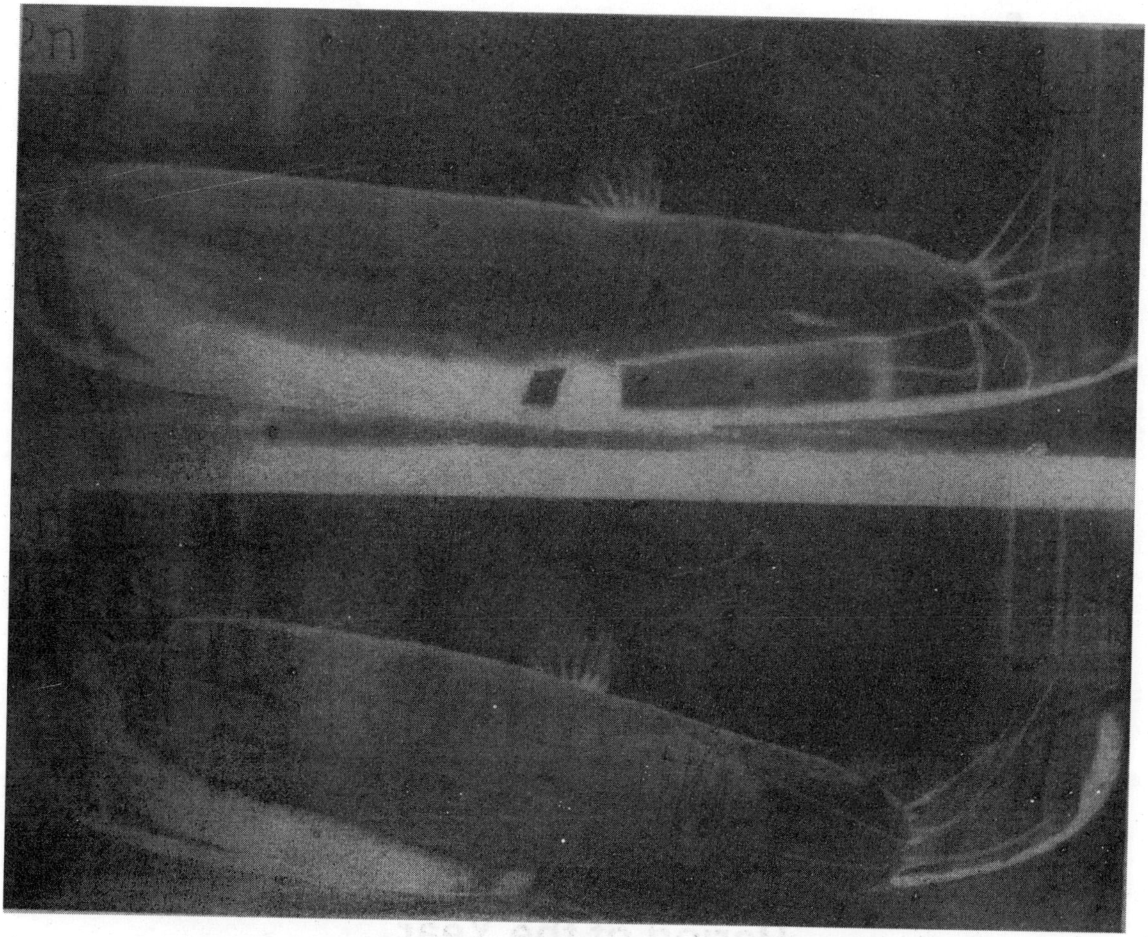

Figure 19.2: Photographs Showing Changes in Growth Patterns in a Triploid Catfish (*H. fossilis*) in Comparison to Normal Diploid Counterpart

endoskeletal changes with alterations in the size of air-sac were also noticed in triploid catfish (Tiwary and Ray, 2004). In addition to these findings, the triploid catfish also exhibited a unique brain-pituiatry-gonad axis, affecting at each level of endocrine hierarchy (Tiwary *et al.*, 2005). The size and number of immunoreactive gonadotropin releasing hormone (GnRH)-cells in the pre-optic area of brain were decreased in triploid catfish resulting in low expression of GnRH decapeptides as well (Tiwary *et al.*, 2002a). Concomitantly, plasma levels of maturational gonadotropin (GtH-II) and sex steroids were also decreased in both sexes of triploids in comparison to their diploid full-sibs (Tiwary *et al.*, 2001). All these adverse effects on endocrine hierarchy of triploid fish led to impaired testicular and ovarian development in males and females respectively (Tiwary *et al.*, 2000). Thus this study on Indian catfish is an example of synergy between genetic and endocrine biotechnology. It is expected that other species of aquacultural interest in India will be chosen for further development of new triploids with better growth potential. Chromosome manipulation also seems to be a cost-effective and eco-friendly

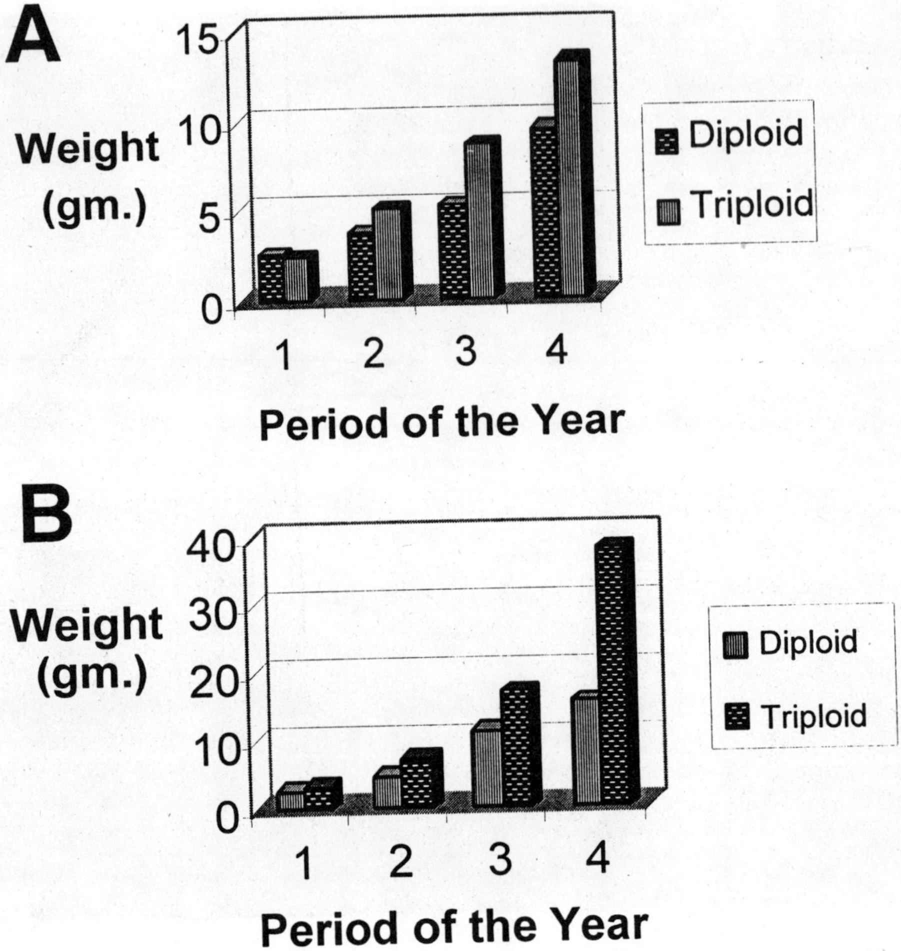

Figure 19.3: The Growth Performance in Terms of Weight Gain in Triploid Males (A) and Females (B) in Comparison to Respective Diploids During Four Periods of a Year. The 1st period indicates August-October, 2nd indicates November-January, 3rd indicates February-April, and 4th indicates May-July months of the year.

technology in comparison to gene transfer (transgenesis). Triploids being sterile pose less risk for genetic pollution of great fish biodiversity of in India.

Thorgaard *et al.* (1981) reported for the first time on successful production of viable tetraploids. Few more teraploids were also produced in seven different species (Pandian and Koteeswaran, 1998). There are reports on production of pentaploids (5n), hexaploids (6n) and heptaploids (7n) in loach (*Misgurnus anguillicaudatus*) (Pandian and Koteeswaran, 1998). Tetraploids are also important for aquaculture because they produce triploids when crossed with diploids.

Transgenesis

Transgenesis is a process in which heterologous DNA is artificially introduced and stably integrated in the genome of an organism. Fish are suitable animal for transgenic study due to large numbers of eggs laid by a single female and external fertilization. The rearing of fish embryo is also simple and cost-effective. Transgenic fish transmit the integrated gene to the next generation in a Mendelian fashion. Experiments with transgenic fish have shown that economically important traits such as enhanced growth rate, disease resistance and increased tolerance to environmental stressors, can be induced through transgenesis. Thus transgenic fish are of great commercial interest owing to their potential to shorten production cycles and increase food production. Several genes have been identified and transferred into different aquatic species (FAO, 2000), designed for human consumption, biopharming, ornamental purposes and industrial applications. Some of the examples are as follows:

1. Growth hormone (GH) gene responsible for increased growth and feed conversion efficiency.
2. Anti-freeze protein (AFP) gene for increased cold tolerance.
3. Lysozyme genes related to generation of immunity from pathogen.
4. Human interferon gene for increased disease resistance.
5. Prolactin (gene), a protein hormone regulating hatching, osmoregulation, behavior and metabolism.
6. Production (biopharming) of human insulin or blood clotting factors in fish (bioreactor) by imposition of the respective genes.

Since 1985, a wide range of transgenic fish has been produced by introducing homologous or heterologous transgenes into unfertilized or newly fertilized eggs through microinjection and electroporation. Different promoter and enhancer constructs have been tested to ensure expression of desirable genes transferred into fish embryo. The success rate in incorporation of microinjected DNA varies as much as, from 1-75 per cent (Chen and Powers, 1990) depending on the fish used and laboratories handling the technique. However, transgenes showed random genome integration with some mosaicism and episomic expression. The use of inducible gene systems, that control temporal and tissue expression and of gene targeting methodologies, based on homologous recombination is desirable to control the expression, efficiency of integration, and locus of transgenes into fish genomes. The use of such system would need further development of stem cell or nuclear transplant technologies in fish (Rocha *et al.*, 2004). Cell-mediated gene transfer has been applied to establish transgenic lines of zebrafish (Fan and Collodi, 2002). Transgenesis in zebrafish was induced by retroviral infection of *in vitro* cultured sperm (Kurita *et al.*, 2004). The particle gun was used effectively to generate transgenic medaka (*Oryzias latipes*) with green fluorescent protein (GFP) (Kinoshita *et al.*, 2003).The enhanced growth performance of transgenic fish has proven to be significant in at least 14 species till date (Aerni, 2004). Transgenic fish showing significant growth enhancement were produced in common carp (*Cyprinus carpio*) (Zhang *et al.*, 1990; Chen *et al.*, 1993), northern pike (*Esox lucinus*) (Gross *et al.*, 1992), Atlantic salmon (*Salmo salar*) (Du *et al.*, 1992), coho salmon (*Oncorhynchus kisutch*), rainbow trout (*Oncorhynchus mykiss*) and cutthroat trout (*Oncorhynchus clarki lewisi*) (Devlin *et al.*, 1994,1995); and tilapia (*Oreochromis niloticus*) (Rahman *et al.*, 1998). The better growth in the transgenics was a variable factor depending upon the fish species and gene constructs used (Sin, 1997). Transgenic Atlantic salmon showed an average two to six fold increase in growth rate. Transgenic coho salmon carrying sockey salmon growth hormone (GH) gene constructs exhibited average growth rate of over 11-fold greater than non-transgenic full-sibs (Devlin *et al.*, 1995). Common carp with rainbow trout

GH-cDNA performed 20-40 per cent better in growth rate than its non-transgenic counterpart (Chen *et al.*, 1993). The transgenic tilapia (*Oreochromis niloticus*) also showed three fold growth enhancement after incorporating Chinook salmon growth hormone (GH) gene with ocean pout anti-freeze promoter. In India, Venugopal *et al.* (2004) generated transgenic rohu (*Labeo rohita*) having endogenous GH-cDNA with two promoters (CMV and beta-actin). The gene transfer efficiency was 25 per cent and growth enhancement in transgenics was 4 to 5.8-fold more than non-transgenics even with less food intake. Transgenic carp revealed more nutritional properties than their non transgenic siblings (Dunham *et al.*, 2002). The total protein with vital amino acids like aspartic acid, Cystine, glutamic acid, histidine, lysine and threonine were significantly higher in transgenic carp. Conversely, total fat and moisture content were significantly low in the same fish (Chatakondi *et al.*, 1995). None of the abovementioned transgenic fish has been approved for aquaculture. Transgenic salmon carrying Chinook salmon GH gene might be the first transgenic animal approved for food in USA. The growth rate of this farmed fish can be increased by four to six-fold with concomitant decrease in food intake by up to 2.5 per cent per unit of output (Hew *et al.*, 1995). There is a growing concern and speculation over the environmental impact regarding introduction of transgenic fish for aquaculture such as competition of transgenic stocks with wild population, introgression of transgene into wild gene pools and increased predation of transgenic fish on prey population. Sundstorm *et al.* (2004) estimated relative fitness of GH-transgenic fish in comparison to non-transgenic conspecifics. They observed that predation pressure on newly hatched transgenic fry was much higher than non-transgenic one and more magnified under less food abundance. Similarly, Devlin *et al.* (2004) noticed dominant behaviour by transgenic coho salmon when reared with their non-transgenic full-sibs. However, they also experienced early population crashes or complete extinctions in comparison to non-transgenics. Thus the environmental concerns of introduction of a transgenic fish become much more relevant in a developing country like India which is endowed with rich biodiversity and less legal constraints. However, the biological risk can be minimized by making transgenic fish sterile through induction of triploidy (Aerni, 2004).

Transgenic fish may be used as a bioreactor for the production of functional recombinant proteins. Goldfish luteinizing hormone (gfLH) has been expressed in four-day old transgenic rainbow trout (Morita *et al.*, 2004). Similarly, human coagulation factor VII (hF VII) was expressed in zebrafish (*Danio rerio*), African catfish (*Clarias gariepinus*) and tilapia with the help of cytomegalovirus (CMV) promoter (Hwang *et al.*, 2004).

Sex Reversal

Sex reversal can be induced in a fish by hormonal treatment to produce monosex population for aquaculture (Hunter and Donadson, 1983; Pandian and Sheela, 1995). Growth is a sexually dimorphic character in fish: males growing bigger than females and vice-versa. The faster growing sex may be desirable for improving productivity in aquaculture. Androgens normally induce male and estrogens induce female sex in fish after treatment. However, androgens after being aromatized to estrogen may induce all-female populations when used at higher doses (paradoxical feminization; Tiwary *et al.*, 1998). 17 α–methyltestosterone (MT), is the most widely used synthetic androgen and other synthetic and natural steroids such as mibolerone, 19-nor-ethynyl-testosterone and testosterone are used for treatment in fish. Similarly, Estradiol-17β is the most favoured estrogen followed by estrone, estriol, diethylstilbesterol (DES). Treatment methods include immersion of embryonic and juvenile stages, through diet, and silastic implantation of the steroid materials. The endocrine treatment through diet is the most accepted mode of administration. The best period for the hormonal treatment is the labile period of a particular species. Therefore, the labile period must be precisely delineated for making the process of sex reversal most effective. Endocrine treatment not only induces complete sex reversal but also sterility when used at higher doses. Sterile fishes produced by hormone treatment showed better survival and growth in *Cyprinus carpio* (Ali and Rao, 1989), *Oncorhynchus tshawytscha* (Barker *et al.*, 1988), *Oreochromis mossambicus* (Guerrero and Guerrero, 1993). The juveniles of *Poecilia sphenops* when treated with 17α-MT showed accelerated growth (Pandian and Sheela, 1995). Common carp (*Cyprinus carpio*) showed two to three-fold positive growth in hormone treated group in comparison to control group (Rao and Rao, 1983). The Indian catfish (*H. fossilis*) has a bigger female than their male counterpart (Dutta Munshi and Hughes, 1992). Thus, production of all-female population will be of great help to aquaculture. Tiwary *et al.* (1998) have demonstrated 100 per cent feminization when hatchlings of this catfish (*H. fossilis*) were treated with testosterone propionate. Therefore, the desirable sex may be induced in various species of aqua cultural interest. As some inducers like diethylstilbesterol (DES) are known endocrine disrupters in animals, treatment must be undertaken in controlled conditions and waste water must be disposed properly. Thus sex reversal mediated by endocrine biotechnology has vast potential for development of aquaculture in India.

Reproductive Manipulation

The removal of seasonal limitations of spawning for year round breeding is a pre-requisite for the full domestication of a fish (Purdom, 1993). Photoperiod and temperature are the most important environmental cues for triggering the gonadal maturation process in fish (Lam, 1982). However, manipulation of these environmental variables is expensive, not always reliable and application is not possible in all locations. Thus an alternative method of endocrine manipulation is being developed to overcome the physiological constraints to gonadal recrudescence. Testosterone exerts a positive feedback on the GnRH secreting system in the brain and thereby triggers precocious maturation in the juvenile rainbow trout (*Oncorhynchus mykiss*, Crim and Evans, 1982). Testosterone together with GnRH analogues was able to induce precocious ovarian maturation in milkfish (*Chanos chanos*, Marte *et al.*, 1988).

Indian catfish (*H. fossilis*) attains gonadal maturity at the end of its first year, when females measure about 12 cm in length (Dutta Munshi and Hughes, 1992). Seeds of this species are only available during a brief period of two months (June-July) of a year. As a result, this fish is less favoured by fish farmers due to non-availability of seeds during major part of a year. Thus an attempt has been made to prepone the spawning season of this catfish with help of endocrine biotechnology. When hatchlings of catfish were exposed to testosterone propionate at the dose of 300μg/litre of water for one month, about 70 per cent of all-female population showed precocious ovarian maturation three months ahead of normal spawning season (Figure 19.4; Tiwary *et al.*, 1998). The endocrine mechanism behind this phenomenon seems to be direct stimulation of GnRH cells in the preoptic area of the brain for elevated plasma levels of GtH-II and sex steroids (Figure 19.5; Tiwary *et al.*, 2002b). Similar studies on other economically important species may help in breaking the seasonal barrier to reproduction for aquaplosion in India. Moreover, the method should be easy, cost-effective and can be used by an innovative farmer. Therefore, controlling the spawning season of various culture species under captivity for out-of-season egg production will have a major impact on intensive fish culture.

Conclusions

All the above-mentioned methods are potentially powerful tools for improving the aquacultural production in India. Among all methods, transgenesis seems to be more effective in dramatically increasing the fish production. However, the process is very costly in terms of instrumentation and manpower; with low efficiency and environmental cost of introduction is not fully assessed. Therefore,

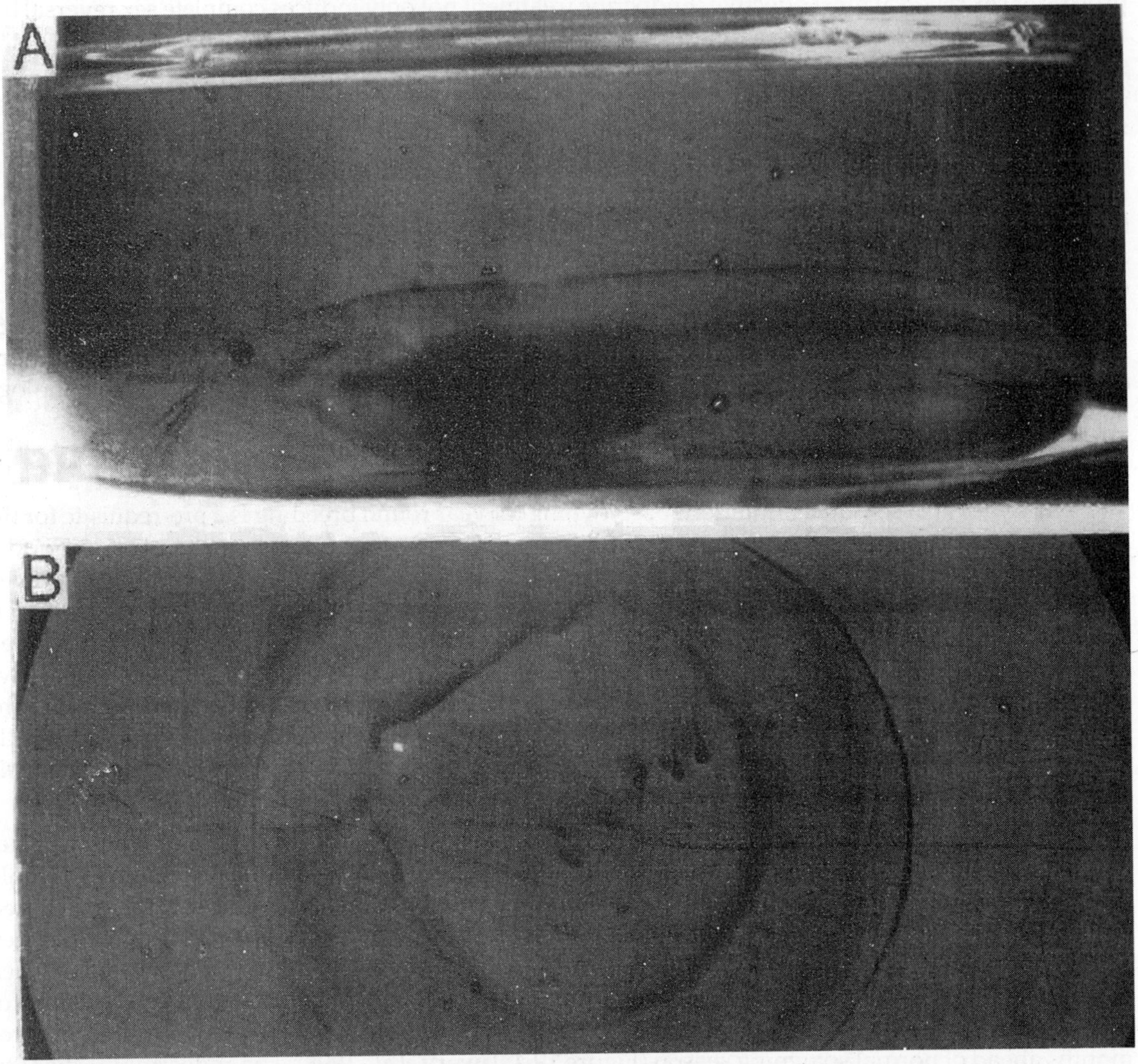

Figure 19.4: The Precociously Matured Female (A) of Indian Catfish (*H. fossilis*) and its Progeny (B)

chromosome manipulation seems to be more feasible method for implementation under prevalent Indian conditions. Similarly, sex reversal and reproductive manipulation are very appropriate technologies for Indian aquaculture. They have enormous potential for production of desirable sex and out-of-season egg production in a fish. Thus the full potential of these technologies on majority of commercial fish is yet to be explored. Thus the genetic and endocrine biotechnology appear as most potential tools for development of aquaculture in Indian conditions and other developing countries as well. It is also expected that these technologies will be utilized for unexplored species of aquaculture importance.

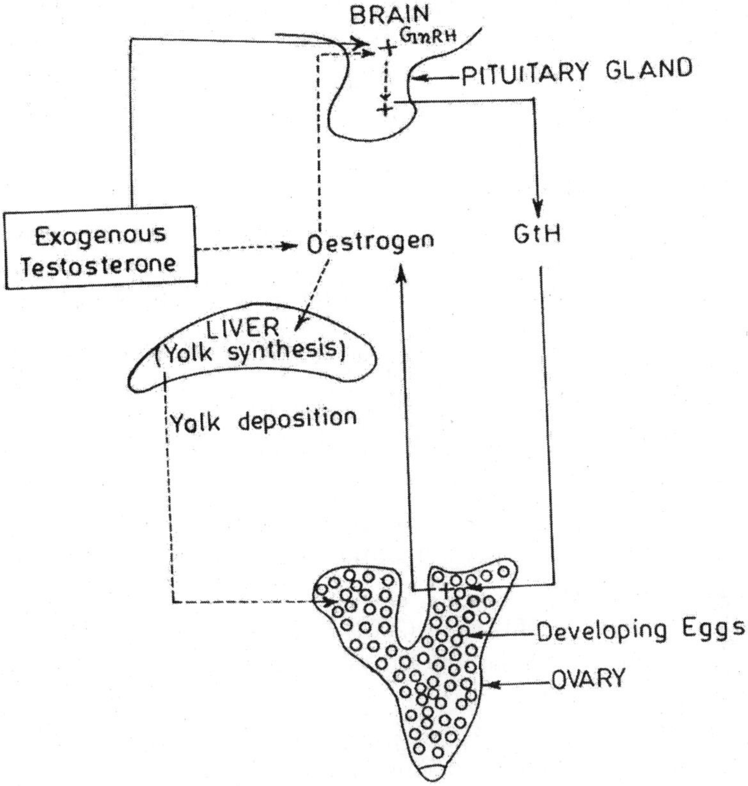

Figure 19.5: Schematic Diagram Showing the Endocrine Mechanism Behind the Precocious Ovarian Maturation

References

Aerni P. (2004). Risk, regulation and innovation: the case of aquaculture and transgenic fish. Aq. Sci., 66, 327-341.

Ali PKMM, and Rao GPS. (1989). Growth improvement in carp, *Cyprinus carpio* (Linnaeus), sterilized with 17α-methyltestosterone. Aquaculture, 76, 157-167.

Anonymous (1982). The Wealth of India, Raw materials, Vol. IV, Supplement Fish and Fisheries, Council of Scientific and Industrial Research, New Delhi, India.

Barker IJ, Solar II, and Dolandson EM. (1988). Musculinization of Chinook salmon (*Oncorhynchus tshawytscha*) by immersion treatments using 17α-methyltestosterone around the time of hatching. Aquaculture, 72, 359-367.

Bramick U, Puckhaber B, Langholz HJ, and Horstgen-Schwark G. (1995). Testing of triploid tilapia (*Oreochromis niloticus*) under tropical pond conditions. Aquaculture, 137, 343-353.

Cassani JR, and Caton WE. (1985). Induced triploidy in grass carp, *Ctenopharyngodon idella* Val. Aquaculture, 46, 37-44.

Chang SL, Chang CF, and Liao IC. (1993). Comparative study on the growth and gonadal development of diploid and triploid tilapia, *Oreochromis aureus*. J. Taiwan Fish. Res., 1, 43-49.

Chatakondi N, Lovell RT, Duncan PL, Hayat M, Chen TT, Powers DA, Weete JD, Cummins K, and Dunham RA. (1995). Body composition of transgenic common carp, *Cyprinus carpio*, containing rainbow trout gene. Aquaculture, 138, 99-109.

Chen TT, and Powers DA. (1990). Transgenic fish. Trends in Biotechnol., 8, 209-215.

Chen TT, Kight K, Lin CM, Powers DA, Hayat M, Chatakondi N, Ramboux AC, Duncan PL, and Dunham RA. (1993). Expression and inheritance of RSVLTR-rt GH 1 complementary DNA in the transgenic common carp, *Cyprinus carpio*. Mol. Marine Biol. Biotechnol., 2, 88-95.

Crim LW, and Evans DM. (1982). Positive testosterone feedback on gonadotrophic hormone in rainbow trout. *In*: Richter CJJ, and Goos HJTh. (Eds.), Proceedings of International Symposium on Reproductive Physiology of Fish, Pudoc, Wageningen, p.23.

Devlin RH, Yesaki TY, Biagi CA, Donaldson EM, Swanson P, and Chan WK. (1994). Extraordinary salmon growth. Nature, 371, 209-210.

Devlin RH, Yesaki TY, Donaldson EM, and Hew, CL. (1995). Transmission and phenotypic effects of an antifreeze/GH gene construct in coho salmon (*Oncorhynhus kisutch*). Aquaculture, 137, 161-169.

Devlin RH, D'Andrade M, Uh M, and Biagi CA. (2004). Population effects of growth hormone transgenic coho salmon depend on food availability and genotype by environment interactions. Proc. Natl. Acad. Sci. USA, 101., 9303-9308.

Donaldson EM. (1997). The role of biotechnology in sustainable aquaculture. *In*: Bardach JE. (Ed.), *Sustainable Aquaculture*, John Wiley and Sons, New York,.

Du SJ, Gong Z, Fletcher GL, Shears MA, King MJ, Idler DR, and Hew CL. (1992). Growth enhancement in transgenic Atlantic salmon by the use of 'all fish' chimeric growth hormone gene construct. Biotechniques, 10, 176-180.

Dunham RA, Chatakondi N, Nichols AJ, Kucuktas H, Chen TT, Powers DA, Weete JD, Cummins K, and Lovell RT. (2002). Effect of rainbow trout growth hormone complementary DNA on body shape, carcass yield, and body composition of F1 and F2 transgenic common carp (*Cyprinus carpio*).Marine Biotechnol., 4, 604-611.

Dutta Munshi JS, and Hughes GM. (1992). *Air-Breathing Fishes of India, Their Structure, Function and Life History*. Oxford and Indian Book House Publishing Company Private Limited, New Delhi.

Fan L, and Collodi P. (2002). Progress towards cell-mediated gene transfer in zebrafish. Functional Genomics and Proteomics, 1, 131-138.

FAO (2000). World Review of Fisheries and Aquaculture, Working document, FAO, Rome

FAOSTAT (2002). Fisheries, 2002, http://faostat.fao.org/faostat/collections?subset= fisheries

Foresti F. (2000). Biotechnology and Fish culture. Hydrobiologia, 420, 45-47.

Galbreath PF, and Thorgaard GH. (1995). Salt-water performance of all female triploid Atlantic salmon. Aquaculture, 138, 77-85.

Gross ML, Schneider JF, Moav N, Moav B, Alvarez C, Myster SH, Liu Z, Hallerman EM, Hackett PB, Guise KS, Ferae AJ, and Kapiscinski AR. (1992). Molecular analysis and growth evaluation of northern pike (*Esox lucius*) microinjected with growth hormone genes. Aquaculture, 103, 253-273.

Guerrero RD, and Guerrero IA. (1993). Effect of oral treatment of mibolerone on sex reversal of *Oreochromis mossambicus*. Asian Fish. Sci., 6, 347-350.

Hew CL, Fletcher GL, and Davis PL. (1995). Transgenic salmon: tailoring the genome for food market. J. Fish Biol., 47 (Suppl. A), 1-19.

Hunter GA, and Donaldson EM. (1983). Hormonal sex control and its application to fish culture. *In*: Hoar WS, Randall DJ, and Donaldson EM. (Eds.), *Fish Physiology*, Vol. IX B, Academic Press, New York, pp. 223-303.

Hussain MG, Rao GPS, Humayun NM, Randall CF, Penman DJ, Kime D, Bromage NR, Myers JM, and McAndrew BJ. (1995). Comparative performance of growth, biochemical composition and endocrine profiles in diploid and triploid tilapia, *Oreochromis niloticus* L. Aquaculture, 138, 87-95.

Hwang G, Muller F, Rahman MA, Williams DW, Murdock PJ, Pasi KJ, Goldspink G, Farahmand H, and Maclean N. (2004). Fish as Bioreactors: Transgenic expression of Human coagulation factor VII in fish embryos. Marine Biotechnol. 6, 485-492.

Johnstone R, Simpson TH, and Youngson AF. (1978). Sex reversal in salmonid culture. Aquaculture, 3, 115-134.

Kerby JH, Eversion JM, Harrell RM, Starling CC, Ravels H, and Geiger JG. (1995). Growth and survival comparisons between diploid and triploid sunshine bass. Aquaculture, 137, 355-358.

Kim DS, Jo YY, and Lee TY. (1994). Induction of triploidy in mud loach (*Misgurnus mizolepis*) and its effect on gonad development and growth. Aquaculture, 120, 263-270.

Kinoshita M, Yamayuchi M, Sasanuma M, Ishikawa Y, Osada T, Inoue K, Wakamatsu Y, and Ozato K. (2003). A transgene and its expression profile are stably transmitted to offspring in transgenic medaka generated by the particle gun method. Zool. Sci., 20, 869-875.

Krasznai Z, and Marian T. (1986). Shock induced triploidy and its effect growth and gonadal development of the European catfish *Silurus glanis* L. J. Fish Biol., 29, 519-527.

Kurita K, Burgess SM, and Sakai N. (2004). Transgenic zebrafish produced by retroviral infection of in vitro-cultured sperm. Proc. Natl. Acad. Sci. USA, 101, 1263-1267.

Lam TJ. (1982). Application of endocrinology to fish culture. Can. J. Fish. Aquatic Sci., 39, 111-137.

Liao IC, and Chao NH. (1997). Developments in aquaculture biotechnology in Taiwan. J. Marine Biotechnol., 5, 16-23.

McCarthy ID, Carter CG, Houlihan DF, Johnstone R, and Mitchell AI. (1996). The performance of all female diploid and triploid Atlantic salmon smolts on transfer together to sea water. J. Fish Biol., 48, 545-548.

McGeachy SA, Benfey TJ, and Friars GW. (1995). Freshwater performance of triploid Atlantic salmon (*Salmo salar*) in New Brunswick aquaculture. Aquaculture, 137, 333-341.

Marte CL, Crim LW, and Sherwood NM. (1988). Induced gonadal maturation and rematuration in milkfish: limited success with chronic administration of testosterone and gonadotrophin-releasing hormone analogues (GnRH-A). Aquaculture, 74, 131-145.

Morita T, Yoshizaki G, Kobayashi M, Watabe S, Takeuchi T. (2004). Fish eggs as bioreactors: the production of bioactive luteinizing hormone in transgenic trout embryos. Transgenic Res., 13, 551-557.

Nanjundappa T, and Varghese TJ. (1988). Effect of diethylstilbesterol on growth and food conversion of common carp, *Cyprinus carpio* (Linn.). Proc. Indian Acad. Sci., B, 98, 85-88.

Pandian TJ, and Sheela SG. (1995). Hormonal induction of sex reversal in fish. Aquaculture, 138, 1-22.

Pandian TJ, and Koteeswaran R. (1998). Ploidy induction and sex control in fish. Hydrobiologia, 384, 167-243.

Purdom CE. (1993). *Genetics and Fish Breeding*, Chapman and Hall, London.

Rahman MA, Mak R, Ayad H, Smith A, and Maclean N. (1998). Expression of a novel piscine growth hormone gene results in growth enhancement in transgenic tilapia (*Oreochromis niloticus*). Transgenic Res., 7, 357-370.

Rao HNS, and Rao GPS. (1983). Hormonal manipulation of sex in the common carp, *Cyprinus carpio* var. *communis* (L.). Aquaculture, 35, 83-88.

Rocha A, Ruiz S, Estepa A, and Coll JM. (2004). Application of inducible and targeted gene strategies to produce transgenic fish: A review. Marine Biotechnol., 6, 118-127.

Schafhauser-Smith D, and Benfey TJ. (2003). The effects of long term estradiol 17 beta treatment on the growth and physiology of female triploid brook trout (*Salvelinus fontinalis*). Gen. Comp. Endocrinol., 131, 9-20.

Sin FYT. (1997). Transgenic fish. Reviews in Fish Biology and Fisheries, 7, 417-441.

Solar II, Donaldson EM, and Hunter GA. (1984). Induction of triploidy in rainbow trout (*Salmo gairdneri* Richardson) by heat shock and investigations on early growth. Aquaculture, 42, 57-67.

Sundstorm LF, Lohmus M, Johnsson JI, Devlin RH. (2004). Growth hormone transgenic salmon pay for growth potential with increased predation mortality. Proc. Royal Soc. London B, 271 (Suppl. 5), S350-352.

Thorgaard GH. (1983). Chromosome set manipulation and sex control in fish. *In*: Hoar WS, Randall DJ, and Donaldson EM. (Eds.), *Fish Physiology*, Vol. IX, Part B, Academic Press, New York, pp. 405-434.

Thorgaard GH, Jazwin ME, and Stier AR. (1981). Polyploidy induced by heat shock in rainbow trout. Trans. Am. Fish. Soc., 110, 546-550.

Tiwary BK, and Ray AK. (2004). Alterations in air sac and skeleton of triploid *Heteropneustes fossilis*. J. Fish Biol., 64, 268-272.

Tiwary BK, Kirubagaran R, and Ray AK. (1997). Induction of triploidy by cold shock in catfish, *Heteropneustes fossilis* (Bloch). Asian Fisheries Science,10, 123-129.

Tiwary BK, Kirubagaran R, and Ray AK. (1998). Testosterone-mediated complete feminization with precocious ovarian maturation in catfish, *Heteropneustes fossilis* (Bloch). Aquaculture Res., 29, 333-336

Tiwary BK, Kirubagaran R, and Ray AK. (1999). Altered body shape as a morphometric indicator of triploidy in Indian catfish, *Heteropneustes fossilis* (Bloch). Aquaculture Res., 30, 907-910.

Tiwary BK, Kirubagaran R, and Ray AK. (2000). Gonadal development in triploid *Heteropneustes fossilis*. J. Fish Biol., 57, 1343-1348.

Tiwary BK, Kirubagaran R, and Ray AK. (2001). Plasma levels of Gonadotropin-II and gonadal sex steroids in triploid catfish, *Heteropneustes fossilis* (Bloch). Fish Physiol. Biochem., 24, 9-14.

Tiwary BK, Kirubagaran R, and Ray, AK. (2002a). Gonadotropin releasing hormone (GnRH) neurones of triploid catfish, *Heteropneustes fossilis* (Bloch): an immunocytochemical study. Comp. Bioch. Physiol., 132A, 375-380.

Tiwary BK, Kirubagaran R, and Ray AK. (2002b). Testosterone triggers the brain-pituitary-gonad axis of juvenile female catfish (*Heteropneustes fossilis* Bloch) for precocious ovarian maturation. Gen. Comp. Endocrinol., 126, 23-29.

Tiwary BK, Kirubagaran R, and Ray AK. (2005). The biology of triploid fish. Rev. Fish Biol. Fish. (in press).

Valenti RJ. (1976). Induced polyploidy in *Tilapia aurea* (Steindachner) by means of temperature shock treatments. J. Fish Biol., 7, 519-528.

Venugopal T, Anathy V, Kirankumar S, and Pandian TJ. (2004). Growth enhancement and food conversion efficiency of transgenic fish *Labeo rohita*. J. Exptl. Zool. A, 301, 477-490.

Withler RE, Beacham TD, Solar II, and Donaldson EM. (1995). Freshwater growth, smolting and marine survival and growth of diploid and triploid coho salmon (*Oncorhynchus kisutch*). Aquaculture, 136, 91-107.

Wolters WR, Libey GS, and Chrisman CL. (1982b). Effects of triploidy on growth and gonadal development of channel catfish. Trans. Am. Fish. Soc., 111, 102-105.

Zhang P, Hayat P, Joycee C, Gonzalez-villasener LI, Lin CM, Dunham RA, Chen TT, and Powers DA. (1990). Gene transfer, expression and inheritance of pRSV-rainbow trout-GH in the common carp, *Cyprinus carpio* (Linn.). Mol. Reprod. Dev., 25, 3-13.

— Section D —
Hormones and Stress

20

Oxidative Stress and the Thyroid Gland: The Role of Melatonin as a Potential Antioxidant

Andrzej Lewinski[*1] *and Malgorzata Karbownik*[2]

[1]*Department of Endocrinology and Metabolic Diseases, Medical University of Lodz; Poland*

[2]*Polish Mother's Memorial Hospital-Research Institute, 281/289, Rzgowska St., 93-338 Lodz, Poland*

ABSTRACT

Free radicals and reactive oxygen species participate in physiological and pathological processes in the thyroid gland. For example, the role of hydrogen peroxide (H_2O_2) is crucial for thyroid hormone biosynthesis. Also other free radicals or reactive species, formed from iodine or tyrosine residues, participate in thyroid hormone synthesis. It is to be emphasized that under physiological conditions, a balance exists between generation and detoxification of free radicals. Moreover, much evidence has been accumulated showing that thyroid diseases, *e.g.*, Graves' disease, non-toxic goitre or thyroid cancer, are accompanied by enhanced oxidative stress. The presence of some antioxidants has been documented in the thyroid gland. In turn, melatonin (N-acetyl-5-methoxytryptamine)–the main secretory product of the pineal gland–is a well-known antioxidant and free radical scavenger, widely distributed in the organism. The relationship between the pineal gland and the thyroid gland has been a subject of intensive research. The abundant to-date's evidence relates mostly to the inhibitory action of melatonin on the thyroid growth processes and secretion. In experimental models on oxidative stress, protective effects

**Corresponding Author*: E-mail: alewin@csk.am.lodz.pl; Phone: 48 42 2711715; Fax: 48 42 2711343

of melatonin against oxidative damage to macromolecules, caused by potential carcinogens, have been documented. It has been shown that melatonin protected against lipid peroxidation, caused by ferrous ions plus hydrogen peroxide (Fenton reaction) *in vitro*, and against lipid peroxidation caused by potassium bromate in porcine or rat thyroid gland, both *in vivo* and *in vitro* Melatonin was also shown to reduce indices of oxidative damage to lipids in different tissues due to either L-thyroxine or iodide treatment. Thus, melatonin seems to be involved in the regulation of oxidative processes in the thyroid.

Keywords: The thyroid gland, Oxidative stress, Free radicals, Antioxidants, Thyroid hormone synthesis, Melatonin.

Introduction

The most basic reaction of oxidative stress is Fenton reaction: $Fe^{2+} + H_2O_2 \rightarrow Fe^{3+} + .OH + OH^-$. Hydroxyl radical (.OH), produced in this reaction, is the most toxic free radical., It is to be recalled that reactive oxygen species (ROS) and free radicals are formed in living organisms–in certain amounts–under physiological conditions. An overproduction of ROS and free radicals results in oxidative stress and can lead to several diseases (Dreher and Junod, 1996; Dröge, 2002).

Melatonin (N-acetyl-5-methoxytryptamine) is produced in the pineal gland, as well as in numerous other tissues and organs (Kvetnoy, 2002). The indoleamine possesses properties of a hormone but also demonstrates other numerous characteristics, allowing its action in cellular compartments of all tissues and organs (Tan *et al.*, 2003). Melatonin is a well-known antioxidant and free radical scavenger, revealing preventive action against oxidative damage in different tissues (Reiter, 1998; Reiter *et al.*, 2000; Hardeland *et al.*, 2001c, 2003; Karbownik and Reiter 2000, 2002).

Pineal-thyroid Relationship

In different animal species, a suppressive effect of melatonin on thyroid growth processes (Lewinski *et al.*, 1984; 1993; 1994; Lewinski and Sewerynek, 1986; Wajs and Lewinski, 1992) and thyroid function (Krotewicz *et al.*, 1992; Krotewicz and Lewinski, 1994; Wright *et al.*, 2000) was shown. In contrast, thyroid hormones reveal mainly stimulatory effects on the pineal, as regards growth and secretory processes of this gland (Lewinski *et al.*, 2002). Experimental evidence on the mutual relations between the thyroid and the pineal gland, also in relation to oxidative stress, has been summarized in numerous review papers (Lewinski, 1990; Lewinski *et al.*, 1996, 1997, 2002; Lewinski and Karbownik, 2002; Karbownik and Lewinski, 2003a).

Experimental Evidence for the Role of Oxidative Stress in the Thyroid Gland

Hydrogen peroxide (H_2O_2)

Hydrogen peroxide is produced in the thyroid gland (Bjorkman and Ekholm, 1984; Raspe and Dumont, 1995) by the NADPH oxidase system of the apical membrane of thyroid follicular cells (thyrocytes) (Dupuy *et al.*, 1999; De Deken *et al.*, 2000). Divalent reduction of oxygen, without superoxide anion radical (O_2^{-}) generation, is involved in this process (Sugawara *et al.*, 2002).

NADPH oxidase is encoded by two genes: human ThOX1 (hThOX1) and human ThOX2 (hThOX2); expression of both genes has been confirmed in thyroid follicular cells. The chromosomal organization of these two genes and the functional characterization of their respective promoter regions have recently been presented (Pachucki *et al.*, 2004). Human ThOX1 gene is composed of 35 exons, and hThOX2 gene–of 34 exons. The promoters of both genes are different from each other and from promoters of other known thyroid-specific genes. The activity of hThOX promoter has been found not restricted to thyroid cells (Pachucki *et al.*, 2004), which is in accordance with the fact that NADPH oxidase is involved in physiological processes in different tissues and cells. Additionally, it is known that both ThOX1 and hThOX2 promoters are not positively controlled by cAMP (Pachucki *et al.*, 2004).

H_2O_2 is an essential factor for thyroid hormone synthesis, acting as an electron acceptor at each step of this process, namely at iodide oxidation and, next, its organification, as well as coupling reaction of iodotyrosines (Karbownik and Lewinski, 2003a). This reactive species is crucial for thyroid peroxidase (TPO) activity. TPO is the key enzyme for thyroid hormone synthesis. It is a heme-dependent protein, thus it contains iron atoms in its structure. On one hand, the increased iron stores in the organism are associated with an increased risk of several diseases, cancer included (Dreher and Junod, 1996; Eaton and Qian, 2002). However, iron is an essential element for metabolism in different tissues and organs, also in the thyroid. It has been shown that an iron deficiency reduces TPO activity, enhances the consequences of iodine deficit, while iron supplementation improves the efficacy of iodine supplementation (Zimmermann and Köhrle, 2002).

Beside the main three steps of thyroid hormone synthesis, H_2O_2 also participates in autocatalytic covalent heme binding to the apoprotein of TPO molecule, thereby stabilizing the activity of the enzyme (Fayadat *et al.*, 1999). It has also been found that ROS and/or free radicals decrease TPO activity in the thyroid gland, without influencing TPOmRNA. Sugawara *et al.* (2002) have speculated that TPO inactivation occurs at the heme-linked histidine residue of the enzyme molecule; this residue plays a critical role for TPO activity due to the presence of heme-derived iron, which–potentially–constitutes a substrate for Fenton reaction.

Thus, while H_2O_2 is necessary for thyroid hormone synthesis, this ROS–when being in excess–may inhibit TPO activity, with a subsequent inhibition of thyroid hormone formation (Sugawara *et al.*, 2002). There are presumptions that oxidative stress may be involved in the pathomechanism of hypothyroidism and goitre formation. Using dog thyroid cells, it has been found that thyrotropin (TSH)–the main hormone stimulating secretory and growth processes in the thyroid–enhances H_2O_2 generation through the cyclic adenosine 3′,5′-monophosphate (cAMP) cascade (Raspe and Dumont 1995). Thus, in any conditions of the increased blood TSH concentration–mainly in case of hypothyroidism and in chronic stimulation of the thyroid gland due to iodine deficiency–the increased synthesis of H_2O_2 must take place with the subsequent enhanced formation of free radicals (especially of.OH).

It is suggested that H_2O_2 participates in Wolff-Chaikoff 's effect, *i.e.* in a phenomenon which relies on the fact that iodides–when in excess–are able to inhibit the process of iodine organification and the subsequent thyroid hormone synthesis. Iodides were shown to strongly inhibit both protein iodination and H_2O_2 generation stimulated by TSH in dog thyroid slices (Corvailain *et al.*, 1988).

Nowadays, it is believed that both apoptosis and necrosis occur in the thyroid gland *via* the mechanism involving H_2O_2, the former process resulting from lower, and the latter–from higher concentration of this species *in vitro* (Riou *et al.*, 1998). Because melatonin directly neutralizes H_2O_2 (Tan *et al.*, 2000) and is the most effective scavenger of.OH (Tan *et al.*, 1993), the indoleamine might prevent pathological processes in the thyroid caused by an excessive amount of H_2O_2.

Reactive Nitrogen Species

Nitric oxide synthase (NOS) catalyzes the formation of free radical–nitric oxide (NO·). An expression of mRNA for the three isoforms of NOS–brain (type I), endothelial (type III), and inducible (type II)–has been detected in the rat thyroid gland (Esteves *et al.*, 1992; Millatt *et al.*, 1993; Colin *et al.*,

1995). Nitric oxide has been found to inhibit cell proliferation in cultured human thyrocytes (Motohashi *et al.*, 1996).

Some data suggest that NO· participates in the regulation of thyroid hormone synthesis. While NO· inhibits TSH-stimulated iodide uptake, probably by stimulation of guanylyl cyclase (GC) activity and cyclic GMP (cGMP) production in calf thyroid (Bocanera *et al.*, 1997), this radical has been found to stimulate TPO activity in monolayer cultures of primary human thyrocytes (Millatt *et al.*, 1998). It is known that melatonin directly neutralizes NO· (Noda *et al.*, 1999) and peroxynitrite anion ($ONOO^-$) (Gilad *et al.*, 1997), the latter being a highly destructive product of the interaction between the $O_2^{\cdot-}$ and NO· Additionally, melatonin has been demonstrated to inhibit the activity of NOS (Reiter, 1998). There is also an indirect evidence that melatonin detoxifies $O_2^{\cdot-}$ (Brömme *et al.*, 1999).

Other Free Radicals

Numerous free radicals and reactive species are undoubtedly involved in the complex process of thyroid hormone synthesis. According to experimental evidence and some other suggestions, the following radicals or reactive species are of special importance: tyrosine free radical (Tyr·), diiodotyrosyl residue radical (DIT·), diiodotyrosyl residue radical in thyroglobulin (Tg-DIT·), iodine radical (I·), iodinium ion (I^+), hypoiodous acid intermediate (IO^-(IOH)), and ascorbate radical (Asc·) (Taurog, 2000).

Antioxidants and Antioxidative Enzymes in the Thyroid Gland

An antioxidative defence system in the thyroid gland comprises both antioxidative enzymes and free radical scavengers. The presence of the following antioxidative enzymes in the thyroid gland has been documented: superoxide dismutase (SOD) (Sugawara *et al.*, 1988; Iwase *et al.*, 1993), glutathione peroxidase (GSH-Px) (Sadani and Nadkarni 1996) and catalase (CAT) (Rhee 1999). Furthermore, a relatively high expression of mRNA encoding for microsomal glutathione S-transferase has been found in the murine thyroid gland (Kelner *et al.*, 2004). It should be stressed here that melatonin has been documented to stimulate the activity of SOD, GSH-Px, glutathione S-transferase, and CAT (reviewed in Reiter 1998; Reiter *et al.*, 2000).

Two enzymes out of mentioned above–GSH-Px and glutathione S-transferase, participate in the synthesis of glutathione (GSH), an intracellular antioxidant, the presence of which has been documented in human thyroid gland (Mano *et al.*, 1998). Additionally, such antioxidants as α- and γ-tocopherols, and coenzyme Q have also been found in human thyroid (Mano *et al.*, 1998), and ascorbic acid has been demonstrated in the hog thyroid (Nakamura and Ohtaki, 1993). As mentioned earlier, ascorbic acid becomes a free radical itself in the process of scavenging free radicals in the thyroid gland (Nakamura and Ohtaki, 1993).

Peroxiredoxins (Prxs), antioxidative proteins involved in the regulation of cell differentiation and proliferation, have been found in human thyroid follicular cells and in FRTL-5 cells (Kim *et al.*, 2000). Using FRTL-5 cells, it has been documented that Prxs are involved in the process of H_2O_2 elimination, when this oxygen species is formed in response to TSH, and that they protect thyroid cells from H_2O_2-induced apoptosis (Kim *et al.*, 2000).

A positive immunostaining with antibodies against melatonin has been found in C cells in the rat thyroid gland (Kvetnoy and Yuzhakow, 1993; Kvetnoy, 2002). It is not clear till now whether melatonin is produced in thyroid follicular cells. However, because melatonin is able to reach any tissues or organs within very short time (Reiter, 1998; Reiter *et al.*, 2000), it is highly probable that the indoleamine is also available in the thyroid gland, when it is required to reveal antioxidative effects.

Oxidative Damage in the Thyroid Gland

The results obtained in humans and in animal models suggest that oxidative damage in the thyroid is accompanied by the increased activities of antioxidative enzymes or increased production of antioxidants, what–probably–represents the defence mechanism (Iwase *et al.*, 1993; Sadani and Nadkarni,1996; Mano *et al.*, 1997).

In case of human thyroid diseases–non-toxic nodular goitre, carcinomas (follicular and papillary) and follicular adenoma–the highest level of lipid peroxidation products was found in carcinomas and it was also increased in follicular adenoma, when compared to control tissue and nodular goitre. Those changes were accompanied by increased activities of antioxidative enzymes, such as SOD, GSH-Px, and CAT, especially in case of carcinomas (Sadani and Nadkarni, 1996).

A much higher amount of Prx I, measured by immunoblotting, has been found in follicular thyroid adenomas and carcinomas than in histopathologically unchanged thyroid tissue; additionally, a significantly higher amount of Prx I has been found in thyroids collected from patients with Graves' disease than in normal thyroid tissue (Yanagawa *et al.*, 1999).

The presence of immunochemical staining of SOD type II (inducible), and of mRNA for this enzyme, was found in the human papillary thyroid carcinoma but not in normal human thyroid tissue (Kitano *et al.*, 1999). On the contrary, a decreased expression of mRNA encoding for CAT, and for copper and zinc SOD were found in anaplastic thyroid carcinoma, when compared to histopathologically unchanged thyroid tissue and to differentiated thyroid tumors (Hasegawa *et al.*, 2003). Thus, in case of advanced neoplastic tumors or, other serious diseases of the thyroid gland, numerous free radicals are generated; that process is followed by an increased production of antioxidants or increased activities of antioxidative enzymes. However, at extremely advanced stages of such diseases, like *e.g.* anaplastic carcinoma, the defence mechanisms remain inactive.

Regarding the protective effects of melatonin against oxidative stress in the thyroid gland, experimental data–unfortunately–have been very scarce. We have recently confirmed a protective action of melatonin against lipid peroxidation in the porcine thyroid gland. Because iron is present in TPO and H_2O_2 is essential for TPO activity, the thyroid gland–under pathological conditions–may be exposed to excessive amounts of either Fe^{2+} or H_2O_2, or both. Ferrous iron and H_2O_2 constitute substrates for Fenton reaction. Using an *in vitro* model of Fenton reaction, we have induced oxidative damage to lipids in homogenates of porcine thyroids; we have found that Fe^{2+} plus H_2O_2 increased–in the Fe^{2+} concentration dependent manner–the level of lipid peroxidation products (malondialdehyde + 4-hydroxyalkenals (MDA+4-HDA)); at the same time, melatonin–in a concentration-dependent manner–decreased the lipid peroxidation induced by Fenton reaction, with the lowest effective concentration of 0.25 mM (Karbownik and Lewinski, 2003b). We would like to stress that the above study is the first one, showing lipid peroxidation induction by Fenton reaction and the protective influence of melatonin against oxidative damage in the thyroid tissue.

In another experimental model, we induced lipid peroxidation in the thyroid gland, both *in vivo* and *in vitro*, using potassium bromate ($KBrO_3$)–a carcinogenic agent. Injection with $KBrO_3$ caused the increase of lipid peroxidation in the rat thyroid gland, which was completely prevented by melatonin, indole-3-propionic acid (IPA) and by PTU (Karbownik *et al.*, 2005). Unfortunately, lipid peroxidation caused by $KBrO_3$ in porcine thyroid homogenates was prevented–in concentration-dependent manner–only by PTU, but not by melatonin or IPA (Karbownik *et al.*, 2005). However, from the three potential antioxidants studied, melatonin is the only one, which could be possibly used for the prevention of cancer initiation.

Thyroid Enlargement: The Role of Oxidative Stress

Several experimental animal data, as well as results of studies in humans, support the hypothesis that oxidative stress is involved in goitre formation. It is supposed that NO· participates in the control of vascularization during goitre formation. In animal model of thiouracil- and low iodine diet-induced goitre, the increased vascularization of the gland was accompanied by an induction of genes encoding for NOS I and NOS III; during goitre involution, the expression for these genes returned to basal values (Colin *et al.*, 1995).

In another study, low iodine diet resulted–as expected–in an increased thyroid weight, increased DNA and protein content, and an increased 3H-thymidine incorporation into DNA of thyroid follicular cells; all those changes were reduced by an administration of an antioxidant–vitamin E (Mutaku *et al.*, 1998). It is worth stressing that the changes in parameters of growth processes after vitamin E treatment were not accompanied by changes in either TSH or thyroid hormone concentrations (Mutaku *et al.*, 1998). These results suggest that vitamin E has a direct antigoitrogenic effect. Further studies performed by the same research group, support the above suggestion. The Authors have observed that vitamin E deficiency resulted in the significant increase in epithelial cell necrosis during goitre development and involution, and in the increased lipid peroxidation and decreased GSH-Px activity in the thyroid (Mutaku *et al.*, 2002).

Studies in humans suggest that iodine deficiency is associated with an increased oxidative stress and decreased antioxidative defence. A decreased SOD activity was observed in endemic goitre tissue (Sugawara *et al.*, 1988). In iodine-deficient children with goitre, the activities of antioxidative enzymes (GSH-Px, CAT and SOD) in erythrocytes, and selenium concentration in plasma and erythrocytes, have been found to be significantly lower than their respective values in non-goitre and non-iodine-deficient control subjects, and even–what is most important–than in non-goitre but iodine-deficient children (Giray *et al.*, 2001). Thus, it seems plausible that low activities of antioxidant enzymes and low selenium concentrations contribute to thyroid enlargement in the iodine-deficient subjects.

In another study, increased damage to genomic DNA (evaluated by the measurement of the concentration of different oxidized bases in peripheral blood) was found in children (aged 15-18 years) with goitre, living in the area of severe or moderate iodine deficiency; DNA damage was accompanied by the decreased blood concentration of free thyroid hormones, decreased concentration of selenium, decreased activity of GSH-Px and SOD in blood erythrocytes, and by decreased ioduria (Giray and Hincal, 2002).

The observation that iodine deficiency resulting in goitre leads to oxidative DNA damage, supports the view on an essential relationship between iodine deficiency and the increased incidence of follicular thyroid carcinoma. It has recently been found that melatonin concentration, measured at night, was significantly higher after the operation performed in patients with very large non-toxic nodular goitre than before the surgery (Karasek *et al.*, 2000). These results, for some reasons–unexpected, could be explained as follows: melatonin might actively be taken up by an enlarged thyroid with a subsequent decrease in blood concentration of the indoleamine before surgery.

Autoantigens of the Thyroid Gland and Oxidative Stress

Thyroglobulin (Tg) and TPO belong to the main thyroid autoantigens. Thyroglobulin is stored extracellularly in the thyroid follicle lumen and is essential for thyroid hormone biosynthesis. During the thyroid hormone synthesis resulting from several oxidative reactions, the multimerized inactive form of Tg (mTg) is produced (Delom *et al.*, 1999). It has recently been found that experimentally-induced oxidative stress participates in mTg fragmentation, producing Tg molecules with a recovered ability to store thyroid hormones; at the same time, the fragmentation of mTg precludes its excessive accumulation in the thyroid and, consequently, the thyroid enlargement (Delom *et al.*, 1999).

Under experimental conditions of Fenton reaction or in the presence of ferric ions (Fe^{3+})–both situations causing oxidative stress–a production of immunoreactive C-terminal fragments (40 kDa) of Tg was observed and that process was accompanied by thyroid hormone synthesis (Duthoit *et al.*, 2000). An excess of H_2O_2 was shown to produce the same 40 kDa immunoreactive fragments of Tg in human thyroid cells. It is worth stressing that both H_2O_2 and iodide have been indispensable to cause Tg fragmentation (Duthoit *et al.*, 2001), what suggests that Tg cleavage occurs during thyroid hormone synthesis. Of great importance is the fact that immunoreactive Tg fragments have been found only in dead cells but not in live ones and that those fragments are able to enter live thyrocytes (Duthoit *et al.*, 2001), what would start autoimmune response. Additionally, it has been found that these 40 kDa immunoreactive fragments of Tg are recognized by Tg antibodies from patients with thyroid disorders, and that they stimulate Tg antiobodies formation in CBA mice model, which suggests that they possess an antigenic and immunogenic character (El Hassani *et al.*, 2004).

As regards TPO, worth mentioning are studies on oxidative stress and myeloperoxidase (MPO), which is an enzyme very similar in structure and properties to TPO. Whole body ionizing radiation (800 cGy) resulted in a decreased level of glutathione, increased levels of lipid peroxidation products and increased levels of MPO (as the index of neutrophil infiltration) in different tissues, the changes which were reduced by a coadministration of melatonin (Sener *et al.*, 2003). It is highly probable that melatonin is also able to modify the level of TPO protein under conditions of excessive oxidative stress. No studies have–till now–been performed, showing that oxidative stress is involved in autoimmunity directed against TPO.

Some other results are worth mentioning, concerning the relationship between oxidative stress and thyroid autoimmunity. It has been documented that free radicals are involved in interleukin-1β-induced glycosaminoglycan production by retroocular fibroblasts and their accumulation in patients with Graves' disease (Lu *et al.*, 1999). Furthermore, it has been shown that $NO^{\cdot}$ is involved in interleukin-1α-induced cytotoxicity in polarized human thyrocytes, suggesting that this free radical may promote the exposure of autoantigens to the immune system (Van den Hove *et al.*, 2002).

The relationship between oxidative stress and thyroid autoimmunity suggests a potential protective role of antioxidants in autoimmunological diseases. However, the application of antioxidants in patients with autoimmunological diseases or patients, predisposed genetically to these disorders, is still the matter of discussion. As far as the influence of melatonin on the immune system is concerned, divergent results exist (Lewinski *et al.*, 1989; Guerrero and Reiter, 2002). However, the detailed effects caused by the indoleamine in patients with autoimmunological diseases, as well as the potential indications and contraindications for the treatment with this substance in such patients, remain to be determined.

The Influence of Thyroid Hormones on Oxidative Processes

Thyroid hormones regulate energy metabolism, among others, by their effects on mitochondria (Goglia *et al.*, 2002), in which free radicals and ROS are produced. Under *in vitro* conditions, thyroid hormones may act as antioxidants; thyroid hormones (3,5,3'-triiodothyronine, T_3; and 3,5,3',5'-tetraiodothyronine (thyroxine), T_4) and their structural analogues (L-thyronine, T_0; 3,5,3'-triiodothyroacetic acid, TA_3; 3,5,3',5'-tetraiodothyroacetic acid, TA_4) were found to reveal antioxidative properties (Hanna *et al.*, 1993; Chomard *et al.*, 1998; Oziol *et al.*, 2001). For example, it was shown that all the above mentioned hormones or their analogues revealed a capacity to scavenge free radicals,.OH and O_2^{-} included (shown for T_3), and that T_0, T_3, and TA_3 effectively prevented the formation of lipid peroxidation products–conjugated dienes and thiobarbituric acid reactive substances–during LDL oxidation (Oziol *et al.*, 2001). In the same study it was documented that the 4'-hydroxydiphenylether structure of thyroid compounds was necessary for their free radical scavenging activity. This observation has been confirmed by antioxidant effects of T_0, which has the structure of thyroid hormones but does not possess any atom of iodine; conversely, monoiodotyrosine (MIT), which possesses the monoiodosubstituted phenolic ring of T_3, does not reveal any antioxidative effects (Oziol *et al.*, 2001).

Thyroid Hormone Excess

Changes in the values of oxidative parameters in different tissues were observed, due to thyroid hormone excess, resulting from either thyroid diseases or treatment with thyroid hormones. In blood samples collected from hyperthyroid patients suffering from Graves' disease, the levels of thiobarbituric acid-reacting substances and conjugated dienes (both being parameters of oxidative stress) and the activities of main antioxidative enzymes (SOD, CAT and GSH-Px) were significantly higher than those in healthy subjects: all those changes were restored following antithyroid drug treatment (Komosinska-Vassev *et al.*, 2000). Similarly, methimazole treatment in hyperthyroid patients suffering from Graves' disease resulted in normalization of all parameters of oxidative stress (H_2O_2 concentration, lipid hydroperoxides, thiobarbituric acid-reactive substances, superoxide dismutase and catalase activities) in patients without infiltrative ophthalmopathy but not in patients with infiltrative ophthalmopathy; the authors suggest that oxidative stress in patients with Graves' disease is influenced not only by thyroid status but also by orbital inflammation (Bednarek *et al.*, 2005). Next, a significant increase in malondialdehyde (MDA) concentration and a slight increase in Schiff's bases and conjugated dienes concentration (all being parameters of oxidative damage to lipids) were found in blood serum collected from hyperthyroid patients with Graves' disease; the treatment with methimazole (an antithyroid drug) brought about a decrease of indices of oxidative stress, accompanied by normalization of thyroid hormone and TSH concentration (Sewerynek *et al.*, 2000).

In mononuclear cells, collected from hyperthyroid patients with Graves' disease, an increased level of 8-oxo-7,8-dihydro-2'-deoxyguanosine (8-oxo-dGuo) (the parameter of DNA oxidative damage) was found; additionally, 8-oxo-dGuo level correlated significantly with free T_4 concentration (Hara *et al.*, 2001). It is presumed that untreated hyperthyroidism contributes to oxidative damage to DNA. An induction of oxidative stress was also observed in experimental models of hyperthyroidism. The treatment of rats with thyroid hormones resulted in generation of O_2^{-} and H_2O_2 in liver mitochondria (Fernandez and Videla, 1993) and of NO· in liver cytosol (Fernandez *et al.*, 1997). Expectedly, T_3-injection increased lipid peroxidation and protein oxidation in the rat liver; it is worth stressing that–in response to thyroid hormone–the damage to lipids preceded the damage to protein molecules, what suggests different susceptibility of target molecules to excessive amounts of thyroid hormones (Tapia *et al.*, 1999).

In animal models of hyper- and hypothyroidism, T_3-treatment in rats brought about a significant increase in lipid peroxidation in the liver, the heart, and skeletal muscles, whereas methimazole administration (causing hypothyroidism) did not influence the level of lipid peroxidation products in those organs (Venditti *et al.*, 1997). Moreover, a decrease in the whole antioxidant capacity of tissues and an increase in GSH-Px and glutathione reductase (GSH-Rd) were observed only in the examined organs in both hyper- and hypothyroid rats (Venditti *et al.*, 1997).

In other animal models of hyper- and hypothyroidism, treatment of *OF1* female mice with L-thyroxine (L-T_4) resulted in the increased sensitivity to lipid peroxidation, increased level of oxidized GSH (GSSG), and increased ratio GSSG/GSH, as well as increased oxidative damage to mitochondrial DNA in the heart; interestingly, a treatment with propylthiouracil (PTU) did not influence oxidative damage to lipids and it decreased GSSG concentration and the GSSG/GSH ratio, as well as reduced oxidative damage to mitochondrial DNA (Gredilla *et al.*, 2001b). The latter finding is of great importance, suggesting protective effects of PTU against oxidative stress. In agreement with that, treatment with L-T_4 but not with PTU increased lipid peroxidation in the skeletal muscle in *OF1* female mice (Gredilla *et al.*, 2001a).

In the study performed in our laboratory, a 2-week-treatment with L-T_4 (100 mg/kg B.W., for 14 days) resulted in an increased concentration of free fractions of both thyroid hormones; a cotreatment with melatonin (5 mg/kg B.W., for 7 days) completely prevented the increase in free T_3-concentration (Sewerynek *et al.*, 1999). Thus, melatonin has abolished the increase in the concentration of a thyroid hormone, which is directly responsible for tissue effects of hyperthyroidism. Furthermore, melatonin injection resulted in decreased basal values of conjugated dienes and Schiff's bases in rat kidney (Sewerynek *et al.*, 1999). It has recently been found that T_3 exerts its mutagenic action through induction of oxidative stress (Djelic and Anderson 2003).

Not only thyroid hormones, but also iodides, being indispensable for thyroid hormone synthesis, may influence oxidative stress in different tissues. The injections with potassium iodide caused the increased level of Schiff's bases in the liver and lung, and these changes have been partially reversed by melatonin (Swierczynska-Machura *et al.*, 2004).

Thyroid Hormone Deficiency

It has been observed that hypothyroidism, resulting from either PTU-treatment or thyroidectomy, causes a decrease in H_2O_2 generation (Swaroop and Ramasarma 1985). Additionally, hypothyroidism in consequence of thyroidectomy, prevented postischemic lipid peroxidation in the rat kidney (Paller, 1986). Thus, it is still unclear if the above-mentioned protective effects of PTU against oxidative stress (Gredilla *et al.*, 2001b) are really related to direct antioxidant action of the drug, or to hypothyroidism, or if they result from both mechanisms jointly. Indeed, PTU has been shown to react directly with.OH and, to a lesser extent, with $O_2^{-\cdot}$, protecting against lipid peroxidation (Hicks *et al.*, 1992), and to prevent the formation of hypochlorous acid (another oxygen species) (Ross *et al.*, 1998). However, it should be expected that hypothyroid state, leads–sooner or later–to oxidative damage of biological molecules.

In agreement with the last assumption, progressive hypothyroidism, observed early in postnatal rats, enhanced oxidative processes; the increased SOD and CAT activities, and decreased GSH level were accompanied by increased amount of.OH, enhanced protein carbonylation and lipid peroxidation in the rat brain (Rahaman *et al.*, 2001). Similarly, the treatment with PTU, causing hypothyroidism, resulted in the increased lipid peroxidation in rat heart and that change was reduced by treatment with T_3 (Chattopadhyay *et al.*, 2003).

In aging humans, the function of several endocrine glands declines progressively, resulting in decreased concentrations of different hormones including melatonin and thyroid hormones. The study, performed in perimenopausal and menopausal women with initial low levels of blood melatonin, revealed that melatonin treatment for 3-6 months resulted in the significant increase in thyroid hormone concentrations (Bellipanni *et al.*, 2001). Thus, melatonin reveals a recovery effect on thyroid function towards a more juvenile pattern of regulation (Bellipanni *et al.*, 2001); it is not excluded that the described effect is a direct one because a similar treatment with melatonin in aging patients did not result in any changes of TSH concentration (Siegrist *et al.*, 2001).

Oxidative Stress Induced by Radioiodine (^{131}I)

Radioiodine (^{131}I)-therapy is a treatment of choice in differentiated thyroid cancer (following total thyroidectomy). The parameters of oxidative stress were evaluated in patients subjected to total thyroidectomy because of thyroid carcinoma, before (at that time patients were hypothyroid) and after ^{131}I treatment (Konukoglu *et al.*, 1998). The level of lipid peroxidation products, measured in blood erythrocytes, significantly increased after thyroidectomy, comparing to values in controls; a further increase in lipid peroxidation was found after ^{131}I treatment. GSH level and GSH-Px and GSH-Rd activities were lower in thyroidectomized patients than those in control subjects; the values of those parameters significantly increased after ^{131}I treatment.

In another study in humans, a treatment with ^{131}I, either in higher (in case of thyroid cancer) or in lower (in case of hyperthyrodism) activities, resulted in the increased levels of isoprostanes (products of oxidative injury to lipids) in blood plasma, serum and urine; the damaging effect of ^{131}I was significantly more pronounced and longer lasting after high-activity therapy (Wolfram *et al.*, 2002).

The above findings are in agreement with numerous data indicating that radiation causes oxidative damage to tissues, while antioxidants, like melatonin, may prevent this phenomenon (reviewed in: Karbownik and Reiter, 2000). Because of the potential role of ionizing radiation in the pathogenesis of thyroid cancer, the studies on protective effects of melatonin against radiation-induced oxidative stress and thyroid cancer seem to be of special value. It is worth mentioning here that melatonin effectively reduced histoenzymological changes in the rat thyroid gland caused by exposure to g-radiation (Kundurovic and Scepovic, 1989).

In conclusion, free radical and reactive species play a significant role in the thyroid gland under both physiological and pathological conditions. Melatonin, being an antioxidant and widely distributed molecule, may modulate physiological processes in the thyroid, and–in the course of thyroid disease–may potentially protect against oxidative damage in the gland and other tissues.

Acknowledgements

Some parts of the cited research studies from our laboratory were supported by the project No. 506-01-042 from the Ministry of Health of Poland.

References

Bednarek J, Wysocki H, and Sowinski J. (2005). Oxidative stress peripheral parameters in Graves' disease: the effect of methimazole treatment in patients with and without infiltrative ophthalmopathy. Clin. Biochem., 38, 13-18.

Bellipanni G, Bianchi P, Pierpaoli W, Bulian D, and Ilyia E. (2001). Effects of melatonin in perimenopausal and menopausal women: a randomized and placebo controlled study. Exp. Gerontol., 36, 297-310.

Bjorkman U, and Ekholm R. (1984). Generation of H_2O_2 in isolated porcine thyroid follicles. Endocrinol., 115, 392-398.

Bocanera LV, Krawiec L, Silberschmidt D, Pignatoro O, Juvenal GJ, Pregliasco LB, and Pisarev MA. (1997). Role of cyclic 3'5'guanosine monophosphate and nitric oxide in the regulation of iodide uptake in calf thyroid cells. J. Endocrinol., 155, 451-457.

Bromme H-J, Ebelt H, Peschke D, and Peschke E. (1999). Alloxan acts as a prooxidant only under reducing conditions: influence of melatonin. Cell Mol. Life Sci., 55, 487-493.

Chattopadhyay S, Zaidi G, Das K and Chainy GB (2003). Effects of hypothyroidism induced by 6-n-propylthiouracil and its reversal by T3 on rat heart superoxide dismutase, catalase and lipid peroxidation. Indian J. Exp. Biol., 41, 846-849.

Chomard P, Seguin C, Loireau A, Autissier N, and Artur Y. (1998). Effects of iodotyrosines, thyronines, iodothyroacetic acids and thyromimetic analogues on in vitro copper-induced oxidation of low-density lipoproteins. Biochem. Pharmacol., 55, 1591-1601.

Colin IM, Nava E, Toussaint D, Maiter DM, VanDenhove MF, Luscher TF, Ketelslegers JM, Denef JF, and Jameson JL. (1995). Expression of nitric oxide synthase isoforms in the thyroid gland: evidence for a role of nitric oxide in vasculatur control during goiter formation. Endocrinol., 136, 5283-5290.

Corvailain B, Van Sande J, and Dumont JE. (1988). Inhibition by iodide binding to proteins: the "Wolff-Chaikoff" effect is caused by inhibition of H_2O_2 generation. Biochem. Biophys. Res. Commun., 154, 1287-1292.

De Deken X, Wang D, Many MC, Costagliola S, Libert F, Vassart G, Dumont JE, and Miot F. (2000). Cloning of two human thyroid cDNAs encoding new members of the NADPH oxidase family. J. Biol. Chem., 275, 23227-23233.

Delom F, Lejeune P-J, Vinet L, Carayon P, and Mallet B. (1999). Involvement of oxidative reactions and extracellular protein chaperons in the rescue of misassembled thyroglobulin in the follicular lumen. Biochem. Biophys. Res. Commun., 255, 438-443.

Djelic N, and Anderson D. (2003). The effect of the antioxidant catalase on oestrogens, triiodothyronine, and noradrenaline in the Comet assay. Teratogenesis, Carcinogen Mutagen, Suppl 2, 69-81.

Dreher D, and Junod AF. (1996). Role of oxygen free radicals in cancer development. Eur. J. Cancer, 32A, 30-38.

Dröge W. (2002). Free radicals in the physiological control of cell function. Physiol. Rev., 82, 47-95.

Dupuy C, Ohayon R, Valent A, Noel-Hudson MS, Deme D, and Virion A. (1999). Purification of a novel flavoprotein involved in the thyroid NADPH oxidase: cloning of the porcine and human cDNAs. J. Biol. Chem., 274, 37265-37269.

Duthoit C, Estienne V, Delom F, Durand-Gorde J-M, Mallet B, Carayon P, and Ruf J. (2000). Production of immunoreactive thyroglobulin C-terminal fragments during thyroid hormone synthesis. Endocrinol., 141, 2518-2525.

Duthoit C, Estienne V, Giraud A, Durand-Gorde J-M, Rasmussen AK, Feldt-Rasmussen U, Carayon P, and Ruf J. (2001). Hydrogen peroxide-induced production of a 40 kDa immunoreactive thyroglobulin fragment in human thyroid cells: the onset of thyroid autoimmunity. Biochem. J., 360, 557-562.

Eaton JW, and Qian M. (2002). Molecular bases of cellular iron toxicity. Free Radical Biol. Med., 32, 833-840.

El Hassani RA, Estienne V, Blanchin S, Durand-Gorde JM, Mallet B, De Micco C, Carayon P, Lalaoui K, and Ruf J. (2004). Antigenicity and immunogenicity of the C-terminal peptide of human thyroglobulin. Peptides, 25, 1021-1029.

Esteves RZ, van Sande J, and Dumont JE. (1992). Nitric oxide as a signal in thyroid. Mol. Cell Endocrinol., 90, R1-3.

Fayadat L, Niccoli-Sire P, Lanet J, and Franc J-L. (1999). Role of heme in intracellular trafficking of thyroperoxidase and involvement of H_2O_2 generated at the apical surface of thyroid cells in autocatalytic covalent heme binding. J. Biol. Chem., 274, 10533-10538.

Fernandez V, Cornejo P, Tapia G, and Videla LA. (1997). Influence of hyperthyroidism on the activity of liver nitric oxide synthase in the rat. Nitric Oxide, 1, 463-468.

Fernandez V, and Videla LA. (1993). Influence of hyperthyroidism on superoxide radical and hydrogen peroxide production by rat liver submitochondrial particles. Free Radical Res. Commun., 18, 329-335.

Gilad E, Cuzzocrea S, Zingarelli B, Salzman AL, and Szabo C. (1997). Melatonin is a scavenger of peroxynitrite. Life Sci., 60, PL169-74.

Giray B, Hincal F, Tezic T, Okten A, and Gedik Y. (2001). Status of selenium and antioxidant enzymes of goitrous children is lower than healthy controls and nongoitrous children with high iodine deficiency. Biol Trace Element Res., 82, 35-52.

Giray B, and Hincal F. (2002). Oxidative DNA base damage, antioxidant enzyme activities and selenium status in highly iodine-deficient goitrous children. Free Radical Res., 36, 55-62.

Goglia F, Silvestri E, and Lanni A. (2002). Thyroid hormones and mitochondria. Bioscience Reports, 22, 17-32.

Gredilla R, Barja G, and Lopez-Torres M. (2001a). Thyroid hormone-induced oxidative damage on lipids, glutathione and DNA in the mouse heart. Free Radical Res., 35, 417-425.

Gredilla R, Torres ML, Portero-Otin M, Pamplona R, and Barja G. (2001b). Influence of hyper- and hypothyroidism on lipid peroxidation, unsaturation of phospholipids, glutathione system and oxidative damage to nuclear and mitochondrial DNA in mice skeletal muscle. Mol. Cell Biochem., 221, 41-48.

Guerrero JM, and Reiter RJ. (2002). Melatonin-immune system relationship. Curr. Topics Med. Chem., 2, 167-179.

Hanna AN, Feller DR, Witiak DT, and Newman HAI. (1993). Inhibition of low density lipoprotein oxidation by thyronines and probucol. Biochem. Pharmacol., 45, 753-762.

Hara H, Sato R, and Ban Y. (2001). Production of 8-OHdG and cytochrome c by cultured human mononuclear cells in patients with autoimmune thyroid disease. Endocr. J., 48, 671-675.

Hardeland R, Poeggeler B, Niebergall R, and Zelosko V. (2003). Oxidation of melatonin by carbonate radicals and chemiluminescence emitted during pyrrole ring cleavage. J. Pineal Res., 34, 17-25.

Hasegawa Y, Takano T, Miyauchi A, Matsuzuka F, Yoshida H, Kuma K, and Amino N. (2003). Decreased expression of catalase mRNA in thyroid anaplastic carcinoma. Jpn. J. Clin. Oncol., 33, 6-9.

Hicks M, Wong LS, and Day RO. (1992). Antioxidant activity of propylthiouracil. Biochem. Pharmacol., 43, 439-444.

Iwase K, Nagasaka A, Kato K, Ohtani S, Tsujimura T, Inagaki A, Jimbo S, Nakai QA, Masunaga R, Hamada M, Mano T, Kotake M, and Miura K. (1993). Localization of Cu/Zn and Mn superoxide dismutase in various thyroid disorders. Acta Endocrinol. (Copen.), 129, 573-578.

Karasek M, Stankiewicz A, Bandurska-Stankiewicz E, Zylinska K, Pawlikowski M, and Kuzdak K. (2000). Melatonin concentrations in patients with large goiter before and after surgery. Neuroendocrinol. Lett., 21, 437-439.

Karbownik M, and Lewiñski A. (2003a). The role of oxidative stress in physiological and pathological processes in the thyroid gland; possible involvement in pineal-thyroid interactions. Neuroendocrinol. Lett., 24, 293-303.

Karbownik M, and Lewiñski A. (2003b). Melatonin reduces Fenton reaction-induced lipid peroxidation in porcine thyroid tissue. J. Cell Biochem., 90, 806-811.

Karbownik M, Lewinski A, and Reiter RJ. (2001). Anticarcinogenic actions of melatonin which involve antioxidative processes: Comparison with other antioxidants. Int. J. Biochem. Cell Biol., 33, 735-753.

Karbownik M, and Reiter RJ. (2000). Antioxidative effects of melatonin in protection against cellular damage caused by ionizing radiation. Proc. Soc. Exp. Biol. Med., 225, 9-22.

Karbownik M, and Reiter RJ. (2002). Melatonin protects against oxidative stress caused by d-aminolevulinic acid: Implication for cancer reduction. Cancer Inv., 20, 276-286.

Karbownik M, Stasiak M, Zasada K, Zygmunt A, and Lewinski A. (2005). Comparison of potential protective effects of melatonin, indole-3-propionic acid and propylthiouracil against lipid peroxidation caused by potassium bromate in the thyroid gland. J. Cell. Biochem., 95, 131-138.

Kelner MJ, Bagnell RD, and Morgenstern R. (2004). Structural organization of the murine microsomal glutathione S-transferase gene (MGST1) from the 129/SvJ strain: identification of the promoter region and a comprehensive examination of tissue expression. Biochem. Biophys. Acta, 1678, 163-169.

Kim H, Lee T-H, Park ES, Suh JM, Park SJ, Chung HK, Kwon O-Y, Kim YK, Ro HK, and Shong M. (2000). Role of peroxiredoxins in regulating intracellular hydrogen peroxide and hydrogen peroxide-induced apoptosis in thyroid cells. J. Biol. Chem., 275, 18266-18270.

Kitano H, Kitanishi T, Nakanishi Y, Suzuki M, Taleuchi E, Yazawa Y, Kitajima K, Kimura H, and Tooyama I. (1999). Expression of inducible nitric oxide synthase in human thyroid papillary carcinoma. Thyroid, 9,113-117.

Komosinska-Vassev K, Olczyk K, Kucharz EJ, Marcisz C, Winsz-Szczotka K, and Kotulska A. (2000). Free radical activity and antioxidant defense mechanism in patients with hyperthyroidism due to Graves' disease during therapy. Clin. Chim. Acta, 300, 107-117.

Konukoglu D, Hatemi HH, Arikan S, Demir M, and Akcay T. (1998). Radioiodine treatment and oxidative stress in thyroidectomised patients for differentiated thyroid cancers. Pharmacol. Res., 38, 311-315.

Krotewicz M, Lewinski A, and Wajs E. (1992). The inhibitory effect of late afternoon melatonin injections, but not of melatonin-containing subcutaneous implants, on thyroid hormone secretion in male Wistar rats. Neuroendocrinol. Lett., 14, 405-411.

Krotewicz M and Lewinski A. (1994). Effects of pinealectomy and of late afternoon injections of pineal indole substances on thyroid hormone secretion in male Wistar rats. Biochem. Lett., 50, 101-107.

Kundurovic Z, and Scepovic M. (1989). Histoenzymological reactions of the thyroid gland in irradiated and previously melatonin-treated irradiated rats. Acta Med. Iugosl., 43, 337-347.

Kvetnoy IM, (2002). Extrapineal melatonin in pathology: new perspectives for diagnosis, prognosis, and treatment of illness. Neuroendocrinol. Lett., 23, (Suppl 1) 92-96.

Kvetnoy, IM, and Yuzhakov J. (1993). Extrapineal melatonin: advances in microscopical identification of hormones in endocrine and nonendocrine cells. Microscopy Anal., 21, 27-29.

Lewinski A. (1990). Some aspects of the pineal-thyroid interrelationship and their possible involvement in the regulation of function and growth of these two glands. In: Reiter RJ, and Lukaszyk A. (Eds.), *Advances in Pineal Research.* Vol. IV,. John Libbey and Company Ltd., London, pp 175-188.

Lewinski A, and Karbownik M. (2002). Melatonin and the thyroid gland. Neuroendocrinol. Lett., 23, (Suppl. 1), 73-78.

Lewinski A, Sewerynek E, and Karbownik M. (2002). Melatonin from the past into the future–Our own experience. *In*: Haldar C, Singaravel M, and Maitra SK. (Eds.), *Treatise on Pineal Gland and Melatonin*, Science Publishers Inc., Enfield (NH), USA, and Plymouth, UK, pp 157-175.

Lewinski A, and Sewerynek E. (1986). Melatonin inhibits the basal and TSH-stimulated mitotic activity of thyroid follicular cells *in vivo* and in organ culture. J. Pineal Res., 3, 291-299.

Lewinski A, Vaughan MK, Champney TH, Reiter RJ, and Smith NKR. (1984). Dark exposure inhibits the mitotic activity of thyroid follicular cells in male mice with intact pineal. Experientia, 40, 1284-1285.

Lewinski A, Wajs E, Klencki M, Karbownik M, Gesing A, Sewerynek E, Slowinska-Klencka D, Skowronska-Jozwiak E, Bilinski P, and Krotewicz M. (1997). Pineal-thyroid interrelationships update. *In*: Webb SM, Puig-Domingo M, Moller M, and Pevet E. (Eds.), *Pineal Update from Molecular Mechanisms to Clinical Implications*, PJD Publications Limited, Westbury, New York, pp 173-181.

Lewinski A, Wajs E, Klencki M, Krotewicz M, and Bilinski P. (1996). Melatonin-induced inhibition of thyroid growth processes and possible mechanisms of that phenomenon. *In*: Haldar C. (Ed.), *Recent Researches in Biology,* Vol. 1 *The Pineal Gland: Its Molecular Signals*, Hindustan Publishing Corporation, New Delhi, pp 27-32.

Lewinski A, Wajs E, and Krotewicz M. (1993). Melatonin and other indolic substances: their influence on thyroid growth and secretion. *In*: Touitou Y, Arendt J, and Pevet P. (Eds.), *Melatonin and the Pineal Gland–From Basic Science to Clinical Application*, Excerpta Medica, Amsterdam, pp 265-268.

Lewinski A, Wajs E, Modrzejewska H, Klencki M, Karbownik M, and Greger J. (1994). Inhibitory influence of melatonin on thymidine kinase activity in the rat thyroid lobes incubated *in vitro.* Neuroendocrinol. Lett., 16, 221- 226.

Lewinski A, Zelazowski P, Sewerynek E, Zerek-Melen G, Szkudlinski M, and Zelazowska E. (1989). Melatonin-induced suppression of human lymphocyte natural killer activity in vitro. J. Pineal Res., 7, 153-164.

Lu R, Wang P, Wartofsky L, Sutton BD, Zweier JL, Bahn RS, Garrity J, and Burman KD. (1999). Oxygen free radicals in interleukin-1beta-induced glycosaminoglycan production by retro-ocular fibroblasts from normal subjects and Graves' ophthalmopathy patients. Thyroid, 9, 297-303.

Mano T, Iwase K, Hayashi R, Hayakawa N, Uchimura K, Makino M, Nagata M, Sawai Y, Oda N, Hamada M, Aono T, Nakai A, Nagasaka A, and Itoh M. (1998). Vitamin E and coenzyme Q concentrations in the thyroid tissues of patients with various thyroid disorders. Am. J. Med. Sci., 315, 230-232.

Mano T, Shinohara R, Iwase K, Kotake M, Hamada M, Uchimura K, Hayakawa N, Hayashi R, Nakai A, Ishizuki Y, and Nagasaka A. (1997). Changes in free radical scavengers and lipid peroxide in thyroid glands of various thyroid disorders. Horm. Metab. Res., 29, 351-354.

Millatt LJ, Jackson R, Williams BC, and Whitley GStJ. (1993). Nitric oxide stimulates cyclic GMP in human thyrocytes. J. Mol. Endocrinol., 10, 163-169.

Millatt LJ, Johnstone AP, and Whitley GS. (1998). Nitric oxide enhances thyroid peroxidase activity in primary human thyrocytes. Life Sci., 63, PL373-380.

Motohashi S, Kasai K, Banba N, Hattori Y, and Shimoda S. (1996). Nitric oxide inhibits cell growth in cultured human thyrocytes. Life Sci., 59, PL227-234.

Mutaku JF, Many MC, Colin I, Denef JF, and van den Hove MF. (1998). Antigoitrogenic effect of combined supplementation with dl-alpha-tocopherol, ascorbic acid and beta-carotene and of dl-alpha-tocopherol alone in the rat. J. Endocrinol., 156, 551-561.

Mutaku JF, Poma JF, Many MC, Denef JF, and Van Den Hove MF. (2002). Cell necrosis and apoptosis are differentially regulated during goitre development and iodine-induced involution. J. Endocrinol., 172, 375-386.

Nakamura M, and Ohtaki S. (1993). Formation and reduction of ascorbate radicals by hog thyroid microsomes. Arch. Biochem. Biophys., 305, 84-90.

Noda Y, Mori A, Liburdy R, and Packer L. (1999). Melatonin and its precursors scavenge nitric oxide. J. Pineal Res., 27, 159-163.

Oziol L, Faure P, Vergely C, Rochette L, Artur Y, and Chomard P. (2001). *In vitro* free radical scavenging capacity of thyroid hormones and structural analogues. J. Endocrinol., 170, 197-206.

Paller MS. (1986). Hypothyroidism protects against free radical damage in ischemic acute renal failure. Kidney Intern., 29, 1162-1166.

Pachucki J, Wang D, Christophe D, and Miot F. (2004). Structural and functional characterization of the two human ThOX/Duox genes and their 5'-flanking regions. Mol. Cell Endocrinol., 214, 53-62.

Rahaman SO, Ghosh S, Mohanakumar KP, Das S, and Sarkar PK. (2001). Hypothyroidism in the developing rat brain is associated with marked oxidative stress and aberrant intraneuronal accumulation of neurofilaments. Neurosci. Res., 40, 273-279.

Raspe E and Dumont JE (1995). Tonic modulation of dog thyrocyte H_2O_2 generation and I^- uptake by thyrotropin through the cyclic adenosine 3',5'-monophasphate cascade. Endocrinol., 135, 965-973.

Reiter RJ, Tan DX, Qi W, Manchester LC, Karbownik M, and Calvo JR. (2000). Pharmacology and physiology of melatonin in the reduction of oxidative stress in vivo. Biol. Signals Receptors, 9, 160-171.

Reiter RJ. (1998). Oxidative damage in the central nervous system: Protection by melatonin. Prog. Neurobiol., 56, 359-384.

Rhee SG. (1999). Redox signaling: hydrogen peroxide as intracellular messenger. Exp. Mol. Med., 31, 53-59.

Riou C, Remy C, Rabilloud R, Rousset B, and Fonlupt P. (1998). H_2O_2 induces apoptosis of pig thyrocytes in culture. J. Endocrinol. 156, 315-322.

Ross AD, Dey I, Janes N, and Israel Y. (1998). Effect of antithyroid drugs on hydroxyl radical formation and α-1-proteinase inhibitor inactivation by neutrophils: therapeutic implications. J. Pharmacol. Exp. Therap., 285, 1233-1238.

Sadani GR, and Nadkarni GD. (1996). Role of tissue antioxidant defence in thyroid cancers. Cancer Lett., 109, 231-235.

Sener G, Jahovic N, Tosun O, Atasoy BM, and Yegen BC. (2003). Melatonin ameliorates ionizing radiation-induced oxidative organ damage in rats. Life Sci., 74, 563-572.

Sewerynek E, Wiktorska J, and Lewinski A. (1999). Effects of melatonin on the oxidative stress induced by thyrotoxicosis in rats. Neuroendocrinol. Lett., 20, 157-161.

Sewerynek E, Wiktorska J, Nowak D, and Lewinski A. (2000). Methimazole protection against oxidative stress induced by hyperthyroidism in Graves' disease. Endocrine Regul., 34, 83-89.

Siegrist C, Benedetti C, Orlando A, Beltran JM, Tuchscherr L, and Noseda CM. (2001). Lack of changes in serum prolactin, FSH, TSH, and estradiol after melatonin treatment in doses that improve sleep and reduce benzodiazepine consumption in sleep-disturbed, middle-aged, and elderly patients. J. Pineal Res., 30, 34-42.

Sugawara M, Kita T, Lee ED, Takamatsu J, Hagen GA, Kuma K, and Medeiros-Neto GA. (1988). Deficiency of thyroid superoxide dismutase in endemic goiter tissue. J Clin Endocrinol. Metab. 67, 1156-1161.

Sugawara M, Sugawara Y, Wen K, and Giulivi C. (2002). Generation of oxygen free radicals in thyroid cells and inhibition of thyroid peroxidase. Exp. Biol. Med., 227, 141-146.

Swaroop A, and Ramasarma T. (1985). Heat exposure and hypothyroid conditions decrease hydrogen peroxide generation in liver mitochondria. Biochem. J., 226, 403-408.

Swierczynska-Machura D, Lewinski A, and Sewerynek E. (2004). Melatonin effects on Schiff's base levels induced by iodide administration in rats. Neuroendocrinol. Lett., 25, 70-74.

Tan DX, Chen LD, Poeggeler B, Manchester LC, and Reiter RJ. (1993). Melatonin: a potent, endogenous hydroxyl radical scavenger. Endocrine J., 1, 57-60.

Tan DX, Manchester LC, Hardeland R, Lopez-Burillo S, Mayo JC, Sainz RM, and Reiter RJ. (2003). Melatonin: a hormone, a tissue factor, an autocoid, a paracoid, and an antioxidant vitamin. J. Pineal Res., 34, 75-78.

Tan DX, Manchester LC, Reiter RJ, Plummer BF, Limson J, Weintraub ST, and Qi W. (2000). Melatonin directly scavenges hydrogen peroxide: a potentially new metabolic pathway of melatonin biotransformation. Free Radical Biol. Med., 29, 1177-1185.

Tapia G, Cornejo P, Fernandez V, and Videla LA. (1999). Protein oxidation in thyroid hormone-induced liver oxidative stress: relation to lipid peroxidation. Toxicol. Lett., 106, 209-214.

Taurog A. (2000). Hormone synthesis. *In*: Braverman LE, and Utiger RD. (Eds.), *The Thyroid*, Lippincott Williams and Wilkins, Philadelphia, pp 52-90.

Van den Hove M-F, Stenoiu MS, Croizet K, Couvreur M, Courtoy PJ, Devuyst O, and Colin IM. (2002). Nitric oxide is involved in interleukin-1α-induced cytotoxicity in polarised human thyrocytes. J. Endocrinol., 173, 177-185.

Venditti P, Balestrieri M, Di Meo S, and De Leo T. (1997). Effect of thyroid state on lipid peroxidation, antioxidant defences, and susceptibility to oxidative stress in rat tissues. J. Endocrinol., 155, 151-157.

Wajs E, and Lewinski A. (1992). Inhibitory influence of late-afternoon melatonin injections and the counter-inhibitory effect of melatonin pellets on thyroid growth processes in male Wistar rats; comparison with effects of other indole substances. J. Pineal Res., 13, 158-166.

Wolfram RM, Budinsky AC, Palumbo B, Palumbo R, and Sinzinger H. (2002). Radioiodine therapy induces dose-dependent in vivo oxidation injury: evidence by increased isoprostane 8-epi-PGF(2 alpha). J. Nuclear Med., 43, 1254-1258.

Wright ML, Cuthbert KL, Donohue MJ, Solano SD, and Proctor KL. (2000). Direct influence of melatonin on the thyroid and comparison with prolactin. J. Exp. Zool., 286, 625-631.

Yanagawa T, Ishikawa T, Ishii T, Tabuchi K, Iwasa S, Bannai S, Omura K, Suzuki H, and Yoshida H. (1999). Peroxiredoxin I expression in human thyroid tumors. Cancer Lett., 145, 127-132

Zimmermann M, and Köhrle J. (2002). The impact of iron and selenium deficiencies on iodine and thyroid metabolism: biochemistry and relevance to public health. Thyroid, 12, 867-878.

21

Oxidative Stress and Male Reproduction

B. Saradha and P.P. Mathur*

Department of Biochemistry and Molecular Biology, School of Life Sciences, Pondicherry University, Pondicherry – 605 014

ABSTRACT

There is a growing concern regarding the decreasing trends in male reproduction, reported in terms of sperm quality and quantity. In the past two decades, many studies have focused on the role of oxidative stress in male infertility. Free radical scavengers and antioxidants handle the key mediators of oxidative stress, the reactive oxygen species, which are produced during normal course of metabolism. Reactive oxygen species (ROS) in limited amounts are important mediators of normal sperm function, such as signal transduction mechanisms that affect fertility. However, in excess can bring about oxidative damage to cellular components, with unsaturated fatty acid being the most venerable molecules. Mammalian spermatozoa, being rich in polyunsaturated fatty acids, are more susceptible to ROS attack, resulting in decrease in sperm motility and viability leading to infertility. The potential of environmental contaminants to generate ROS and create oxidative damage is the focus research in the past two decades. The present review will focus on few toxicology studies that have addressed the possible relationship between reproductive toxicity and exposure to chemicals that generates ROS.

Introduction

There is a growing concern regarding the decreasing trends in male reproduction, reported in terms of sperm quality and quantity. The report on meta-analysis of 61 sperm count studies published

**Corresponding Author*. E-mail: ppmathur@hotmail.com; Tel: + 91-413-2655212; Fax: +91-413-2655211

in 1992 which showed a nearly 50 per cent decline in sperm count between 1940 and 1990 world wide, created an alarming concern over the declining trends in male reproduction (Carlsen *et al.*, 1992). During the same period, the incidence of testicular cancer had markedly increased in many countries (Forman and Moller, 1994). These and other observations provided a clue that this observed decline in male reproductive health might be caused by some common environmental factor. Subsequently, several studies pointed out the correlation between the environmental exposure and reproductive disorders such as testicular cancer, crptochidism, infertility in both wildlife as well as humans (Colborn *et al.*, 1993). These reports further triggered the interest of many scientists to look into the underlying mechanism through which these environmental contaminants cause adverse effects on male reproduction. The one, which has gained greater interest in last two decades, is the potential of environmental contaminants to generate ROS and role of associated oxidative stress on male reproduction (Bagchi and Stohs, 1992; Sikka *et al.*, 1995). Occupational exposure to many environmental chemicals has been reported to induce oxidative stress in biological system. Several studies have conducted to determine whether oxidative stress resulting from various environmental chemicals is associated with their toxic effect. The present review will focus on few recent toxicology studies that have addressed the possible relationship between reproductive toxicity and exposure to chemicals that generates ROS. However, to better understand the nature of oxidative stress, some basic principles of free radicals/reactive oxygen species (ROS) production, their physiological and pathological role in male reproduction will also be discussed in brief.

Free Radicals and Oxidative Stress

All aerobic cells are exposed to ROS, but when the level of exposure increases, oxidative stress occurs, which results in increased cellular damage by oxidants. Radicals of oxygen (superoxide anion, hydroxyl radical and peroxy radicals), reactive non-radical oxygen species such as hydrogen peroxide (H_2O_2) and singlet oxygen, nitrogen species ($NO^{\cdot}$) make up the variety of reactive molecule that constitute an oxidative stress to cells (Younes, 1999). Much of the reactive species are free radicals characterized by the presence of one or more unpaired electrons, therefore, highly unstable and reactive. In the process seeking of stability, they attack nearby molecules to obtain electron, resulting in structural and function damage to molecules. If the free radicals are not inactivated, their chemical reactivity can damage cellular macromolecules, including lipids, proteins, carbohydrates and nucleic acid (Farber, 1994). The free radicals are implicated in pathophysiology of many diseases like Alzheimer's disease (Butterfield and Launderback, 2002), cataracts (Spector, 1995), atherosclerosis (Halliwell, 1999), reperfusion injury, Parkinson's diseases (Olanow, 1990), arthritis, and rheumatic disorders and many others (Kehrer, 1993). Radicals of different type are also are potentially involved in both initiation and promotion in multistage cancer development.

The sources of free radicals and other reactive oxygen species are categorized into the internal sources *i.e.* from normal cellular metabolism and the external sources, such as exposure to drugs, radiation, smoking, pesticides and other industrial chemicals. The free radicals are continuously formed in the cell as a consequence of both enzymatic and non-enzymatic reactions. Within normal limits of production, free radicals may have many physiological roles. In relation male reproduction, ROS serves as an important mediator in signal transduction mechanisms, regulation of sperm capacitation and facilitation of acrosome reaction and spermatozoon-oocyte attachment (Holland *et al.*, 1982; De Lamirande and Gagnon, 1993)

There are numerous cellular defenses, which under normal metabolic conditions, regulate the level of ROS and protect against the ill effects of free radicals. These defenses can be classified into two categories: primary and secondary defenses. The primary defenses are the antioxidant scavenging enzymes and the secondary includes antioxidant Vitamins, such as vitamin E, vitamin C and beta-carotene, as well as glutathione and uric acid. The balance between the ROS-generating systems and the enzymatic or non-enzymatic scavengers in the body has important health implication. Therefore, any factor whether internal or external, which leads to excessive generation of radicals or decrease in the levels of the antioxidants, results in a condition of oxidative stress, which may cause chronic and permanent damage.

Environmental Contaminant Induced Oxidative Stress in Male Reproduction

There is now unequivocal evidence that a wide variety of chemicals capable of disturbing male reproductive function are present in the environment (Wong *et al.*, 2003). These chemical ranges from components in smoke to industrial chemicals. Some of the environmental contaminants have been shown to affect estrogen metabolism by activating a pathway leading to the production of 16α-hydroxyestrone (16α-OHE1), a potent estrogen and genotoxic compound. Bradlow *et al.* (1995) demonstrated the effects of various pesticides on metabolism of estradiol to 2-hydroxyestrone (2-OHE1) and 16α-OHE1 in MCF-7 human breast cancer cells and concluded that the ratio of 16α-OHE1/2-OHE1 as a biological marker of breast cancer risk. Polychlorinated biphenyls (PCBs) and other weakly estrogenic pesticides were found to induce higher 16α-OHE1/2-OHE1 ratio. Altered ratio of these metabolites may cause disturbance in male reproductive health also.

Pesticides and many oxidatively reactive compounds, such as those found in cigarette smoke, alcohol, and air pollution, have all been shown to induce the formation of harmful ROS by affecting the pro-oxidant and antioxidant balance which is vital for normal biological functioning of cells (Abdollahi *et al.*, 2004; Bagchi and Stohs, 1992; Sinclair, 2000). One of the significant developments in past two decades has been the discovery of ROS and role of associated oxidative stress in the etiology of defective sperm function and male infertility (Sikka *et al.*, 1995). Elevation in the levels of ROS, results in oxidation of cellular components, with polyunsaturated fatty acid being the more vulnerable molecules. As the cytoplasm of the spermatozoa is extremely limited in volume and localization, the PUFA that are bound in the sperm plasma membrane are very susceptible to ROS attack resulting in decrease in sperm motility and viability leading to infertility (Aitken and Clarkson, 1987). The effect of environmental contaminants on male reproduction is thereby reflected in sperm quality and quantity. Therefore, subsequent discussions on environment contaminants on male reproduction are subdivided into its effect on testicular and epididymal phase of sperm maturation.

Effect of Environmental Contaminants on Testis

Testis has two major functions–germ cell production and steroidogenesis, which are controlled by gonadotropins. Germ cells continuously proliferate and differentiate to supply a large amount of mature spermatozoa. During this process, upto 75 per cent of the spermatozoa are estimated to undergo apoptotic cell death in the testis of adult male animals (Print and Loveland, 2000). In addition to spontaneous cell death, the process of spermatogenesis is highly susceptible to stressful stimuli (Izu *et al.*, 2004). Expression of stress response genes in germ cells during spermatogenesis has been reported in isolated rat germ cells (Aguilar-Mahecha *et al.*, 2001). During germ cell development, different spermatogenic cell types show a remarkable variation in their susceptibility to stressful stimuli. A variety of cellular mechanisms such as DNA repair (Sancar, 1995), the heat shock response (Craig, 1985), apoptosis (Cosulich and Clarke, 1996), and oxidative stress response (Gutteridge and Halliwell, 2000) were triggered after exposure to different types of stressors. Many toxicants have been reported to block the spermatogenesis at various stages resulting in decreased sperm count (Bendsen *et al.*, 2001; Toyama and Yuasa1, 2004). As normal male fertility relies on normal spermatogenesis, block at any stage could result in infertility.

Currently, it is well appreciated that damage to testicular male germ cells induced by various xenobiotics or ROS can result in testicular dysfunction, leading to infertility. Doreswamy *et al.* (2004) has demonstrated the induction of oxidative stress by multiple doses of nickel chloride in testis of mice being associated with significant DNA damage and apoptosis. The study also demonstrated the propensity of nickel chloride to cause stage specific genotoxic effects measured in terms of higher percentage of sperm head abnormalities and dominant lethal–type mutations discernible only specific post treatment weeks. These data are consistent with earlier observations that many germ cell mutagens induce genotoxic effect by producing higher incidence of abnormal sperms in mice (Ehling, 1977).

Leydig cells are the principle cells involved in secretion of androgens, notably testosterone. The roles of gonadotropins as endocrine regulator of spermatogenesis are well known (Sharpe, 1994). LH induced Leydig cell secretion of androgens, namely testosterone, are involved in regulation of spermatogenesis by targeting androgen receptors in the seminiferous epithelium. FSH targets receptors within the Sertoli cell to regulate spermatogenesis by stimulating the production of numerous Sertoli cell factors. Therefore, any factor affecting the Leydig cell or Sertoli cell viability in turn can interrupt the endocrine regulation of spermatogenesis and consequently affects the reproductive performance.

Intraperitoneal exposure to Aroclor 1254, a commercial mixture of polychlorinated biphenyls (PCB), at a dose of 2 mg/kg bw/day for 30 days resulted in increased lipid peroxidation and decreased specific activities of antioxidant enzymes indicating induction of oxidative stress in Sertoli cells of treated rats (Senthil Kumar *et al.*, 2004). Exposure of primary cultured Leydig cell *in vitro*, to cadmium at the concentrations of 10 mM, caused increased oxidative DNA damage in exposed Leydig cells resulting in decreased viability of cell and hence the testosterone secretion (Yang *et al.*, 2003). Similar induction of oxidative stress, DNA damage and impairment of antioxidant cellular defenses in primary cultures of rat Leydig cells has also been reported with a solvent, 2-Bromopropane (Wu *et al.*, 2002). The high levels of corticosterone (CORT) associated with stress are known to induce apoptosis in Leydig cells. The activation of Fas system, cleavage of procaspase-3, loss of mitochondrial membrane potential and increased ROS generation are reported to be the possible mechanisms involved in CORT-induced Leydig cell death (Gao *et al.*, 2003). The decline in testosterone production following contaminants exposure may be in part due to stress-induced effect of CORT on Leydig cells.

Many toxicants have been shown to induce lipid peroxidation in the testis (ethanol, cadmium, and cigarette smoke) indicating the susceptibility of this organ to reactive oxygen species (Rosenblum *et al.*, 1985; Peltola *et al.*, 1994). However, the susceptibility pattern varies with the age of the animal, dose and duration of exposure to the toxicant. In facts, many reports have indicated that immature testis is more prone to deleterious effects of environmental contaminants in comparison to mature ones (Samanta and Chainy, 1997).

Sub-acute exposure to benzo(a)pyrene (75 mg BaP/m3) via nose-only inhalation, 4 h daily for 10 days resulted in decline in plasma testosterone concentrations in rats, thereby contributing to impaired testicular endocrine function (Inyang *et al.*, 2003). The observed benzo(a)pyrene–induced decrease in the circulating concentrations of testosterone have been proposed to be resulted from sequestration of this compound by high-density lipoproteins that are essential for steroid hormone biosynthesis (Polyakov *et al.*, 1996) or due to the induction of the liver cytochrome P450s that are necessary for detoxification of the polycyclic aromatic hydrocarbons (PAHs) compound (Baird and Ralston, 1997). Recently aryl hydrocarbon receptor mediated induction of xanthine oxidase and xanthine dehydrogenase system has been related to oxidative stress in vivo in the 2,3,7,8-tetrachlorodibenzo-p-dioxin (TCDD)-treated rats (Sugihara *et al.*, 2001).

The most consistent biochemical effect of most of the pesticides has been shown to be the induction of Cytochrome P-450 system (CYP) associated monooxygenase and related protein. These monooxygenases catalyzes the metabolism of steroids, fatty acids as well as xenobiotics such as drugs and pesticides. Thus, exposure to environmental inducers of CYPs may profoundly modulate the testosterone metabolism and hence the final concentration of testosterone in organisms exposed to these chemicals (Wilson and LeBlanc, 1998). Several contaminants have been reported to exhibit receptor as well as non-receptor mediated induction of several CYP genes, thereby significantly increasing testosterone metabolism (Dai *et al.*, 2001). The ability of CYP system to induce ROS has been reported (Bondy and Naderi, 1994). Furthermore, Peltola *et al.*, reported that physiological lutuneizing hormone (LH) action in the rat testis caused lipid peroxidation and maintained high activities of peroxide-metabolizing enzymes in the interstitial tissue. The steroidogenic steps regulated by the P450 enzymes were the most likely sites of ROS generation. These free radicals are produced as a result of electron leakage due to interaction of steroid products or other pseudosubstrate with the enzymes. The inability of the pseudosubstrate to be oxygenated promotes the release of ROS (Peltola *et al.*, 1996). Several environmental contaminants are reported to produce ROS through the above-mentioned mechanism and thereby produce oxidative changes in testis. Studies from our laboratory have shown that, exposure to lindane (5 mg/kg body weight/day) orally for 30 days can induce oxidative stress and impair steroidogenesis in adult rat testis (Sujatha *et al.*, 2001). Increased levels of hydrogen peroxide generation associated with decrease in activity of antioxidant enzymes superoxide dismutase (SOD), catalase and glutathione reductase indicated induction of oxidative stress by lindane. The activities of steroidogenic enzymes 3β-hydroxysteroid dehydrogenase and 17β-hydroxysteroid dehydrogenase were decreased indicating reduced steroidogenesis. Decreased activity of steroidogenic enzyme and SOD, which has been shown to act as an alternate regulatory switch in testicular steroidogenesis (Kumar *et al.*, 1990) have been linked to the possible role of ROS in reduced steroidogenesis.

Mitochondrial respiration is the main biological source of superoxide anion radicals under physiological conditions. Free radicals generated in the subcellular compartments are efficiently scavenged by the antioxidant enzymes of the corresponding cellular compartments. The subcellular membranes have been reported to undergo permeability changes following enhanced lipid peroxidation and glutathione depletion leading to disruptions of testicular function (Chance *et al.*, 1979). Similar impairment of antioxidant system in mitochondrial and microsome-rich fractions of rat testis has been noted with methoxycholr, TCDD and nonylphenol (Latchoumycandane *et al.*, 2002; Latchoumycandane and Mathur, 2002; Chitra and Mathur, 2004). Environmental contaminants, such as mercuric chloride, have been shown to produce dose-dependent and differential influence over plasma membrane enzymes in the testis of rat, thereby affecting the physico-chemical properties of testicular membranes (Ramalingam and Vimaladevi, 2002). As testis is the major site for testosterone production, which plays a crucial role in the development of secondary sexual characteristics and in initiation as well as regulation of spermatogenesis, androgen deprivation due to exposure to environmental toxicants have the potential to interrupt the endocrine regulation of spermatogenesis and consequently affects the reproductive performance.

Effect of Environmental Contaminants on Epididymis

After completing the testicular phase of maturation, spermatids as well as their surrounding fluid microenvironment enters the epididymis, which through its very active secretory and reabsortive functions creates sequential changes in composition of luminal fluid throughout its length (Brooks, 1983). Epididymal sperm maturation involves a complex cascade of changes involving the remodeling of the sperm surface, the induction of chromatin condensation, the acquisition of movement, and development of the potential for capacitation. Correlations between the highly regionalized epididymal activities and sperm characteristics linked to their survival and fertility potential have been reviewed by Dacheux *et al.* (2003). During epididymal transit, the sperm concentration reaches nearly 10^{10} cells/ml at the same time as sperm motility and metabolism increase, with the threat of oxidative stress for gametes (Taylor, 2001). Several antioxidant systems have been identified in the epididymal environment and they are believed to protect sperm during maturation and storage (Vernet *et al.*, 2004). The existence of an epididymal-specific form of glutathione peroxidase and a high concentration of SOD within the epididymis of rat has been reported (Perry *et al.*, 1992, 1993). Recent experiments utilizing golden hamsters demonstrated that removal of the male accessory sex glands, which secrete a variety of antioxidants, results in increased oxidative damage in the form of DNA fragmentation, most notably in the less mature sperm in the caput epididymis (Chen *et al.*, 2003). Removal of accessory sex gland also resulted in delayed pronuclear formation at fertilization and increased embryonic wastage, which indicates the importance of accessory sex gland secretions in protecting sperm during the epididymal maturation process. Epididymal antioxidant status and degree of oxidative stress experienced by sperm therein may be important determinants of male fertility.

For the epididymal sperm maturation, the timing and the cellular events are linked exclusively to the transit of the sperm through the different parts of the epididymal tubule. Several environmentally relevant, endocrine-disruptive chemicals including methoxychlor, vinclozolin, PCBs, TCDD and chloroethylmethanesulphonate (CEMS), an alkylating agent, have been reported to decrease the sperm density of cauda epididymis significantly with little or no decrease in testicular sperm number (Klinefelter and Suarez, 1997). It has been speculated that CEMS and other environmental contaminants mentioned above result in some form of androgen deprivation, leading to decrease in epididymal sperm transit time. Toxicant-induced accelerated sperm transition through the epididymis not only affects the number of sperm available for ejaculation, but also the quality by compromising the process of epididymal sperm maturation, and in turn, the fertility of sperm. Androgen deprivation has also been reported to induce a wave of apoptotic cell death in the epididymis, subsequently hampering the epididymal sperm maturation and fertility (Fan and Robaire, 1998).

The assessment of sperm parameters is considered an important aspect of identifying and characterizing male reproductive toxicants. Various environmental contaminants have been shown to induce oxidative stress in sperm. Oxidative stress is known to play a major role in the aetiology of defective sperm function via the mechanism involving the induction of peroxidative damage to the plasma membrane (Aiken *et al.*, 1987). LPO constitutes a free radical oxidation process in which polyunsaturated fatty acid of the cell membrane decompose to yield, among others, highly reactive lipid hydroperoxides, H_2O_2, hydroxyl radicals and malondialdehyde (MDA). MDA has been demonstrated to cause cross-linking and polymerization of membrane components and may contribute to mutagenic, genotoxic and carcinogenic effects. Such peroxidative damage to the sperm plasma membrane leads to a loss of membrane fluidity and integrity as a result of which the spermatozoa lose their competence to participate in the membrane fusion events associated with fertilization (Storey, 1997).

High concentrations of toluene appear to have direct effects on epididymal spermatozoa without testicular effects (Ono *et al.*, 1999). Exposure of seven-week-old male Sprague–Dawley rats to toluene vapor inhalation (6000 ppm; 2h/day daily for 5 days) has been shown to cause significant reduction in the epididymal sperm counts, sperm motility, sperm quality and in vitro penetrating ability to zona-free hamster eggs, while no exposure-related changes in the testes weight or spermatogenesis were detected. Since there were no significant changes observed in serum testosterone levels following 1 month of exposure to 6000 ppm toluene, androgen independent effects on epididymal sperm have been suggested. It has been speculated that, these effects of toluene may be mediated through peroxidative damages to sperm.

Studies from our laboratory have shown that, nonylphenol, degradation product of alkyl- phenol polyethoxylate, can induce oxidative stress in epididymal sperm of rats (Chitra *et al.*, 2002). Nonylphenol when administered orally to male rats at the dose levels of 1, 10 and 100 mg/Kg body weight per day for 45 days caused a significant reduction in the epididymal sperm motility and sperm count in a dose-dependent manner. The activities of superoxide dismutase, catalase, glutathione reductase and glutathione peroxidase were also decreased, while the levels of hydrogen peroxide and lipid peroxidation increased significantly leading to induction of oxidative stress in sperm. Similar effect was observed with methoxychlor on antioxidant system of goat epididymal sperm (in *vitro*) (Gangadharan *et al.*, 2001).

In addition peroxidative damage to the sperm plasma membrane, ROS is known to affect the integrity of sperm genome, inducing high frequencies of single and double DNA strand breaks (Hughes *et al.*, 1996). Cigarette smoking has been proved to induce seminal oxidative stress affecting both the sperm function as well as sperm DNA integrity (Saleh *et al.*, 2002). Substantial oxidative modification in sperm DNA in terms of the oxidized deoxynucleotide and associated male infertility has been demonstrated in smokers (Shen and Ong, 2000). Sun *et al.*, observed a negative correlation between DNA fragmentation and sperm quality as reflected by sperm motility, morphology and count. The degree of sperm chromatin condensation appears to be a sensitive and functional indicator of male reproductive function. The sperm nuclear chromatin undergoes condensation during epididymal maturation; thereby the DNA of normal spermatozoa is relatively resistant to denaturation by acid treatment. Arvindan *et al.* (1997) found a strong correlation between DNA strand break and susceptibility of spermatozoa to low pH-induced DNA denaturation, which is, in turn negatively correlated with the fertilizing potential of spermatozoa. Occupational exposure to organophosphorous pesticides in Mexican agricultural worker has been reported to alter chromatin structure as revealed by increased susceptibility to in situ acid-induced denaturation (Sanchez-Pen *et al.*, 2004). Giri *et al.* (2002) reported the genotoxic potential of carbosulfan as revealed by chromatin abberation, bone marrow micronucleus and sperm head abnormality assays in Swiss albino mice. Similar results have been reported for malathion, another organophosphorous pesticide. *In vitro* experiments carried out by Contreras and Badilla (1999) demonstrated the direct effects of parathion and its metabolite paraoxon on human sperm viability, chromatin condensation and fertilization ability.

Conclusions

All reported studies in human and animals support that the induction of oxidative stress as a possible mechanism underlying the adverse effects of environmental contaminants on male reproductive function. Data from our laboratory shows decreased activity of steroidogenic enzyme and SOD in testis following the exposure of adult rats to lindane, which could be linked to the possible role of ROS in reduced steroidogenesis. It is clear from several studies that when a spermatozoon is subjected to oxidative stress, both the plasma membrane and the integrity of the DNA will come under attack. Although small physiological amounts of ROS play important role in normal reproductive function, high amount mediate wide range of pathological effect that affect fertility.

Acknowledgements

The authors thank the staff of Bioinformatics center, Pondicherry University, Pondicherry for providing various facilities. B. Saradha acknowledges the Indian Council of Medical Research, New Delhi, India for a Junior Research Fellowship. P. P. Mathur acknowledges the receipt of financial support from the Department of Science and Technology, Government of India.

References

Abdollahi M, Ranjbar A, Shadnia S, Nikfar S, and Rezale A. (2004). Pesticides and oxidative stress: a review. Med. Sci. Monit., 10, RA141-147.

Aguilar-Mahecha A, Hales BF, and Robaire B. (2001). Expression of stress response genes in germ cells during spermatogenesis. Biol. Reprod., 65, 119-127.

Aitken RJ, and Clarkson JS. (1987). Cellular basis of defective sperm function and its association with the genesis of reactive oxygen species by human spermatozoa. J. Reprod. Fertil., 81, 459-469.

Aravindan GR, Bjordahl J, Jost LK, and Evenson DP. (1997). Susceptibility of human sperm to in situ DNA denaturation is strongly correlated with DNA strand breaks identified by single-cell electrophoresis. Exp. Cell Res., 236, 231-237.

Bagchi M, and Stohs SJ. (1992). *In vitro* induction of reactive oxygen species by 2,3,7,8-tetrachlorodibenzo-p-dioxin, endrin and lindane in rat peritoneal macrophages, and hepatic mitochondria and microsomes. Free Rad. Biol. Med., 14, 11-18.

Baird WM, and Ralston SL. (1997). Carcinogenic polycyclic aromatic hydrocarbons. *In*: Bowden, G. T., Fischer, S. M. (Eds.), *Comprehensive Toxicology*, Vol. 12. *Chemical Carcinogens and Anticarcinogens.* Pergamon Press, New York, pp. 171-200.

Bendsen E, Laursen SB, Olesen C, Westergaard LG, Andersen CY, and Byskov AG. (2001). Effect of 4-octylphenol on germ cell number in cultured fetal human gonads. Hum. Reprod., 16, 236-243.

Bondy SC, and Naderi S, (1994). Contribution of hepatic cytochrome P450 systems to the generation of reactive oxygen species. Biochem. Pharmacol., 48, 155-159.

Bradlow HL, Davis DL, Lin G, Sepkovic D, and Tiwari R. (1995). Effects of pesticides on the ratio of 16α/2-hydroxyestrone: a biologic marker of breast cancer risk. Environ. Health Perspect., 103, 147-150.

Brooks DE. (1983). Epididymal functions and their hormonal regulation. Aust. J. Biol. Sci., 36, 205-221.

Butterfield DA, and Launderback CM. (2002). Lipid peroxidation and protein oxidation in Alzheimer's disease brain: potential causes and consequences involving amyloid beta-peptide associated free radical oxidative stress. Free Rad. Biol. Med., 32, 1050-1060.

Carlsen E, Giwercman A, Keiding N, and Skakkebaek NE.(1992). Evidence for the decreasing quality of semen during the past 50 years. Br. Med. J., 305, 609-613.

Chance B, Sies H, and Boveris A. (1979). Hydroperoxide metabolism in mammalian organs. Physiol. Rev., 59, 527–534.

Chen H, Chow PH, Cheng SK, Cheung AL, Cheng LY, and O WS. (2003). Male genital tract antioxidant enzymes: their source, function in the female, and ability to preserve sperm DNA integrity in the golden hamster. J. Androl., 24, 704-711.

Chitra KC, and Mathur PP. (2004). Vitamin E prevents nonylphenol-induced oxidative stress in testis of rats. Indian J. Exp. Biol., 42, 220-223.

Chitra KC, Latchoumycandane C, and Mathur PP. (2002). Effect of nonylphenol on the antioxidant system in epididymal sperm of rats. Arch. Toxicol., 76, 545-551.

Colborn T, Vom Saal FS, and Soto AM. (1993). Developmental effects of endocrine- disrupting chemical in wildlife and humans. Environ. Health Perspect., 101, 378-384.

Contreras HR, and Badilla J. (1999). Morphofunctional disturbance of human sperm after incubation with organophosphorate pesticides. Biocell, 23, 135-141.

Craig EA. (1985). The heat shock response. Crit. Rev. Biochem., 18, 239-280.

Dacheux JL, Gatti JL, and Dacheux F. (2003). Contribution of epididymal secretory proteins for spermatozoa maturation. Microsc. Res. Tech., 61, 7-17.

Dai D, Cao Y, Falls G, Levi EP, Hodgson E, and Rose RL. (2001). Modulation of Mouse P450 Isoforms CYP1A2, CYP2B10, CYP2E1, and CYP3A by the Environmental Chemicals Mirex, 2,2-Bis (*p*-chlorophenyl)-1,1-dichloroethylene, Vinclozolin, and Flutamide. Pesticide Biochem. Physiol., 70, 127-141.

De Lamirande E, and Gagnon C. (1993). A positive role for the superoxide anion in triggering hyperactivation and capacitation of human spermatozoa. Int. J. Androl., 16, 21-25.

Doreswamy K, Shrilatha B, Rajeshkumar T, and Muralidhara (2004). Nickle-induced oxidative stress in testis of mice: evidence of DNA damage and genotoxic effects. J. Androl., 25, 996-1003.

Fan X, and Robaire B. (1998). Orchidectomy induces a wave of apoptotic cell death in the epididymis. Endocrinology,139, 2128-2136.

Farber JL. (1994). Mechanisms of cell injury by activated oxygen species. Environ. Health Perspect., 102, 17-24.

Forman D, and Moller H. (1994). Testicular cancer. Cancer Surv., 19, 323-341.

Gangadharan B, Murugan MA, and Mathur PP. (2001). Effect of methoxychlor on antioxidant system of goat epididymal sperm *in vitro*. Asian J. Androl., 3, 285-288.

Gao HB, Tong MH, Hu YQ, You HY, Guo QS, Ge RS, and Hardy MP. (2003). Mechanisms of glucocorticoid-induced Leydig cell apoptosis. Mol. Cell Endocrinol., 199, 153-163.

Giri S, Giri A, Sharma GD, and Prasad, SB. (2002). Mutagenic effects of carbosulfan, a carbamate pesticide. Mutat. Res., 519, 75-82.

Gutteridge JM, and Halliwell B. (2000). Free radicals and antioxidants in the year 2000. A historical look to the future. Ann. N. Y. Acad. Sci., 899, 136-147.

Halliwell B.(1989). Current status review: free radicals, reactive oxygen species and human disease. A critical evaluation with special reference to atherosclerosis. Br. J. Exp. Pathol., 70, 737-757.

Holland MK, Alvarez JG, and Storey BT.(1982). Production of superoxide and activity of superoxide dismutase in rabbit epididymal spermatozoa. Biol. Reprod., 27, 1109-1118.

Hughes CM, Lewis SEM, McKelvey-Martin VJ, and Thompsen W. (1996). A comparison of baseline and induced DNA damage in human spermatozoa from fertile and infertile men, using a modified comet assay. Mol. Reprod. Dev., 2, 613-619.

Inyang F, Ramesha A, Kopsombut P, Niaz M S, Hooda DB, Nyandaa AM, and Archibong AE. (2003). Disruption of testicular steroidogenesis and epididymal function by inhaled benzo(a)pyrene. Reprod. Toxicol., 17, 527-537.

Izu H, Inouye S, Fujimoto M, Shiraishi K, Naito K, and Nakai K. (2004). Heat shock transcriptional factor 1 is involved in quality- control mechanisms in male germ cells. Biol. of Reprod., 17, 18-24.

Kehrer JP.(1993). Free radicals as mediators in tissue injury and disease. Crit. Rev. Toxicol., 23, 21-48.

Klinefelter GR, and Suarez JD. (1997). Toxicant-induced acceleration of epididymal sperm transit: Androgen-dependent proteins may be involved. Reprod. Toxicol., 11, 511-519.

Kumar GP, Seerwani N, Laloraya M, Nivsarkar M, Verma S, and Singh A. (1990). Superoxide dismutase as a regulatory switch in mammalian testicular steroidogenesis. Biochem. Biophys. Res. Commun., 173, 302-308.

Latchoumycandane C, and Mathur PP. (2002). Effect of methoxychlor on the antioxidant system in mitochondrial and microsome-rich fractions of rat testis. Toxicology,176, 67-75.

Latchoumycandane C, Chitra KC, and Mathur PP. (2002). The effect of 2,3,7,8-tetrachlorodibenzo-p-dioxin on mitochondrial and microsomal fractions of rat testis. Toxicology, 171, 127-135.

Olanow CW. (1990). Oxidation reaction in Parkinson's disease. Neurology, 40, 32-37.

Ono A, Kawashima K, Sekita K, Hirose A, Ogawa Y, Saito M, Naito K, Yasuhara K, Kaneko T, Fuanya T, Inoue T, and Kurokawa, Y. (1999). Toluene inhalation induced epididymal sperm dysfunction in rats. Toxicology, 139, 193-205.

Peltola V, Huhtaniemi I, Metsa-ketela T, and Ahoutapa M. (1996). Induction of lipid peroxidation during steroidogenesis in the rat testis. Endocrinology, 137, 105-112.

Peltola V, Mantyla E, Hutaniemi I, and Ahutupa M. (1994). Lipid peroxidation and antioxidant enzyme activities in the testis after cigarette smoke inhalation or administration of polycholorinated biphenyls or polycholorinated napthalenes. J. Androl., 15, 353-362.

Perry AC, Jones R, Niang LS, Jackson RM, and Hall L. (1992). Genetic evidence for an androgen regulated epididymal secretory glutamine peroxidase where transport does not contain a selenocystein codon. Biochem. J., 285, 863-870.

Perry AC, Jones R, and Hall L. (1993). Isolation and characterization of rat cDNA clone encoding a secreted superoxide dismutase reveals the epididymis to be a major site of its expression. Biochem. J., 293, 21-25.

Polyakov LM, Chasovskikh MI, and Panin LE. (1996). Binding and transport of benzo(*a*)pyrene by blood plasma lipoproteins: the possible role of apolipoprotein B in this process. Bioconjug. Chem., 7, 396-400.

Print CG, and Loveland KL. (2000). Germ cell suicide: new insights into apoptosis during spermatogenesis. Bioessays, 22, 423–430.

Ramalingam V, and Vimaladevi V. (2002). Effect of mercuric chloride on membrane-bound enzymes in rat testis. Asian J. Androl., 4, 309-311.

Rosenblum I, Gavaler JS, and Van Thiel DH. (1985). Lipid peroxidation: a mechanism for ethanol associated testicular injury in rats. Endocrinol., 116, 311-318.

Saleh RA, Agarwal A, Sharma RK, Nelson DR, and Thomas AJ. (2002). Effect of cigarette smoking on levels of seminal oxidative stress in infertile men: a prospective study. Fertil. Steril., 78, 491-499.

Samanata L, and Chainy GBN. (1997). Comparison of hexachlorocyclohexane-induced oxidative stress in the testis of immature and adult rats. Comp. Biochem. Physiol., 3, 319-327.

Sancar A. (1995). DNA repair in humans. Annu. Rev. Genet., 29, 69-105.

Sanchez-Pen LC, Reyes BE, Lopez-Carrillo L, Recio R, Moran-Martynez J, Cebrian ME, and Quintanilla-Vegaa B. (2004). Organophosphorous pesticide exposure alters sperm chromatin structure in Mexican agricultural workers. Toxicol. Appl. Pharmacol., 196, 108-113.

Senthil Kumar J, Banudevi S, Sharmila M, Murugesan P, Srinivasan N, Balasubramanian K, Aruldas MM, and Arunakaran J. (2004). Effects of vitamin C and E on PCB (Aroclor 1254) induced oxidative stress, androgen binding protein and lactate in rat Sertoli cells. Reprod. Toxicol., 19, 201-208.

Sharpe RM. (1994). Regulation of spermatogenesis. *In*: Knobil E, and Neill JD. (Eds.), *The Physiology of Reproduction*, Second edition. Raven Press Ltd, New York, pp.1363-1434.

Shen H, and Ong C. (2000). Detection of oxidative DNA damage in human sperm and its association with sperm function and male infertility. Free Radic. Biol. Med., 28, 529-536.

Sikka SC, Rajasekaran M, and Hell Strom WJG. (1995). Role of oxidative stress and oxidants in male infertility. J. Androl., 16, 464-468.

Sinclair S. (2000). Male infertility: nutritional and environmental considerations. Altern. Med. Rev., 5, 28-38.

Spector A. (1995). Oxidative stress- induced cataract: mechanism of action. FASEB J., 9, 1173-1182.

Storey BT. (1997). Biochemistry of the induction and prevention of lipoperoxidative damage in human spermatozoa. Mol. Hum. Reprod., 3, 203-214.

Sugihara K, Kitamura S, Yamada T, Ohta S, Yamashita K, Yasuda M, and Fujii-Kuriyama Y.(2001). Aryl hydrocarbon receptor (AhR)-mediated induction of xanthine oxidase/xanthine dehydrogenase activity by 2,3,7,8-tetrachlorodibenzo-*p*-dioxin. Biochem. Biophys. Res. Comm., 281, 1093-1099.

Sujatha R, Chitra KC, Latchoumycandane C, and Mathur PP. (2001). Effect of lindane on testicular antioxidant system and steroidogenic enzymes in adult rats. Asian J. Androl., 3, 135-138.

Sun JG, Jurisicova A, and Casper RF. (1997). Detection of deoxyribonucleic acid fragmentation in human sperm: correlation with fertilization in vitro. Biol. Reprod., 56, 602-607.

Taylor CT. (2001). Antioxidants and reactive oxygen species in human fertility. Environ. Toxicol. Pharmacol., 10, 189-198.

Toyama Y, and Yuasal S. (2004). Effects of neonatal administration of 17β-estradiol, β-estradiol 3-benzoate, or bisphenol A on mouse and rat spermatogenesis. Reprod. Toxicol., 19, 181-188.

Vernet P, Aitken RJ, and Drevet JR. (2004). Antioxidant strategies in the epididymis. Mol. Cell Endocrinol., 216, 31-39.

Wilson VS, and LeBlanc GA. (1998). Endosulfan elevates testosterone biotransformation and clearance in CD-1 mice. Toxicol. Appl. Pharmacol., 148, 158-168.

Wong WY, Zielhuis GA, Thomas CMG, Merkus HMWM, and Steegers-Theunissen RPM. (2003). New evidence of the influence of exogenous and endogenous factors on sperm count in man. Eur. J. Obstet. Gynecol. Reprod. Biol.,110, 49-54.

Wu X, Faqi AS, Yang J, Pang B, Ding X, Jiang X, and Chahoud I. (2002). 2-Bromopropane induces DNA damage, impairs functional antioxidant cellular defenses, and enhances the lipid peroxidation process in primary cultures of rat Leydig cells. Reprod. Toxicol., 16, 379-384.

Yang JM, Arnush M, Chenc OY, Wu XD, Pang B, and Jiang XZ. (2003). Cadmium-induced damage to primary cultures of rat Leydig cells. Reprod. Toxicol., 17, 553-560.

Younes M. (1999). Free radicals and Reactive Oxygen Species, *In*: Margaurdt H, Schafer SG, and McClellan R, and Welsch F. (Eds.), *Toxicology*, Academic Press, New York, pp-111-125.

22

Endocrine Regulation of Metabolism, Oxidative Stress and Reproduction: Physiological Implications of Functional Interactions

O.V. Oommen,[1] F. Sunny,[2] M. Smita[1], J.M. George[1], P. Sreejith[1], R.S. Beyo[1], L. Divya[1], A.S. Vijayasree[1], M. Manju[1], C. Johnson[1] and M.A. Akbarsha[3]*

[1]*Department of Zoology, University of Kerala. Kariavattom, Thiruvananthapuram – 695 581*

[2]*Department of Zoology, University College, Thiruvananthapuram – 695 034*

[3]*Department of Animal Science, Bharathidasan University, Tiruchirappalli – 620 024, Tamil Nadu*

ABSTRACT

Vertebrates have a number of metabolic pathways to meet the various requirements for a successful survival. For more than a century, it has been known that thyroid hormones control the metabolic rate of mammalian tissues and a parallel situation is reported in sub mammalian species also. These hormones influence different life processes such as metabolism, growth, differentiation, reproduction and behaviour. To serve these functions, thyroid hormones play roles that are permissive or synergistic with other hormones. The endocrine control of oxidative metabolism in fish, amphibian and reptiles, both *in vivo* and *in vitro* has been evaluated in our laboratory for the last 25 years. Currently our focus is on the regulation of the antioxidant

**Corresponding Author.* E-mail: oommenvo@yahoo.com

defense system, and its regulation by hormones and other plant derived antioxidants. Plants produce secondary metabolites of which steroid compounds form a major group of chemicals. We have established a non-genomic effect of steroid hormones on various metabolic pathways employing transcription inhibitors, in addition to the classic genomic effect. Therefore, we have inferred that the genomic and non-genomic actions of steroids interact to produce the synergistic effects. The combination of the dual mechanisms endows steroids with considerable diversity, range and power in regulating various physiological functions in fishes. A possible cross talk of rapid steroid signaling with genomic actions is hypothesized. Recently, we have established the sequence of spermatogenesis in a less known amphibian group, Caecilians, employing electron microscopy for the first time to establish the various stages. We have also evidence to suggest that the mullerian gland in Caecilians could be the progenitor of accessory sex glands, a preparation for internal fertilization and a prelude to terrestrialisation.

Introduction

The beginning of the molecular endocrinology era was in the 1950's with the determination of the amino acid sequences of the nonapeptides vasopressin and oxytocin by du Vigneaud and his co-workers. (Popenos and du Vigneaud, 1954, Pierce and du Vigneaud, 1950). In the ensuing 25 years, the amino acid sequences of approximately 30 different polypeptide hormones and regulatory peptides were established through the tedious approach of the isolation and purification of scarce peptides from large amounts of endocrine tissues. The last major break through in endocrinology has been the application of techniques of molecular biology that are more powerful and easier to use, in general than those of polypeptide biochemistry. In an amazingly short time (< 10 yrs), the genes for a number of receptors as well as prohormones have been isolated and sequenced. Not too surprisingly, the nucleotide sequences of some oncogenes producing abnormal growth show some similarity to those of receptors for growth factors regulating normal growth.

Endocrinology thus might greatly contribute to the development of regulated transgene expression system. Regulated endocrine axes might ideally represent a paradigm of any gene switch system and knowledge in the field of endocrinology should be translated in the set-up and refinement of these gene expression tools. Potential application will range from oncology to cardiovascular diseases, hormone deficiencies and inherited diseases.

Vertebrates have evolved a number of metabolic pathways to meet various requirements for a successful survival. For a century, it has been known that thyroid hormones control the metabolic rate of mammalian tissues which is thoroughly investigated and continues to attract attention of endocrinologists, despite the appearance of scores of papers in the last 100 years of thyroid research, ever since Magnus Levey's experiments in 1895. Thyroid hormones (TH) influence different life processes such as metabolism, growth, differentiation, reproduction and behaviour. To serve these functions TH may play roles that are permissive or synergistic with other hormones. The oxidative metabolism is responsible for cellular oxidation of metabolic substrates and production of energy which is conserved in the form of ATP. The entire process of biological oxidation involves a number of enzymes, metabolic substrates and utilizes molecular oxygen at the mitochondrial level.Therefore, respiration or oxygen uptake acts as an index of oxidative metabolism or biological oxidation. This in turn reflects the physiological status of an individual organism. The endocrine control of oxidative metabolism in fish, amphibian and reptiles has been the subject of numerous studies. Administration of T_3 and T_4 caused an increase in plasma concentration of the same and altered the total mitochondrial nucleic acid, emphasizing the role of TH in teleost metabolism (Oommen and Matty, 1991). Studies from our laboratory on *Anabas testudineus* have revealed a dose dependent and time dependent stimulatory role on intermediary metabolism. Administration of different doses of T_3 stimulated the activities of the enzymes, alanine amino transferase (ALT), aspartate amino transferase (AST) and glucose-6-phophatase (G-6-Pase) after short term (hours) and a long term (days) treatment. This may mean a glucogenic action of T_3 in fish (Nair and Oommen, 1997). Thyroid hormone *in vivo* and *in vitro* is known to stimulate the activities of cytochrome oxidase, α-glycerophosphate dehydrogenase and succinate dehydrogenase in teleosts (Peter and Oommen, 1989a, b, c, 1991, 1993). The increased mitochondrial respiration in *Anabas* may partly be due to an increase in the respiratory chain components of hepatic mitochondria (Peter and Oommen, 1987). In the life cycle of amphibians several periods of thyroidal activity have been observed. In earlier studies, we have measured the plasma concentration of T_3 and T_4 employing RIA in the tadpoles of a little known tropical amphibian, *Rana curtipes*, for the first time (Valamparampil and Oommen, 1997). Both T_3 and T_4 become detectable in the late prometamorphic stage. A gradual increase in T_3 and T_4 is observed during this period. A sharp rise in hormone levels is apparent at the onset of metamorphic climax. Thereafter, the circulating hormones (particularly T_4) decrease rapidly and reach levels similar to prometamorphosis. This kind of a variation in thyroid hormone concentration at various stages of metamorphosis is essential for the normal development of a frog. Our studies also point out the possibility of peripheral conversion of T_4 and T_3 and T_3 appears to be the most potent from of TH in this amphibian species. Adult amphibians have received very little attention with regards to seasonal changes of plasma concentration of thyroid hormones. The periods of maximal thyroid activity are associated with breeding or preparation for breeding in many species. Reports from our laboratory have also established that TH (T_4 and T_3) exert a dose responsive stimulatory effect on the activities of hepatic mitochondrial oxidative enzymes in an apodan amphibian, *Gegenophis carnous* (Sutharam *et al.*, 1990).

Thyroid hormones are well known to elicit an amazingly diverse set of developmental and physiological responses in different tissues of higher vertebrates and appear to have the same function in the maintenance and stimulation of the metabolic rate in reptiles too. In *Calotes versicolor*, thyroidectomy decreased and thyroxine accelerated oxygen consumption of liver and muscle *in vivo* and *in vitro* (Oommen, 1981). It is clear from our report that thyroidectomy does not change the glycemic level, T_4 and T_3 administration in thyroidectomised lizards produces hyperglycemia (Jacob and Oommen, 1990). In a subsequent study on the thyroid-gonad interrealationship on the intermediary metabolism in a reptile, we have reported that both T_4 and testosterone are effective in stimulating intermediary metabolism in lizards, irrespective of the removal of either thyroid or testis in *C. versicolor* and exhibit a direct interrelationship which may involve gonadotropins and thyrotropin. The decrease of hepatic cholesterol in alloxan-pre treated lizards after T_4 administration emphasizes that the influence of thyroid gland on lipolysis is independent of the actions of insulin (Jacob and Oommen, 1992). Thus the effects of thyroid hormone represent complex integration of pathways both at the cellular level and in terms of whole-animal physiology.

Endocrine Regulation of Ionic Metabolism: Role of Insulin and Melatonin

Ion and acid-base regulation is critical to the normal function of most metabolic and physiological processes. Na^+K^+ ATPase (sodium pump) and Ca^{2+} ATPase are universal membrane bound enzymes in eukaryotic cells. They serve for a general "house keeping" function by establishing the important ionic gradients that underlie the transport of ions and nutrients across the cell membranes to maintain intracellular homeostasis (Lechene, 1988, Sweadner and Goldin, 1980). Na^+K^+ATPase also provide the driving force for vectorial Na^+ transport in a variety of vertebrate osmoregulatory epithelia, including kidney tubules, bladder, intestine, nasal salt glands in reptiles and birds, and rectal gland in sharks.

There are abundant evidences that the activity of these enzymes is regulated by a number of hormones (Leena and Oommen, 2000., Peter *et al.*, 2000). Although Na^+K^+ ATPase and Ca^{2+}ATPase are different enzymes, their mechanism of action have more features in common (Graham *et al.*, 1993). We, therefore, undertook a study to evaluate the short term effect of insulin, a major anabolic hormone, on ionic metabolism in brain and branchial tissue of *Anabas testudineus.* Short-term effects of insulin on total brain and branchial Na^+K^+ATPase, Ca^{2+}ATPase and Na^+, K^+, and Ca^{2+}ions in the gill were investigated (Vijayasree *et al.*, 2005).

The short term *in vivo* administration of alloxan and insulin markedly influenced the activity of Na^+ K^+ATPase and Ca^{2+}ATPase in the brain and gill of *A. testudineus.* The major function of insulin is the control of whole body glucose homeostasis. Alloxan administration resulted in hyperglycemia evidenced by increased level of blood glucose. This may be due to insulin deficiency caused by the impairment of β-cell by alloxan. In teleosts, alloxan treatment resulted in hyperglycemia due to β-cell depletion (Bhatt and Bora, 1981). Insulin administration to alloxan pretreated fish reversed the effect of this β-cell cytotoxin (Nair, 1996). The results showed a stimulatory effect on the total brain Ca^{2+}ATPase after alloxan injection. It is assumed that an increase in Ca^{2+}ATPase may account for an increased amount of intracellular calcium required for biochemical events taking place inside the cells. Branchial Na^+K^+ATPase was significantly stimulated while Ca^{2+}ATPase inhibited after alloxan treatment. Na^+ in the gill significantly increased, Ca^{2+} content showed a reversal while K^+ showed no change. This suggests that alloxan exerts its inhibitory effect on the ATP-driven Ca^{2+} transport via its action on the Ca^{2+} pump protein rather than the membrane permeability to Ca^{2+}. The increased activity of brain Na^+K^+ATPase at 3 and 24 hr by insulin on alloxan pretreated fish may account for the stimulated co-transport of glucose and its utilization for energy requirements and the excitatory action on neurons in the brain. The stimulated Ca^{2+}ATPase activity in the brain may be due to increased intracellular calcium. At 24 hr, the activity of branchial Na^+ K^+ATPase and Ca^{2+}ATPase in alloxan pretreated specimens were significantly stimulated by insulin. This may be due to increased synthesis of these enzyme units. Administration of insulin 1U/fish in normal fish significantly inhibited the activities of brain and branchial Na^+K^+ATPase while Ca^{2+}ATPase showed a stimulatory effect at 3 and 24 hr compared to control. Inhibition of total branchial Ca^{2+}ATPase by insulin may be due to increased Ca^{2+} concentration. Na^+ and Ca^{2+} ions in the gill were significantly elevated and K^+ ion reduced at 3, 6 and 24 hr. This may have occurred as a result of increased activity of Na^+K^+ATPase and Ca^{2+}ATPase. The stimulation of ionic uptake corroborates earlier work that insulin- mediated translocation of sodium pump from intracellular stores to the cell surface is considered to be the main mechanism of pump stimulation in skeletal muscle (Evans, 2002., Shulman and Fox, 1996).

Significant elevation of protein content in the brain and gill substantiates the anabolic role of insulin. Insulin treatment gradually decreased the glucose level at 3, 6 and 24 hr confirming the antidiabetic effect of insulin in alloxan pretreated groups. But insulin treatment alone caused a hyperglycemic effect at 3 and 6 hr, followed by a decrease at 24 hr. Since glucose has been shown to have an effect on ATPase activity, it may be inferred that ATPase is controlled primarily by insulin along with other hormones (Yamashita *et al.*, 1992). Owada *et al.* (1999) demonstrated that glucose–induced inhibition of Na^+ K^+ATPase activity in pancreatic β-cells is mediated by Protein kinase C (PKC). PKC is a central signaling molecule in insulin action, and atypical PKCs are activated by the lipid products of phosphatidyl inositol 3-kinase (Sweeney *et al.*, 2001). The possible mechanism by which insulin controls Na^+K^+ATPase activity appears to be tissue specific and include reversible covalent modification of catalytic subunits, activation by a rise in intracellular Na^+ concentration, altered Na^+ sensitivity and changes in subunit gene or protein expression (Sweeney and Klip, 1998).

A need for rapid switches in pump activity arises in response to variable salt intake, stress or exercise. Thus insulin can elicit both rapid and sustained changes, reflecting adaptation to metabolic need. Short term insulin mediated sodium pump stimulation can also be secondary to an increase in the cytoplasmic Na^+ concentration (Ewart and Klip, 1995). These further confirm our observation in *A. testudineus* These data are consistent with the hypothesis that insulin performs a role in hydro mineral regulation in freshwater teleosts. Clearly, future studies are required to elucidate whether insulin controls the expression of any of the pump subunits in brain and gill tissue, in teleosts.(Sweeney *et al.*, 1998).

The pineal organ of ectotherms is a part of a circadian pacemaker system transducing photothermal environmental inputs into a neurochemical signal (Firth and Kennaway, 1989). Within the field of endocrinology, the pineal hormone, melatonin has emerged as the top multifunctional hormone and its wide distribution (Kvetnoy, 1999., Kvetnoy *et al.*, 2000) emphasizes its key role as an intracellular neuroendocrine regulator and co-ordinator of many complex and interrelated biological processes. Any functional relationship between time period and ionic metabolism in response to melatonin treatment is yet a topic of speculation. Hence, the present study evaluates and compares the time and dose dependent response to melatonin by total brain and gill Na^+, K^+ ATPase, Ca^{2+}ATPase and also by inorganic ions like Na^+, K^+ and Ca^{2+} in a fresh water teleost *A. testudineus* (Bloch).

Melatonin (MEL) administered at doses of 0.20, 0.40 and 0.60 mg/g bwt inhibited the Na^+K^+ ATPase activity of both brain and gill tissues.The activity of Ca^{2+} ATPase in brain and gills also showed significant inhibition. The data of the present study have revealed an inhibitory action of the pineal hormone, melatonin on total brain and gill Na^+, K^+ATPase activity (Divya *et al.*, unpublished observations from authors' lab). It has been reported that in adult male rose-ringed parakeets, administration of melatonin at a dose of 25mg/100 gm bwt for 30 consecutive days resulted in glycemia (Maitra *et al.*, 2000). Melatonin has been shown to suppress blood glucose levels in rats (Baydas *et al.*, 2002). Hypoglycemic effect of melatonin has been demonstrated in the liver of goldfish where a decrease in plasma glucose and an increase in liver glycogen have been observed after melatonin injection (Delhaunty *et al.*, 1978). Na^+ coupled glucose co-transport was reported in the mucosa of the pigmented conjunctiva of rabbit (Horbie *et al.*, 1997). Melatonin is known to elevate gluconeogenesis in the liver and changes in plasma glucose level itself in turn affect hepatic melatonin binding (Choy *et al.*, 2001). The decreased activity of Na^+K^+ATPase by MEL may account for inhibition of glucose co-transport and its utilization for energy requirement and thus contributes to inhibitory action on neurons in the brain of *A. testudineus.* The ion gradients formed by the enzyme are necessary for Na^+ coupled transport of nutrients into cells for osmotic balance, cell volume regulation and restoration of membrane potential in excitable tissues (Gloor, 1997). A direct consequence of inhibition of Na^+K^+ATPase is an increase in concentration of intracellular Na^+ and depletion of K^+. Increase in intracellular Na^+ in turn raises intracellular concentration of Ca^{2+} by stimulating Na^+Ca^{2+} exchanges. The variation in ionic concentration in response to different time periods also suggests an involvement of melatonin in Na^+Ca^{2+} pump dynamics and the transport of ions.

Control groups of both brain and gill tissues which received vehicle injection at two different times of a day (6.00 and 18.00 hr) themselves showed significant differences, with the evening groups in both brain and gills showing an increased Na^+K^+ and Ca^{2+}ATPase. Leena and Oommen (2001) have earlier reported a circadian variation in activities of lipogenic enzymes (malic enzyme, glucose-6-phosphatedehydrogenase and isocitrate dehydrogenase, with significantly higher values in evening than morning. The remodeling of biological lipid composition in response to changes in temperature is perhaps the most recurrent cellular adaptation of ectotherms to alterations in growth and is believed to be responsible for thermal compensation of membrane function (Carvey and Hazel, 1989).

Trout acclimated to low temperature exhibited increased turnover rates of cytochrome C oxidase, erythrocyte Na^+K^+ATPase and increased membrane fluidity (Gibbs, 1998). Melatonin (0.40 mg/g bwt) administered in the evening for duration of 12 hrs produced an inhibitory effect in brain and gill Na^+K^+ and Ca^{2+} ATPase.

Evening melatonin treatment (0.40 µg/gm bwt) produced an inhibitory effect in both brain and gill Na^+K^+ATPase and Ca^{2+}ATPase. Administration of melatonin in the evening for 12 hours produced a significant decrease in brain and gill Na^+K^+ATPases. The effect of melatonin is believed to be mediated by membrane bound receptors. When injected late in the light phase, melatonin was effective as short photoperiod in inducing gonadal atrophy whereas melatonin injections at other times were ineffective (Stetson and Watson–Whitmyre, 1984). The interpretation was that melatonin administered in the afternoon advanced the circadian sensitive phase, so that it was then coincident with endogenous peak (Reiter, 1987).This observation is related to the hypothesis that the coincidence of the injection of melatonin with a phase of sensitivity was a deciding factor for the appearance of a physiological effect (Tamarkin *et al.*, 1976., Reiter, 1987). Na^+, K^+ and Ca^{2+} ions of the gills were significantly reduced in evening treated groups as a result of reduced pump activity.

Melatonin treatment for 12 hrs in the morning had null effect on brain Na^+K^+ATPase. The early morning injection probably coupled to the endogenous evening melatonin synthesis. As melatonin is chronically available, it binds all available melatonin receptors, making the animal refractory to the action of additional melatonin. The activity of Ca^{2+}ATPase was found to be inhibited. Melatonin being lipid soluble, it can readily enter cells, where in the cytosol, it may form a complex with calmodulin.The inhibition of cAMP through melatonin receptor is a common pathway like in the rat neonatal pars distalis, the main effect of melatonin appears to be the inhibition of activated second messenger pathway as melatonin dose dependently inhibits forskolin-stimulated cAMP production. Similar treatment in gills stimulated both Na^+K^+ATPase and Ca^{2+}ATPase. It has also been shown that circulating levels of melatonin are increased significantly in homing pigeons after flight so as to reduce flight-induced hyperthemia (John *et al.*, 1993., George, 1997). Melatonin administration produced a 2-3 °C decline in temperature in mice. In both trout and gold fish Na^+K^+ATPase activity fell sharply with increases in acclimation temperature. In rainbow trout gill, Na^+ K^+ATPase activity was elevated than that of kidney suggesting that these animals may rely more on branchial uptake of ions at higher temperature (Houston *et al.*, 1982). In both trout and gold fish Na^+ K^+ATPase activity fell sharply with increases in acclimation temperature. This phenomenon was believed to be the result of temperature related changes in membrane lipid composition rather than alterations in enzyme type or amount (Hazel and Prosser, 1974). The present study confirms the circadian activity of melatonin and is the first report on its regulatory role on the ionic turnover (Na^+, K^+ and Ca^{2+}) and enzyme activity in the gills and brain of *Anabas testudineus.* The precise mechanism of action of MEL in teleosts remains to be understood.

Hormonal Regulation of Antioxidants in Combating Oxidative Stress

Thyroid hormone T_3 is known to be associated with increased oxidation capacity, lipid peroxidation and susceptibility to oxidative stress and decreased levels in rat liver mitochondrial population (Venditti *et al.*, 1999). Thyroid hormones are known to influence cellular metabolism including oxygen consumption (Peter and Oommen, 1993., Freak and Oppenheimer, 1995), oxidative phosphorylation and proton leak (Harper and Brand, 1994). We have conducted short term *in vivo* and *in vitro* experiments to study the regulatory effect of triiodothyronine in the antioxidant defense system in the liver of a teleost, *Anabas testudineus.*

Our results have shown that lipid peroxidation products such as MDA and CD decreased in 6-PTU treated group compared to control. This is due to the suppression of lipid peroxidation by the thyroid hormone inhibitor, 6-PTU. Injection of thyroid hormone did not cause any significant change in MDA concentration, but increased the concentration of CD in 15 min thyroid hormone injected fish when compared to 6-PTU treated groups. The rise in CD is due to the increased lipid peroxidation, induced by thyroid hormone. MDA forms only after a certain period as it is the end product of lipid peroxidation.The activity of SOD and catalase were increased in 6-PTU treated specimen and thyroid hormone decreased their activity. This clearly reveals that 6-PTU produces reactive oxygen species, which in turn, enhances defense mechanism through the activation of enzymes. The increased production of hydrogen peroxides as a result of the removal of reactive oxygen species by SOD may have stimulated the catalase activity in this study. Glutathione reductase and GPx activity increased in 6-PTU treated group and hormone injection had varied effects. Glutathione content in the liver was not altered by 6-PTU treatment and T_3 injection decreased its amount in 30 min and at 60 min it was increased. The decrease in glutathione content and an increase after T_3 administration suggests the availability of substrate for increased GPx activity by T_3 treatment in 6-PTU injected fish specimens. Injection of thyroid hormone in normal fish decreased MDA and CD concentration and increased the activity of SOD, catalase and GPx. This helps to suggest that both hypothyroidism and hyperthyroidism alter lipid peroxidation and antioxidant defense mechanism in *A. testudineus.* A long–term effect of T3 *in vivo* on these processes in *Anabas* was earlier reported by Varghese *et al.*, 2001.

The fast *in vivo* action of T_3 on antioxidant activity is substantiated by similar *in vitro* results. Hyperthyroidism resulted in marked increase in antioxidant enzyme such as catalase, GPx and also non-enzymatic glutathione content in the *in vitro* study. The change in lipid peroxidation may be due to the significant change in oxidative stress in mitochondria by altered thyroid status of the body. Triiodothyronine, thus, is capable of combating the experiment-induced free radical formation by activating the enzyme systems. It appears to be a non-genomic effect of thyroid hormone as is reported for steroid hormones in a teleost (Sunny *et al.*, 2002). However, the precise mechanism of action remains to be understood.

Nuclear Transcription Factor (NF-κB): Role in Oxidative Stress and Cancer

The NF-KB transcription family represents an important group of regulators of a broad range of genes involved in cellular responses to inflammation and stress signals. This family contains p50, p52, p65 (Rel A), c-Rel and Rel B proteins (Aggarwal *et al.*, 2004). In most cell types, inactive NF-êB complexes are sequestered in the cytoplasm via their interaction with inhibitory proteins known as Ikappa Bs (IêBs). In response to multiple stimuli, including cytokines, viral and bacterial pathogens and stress–inducing agents, the latent cytoplasmic NF-êB/IêBα complex gets activated by phosphorylation of IêB (Karin and Delhase, 2000., Karin and Lin., 2002, Kumar *et al.*, 2003, and Kumar and Boriek, 2003). The physiological functions of various proteins of NF-êB signaling pathway have been studied by targeted disruption of individual gene. The gene targeting experiments have also revealed the importance of other key components of NF-êB signaling pathways in mouse development (Senftleben and Karin, 2002). Due to its critical role in cell survival, cell adhesion, inflammation, differentiation and cell growth, NFê-B has been implicated in carcinogenesis (Garg and Aggarwal, 2002). It was recently shown that oxidative stress stimulated the activity of NF-êB whereas corticotropin releasing hormone inhibited both constitutive and oxidative stress induced NF-êB, indicating that this transcription factor may play a critical role in stress response (Karalis *et al.*, 2004). Many antioxidants present in the typical diet, such as curcumin, catechins, genistein, capsaicin, resveratrol, caffeic acid phenethyl ester (CAPE) have been shown to suppress the activation of the NF-êB signaling pathway and thus is a good target for both prevention and therapy of cancer (Kumar *et al.*, 1998, Natarajan *et al.*, 1996, Singh *et al.*, 1996, Aggarwal *et al.*, 2004).

Plant Products and Steroids: A Cross Talk

Plants produce several secondary metabolites, of which steroid compounds form a major group of chemicals. Recently there is an explosion of research concerning the health-benefits of phytochemicals in food and medicinal plants. The present study examined the ability of allitin and raw garlic to exert an antilipogenic activity in a teleost, *Anabas testudineus*. Garlic (*Allium sativum*), a member of the Lily family is one of the widely researched medicinal plants. Allitin is an analogue of allicin which is one of the major components of freshly crushed garlic homogenate. It contains 98 per cent diallyl disulphide and 2 per cent diallyl trisulphide. It is synthetically prepared from allyl thiol by oxidation with iodine. It is purified by column chromatography and vacuum distillation.Experiment was done in 2 sections. In the first section, fish was administered with allitin, intraperitonially at a concentration of 10mg/kg bwt for 10, 20, and 30 days, using physiological saline as vehicle. In the second section, raw garlic was given in conjunction with the standard fish feed at concentration equal to human consumption of 10, 20 and 40 garlic per day for 45 days. The controls were given standard fish feed.

The overall evaluation of results confirm the lipid lowering properties of garlic and/or its derivatives especially diallyldisulphide (DADS), the major constituent of allitin, for the first time in a teleost (Johnson *et al.*, 2003).The results obtained from this study confirm a decrease in the activity of lipogenic enzyme after allitin administration. Another finding with allitin was the reduction in both HDL and LDL levels; however, there was a marked increase in HDL and decrease in 10 and 20 day doses when compared to their controls. With the supplementation of raw garlic, HDL levels showed a significant increase with a corresponding decrease in LDL. Also, garlic, when given through diet, reduced the cholesterol levels in the present experimental fish. The consumption of garlic tends to show a decrease in the plasma cholesterol and LDL levels with maximum effect of 40 gm dose/day. The 40gm garlic administration dose showed a promising result with low plasma cholesterol and high HDL levels. The probable explanation for reduction in the hepatic total lipid along with cholesterol in fish administered with allitin at all the 3 doses (10mg/kg bwt/day for 10, 20 and 30 days respectively) may be due to the deposition of lipids in other tissues or the conversion to other sources of stored forms in the animal body. The reduction in the lipid profiles substantiated the decrease in total hepatic lipid level. The total lipid content in garlic fed fish confirmed an inverse relationship between the liver and muscle. Contradictory to the influence of allitin, where there was a direct relationship between the liver lipid and blood cholesterol, feeding the fish with garlic increased the hepatic and decreased the muscle total lipid content along with blood cholesterol. So the reduction in plasma cholesterol may be interpreted on the basis of conversion of serum cholesterol into hepatic lipid. The decrease in the serum LDL levels and total lipid level in muscle tissue encourages using garlic as a therapeutic agent. These finding support the fact that garlic is effective in various ailments of cardiovascular diseases because of its ability to reduce serum cholesterol. Therefore, it may be concluded that allitin, an analog of allicin and raw garlic demonstrates a cholesterol lowering effect in a fish (Johnson, 2004.) This study indicates that raw garlic feeding and allitin administration produced a hypocholestrolemic effect as evidenced by changes in the activity of lipogenic enzymes and lipid profile. It will be of interest to include garlic products in aquaculture fish feed and to have fish with low LDL and cholesterol and high HDL for human consumption.

Dual Action of Steroid Hormones in Teleost

Steroid hormones bind to specific receptors, which are intracellular transcription factors and exert positive or negative effects on the expression of target genes (Beato and Klug, 2000). Numerous studies have implicated the role of glucocorticoids in the regulation of hepatic intermediary metabolism (Mommsen *et al.*, 1999; Vijayan *et al.*, 1997; Sunny *et al.*, 2002) and osmoregulation (McCormick, 1995., Sunny and Oommen, 2000, 2001) in teleosts. Sex steroids, apart from their role in reproduction and control of secondary sexual characters, are renowned for their effect on intermediary metabolism in vertebrates particularly, in mammals (Swinnen *et al.*, 1997., Cinci *et al.*, 2000). These effects of steroids are characterized by a specific delay in hormone action and sensitivity towards transcription and translation inhibitors. In addition to the well-known genomic action, increasing evidence for rapid, non-genomic steroid effects has been demonstrated for virtually all groups of steroids (Falkenstein *et al.*, 2000; Borski 2000; Borski *et al.*, 2001., Sunny and Oommen, 2001, 2002). Cortisol and other glucocorticoids are known for their rapid influence on various physiological activities like hormone secretion from pituitary (Borski *et al.*, 2001), neuronal excitability (Venero and Borrel, 1999), behavioral response (Orchinik, 1991), lipogenic enzymes (Sunny *et al.*, 2002) and enzymes of osmoregulation (Sunny and Oommen, 2002). Similar to other steroids, rapid effects of testosterone on intracellular calcium ions have been observed in human granulosa luteinizing cells (Machelon *et al.*, 1998), T cells (Benten *et al.*, 1999a), mouse IC-21 macrophages (Benten *et al.*, 1999b) and in freshly isolated Sertoli cells (Gorczynska and Handelsman, 1995). Testosterone produced a rapid dose dependent inhibition of chloride secretion in cultured rat efferent duct cells (Leung *et al.*, 2001) and on lipogenic enzymes in fish (Sunny *et al.*, 2003). Non-genomic action of steroids on cellular function is mediated by conventional second messenger cascades (Wehling *et al.*, 1996) and free intra cellular Ca^{2+} (Wehling *et al.*, 1990, Sunny and Oommen, 2002).

The present review intends to analyze the long term [*in vivo*] and short term [*in vivo* and *in vitro*] effects of steroids on the activity of hepatic ME, G6PDH, ICDH, G-6-Pase, AST and ALT, and branchial Na^+K^+ and Ca^{2+}ATPase in a fresh water fish, *Oreochromis mossambicus*. The modes of action of hormones were identified by administrating actinomycin D-the transcription inhibitor and puromycin–the translation inhibitor prior to hormones' treatment.

Long-term *in vivo* Study

In long term *in vivo* study, exogenous administration of DES and testosterone stimulated the activity of hepatic lipogenic enzymes–ME, G6PDH and ICDH and cortisol, corticosterone and progesterone inhibited the activity of all the lipogenic enzymes in *O. mossambicus*. The activity of G-6-Pase, AST and ALT was significantly inhibited by DES and stimulated by glucocorticoids, progesterone and testosterone. Exogenous administration of multiple doses of cortisol, corticosterone, and DES enhanced branchial Na^+K^+ ATPase activity in *O. mossambicus*. Sunny and Oommen (2004) also reported that *in vivo* administration of cortisol conspicuously increased the number of shallow basin chloride cells and caused noticeable changes in the microridges of pavement cell right from 24 hr treatment onwards. Among the steroids studied, only progesterone exhibited a biphasic effect on Na^+K^+ ATPase activity, lowest and middle doses significantly stimulated while highest dose significantly inhibited the enzyme activity. The lowest and middle doses of testosterone responded positively while the highest dose did not modify Na^+K^+ATPase activity. Administration of cortisol, corticosterone, DES and testosterone significantly enhanced the activity of branchial Ca^{2+}ATPase in a dose dependent manner. While only the middle and highest doses of progesterone increased Ca^{2+}ATPase, the lowest dose did not evoke any response on its activity. Among the steroid hormones, progesterone has minimum stimulatory effect and cortisol has maximum stimulatory effect on both ATPase activities in *O. mossambicus*. When protein synthesis inhibitors actinomycin D or puromycin was administered prior to hormones' treatment, the activity of all the enzymes studied was either completely or partially inhibited when compared to hormone treated groups irrespective of their nature of activity. The complete or partial inhibition of ATPases' activity by the inhibitors suggest genomic mode of action for all the hormones studied (Figure 22.1).

Short-term *in vivo* Study

Exogenous administration of glucocorticoids, testosterone and DES *in vivo* produced significant increase in the activity of branchial Na^+K^+ ATPase and Ca^{2+} ATPase in *O. mossambicus*. Cortisol significantly inhibited ME, G6PDH and ICDH as early as 30 min *in vivo* and continued to be decreased at 60 and 120 min, the maximum inhibition being noticed at 30 min. While testosterone significantly stimulated the activity of these enzymes as early as 30 min *in vivo*. The rapid action of steroids on lipogenic enzymes and branchial ATPases' activity was not inhibited by 10^{-6} M actinomycin D *in vitro* when treated prior to hormones suggesting non-genomic action.

Short-term *in vitro* Study

Short term *in vitro* incubation of gill cells with of 10^{-6} M glucocorticoids, testosterone and DES significantly increased Na^+K^+ ATPase activity as early as 5 min. Incubation of liver cell suspensions with 10^{-6}M cortisol also inhibited the activity of all three lipogenic enzymes' activity as early as 5 min *in vitro*. Actinomycin D treatment did not interfere with the inhibiting effect of cortisol on enzyme activities when measured at 10 min in the *in vitro* system. The transcriptional inhibitor appeared to partially block the effect of cortisol *in vivo*. Incubation of 10^{-6}M testosterone significantly stimulated the activity of ME, G6PDH and ICDH in the liver of *O. mossambicus* as early as 5 min *in vitro*. The rapid

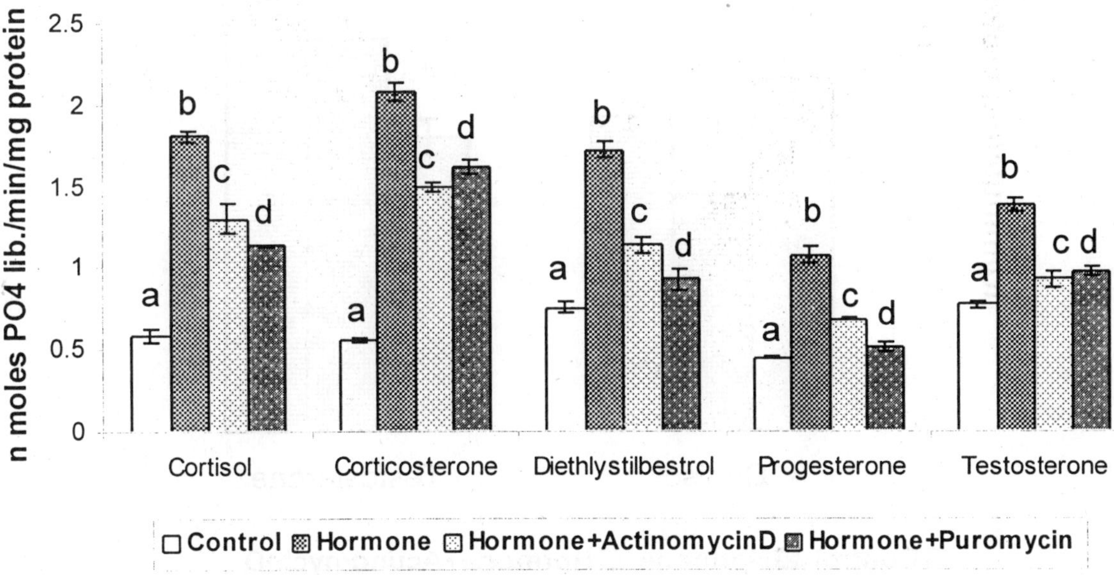

Figure 22.1: Genomic Action of Steroids on Branchial Na^+ K^+ ATPase Activity
Each histogram represents mean ± SEM of 6 animals.

action of hormones on enzymes was not inhibited by 10^{-6} M actinomycin-D *in vitro* when treated prior to hormones suggesting non-genomic action (Figure 22.2).

The study long-term and short-term studies clearly demonstrated two separate modes of action for steroid hormones in *O. mossambicus*. In long term treatment the activity of enzymes' was sensitive to transcription and translation inhibitors suggesting genomic action. While in short term study the enzymes rapidly responded to the hormones as early as 30 min *in vivo* and 5 min *in vitro*. The activity of enzymes' was in- sensitive to the inhibitors suggesting a rapid non-genomic action.

Cortisol and corticosterone significantly enhanced the hepatic gluconeogenic enzyme G-6-Pase and the aminotransferases ALT and AST. Glucose-6-phosphatase catalyses a critical step in gluconeogenesis (Nordlie *et al.*, 1993), and plays an important role in the regulation of glucose homeostasis. Glucocorticoids considerably enhanced hepatic G-6-Pase activity, which was in agreement with similar studies by Voice *et al.* (1997) and Plewka *et al.* (2000) in dexamethasone treated rats. The elevated aminotransferases' activities, especially of ALT in glucocorticoids' treated fish specify a higher liver capacity for amino acid metabolism. The enhanced AST and ALT activity in the present study support same kind of earlier studies on fish (Vijayan *et al.*, 1994; 1996; Oommen and Kaippallil,

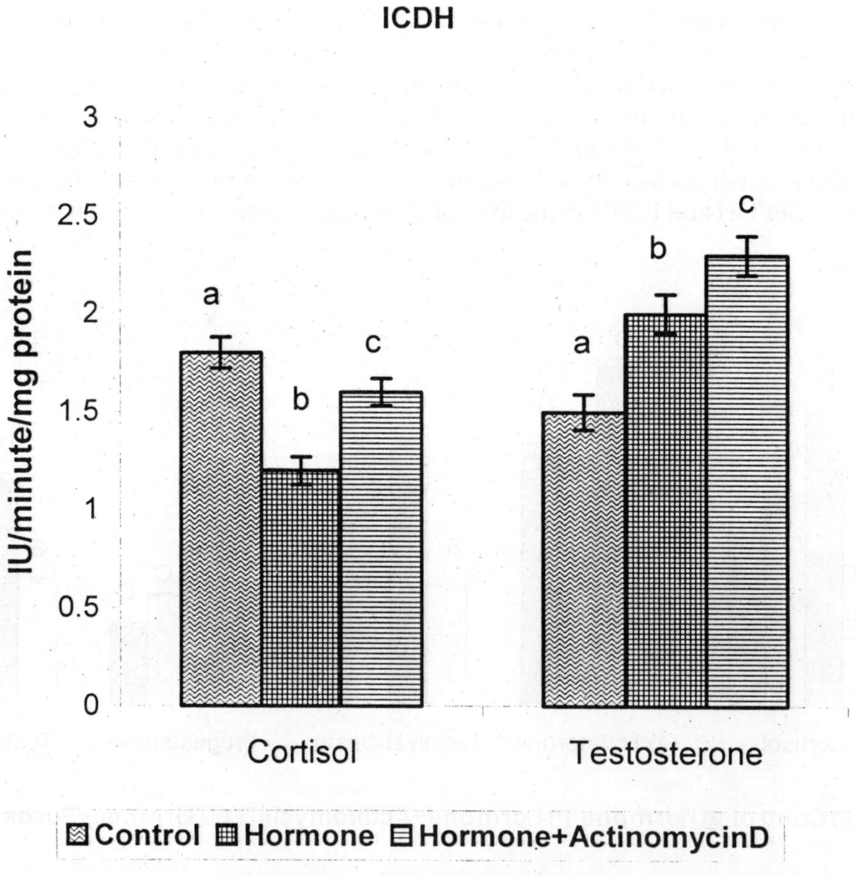

Figure 22.2: Non-genomic Action of Steroids on Lipogenic Enzymes. Groups with different letter headings are significantly different (P<0.05).

1995). Cortisol-induced gluconeogenesis may be important for long term maintenance of glucose levels during stress in teleost (Vijayan *et al.*, 1994).

The specific activities of all the three lipogenic enzymes–ME, G6PDH and ICDH in *O. mossambicus* were inhibited by long-term exogenous administration of cortisol and corticosterone. It is important that liver and adipose tissue possess an active lipogenesis and it is stated that the pentose phosphate pathway may be the major route for the break down of glucose in fish (Fidue *et al.*, 1983). Lipolysis is promoted by a number of hormones, including cortisol, thyroid hormones, prolactin, GH, ACTH, and somatostatins (Sheridan and Kao, 1998). Cortisol and corticosterone generally seem to be lipolytic in anurans during metamorphosis. The *de novo* biosynthesis of glucose by G-6-Pase, AST, and ALT under the influence of cortisol confirms a lipolytic effect of glucocorticoids in this teleost. In support for this view, Vijayan *et al.* (1996) has reported an increased capacity for fatty acids oxidation in sea raven treated with cortisol, cortisol treatment was also reported to mobilize lipids in fish (Sheridan, 1986) and it facilitated the utilization of glycerol for gluconeogenesis in teleosts (Vijayan *et al.*, 1991). Treatment with all steroids significantly stimulated branchial Na^+K^+ and Ca^{2+}ATPases' showing that not only interrenal but also gonadal steroids play a critical role on the enzymes of osmoregulation in fresh water fish *O. mossambicus*.

The genomic mode of steroid action, which has been widely accepted for the past four decades can satisfactorily account for this type of slow long-term action of steroids, which are sensitive to protein synthesis inhibitors. Earlier reports from this laboratory confirmed the inhibitory role of actinomycin D and puromycin alone and together with GCs on AST, ALT and RNA content in a teleost and a reptile (Oommen and Kaippallil, 1995). Reduction in enzymes' activity and RNA content compared to hormones' treatment convincingly established the fact that the inhibition occurred was due to the action of protein synthesis inhibitors. It is assumed that a similar situation on enzyme synthesis would have prevailed in the present study with these inhibitors. In the present study, actinomycin D or puromycin when treated together with the hormones, completely or partially inhibited the activity of all the enzymes studied. The inhibition by the inhibitors has plausibly confirmed the genomic action of steroids. This shows that the inhibitors were able to mask a part of the action of hormones, which confirms that the action of hormones is sensitive to that of inhibitors. This finding helped to assume that in tilapia the long-term action of steroids on enzymes of intermediary metabolism is taking place through the stimulation of protein synthesis suggesting a genomic action.

The stimulatory effect of steroids on branchial ATPases' activity and the inhibitory and stimulatory control on lipogenic enzyme activity by cortisol and testosterone, in short term study, is rapid, time dependent, specific, partially sensitive and in-sensitive to transcription inhibitor, actinomycin D. It would appear that these steroids may act independent of gene transcription or in a non-genomic fashion to regulate osmoregulation and lipid metabolism in fish. The opposing action of cortisol and testosterone on lipogenic enzymes indicates that they are not only working in a catabolic and anabolic fashion, respectively, but also suggests their actions are specific. That is, the actions of cortisol are not one that would simply reflect an inhibition by any steroid, since testosterone had the opposite effect.

It appears that steroids specifically control the key enzymes involved in carbohydrate, protein and lipid metabolism in the liver of *O. mossambicus* to suggest a general influence of steroids on intermediary metabolism. The activities of lipogenic and gluconeogenic enzymes and branchial ATPases' are either stimulated or inhibited following the administration of steroids. The long term *in vivo* i.p. injection of steroids intensely reveals that glucocorticoids, testosterone and progesterone are hyperglycemic; DES is hypoglycemic, testosterone and DES lipogenic, and glucocorticoids and progesterone antilipogenic (lipolytic) in tilapia. It is also established that amino acid catabolism, mostly that of alanine may be a major source of substrate for gluconeogenesis.

It is evident that steroid hormones influenced intermediary metabolism and osmoregulation in this fish by two different mechanisms of action. The new proposed mechanism of action of steroid hormones should not be viewed as a challenge to the classical genomic mechanism. These diverse mechanisms prove an integrated action of steroid hormones, which may be rapid after short duration or delayed after prolonged duration to address the physiological needs of the fish. A possible cross talk of rapid steroid signaling with genomic actions has been hypothesized. Steroid hormones may act through both mechanisms simultaneously, and these mechanisms appear to be important co-mediators of the wide array of cellular steroid effects. This is in full agreement with the dual modes of actions of steroids reported on the inhibition of liver lysosomal β- glucuronidase in rat (Hinz and Hirschelmann, 2000) and on the regulatory role of nitric oxide on HeLa cells (Marino *et al.*, 2001).

From the present study it can be inferred that the genomic and non-genomic actions of steroids interact to produce synergistic effects. The combination of the dual mechanisms endows steroids with considerable diversity, range and power in regulating various physiological functions in *O. mossambicus*. A possible cross talk of rapid steroid signaling with genomic actions has been hypothesized (Figure 22.3). Steroid hormones may act through both mechanisms simultaneously, and these mechanisms appear to be important co-mediators of the wide array of cellular steroid effects. In view of recent developments, it may be assumed that membrane-initiated signals appear to be the primary response of the target cell to steroid hormones and may be prerequisite to subsequent genomic activation.

Functions of Steroid Hormones and AMH in Caecilian Reproduction

Within the modern amphibia, several trends in reproductive evolution are evident. All three extant anurans, urodeles, and apodans show a reduction in the use of aquatic habitat with a tendency towards terrestrial development. This trend is accompanied by greater reliance on internal fertilization. Apodans are a unique group of vertebrates that practice internal fertilization among amphibians. Several species are viviparous. Published reports on seasonal variations in caecilians are limited to a very few species. The caecilians have evolved a unique pattern of cystic spermatogenesis in which cysts representing different stages in spermatogenesis co-exist in a testis lobule. We observed seasonal variations of testicular activity with respect to spermatogenesis and spermiogenesis in caecilians *Ichthyophis tricolor* and *Uraeotyphlus narayani* abundant in the Western Ghats of Kerala. The phases of spermatogenic activity namely, active spermatogenesis (July–November), early regression (December-March) and spermatogenic quiescence (April-June) are identified in *Ichthyophis tricolor*.

During spermatogenesis, whether it is the cystic as in the case of fishes and amphibians or non-cystic as in amniotic vertebrates, germ cells of the testis are intimately associated with the somatic Sertoli cells. In cystic spermatogenesis, spermatogenic cysts are produced when follicle cells are associated with the primary spermatogonia. The primary spermatogonia develop synchronously, in close association with the somatic follicle cells that surround the cyst/follicle (Saez *et al.*, 1990). On the other hand, in amniotes spermatogenesis occurs in the convoluted seminiferous tubules, which are lined by a permanent epithelium in which germ cell transformation occurs as a wave along the length of the tubule. A single Sertoli cell is in contact with several germ cells at different stages of differentiation (Grier, 1993). Sertoli cells in the caecilian testis, therefore, suggest an origin and association with the germ cells in a different manner than the Sertoli cells in the amniotes where there is cystic spermatogenesis in the strict sense. However, we found that Sertoli cells in the caecilian testis are also dynamic and that newer cells arise in parafollicular cells, which later acquire a relationship with the

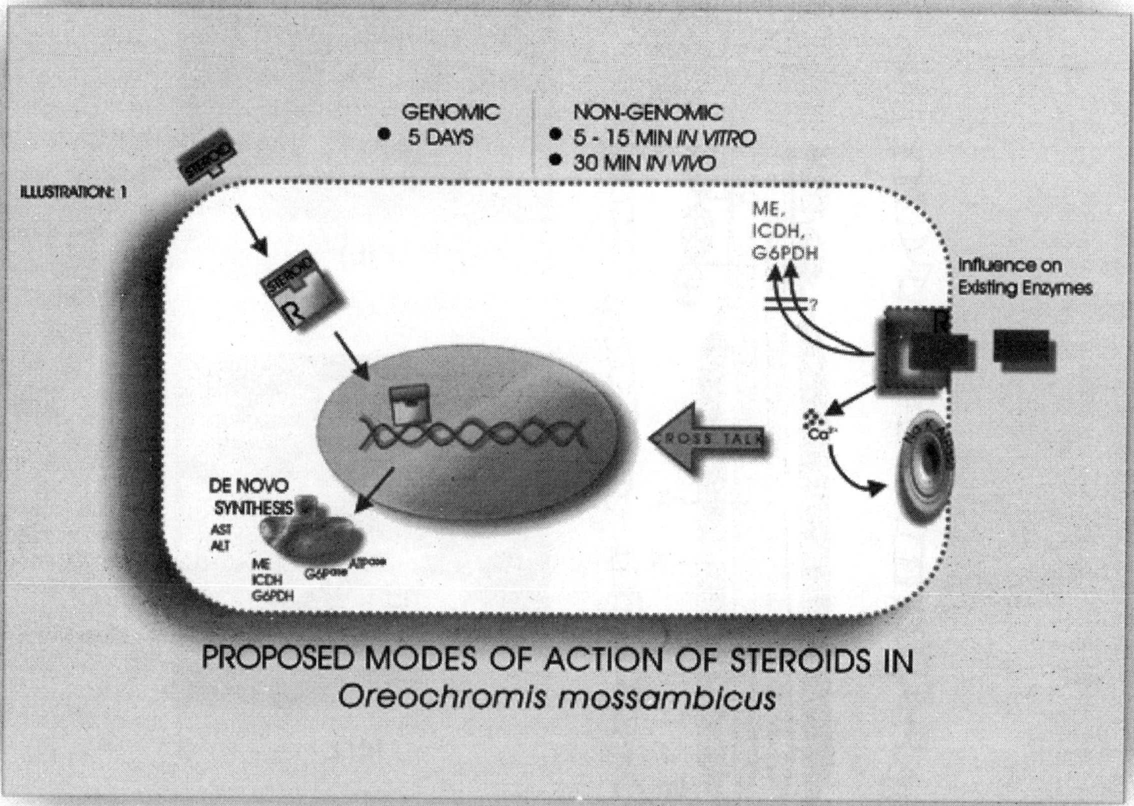

Figure 22.3: Proposed Modes of Steroid Action

wall of the tubule (Smita *et al.*, 2003, 2004). Sertoli cells in caecilian testicular lobules are extremely tall and can meet at various points in the core of the tubule, providing support for the developing germ cells until spermiation (Figure 22.4b).

It is an established fact that either interstitial or peritubular Leydig cells are present in the testis of majority of the vertebrates and they secrete androgens. These hormones are responsible for development, maintenance and regulations of the secondary sexual characters, accessory reproductive organs and to a great extent, spermatogenesis. Several authors have reported the occurrence of Leydig cells in the interstitial tissues of the caecilian testis (Exbrayat, 1992). In these studies Leydig cells are examined from histological perspective (Figure 22.4c). Various changes in the structural organization of Leydig cells during differentiation and maturation of mammalian testis are clearly indicative of some changes in the function involving steroid hormone synthesis (Robinovici and Jaffe, 1990). The prominent zones of loosely organized tubular endoplasmic reticulum generally occur near the nucleus. Scattered patches of polyribosomes can be seen on some area of its surface and they may be involved in the synthesis of some proteins, especially steroidogenic enzyme. We subjected *Ichthyophis tricolor* to a more systematic and computer assisted histometric analysis of the interstitial tissue and biochemical determination of testicular hydroxy steroid dehydrogenase activity during different phases of

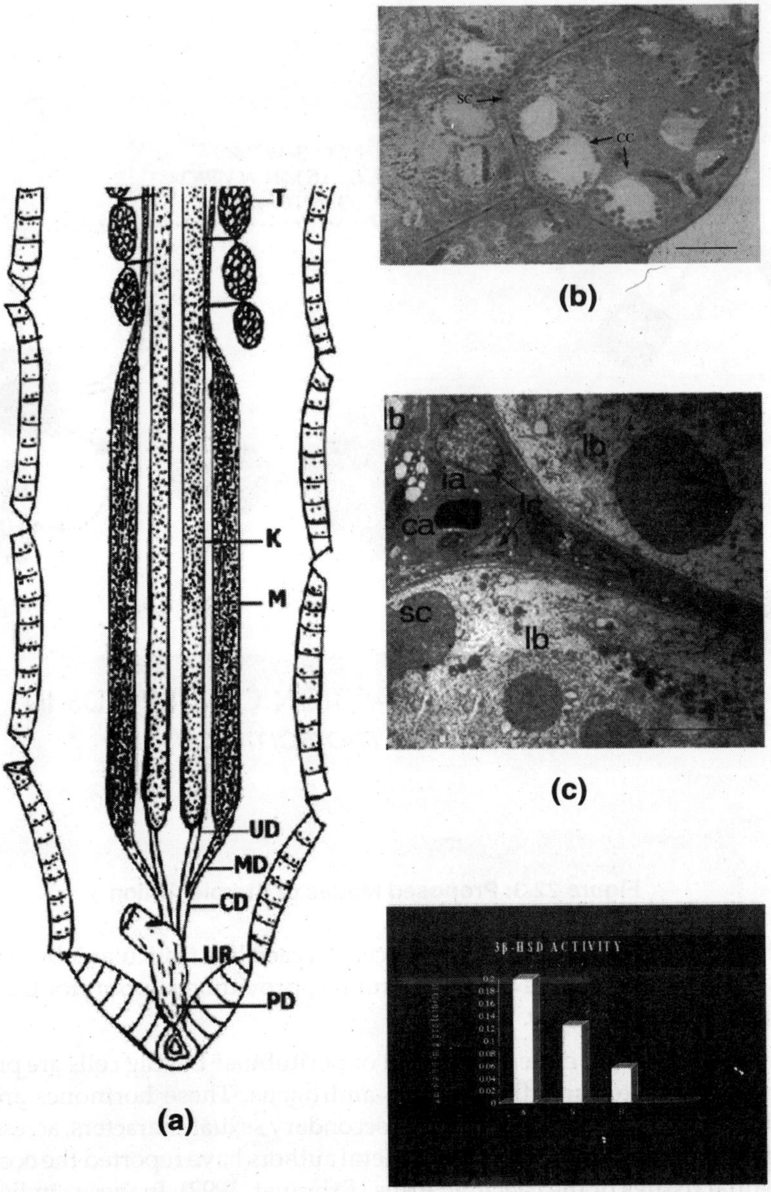

Figure 22.4(a) Drawing of the male urinogenital system of *Uraeotyphlus narayani* showing testis lobes (T), kidney (K), Mullerian gland (M), urinogenital duct (UD), Mullerian duct (MD), common duct (CD), urodeum (UR) and phallodeum (PD). Actual size; (b) Testis of *Ichthyophis tricolor* in active spermatogenesis showing hexagonal lobules containing the cell cyst (CC) and the Sertoli cells (SC) with basal nucleus. L M. Toluidine blue stained semi thin section. Scale bar = 200 μm; (c) TEM of a testis lobe showing three lobules (lb) and the interstitium (ia) lying outside the lobules. The lobules are bounded by myoid cells (arrow). The interstitium contains Leydig cells (lc), Sertoli cells (SC) and a capillary (ca). Scale bar = 20 μm; (d) 3 β-HSD activity in testis of *Ichthyophis tricolor*

spermatogenic cycle (Smita *et al.*, 2005). Activity of testicular 3β-HSD was high in animals collected in November (active spermatogenesis), at an intermediate level in March (regressive but spermiating) and lowest in April (spermatogenically quiescent) (Figure 22.4d).

The relative functions of Leydig cells and Sertoli cells in supporting and regulating spermatogenesis are yet to be defined more precisely at the molecular level. The embryonic Sertoli cells indicate that they have all organelles which are indicative of synthesis of proteins. One of these proteins is the regulatory anti-Mullerian hormone which is well known to be synthesized by the Sertoli cells. The regression of the Mullerian duct occurring shortly after testicular differentiation is due to the effect of AMH (Blanchard and Josso, 1974; Josso *et al.*, 1977). As far as caecilians are concerned, the Mullerian duct is retained in the adult males as a functional glandular structure and it was first reported by Spengal (1876). This retention is an enigmatic aspect of caecilian biology. In all other vertebrate males, the Mullerian duct regresses in the embryonic stage itself in response to the anti Mullerian hormone (AMH) secreted by the Sertoli cells of the embryonic testis and any further development of the adult is aborted. Therefore, the retention of the Mullerian duct in the adult caecilians raises a question as to whether the Sertoli cells of the embryonic testis do not secrete MIH at all. The mechanism of sperm transfer in the context of internal fertilization in the caecilians is different. The spermatozoa are directly ejaculated into the female tract making use of the eversible phallodeum as the phallus (Wake, 1977). Thus a medium suspend the spermatozoa as in amniotic vertebrates is imminent. But in the organs in the higher vertebrates concerned with the secretion of the seminal plasma *viz.*, the sexual segment of kidney (Bishop, 1959, Sarkar and Sivanandappa, 1989), epididymis, (Depegis and Dufarue, 1977., Robaire and Hermo, 1988) and Ampulla ductus efferentis (Akbarsha and Meeran, 1995, Daisy *et al.*, 2000) as in the case of reptiles, and the prostate gland, seminal vesicles (Luke and Coffey, 1994) and Ampulla ducts deferentes (Setchell *et al.*, 1994) as in the case of mammals are absent in the caecilians.

The Mullerian gland is the only glandular structure anatomically connected with the urinogenital system; therefore hypothetically, it would be the candidate organ to contribute the seminal plasma. The Mullerian duct of *Uraeotyphlus narayani* on each side is an elongated organ, white in color and extends up to the level of the transverse septum along side of the kidney, parallel to the testis lobes (Figure 22.4a). The posterior one third of the duct from below the level of the testis lobes is cylindrical organ, the Mullerian gland and in the front it extends as a connective tissue strand. Mullerian duct narrows at the posterior end and joins the urinogenital duct to form a common duct that opens into cloaca (George *et al.*, 2004a, 2005). It is a clear indication that the secretion of male Mullerian gland would potentially mix with sperm before arriving at the cloaca, strengthening the concept of Tonutti, (1931), Wake (1977, 1981) and Exbryat (1985, 1986) that male Mullerian gland of caecilian would provide a medium in which spermatozoa are suspended in and even provide nutritional support to the sperm. Therefore, we support the hypothesis of Wake (1981) that the male Mullerian gland of caecilian is an attempt towards secretion of seminal plasma, a requirement for internal fertilization, which in the mammals is taken care of by the prostate gland and seminal vesicles. Thus, the Mullerian gland of caecilians is proved to be the male accessory reproductive gland (George *et al.*, 2004b). Tonutti (1931), for the first time, suggested that the male Mullerian gland might be a functional analogue of mammalian prostate gland. The studies of the protein profile of Mullerian gland secretion and the antigenic homogeneity of some of these proteins with secretory proteins of male accessory reproductive glands of amniotes indicate an evolutionary relationship between caecilian male Mullerian gland and amniote male accessory gland.

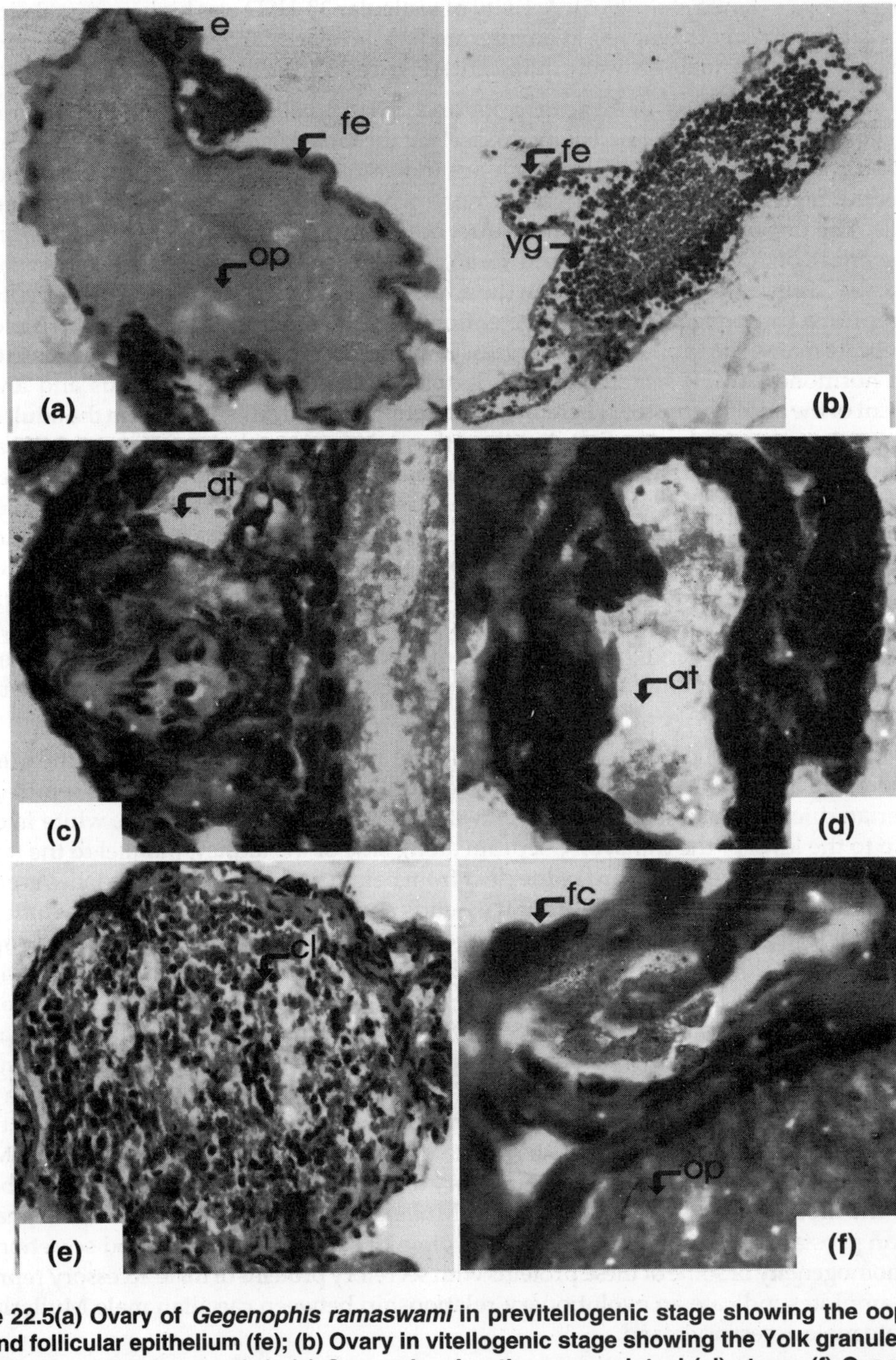

Figure 22.5(a) Ovary of *Gegenophis ramaswami* in previtellogenic stage showing the ooplasm (op) and follicular epithelium (fe); (b) Ovary in vitellogenic stage showing the Yolk granules(yg); (c, d.) Ovary in atretic stage(at); (e) Ovary showing the corpus luteal (cl) stage; (f) Ovary with follicular cells (fc).

Fertilization in caecilians occur in the upper part of the oviduct and the fertilized eggs are either laid in burrows or as may prove to be the case of most species, are retained in the oviducts until the developing larvae have completed metamorphosis (Norris, 1996). Ovarian development is typical to that of amphibians generally (Lofts, 1974). Oviparous caecilians exhibit reproductive pattern, where spermatogenesis and ovarian follicular development appear to be synchronous and dependent on season. The ovaries are paired, elongate sac-like structures. Developing ova in the follicles are associated with the thin wall of the ovary. Since the ovary of the caecilian is elongate and developing eggs are strung out, and do not develop typically in a compact ovary body, the ovary tissue seems to serve to great degree as a binding support for the ova. Depending up on the phase of the reproductive cycle in the ovaries, oogonia, previtellogenic oocytes of different sizes, vitellogenic and fully grown yolky oocytes, atretic follicles and corpus lutea are identified (Figure 22.5a-f). However, large yolky oocytes are present in greater number only during the month of July to November. It was found that granulosa cells and thecal cells contain 3β-HSD activity and hence, they may be the source of circulating ovarian steroids. At the end of the breeding season the ovary contains young follicles and will become the next crop of mature oocytes, numerous cell nests will become the young follicles of the next vitellogenic period, and primary germ cells that will give rise to new cell nests for future generation of oocytes (Norris, 1996).

Fat bodies play an important role and they vary in their weight in an inverse ratio with that of the ovarian weight. This suggests that the fat bodies seem to have supporting and not an obligatory role in vitellogenesis (Saidapur, 1989). The viviparous species, with their lengthy gestation periods and provision of nutrient material after the yolk is exhausted, reflect a very different mode of maternal investment, where contribution of maternal energy through oviductal secretion to support the developing young is also considerable. Publications from the author's laboratory revived interest in amphibian reproductive biology after a gap of 60 years in India, since the pioneering work of Seshachar in the 1930s. However, the hormonal regulation in caecilian amphibians still remains inconclusive.

Acknowledgements

Financial assistance from UGC to Oommen O.V (F.3.33/2002 (SR-II), DST (SP/SO/C12/2000), KCSCTE (SARD facility) and DST-FIST Programme is gratefully acknowledged.

References

Aggarwal BB, Takada Y, and Oommen OV. (2004). From chemoprevention to chemotherapy: common targets and common goals. Expert Opinion, 13, 1327-1338.

Aggarwal BB, Takada Y, Shishodia S, Gutierrez AM, Oommen OV, Ichikawa H, Baba Y, and Kumar A. (2004). Nuclear transcription factor NF-kappa B: Role in biology and medicine. Indian J. Exp. Biol., 42, 341-353.

Akbarsha MA, and Meeran MM. (1995). Occurrence of ampulla in the ductus deferentis of *Calotes versicolor*. Daudin. J. Morphol., 225, 261–268.

Baydas G, Canatan H, and Turkoglu A. (2002). Comparative analysis of the protective effect of melatonin and vitamin E on streptozotocin induced diabetes mellitus. J. Pineal. Res., 32, 225-230.

Beato M, and Klug J, (2000). Steroid hormone receptors: An update. Hum. Reprod. Update, 6, 225-236.

Benten WP, Lieberherr M, Giese G, Wrehlke C, Stamm O, Sekeris CE, Mossmann H, and Wunderich F. (1999a). Functional testosterone receptors in plasma membranes of T cells. FASEB. J., 13, 113-133.

Benten WP, Lieberherr M, Stamm O, Wrehlke C, Guo Z, and Wunderich F. (1999b). Testosterone signaling through internalizable surface receptors in androgen receptor frees macrophages. Mol. Biol. Cell., 10, 3113-3123.

Bhatt SD, and Bora PS. (1981). Response of blood glucose, hepatic glycogen and pancreatic islets to treatment of sulphydryl inhibitors in *Clarias batrachus* (L). J. Bio. Sci., 3, 259-268.

Bishop JE. (1959). A histological and histochemical study of the kidney tubule of the common garter snake *Thamnophis* with special reference to the sexual segment in the male. J. Morphol., 104, 307–357.

Blanchard MG, and Josso N. (1974). Source of the anti- Mullerian hormone synthesized by the fetal testes: Mullerian inhibitory activity of fetal bovine Sertoli cells in tissue culture. Pediatr. Res., 8, 968-971.

Borski RJ. (2000). Non-genomic membrane actions of glucocorticoids in vertebrates. Trends. Endocrinol. Metab., 11, 427-436.

Borski RJ, Hyde JN, Fruchtman S, and Tsai WS. (2001). Cortisol suppresses prolactin release through a non-genomic mechanism involving interactions with the plasma membrane. Comp. Biochem. Physiol. Part B, Biochem. Mol. Biol., 129, 533-541.

Carvey C, Hazel JR. (1989). Diurnal variation in membrane in lipid composition of sonoran desert teleost, J. Exp. Biol., 147, 375-391.

Choy E, Poon AMS, and Pang SF. (2001). Effects of melatonin in blood glucose and melatonin receptors in mouse hepatocytes. Biological Signals and Receptors, 10, 367-379.

Cinci G, Guerranti R, Pagani R, Carlucci F, Terzuoli L, Rosi F, and Marinello E. (2000). Fatty acid composition of phospholipids, triglycerides and cholesterol in serum of castrated and estradiol treated rats. Life Sci., 66, 1647-1664.

Daisy P, Meeran MM, and Akbarsha MA. (2000). Histological variation along the vas deferens of the Indian garden lizard, *Calotes versicolor* Daudin with special reference to the ampulla ductus deferentis. J. Endocrinol. Reprod., 4, 74–88.

Delahunty GG, Bauer M, Prack de Vlaming Vl. (1978). Effects of pinealectomy and melatonin treatment on liver and plasma metabolites in gold fish, *Carassius auratus*. Gen. Comp. Endocrinol., 35, 99-104.

Depeiges A, and Dufaure JP. (1977). Secretory activity of the lizard epididymis and its control by testosterone. Gen. Comp. Endocrinol., 33, 473–479.

Evans DH. (2002). Cell signaling and ion transport across the fish gill epithelium. J. Exp. Zool., 293, 336-347.

Ewart SH, and Klip A. (1995). Hormonal regulation of the Na^+K^+ATPase: mechanisms underlying rapid and sustained changes in pump activity. Am. J. Physiol. Cell Physiol., 269, 295-311.

Exbrayat JM. (1985). Cycle des canaux de Mu¨ller chez le ma^ le adulte de Typhlonectes compressicaudus (Dumeril et Bibron, 1841), Amphibian apode. Comptes. Rendus. Des. Sciences Paris, 301, 507–512.

Exbrayat JM. (1986). Quelques aspects de la biologie de la reproduction chez *Typhlonectes compressicaudus*, Amphibian apode. Doctorates Sciences naturelles. Universite Paris VI.

Exbrayat JM. (1992). Appareils genitaux et reproduction chez les amphibians gymnophiones. Bull. Soc. Zool. Fr., 117, 291–296.

Falkenstein E, Hanns-Chritian T, Christ M, Feuring M, and Wehling M. (2000). Multiple actions of steroid hormones-A focus on rapid, non-genomic effects. Pharmacol. Rev., 52, 513-555.

Fidue MD, Soler G, and Ruiz-Amil M. (1983). Nutritional regulation of glycolysis in rainbow trout (*Salmo gairdneri*), Comp. Biochem. Physiol., 74B, 795-799.

Firth BT, and Kennaway DJ. (1989). Thermo period and photoperiod interact to affect the phase of the plasma melatonin rhythm in the lizard, *Tiliqua rogosa*. Neuroscience letters, 106, 125-130.

Freak HC, and Oppenheimer JH. (1995). Thermogenesis and thyroid function. Annu. Rev. Nurtr., 15, 183- 192.

Garg A, and Aggarwal BB. (2002). Nuclear transcription factor-êB as a target for cancer drug development. Leukemia 16, 1053-1068.

George JC. (1997). Thermoregulatory and metabolic responses to cold exposure in birds. *In*: Maitra, S.K. (Ed.), *Frontiers in Environmental and Metabolic Endocrinology*. The University of Burdwan Publications, Burdwan, India, pp. 135-144.

George JM, Smita M, Oommen OV, and Akbarsha MA. (2004a). Anatomy, histology and ultrastructure of male Mullerian gland of *Uraeotyphlus narayani* (Amphibia: Gymnophiona). J. Morphol., 260, 33–56.

George JM, Smita M, Kadalmani B, Girija R, Oommen OV, and Akbarsha MA. (2004b). Secretory and basal cells of the epithelium of the tubular glands in the male Mullerian gland in the Caecilian, *Uraeotyphlus narayani* (Amphibia: Gymnophiona). J. Morphol., 262, 760-769.

George JM, Smita M, Kadalmani B, Girija R, Oommen OV, and Akbarsha MA. (2005). Contribution of Caecilian (Amphibia: Gymnophiona) male Mullerian gland to motility of sperm: A study in *Uraeotyphlus narayani*. J. Morphol., 263, 227-237.

Gibbs AG. (1998). The role of lipid physical properties in lipid barriers. Amer. Zool., 38, 268-279.

Gloor SM. (1997). Relevance of $Na^{+}K^{+}$ ATPase to local extra cellular potassium homeostasis and modulation of synaptic transmission. FEBS letters, 412, 1-4.

Gorczynska E, and Handelsman DJ. (1995). Androgens rapidly increase the cytosolic calcium concentration in Sertoli cells. Endocrinology, 136, 2052-2059.

Graham E, Mishra OP, and Delivoria-Papadopoulas M. (1993). Brain cell membrane $Na^{+}K^{+}$ ATPase modification following hypoxia in the guinea foetus. Neurosci. Lett., 153, 93-97.

Grier HJ. (1993). Comparative organization of Sertoli cells including the Sertoli cell barrier. *In*: Russell, L.D., Griswold, M.D. (Eds.), *The Sertoli cell*. Cache River Press, Clearwater FL, pp. 704-739.

Harper ME, and Brand MD. (1994). Hyperthyroidism stimulates mitochondrial proton leak and ATP turnover in rat hepatocytes but does not change the overall kinetics of substrate oxidation reaction. Can. J. Physiol. Pharmacol., 72, 899-908.

Hazel JR, and Prosser CL. (1974). Molecular mechanism of temperature compensation in poikilotherms. Physiol. Rev., 54, 620-677.

Hinz H, and Hirschelmann R. (2000). Dexamethasone megadoses stabilize rat liver lysosomal membranes by non-genomic and genomic events. Pharm. Res., 17, 1489-1493.

Horbie Y, Hosoya K, Kim KJ, and Lee VH. (1997). Kinetic evidence for Na^+, glucose co transport in the pigmented rabbit conjunctiva. Curr. Eye. Res., 16, 1050-1055.

Houston AH, and Mearow KM. (1982). Branchial and renal Na^+K^+ ATPase and carbonic anhydrase activities in a eurythermal freshwater teleost *Carassius auratus*, J. Comp. Biochem. Physiol., 71, 175-180.

Jacob V, and Oommen OV. (1990). Intermediary metabolism in a lizard, *Calotes versicolor*: Role of thyroid hormones. Gen. Comp. Endocrinol., 77, 324-336.

Jacob V, and Oommen OV. (1992). Role of thyroid hormone in intermediary metabolism of propranolol or alloxan- treated *Calotes versicolor*. J. Exp. Zool., 262, 263–270.

John TM, George JC, Yie SM, and Brown GM. (1993). Flight induced increase in circulating levels of melatonin in the homing pigeon. Comp. Biochem. Physiol.,106A, 645-648.

Johnson C. (2004). Influence of certain plant extract on the growth and metabolism of teleosts (*Anabas testudineus* and *Labeo rohita*) and a mammal, (*Rattus norvegicus*). Ph.D thesis, University of Kerala, Thiruvananthapuram. pp. 1-30.

Johnson C, Banerji A, and Oommen OV. (2003). A comparative study of allitin and garlic on lipid turnover in a teleost, *Anabas testudineus* (Bloch). Indian J. Exp. Biol., 41, 242-247.

Josso N, Picard JY, and Tran D. (1977). The antimullerian hormone. Recent Prog. Hormone Res., 33,117-160.

Karalis KP, Venihaki M, Zhao J, Van vlerken LE, and Chandras C. (2004). NF-kappa B participates in the corticotropin- releasing, hormone- induced regulation of the pituitary pro opiomelanocortin gene. J. Biol.Chem., 279, 10837-10840.

Karin M, and Delhase M. (2000). The Ikappa B kinase (IKK) and NF-kappaB: Key elements of proinflammatory signaling, Semin. Immunol., 12, 85-98.

Karin M, and Lin A. (2002). NF-kappaB at the crossroads of life and death. Nat. Immunol., 3, 221-227.

Kumar A, and Boriek AM. (2003). Mechanical stress activates the nuclear factor–kappaB pathway in skeletal muscle fibers: A possible role in Duchenne muscular dystrophy. FASEB. J., 17, 386-396.

Kumar A, Dhawan S, Hardegen NJ, and Aggarwal BB. (1998). Curcumin (Diferuloyl methane) inhibition of tumor necrosis factor (TNF)–mediated adhesion of monocytes to endothelial cells by suppression of cell surface expression of adhesion molecules and of nuclear factor-êB activation. Biochem. Pharmacol., 55, 775-783.

Kvetnoy IM. (1999). Extra pineal melatonin location and function within diffuse neuroendocrine system. Histochem., 31, 1-12.

Kvetnoy IM, Reiter RJ, and Khavinson VKH. (2000). Claude Bernard was right: Hormones may be produced by non-endocrine cells. Neuroendocrinology letters 21, 173-174.

Lechene C. (1988). Physiological role of the Na-K pump. *In*: Skou JC, Norby JG, Maunsbach AB, and Esmann M. (Eds.), *Progress in Clinical and Biological Research; The Na^+K^+ pump. Cellular Aspects.* Liss., New York, 268B, pp. 171-194.

Leena S, and Oommen OV. (2000). Hormonal control on enzymes of osmoregulation in a teleost, *Anabas testudineus* (Bloch): an *in vivo* and *in vitro* study. Endocrine Res., 26, 169-187.

Leena S, and Oommen OV. (2001). Growth hormone and prolactin action in a teleost, *Anabas testudineus* (Bloch): Effect of time of administration. Biol. Rhythm. Research, 32, 501-510.

Leung GP, Cheng-Chew SB, and Wong PY. (2001). Non-genomic effect of testosterone on chloride secretion in cultured rat efferent duct epithelia. Am. J. Cell Physiol., 280, 1160-1167.

Lofts B. (1974). Physiology of the amphibia. *In*: Lofts B. (Ed.), *Reproduction*. Academic Press, NewYork, pp.107-218.

Luke CM, and Coffey DS. (1994). The male sex accessory tissue: structure, androgen action and physiology. *In*: Knobil, E., Neill, J.D. (Eds), *The Physiology of Reproduction*. Raven Press, New York, pp. 1435–1487.

Machelon V, Nome F, and Tesarik J. (1998). Non-genomic effects of androstenedione on human granulosa luteinizing cells. J. Clin. Endocrinol. Metab., 83, 263-269.

Magnus-Levy A. (1895). Ueber den respiratorischen Gaswechsel unter demy Einfluss der Thyreoidea sowie unter Vreschiedenen pathologisthe. Zustanden. Berl. Klin. Wochenschr., 32, 650-652.

Maitra SK, Dey M, Dey R, Bhattacharya S, and Sengupta A. (2000). Influence of photoperiods on glycemic and adrenal catecholaminergic responses to melatonin administrations in adult male roseringed parakeets, *Psittacula krameri* Neumann. Indian J. Exp. Biol., 38, 1111-1116.

Marino M, Ficca R, Ascenzi P, and Trentalance A. (2001). Nitric oxide inhibits selectively the 17-β estradiol-induced gene expression without affecting non-genomic events in HeLa cells. Biochem. Biophys. Res. Commn., 286, 529- 533.

McCormick SD. (1995). Hormonal control of gill Na^+ K^+ ATPase and chloride cell function. *In*: Hoar WS, Randall DJ, and Farrell AP. (Eds), *Fish Physiology, Cellular and Molecular Approaches to Ionic Regulation*, Academic Press, New York, pp. 285-316.

Mommsen TP, Vijayan MM, and Moon TW. (1999). Cortisol in teleost: dynamics, mechanism of action, and metabolic regulation. Rev. Fish Biol. Fish, 9, 211-268.

Nair P. (1996). Intermediary metabolism in a teleost, *Anabas testudineus*: Regulation by hormones of hypothalamus, pituitary, thyroid and islets of langerhans. Ph.D thesis, University of Kerala, Thiruvananthapuram. pp. 116-129.

Nair P, and Oommen OV. (1997). Role of thyroxine on intermediary metabolism of thiouracil treated, *Anabas testudineus*. Indian J. Exp. Biol., 35, 983-986.

Natarajan K, Singh S, Burke TR Jr, Grunberger D, and Aggarwal BB. (1996). Caffeic acid phenethylester is a potent and specific inhibitor of activation of nuclear transcription factor NF-êB. Proc. Natl. Acad. Sci. USA, 93, 9090-9095.

Nordlie RC, Bode AM, and Foster JD. (1993). Recent advances in hepatic glucose-6-phosphatase regulation and function. Proc. Soc. Exp. Biol. Med., 203, 274-285.

Norris OD. (1996). Vertebrate endocrinology: Amphibians. *In*: Norris OD. (Ed.), *The Reproduction in Female Amphibians*. Academic Press, California, pp. 438-444.

Oommen OV. (1981). Stimulation of oxygen consumption of skeletal muscle and liver of *Calotes versicolor* by thyroid hormones in *vitro*. Comp. Physiol. Ecol., 6, 95-98.

Oommen OV, and Kaippallil JD. (1995). A comparative study on the effects and mechanism of action of glucocorticoids in the teleost, *Anabas testudineus* and the lizard, *Calotes versicolor*. J. Reprod. Biol. Comp. Endocrinol., 7, 1-8.

Oommen OV, and Matty AJ. (1991).The effects of thyroid hormones and starvation on mitochondrial nucleic acid and on plasma proteins of rainbow trout, *Onchorhynchus mykiss*. Gen. Comp. Endocrinol., 83, 468-472.

Orchinik M, Murray TF, and Moore FL. (1991). A corticosteroid receptor in neuronal membrane. Science, 252, 848-850.

Owada S, Larson O, Arkhammar P, Katz IA, Chilabin VA, Berggren OP, and Bertorello MA. (1999). Glucose decreases the Na^+K^+ATPase activity in pancreatic β-cells: An effect mediated via Ca^{2+}-independent phospholipase A_2 and Protein kinase C- dependent phosphorylation of the α-subunit. J. Biol. Chem., 274, 2000-2008.

Peter MCS, and Lock RAC, and Wendelaar Bonga SE. (2000). Evidence for an osmoregulatory role of thyroid hormones in the freshwater Mozambique tilapia, *Oreochromis mossambicus*, Gen. Comp. Endocrinol., 120, 157-167.

Peter MCS, and Oommen OV. (1987). Effect of administration of triiodo-L-thyronine on oxidative metabolism of a teleost, *Anabas testudineus* (Bloch). J. Reprod. Biol. Comp. Endocrinol., 7, 53-60.

Peter MCS, and Oommen OV. (1989a). Oxidative metabolism in a teleost, *Anabas testudineus* (Bloch): Effect of thyroid hormone on hepatic enzyme activities. Gen. Comp. Endocrinol., 73, 96-107.

Peter MCS, and Oommen OV. (1989b). Effect of thyroid hormones on the activities of hepatic enzymes in thiouracil-treated teleost, *Anabas testudineus* (Bloch). Current Sci., 58, 37-38.

Peter MCS, and Oommen OV. (1989c). Effects of thyroid and gonadal hormones *in vitro* on hepatic succinate dehydrogenase activity of the teleost, *Anabas testudineus* (Bloch). Zool. Sci., 6, 185-189.

Peter MCS, and Oommen OV. (1991). Role of thyroid hormones in rotenone-treated teleost; *Anabas testudineus* (Bloch): Effects on oxidative enzyme activities. Indian J. Exp. Biol., 29, 481-483.

Peter MCS, and Oommen OV. (1993). Stimulation of oxidative metabolism by thyroid hormones in propanol/alloxan–treated bony fish, *Anabas testudineus* (Bloch). J. Exp. Zool., 266, 85- 91.

Pierce JG, du Vigneaud V. (1950). Preliminary studies on aminoacid content of a high-potency preparation of the oxytocic hormone of the posterior lobe of the pituitary gland. J. Biol. Chem.,182, 359-366.

Plewka A, Kaminski M, Plewka D, and Nowaczyk G. (2000). Glucose-6-phosphatase and age: biochemical and histochemical studies. Mech. Ageing Dev., 113, 49-59.

Popenos NA, and du Vigneaud V. (1954). A partial sequence of amino acid in performic acid-oxidised vasopressin. J. Biol. Chem., 205, 353-360.

Rabinovici J, and Jaffe RB. (1990). Development and regulation of growth and differentiated function in human and subhuman primate fetal gonads. Endocrine Reviews, 11, 532-557.

Reiter RJ. (1987). The melatonin message, duration versus coincidence hypotheses. Life. Sci., 40, 111-116.

Robaire B, and Hermo L. (1988). Efferent ducts, epididymis and vas deferens: Structure, functions and their regulation. *In*: Knobil E, and Neill JD. (Eds.), *The Physiology of Reproduction*. Raven Press, New York, pp. 999–1080.

Saez FJ, Frail B, and Paniagua R. (1990). Histological and quantitative changes in the annual testicular cycle of *Triturus marmoratus*. Can. J. Zool., 68, 63-72.

Saidapur SK. (1989). Reproductive cycles of amphibians. *In*: Saidapur SK. (Ed.), *The Reproductive Cycles of Indian Vertebrates*. Allied Publishers Ltd., New Delhi, pp.166-224.

Sarkar HBD, and Shivanandappa T. (1989). Reproductive cycles of reptiles. *In*: Saidapur SK. (Ed.), *Reproductive Cycles of Indian Vertebrates*. Allied Publishers Ltd., New Delhi, pp. 224–271.

Senftleben U, and Karin M. (2002). The IKK/NF-kappaB pathway. Crit. Care. Med, 30, S18-S26.

Setchell BP, Maddocks S, and Brooks DE. (1994). Anatomy, vasculature, innervation and fluids of the male reproductive tract. *In:* Knobil E, and Neill JD. (Eds.), *The Physiology of Reproduction*. Raven Press, NewYork, pp. 1063-1175.

Sheridan MA. (1986). Effects of thyroxin, cortisol, growth hormone and prolactin on lipid metabolism of coho salmon, *Oncorhynnchus kisutch* during smoltification. Gen. Comp. Endocrinol., 64, 220-228.

Sheridan, M.A., Kao, Y.H., 1998. Regulation of metamorphosis associated changes in the lipid metabolism of selected vertebrates. Am. Zool. 38, 350-368.

Shulman LM, and Fox D. (1996). Dopamine inhibits mammalian photo receptor Na^+K^+ATPase activity via a selective effect on the α isozyme. Proc. Natl. Acad. Sci. USA, 93, 8034-8039.

Singh S, Natarajan K, and Aggarwal BB. (1996). Capsaicin (8- methyl–N-Vanillyl-6-nonenamide) is a potent inhibitor of nuclear transcription factor–êB activation by diverse agents. J. Immunol., 157, 4412-4420.

Smita M, George JM, Akbarsha MA, and Oommen OV. (2005). Amoeboid cells in spermatogenic cysts of caecilian testis. J. Morphol., 263, 340-355.

Smita M, George JM, Girija R, Akbarsha MA, and Oommen OV. (2004). Spermiogenesis in caecilians, *Ichthyophis tricolor* and *Uraeotyphlus cf. narayani* (Amphibia: Gymnophiona): Analysis by light and transmission electron microscopy. J. Morphol. 262, 484-499.

Smita M, Oommen OV, George JM, and Akbarsha MA. (2003). Sertoli cells in the testis of caecilians, *Ichthyophis tricolor* and *Uraeotyphlus cf. narayani* (Amphibia: Gymnophiona): Light and electron microscopic perspective. J. Morphol., 258, 317–326.

Spengel JW. (1876). Das Urogenital system der Amphibien. I. Theil Der annatomische Bau des Urogenitalsystems. Arbeitenaus demm Zoologyzootom. Institute in Wurzburg. 3, 51-114.

Stetson MH, and Watson-Whitmire M. (1984). The physiology of the pineal and its hormone melatonin in annual reproduction in rodents. *In*: Reiter RJ. (Ed.), *The Pineal Gland*, Raven Press, NewYork, pp.109-153.

Sunny F, Lakshmy PS, and Oommen OV. (2002). Rapid action of cortisol and testosterone on lipogenic enzymes in a fresh water fish *Oreochromis mossambicus*: Short–term *in vivo* and *in vitro* study. Comp. Biochem. Physiol., 131, 297-304.

Sunny F, and Oommen OV. (2000). Steroid hormones regulate enzymes of osmoregulation in a freshwater fish, *Oreochromis mossambicus*. J. Endocrinol. Reprod.4, 63-73.

Sunny F, and Oommen OV. (2001). Rapid action of glucocorticoids on branchial ATPase activity in *Oreochromis mossambicus*: An *in vivo* and *in vitro* study. Comp. Biochem. Physiol. Part B, 130, 323-330.

Sunny F, and Oommen OV. (2002). Rapid action of testosterone and diethylstilbestrol on enzymes of osmoregulation in a freshwater fish, *Oreochromis mossambicus*. Endocrine Res., 28, 69-81.

Sunny F, and Oommen OV. (2003). Genomic effect of glucocorticoids on enzymes of intermediary metabolism in *Oreochromis mossambicus*. Endocrine Res., 29, 119-131.

Sunny F, and Oommen OV. (2004). Scanning electron microscopic evaluation of effect of cortisol on chloride cells of tilapia, *Oreochromis mossambicus*. Indian J. Exp. Biol., 42, 398-402.

Sutharam KK, Peter MCS, and Oommen OV. (1990). Stimulation of oxidative metabolism by thyroid hormones in an apodan amphibian, *Gegenophis carnosus* (Beddome). Gen.comp. Endocrinol., 79, 246-252.

Sweadner KJ, and Goldin SM, (1980). Active transport of sodium and potassium ions: mechanism, function and regulation. N. Engl. J. Med., 302, 777-783.

Sweeney G, and Klip A. (1998). Regulation of the Na^+K^+ATPase by insulin: Why and how? Mol. Cell. Biochem., 182, 121- 133.

Sweeny G, Niu W, Canfield AV, Levenson R, and Klip A. (2001). Insulin increases plasma membrane content and reduces phosphorylation of Na^+K^+ pump α_1-subunit in HEK-293 cells. Am. J. Physiol. Cell. Physiol., 281, 1797-1803.

Swinnen J, Ulrix W, Heyns W, and Verhoeven G. (1997). Coordinate regulation of lipogenic gene expression by androgens: Evidence for a cascade mechanism involving sterol regulatory element binding proteins. Proc. Natl. Acad. Sci. USA, 94, 12975-12980.

Tamarkin L, Westrom WK, Hamill AI, and Goldmann BD. (1976). Effect of melatonin on the reproductive systems of male and female Syrian hamsters, a diurnal rhythm in sensitivity to melatonin. Endocrinol., 99, 1534-1541.

Tonutti E. (1931). Beitrag zur Kenntnis der Gymnophionen XV. Das Genitalsystem. Morphol. Jahrb., 68, 151–292.

Valamparampil TT, Oommen OV. (1997). Triiodothyronine (T_3) and thyroxine (T_4) in *Rana curtipes* during development and metamorphosis. Indian J. Exp. Biol., 35, 1375-1377.

Varghese S, Shameena B, and Oommen OV. (2001). Thyroid hormones regulate lipid peroxidation and antioxidant enzyme activities in *Anabas testudineus* (Bloch). Comp. Biochem. Physiol., 128B, 165-171.

Venditti P, Daniele MC, Masullo P, and Meo SD. (1999). Antioxidant–sensitive Triiodo thyronine effects on characterization of rat liver mitochondrial population. Cell. Physiol. Biochem., 9, 38-52.

Venero C, and Borrell J. (1999). Rapid glucocorticoid effects on excitatory amino acid levels in the hippocampus: A micro dialysis study in freely moving rats. Eur. J. Neuro Sc., 11, 2465-2473.

Vijayan MM, Ballantyne JS, and Leatherland JF. (1991). Cortisol induced changes in some aspects of the intermediary metabolism of *Salvelnius fontinalis*. Gen. Comp. Endocrinol., 82, 476-486.

Vijayan MM, Mommsen TP, Glemet HC, and Moon TW. (1996). Metabolic effects of cortisol treatment in a marine teleost, the sea raven. J. Exp. Biol., 199, 1509-1514.

Vijayan MM, Pereira C, Grau EG, Iwama GK. (1997). Metabolic responses associated with confinement stress in tilapia: The role of cortisol. Comp. Biochem. Physiol., 116, 89-95.

Vijayan MM, Reddy PK, Leatherland JF, Moon TW. (1994). The effects of cortisol on hepatocytes metabolism in rainbow trout: A study using the steroid analogue RU 486. Gen. Comp. Endocrinol., 96, 75-84.

Vijayasree AS, Divya L, Sreejith P, Cyril J, Smita M, and Oommen, OV. (2005). Insulin regulates ionic metabolism in a fresh water teleost, *Anabas testudineus* (Bloch). Indian J. Exp. Biol., 43,702-709.

Voice MW, Webster AP, and Burchell A. (1997). The *in vivo* regulation of liver and kidney glucose-6-phosphatase by dexamethasone. Horm. Metab. Res., 29, 97-100.

Wake MH. (1977). Reproductive biology of caecilians: an evolutionary perspective. *In*: Guttman S, and Taylor D. (Eds.), *The Reproductive Biology of the Amphibia*. Plenum Press, New York, pp. 73–102.

Wake MH. (1981). Structure and function of the male Mullerian gland in caecilians, with comments on its evolutionary significance. J. Herpetol., 126, 291–332.

Wehling M, Bauer M, Ulsenheimer A, Schneider M, Neylon CB, and Christ M. (1996). Non-genomic effects of aldosterone on intracellular pH in vascular smooth muscle cells. Biochem. Biophys. Res. Commun., 223, 181-186.

Wehling M, Kasmayr J, and Theisen K. (1990). Aldosterone influences free intracellular calcium in human mononuclear leukocytes *in vitro*. Cell Calcium, 11, 565-571.

Yamashita T, Umeda F, Hashimoto T, Inoguchi T, Yamauchi T, Mimura K, Watanabe J, and Nawata H. (1992). Effect of glucose on $Na^{+}K^{+}$ ATPase activity in cultured bovine aortic endothelial cell Endocrinol. Jpn., 39, 1-7.

23

Stress Endocrinology and Conservation

*J.F. Cockrem**

Conservation Endocrinology Research Group, Institute of Veterinary, Animal and Biomedical Sciences, Massey University, Palmerston North New Zealand

ABSTRACT

The conservation of animals and their habitats is a major issue for biologists, with the loss of animal species and biodiversity increasing in some areas of the world. Stress responses are natural responses, but stress can be harmful to animals and stress is often a concern in conservation programmes for animals. Stress can be defined as a state in which an animal is responding to a stressor. Stressors are factors that activate the hypothalamo-pituitary-adrenal axis and lead to an increase in glucocorticoid secretion from the adrenal gland. Stressors can be physical or emotional, with marked variations between animals in their glucocorticoid responses to emotional stressors. This individual variation in glucocorticoid responses is associated with similar variation in behavioural responses. Animals that tend to have low responses have a proactive coping style and animals with high responses a reactive style. Animals with reactive coping styles may be more successful than those with proactive styles when habitat changes occur due to global warming. The degree of stress that an animal experiences in different situations can be determined by measurements of baseline glucocorticoid concentrations and glucocorticoid responses to stressors. Measurements can be made in plasma or non-invasively in faecal or urine samples. Studies of stress endocrinology can contribute to conservation across a wide range of areas including stress in animals in captive breeding programmes and in relation to habitat quality, stress responses to environmental stressors, and responses of animals to tourists. The application of stress endocrinology to conservation biology can make significant contributions to the survival of endangered species of animals.

---

*E-mail: J.F.Cockrem@massey.ac.nz

Introduction

Stress is often mentioned in relation to conservation, and there is increasing interest in stress and the conservation of rare and endangered animals. Stress is usually considered to be harmful to animals, but stress responses are natural responses that help animals adjust to difficult situations. Animals can potentially experience stress when in contact with people, in situations where their environment has changed due to human activities and as a consequence of large scale changes associated with global warming. Conservation biologists face challenges ranging from the maintenance of populations of individual species of animals through to retaining viable habitats and ecosystems. Endocrine measures of stress may provide useful tools for conservation managers to meet these challenges.

Stress and Stress Responses

Stress is a subjective term, but many publications that discuss stress in animals do not define stress. When definitions are provided they vary; for example "a non-specific response of the body to any demand made on it" (Selye, 1952; Von Holst, 1998), "complex changes which occur in various plasma or adrenal constituents following an adverse stimulus" (Freeman, 1971), "an animal's defense mechanisms" (Siegel, 1980), "the homeostatic, physiological and behavioural responses detectable in the animal resulting from its interactions with environmental stressors" (Stephens, 1980), and "a stressor is any stimulus (behavioural, environmental or demographic) that provokes a physiological stress response, as measured by an increase in glucocorticoid secretion" (Creel, 2001). The definitions used by the present author in the current treatise: *Stress* –a state in which an animal is responding to a stressor; *Stressor*–any factor that activates the hypothalamo-pituitary-adrenal (HPA) axis and leads to an increase in glucocorticoid secretion from the adrenal gland; *Stress response*–the response of an animals to one or more stressors

Stressors can be classified as physical or emotional. Physical stressors involve signals arising in the body due to an internal or external disturbance of physical or chemical tissue parameters. There is no requirement for an appraisal of the stimulus in relation to stored information in the brain, either learnt or inherited. An example of a physical stressor is a fall in blood glucose concentrations that leads to an activation of the HPA axis and cortisol secretion. Physical stressors are also called physiological, homeostatic or systemic stressors. Emotional stressors involve signals that arise in the brain and are less easy to define than physical stressors. They are also called psychological or neurogenic stressors. These stressors require the appraisal of information in relation to stored information that is either learned or inherited and require processing in the limbic or cortical areas of the brain. An example of an emotional stressor is the sight of a potential predator. Note that conscious appraisal of a stimulus is not needed for it to be perceived as a threat (*i.e.* a stressor) and for a stress response to be initiated.

Stress responses can include a range of physiological systems and behaviours, but the common element is activation of the HPA axis. Corticotropin-releasing factor from the hypothalamus stimulates the release of adrenocorticotropic hormone (ACTH) from the anterior pituitary gland. Arginine vasopressin or arginine vasotocin from the median eminence can also stimulate the release of ACTH. ACTH then stimulates the release of glucocorticoids (cortisol in most mammals and corticosterone in rodents and most other vertebrates) from the adrenal cortex. Glucocorticoids are metabolic hormones that help to maintain blood glucose levels during fasting. They also have a wide range of other actions including effects on intermediary metabolism, immune function, behaviour, electrolyte balance, growth and reproduction. Glucocorticoids released when an animal responds to a stressor help the animal adjust to the stressor (Sapolsky *et al.*, 2000), but if a stress response is prolonged then glucocorticoids may have detrimental effects on the animal.

Individual Variation and Coping Styles

Plasma glucocorticoid concentrations increase when an animal responds to a stressor. Concentrations typically increase rapidly over several minutes then return towards initial concentrations over some minutes or hours once the stressor is no longer present. A typical glucocorticoid response in birds is shown in Figure 23.1. Japanese quail (*Coturnix coturnix japonica*) were subjected to a handling stressor for 15 minutes. Mean plasma corticosterone concentrations increased for 15 min then declined. However, examination of the individual corticosterone responses (Figure 23.2) shows that a statement about mean corticosterone responses does not accurately describe the responses in all the individual birds. Some birds had much lower corticosterone responses than others, with no measurable response at all in one quail. This illustrates the need to consider individual responses when examining the responses of animals to stressors.

Glucocorticoid responses of individual animals are repeatable (Littin and Cockrem, 2001), and animals that have relatively low or high glucocorticoid responses tend to have different behavioural characteristics such as levels of aggression, sociability and fearfulness. Animals with lower glucocorticoid responses to stressors are generally those with a proactive coping style whilst those with higher responses have a reactive coping style. Coping styles are coherent sets of behavioral and physiological responses to common challenges faced by animals (Koolhaas *et al.*, 1999). One type of style is called proactive, active or bold whilst the other is called reactive, passive or shy. Animals with

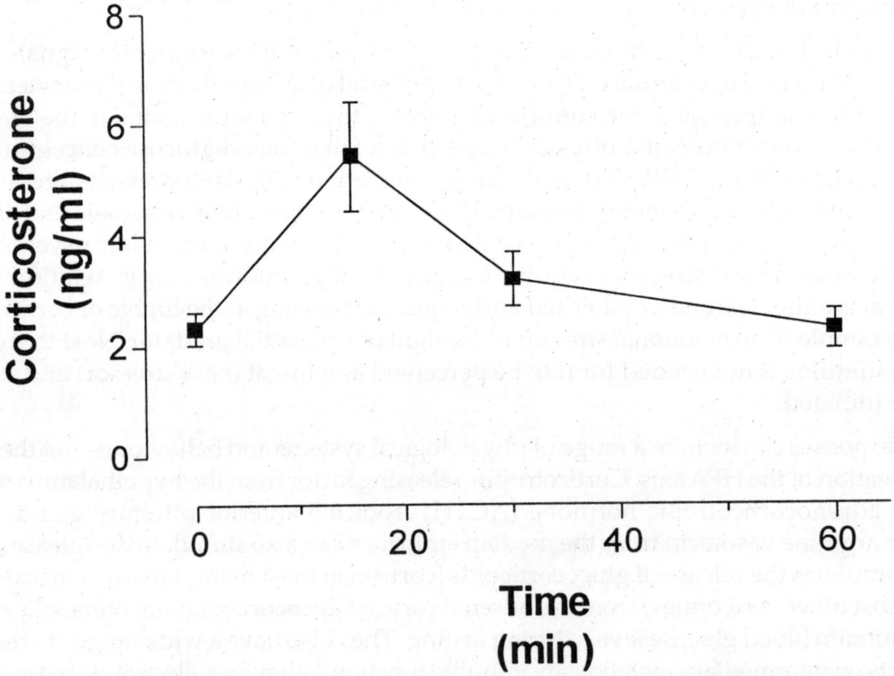

Figure 23.1: Mean Plasma Corticosterone Response to Capture and Handling of Five Japanese Quail. Quail were obtained from a commercial supplier and held individually in cages.

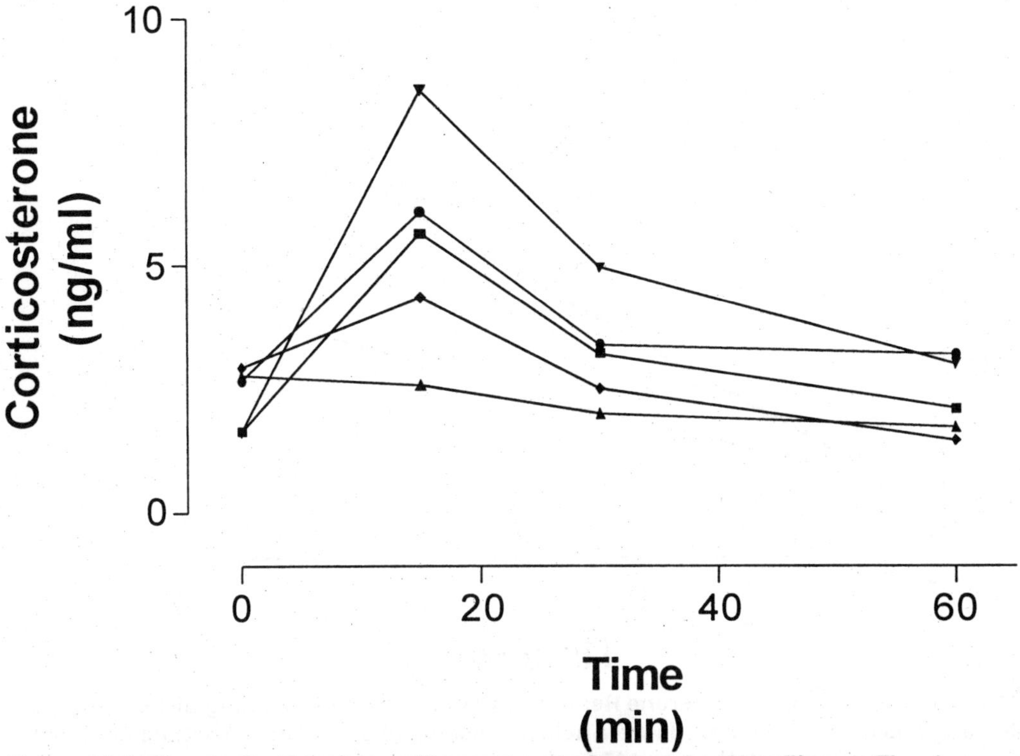

Figure 23.2: Plasma Corticosterone Responses to Capture and Handling on Five Japanese Quail. These are the individual birds for which mean data were presented in Figure 23.1.

proactive coping styles and lower glucocorticoid responses are likely to be more successful in an environment that remains constant than those with reactive styles. Conversely, animals with reactive coping styles may be more successful in a changing environment. This could be important for animals that experience habitat changes as consequences of global warming, with the ability of animals to adapt to these changes dependent on their coping style.

The Measurement of Stress

Stress responses involve increases in plasma glucocorticoid concentrations. The question "Is this animal stressed?" can be expressed in endocrine terms as two questions: "What are the baseline plasma glucocorticoid concentrations in this animal?" and "What is the magnitude and duration of the glucocorticoid response to handling in this animal?" (Cockrem, 1997). Baseline plasma glucocorticoid concentrations indicate the level of activity in the HPA axis when an animal is not responding to an acute stressor. These concentrations do not indicate the responsiveness of an animal to stressors, and it is important to measure glucocorticoid responses as well as basal concentrations. For example, Dunlap and Wingfield (1995) showed that corticosterone responses to handling were more informative than basal corticosterone concentrations for reflecting the extent to which environmental influences were affecting male fence lizards (*Scleoporus occidentalis*). Figure 23.3 illustrates how two groups of Adelie penguins (*Pygoscelis adeliae*) had similar baseline corticosterone concentrations (at 0 min) but markedly different corticosterone responses to a stressor.

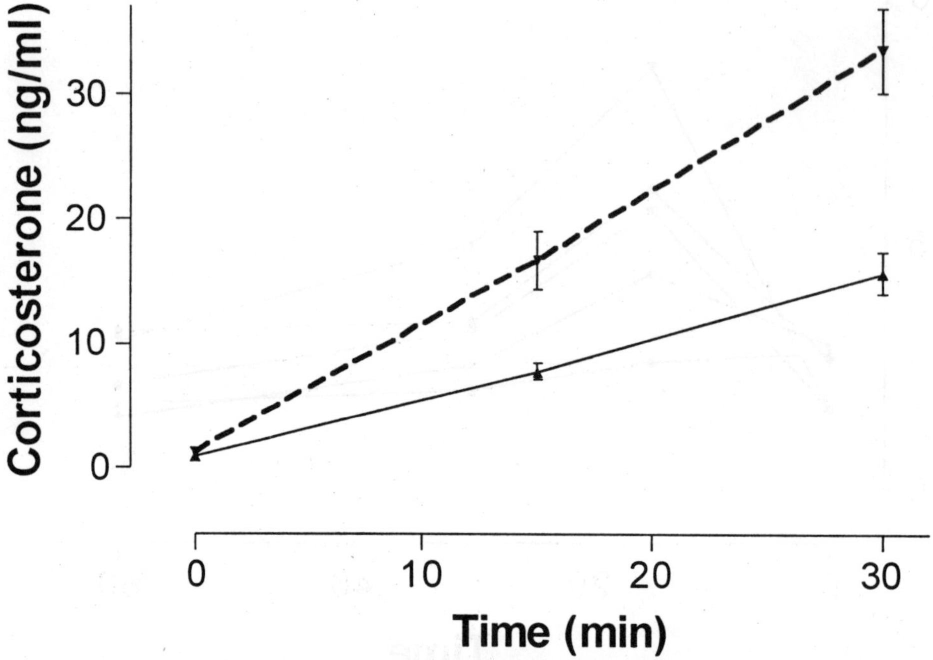

Figure 23.3: Mean Plasma Corticosterone Responses to Capture and Handling of Two Groups of Adelie Penguins in Antarctica. Birds were held in mesh bags during the 60 minute period of confinement. N = 8 and 13 for the two groups of penguins.

Contributions of Stress Endocrinology to Conservation

The newly recognised discipline of conservation endocrinology is defined as "Endocrine studies that can contribute to conservation" (Cockrem and Ishii, 1999; Cockrem *et al.*, 2004). Conservation endocrinology research includes theoretical, diagnostic and management studies, and studies in stress endocrinology can contribute to all of these areas of conservation endocrinology. Theoretical studies address basic questions about the biology of animals such as how do animals respond to emotional stressors. Diagnostic studies address questions asked by conservation managers about the physiological and behavioral status of individual animals or groups of animals. Two examples of diagnostic questions are "Is this animal experiencing stress?" and "What proportion of this population shows a reactive coping style?". Management studies deal directly with practical questions in conservation such as how to avoid negative effects of stress on breeding in animals of an endangered species.

The application of stress endocrinology to the conservation of wild and captive animals can include the measurement of plasma or faecal glucocorticoid concentrations and of glucocorticoid responses to acute stressors to provide a basis for the assessment of the physiological status of animals in particular habitats or in captivity. Conservation managers do not have the luxury of defining the study conditions for endangered species, and the practical application of endocrine methods to directly aid conservation programmes provides a major challenge for stress endocrinologists with interests in conservation biology. Conservation managers need answers to immediate management problems, and are always looking for diagnostic tools that may provide such answers.

It can be tempting to suggest that measurements of plasma glucocorticoid concentrations will indicate whether an animal is "stressed" or not. However, the interpretation of plasma glucocorticoid concentrations in relation to stress is fraught with difficulty and must be approached from a sound understanding of the physiology of the adrenal gland. If plasma glucocorticoid concentrations are elevated in samples collected from one group of animals compared with concentrations in samples collected from another group of animals of the same species it would be unwise to conclude that these differences meant that one group was "stressed". A difference in plasma glucocorticoid concentrations may reflect differences in the normal physiological state of the animals. Conversely, it is possible that animals may indeed be experiencing chronic stress but that basal glucocorticoid concentrations were elevated only transiently during the initiation of the stressor.

Non-invasive methods are especially useful to address diagnostic questions about the physiological status of individual animals in captive breeding programmes. They are widely used in studies of mammals such as descriptions of faecal glucocorticoid responses to stressors in felids (Carlstead *et al.*, 1992), chimpanzees (*Pan troglodytes;* Whitten *et al.*, 1998) and African elephants (*Loxodonta africana;* Ganswindt *et al.*, 2003).

Studies in conservation endocrinology in which glucocorticoid concentrations have been measured include responses of African wild dogs (*Lycaon pictus*) to the attachment of radio collars (Creel *et al.*, 1997), the effects of habitat disturbance on northern spotted owls (*Strix occidentalis caurina;* Wasser *et al.*, 1997), and the effects of pollutants on Magellanic penguins (*Spheniscus magellanicus;* Fowler *et al.*, 1995) and southern toads (*Bufo terrestris;* Hopkins *et al.*, 1997). Corticosterone responses of Magellanic penguins to people were influenced by tourist visits (Fowler, 1999). Corticosterone responses to a stressor were decreased at a tourist site in Galapagos marine iguanas (*Amblyrhynchus eristatus;* Romero and Wikelski, 2002) and increased in hoatzins (*Opisthocomus hoazin*) in juveniles but not chicks (Mullner *et al.*, 2004). Measurements of baseline glucocorticoid concentrations and glucocorticoid responses to stressors may provide indicators of habitat quality (American redstart *Setophaga ruticilla;* Marra and Holberton, 1998) and spotted salamander (*Ambystoma maculatum;* Homan *et al.*, 2003) and food availability (black-legged kittiwake *Rissa tridactyla;* Lanctot *et al.*, 2003). Glucocorticoid responses to environmental stressors as indicators of coping style might also provide some indication of the susceptibility of different species to changes resulting from global warming. Stressors can affect the susceptibility of animals to disease (Miller *et al.*, 2001), and diagnostic studies in stress endocrinology can address questions about the disease status of animals.

The practical conservation environment is very different from the academic one in which scientists choose a study animal or study topic, choose what aspects of the biology of the animal or topic to study, and define according to their own interests how they will study the animal or topic. The direct application of stress endocrinology to conservation biology is a challenge that can lead in future to significant contributions to the survival of endangered species of animals.

Acknowledgements

Research in stress endocrinology described here has been supported by Massey University and Antarctica New Zealand.

References

Carlstead K, Brown JL, Monfort SL, Killens R, and Wildt DE. (1992). Urinary monitoring of adrenal responses to psychological stressors in domestic and nondomestic felids. Zoo. Biol., 11, 165-176.

Cockrem JF. (1997). Contributions of comparative endocrinology to conservation biology. *In*: Kawashima S, and S Kikuyama S. (Eds.), Advances *in Comparative Endocrinology. Proc. XII Intnl. Congrr. Comp. Endocrinol., Yokohama, Japan, November 1997"*. Monduzzi Editore, Bologna, pp. 1695-1700.

Cockrem JF, Adams DC, Bennett EJ, Candy EJ, Chua WH, Henare SJ, Hawke EJ, and Potter MA. (2004). Endocrinology and the conservation of New Zealand birds. *In*: Gordon MS, and Bartol SM. (Eds.), *Experimental Approaches to Conservation Biology"*, Univ. of California Press, Los Angeles, pp. 101-121.

Cockrem JF, and Ishii S. (1999). Conservation endocrinology–a new field of comparative endocrinology. *In*: Kwon HB, Joss JMP, and Ishii S. (Eds.), *Recent Progress in Molecular and Comparative Endocrinology"*, Hormone Research Center, Kwangju, Korea, pp. 413-418.

Creel S. (2001). Social dominance and stress hormones. Trends Ecol. Evol., 16, 491-497.

Creel S, Creel NM, and Monfort SL. (1997). Radiocollaring and stress hormones in African wild dogs. Cons. Biol., 11, 544-548.

Dunlap KD, and Wingfield JC. (1995). External and internal influences on indices of physiological stress. I. Seasonal and population variation in adrenocortical secretion of free-living lizards, *Scleoporus occidentalis*. J. Exp. Zool., 271, 36-46.

Fowler GS. (1999). Behavioral and hormonal responses of Magellanic penguins (*Spheniscus magellanicus*) to tourism and nest site visitation. Biol. Conserv., 90, 143-149.

Fowler GS, Wingfield JC, and Boersma PD. (1995). Hormonal and reproductive effects of low levels of petroleum fouling in Magellanic penguins (*Spheniscus magellanicus*). Auk, 112, 382-389.

Freeman BM. (1971). Stress and the domestic fowl: a physiological appraisal. Worlds Poult. Sci. J., 27, 263-275.

Ganswindt A, Palme R, Heistermann M, Borragan S, and Hodges JK. (2003). Non-invasive assessment of adrenocortical function in the male African elephant (*Loxodonta africana*) and its relation to musth. Gen. Comp. Endocrinol., 134, 156-166.

Homan RN, Regosin JV, Rodrigues DM, Reed JM, Windmiller BS, and Romero LM. (2003). Impacts of varying habitat quality on the physiological stress of spotted salamanders (*Ambystoma maculatum*). Anim. Conserv., 6, 11-18.

Hopkins WA, Mendonca MT, and Congdon JD. (1997). Increased circulating levels of testosterone and corticosterone in southern toads, *Bufo terrestris*, exposed to coal combustion waste. Gen. Comp. Endocrinol., 108, 237-246.

Koolhaas JM, Korte SM, De Boer SF, Van Der Vegt BJ, Van Reenen CG, Hopster H, De Jong IC, Ruis MAW, and Blokhuis HJ. (1999). Coping styles in animals: current status in behavior and stress-physiology. Neurosci. Biobehav. Rev., 23, 925-935.

Lanctot RB, Hatch SA, Gill VA, and Eens M. (2003). Are corticosterone levels a good indicator of food availability and reproductive performance in a kittiwake colony? Horm. Behav., 43, 489-502.

Littin KE, and Cockrem JF. (2001). Individual variation in corticosterone secretion in laying hens. Brit. Poult. Sci., 42, 536-546.

Marra PP, and Holberton RL. (1998). Corticosterone levels as indicators of habitat quality: effects of habitat segregation in a migratory bird during the non-breeding season. Ocealogia, 116, 284-292.

Miller AH, Pearce BD, Ruzek MC, and Biron CA. (2001). Interactions between the hypothalamic-pituitary-adrenal axis and immune system during viral infection: pathways for environmental effects on disease expression. *In*: McEwen BS, (Eds.), *Handbook of Physiology. Section 7: The endocrine system. Volume IV: Coping with the environment: neural and endocrine mechanisms*, Oxford University Press, Oxford, pp. 425-450.

Mullner A, Linsenmair KE, and Wikelski M. (2004). Exposure to ecotourism reduces survival and affects stress response in hoatzin chicks (*Opisthocomus hoazin*). Biol. Conserv., 118, 549-558.

Romero LM, and Wikelski M. (2002). Exposure to tourism reduces stress-induced corticosterone levels in Galápagos marine iguanas. Biol. Cons., 108, 371-374.

Sapolsky RM, Romero LM, and Munck AU. (2000). How do glucocorticoids influence stress responses? Integrating permissive, suppressive, stimulatory, and preparative actions. Endocr. Rev., 21, 55-89.

Selye H. (1952). The story of the adaptation syndrome. Acta, Montreal.

Siegel HS. (1980). Physiological stress in birds. Biosci., 30, 529-534.

Stephens, DB. (1980). Stress and its measurement in domestic animals: a review of behavioural and physiological studies under field and laboratory situations. Adv. Vet. Sci. Comp. Med., 24, 179-210.

Von Holst D. (1998). The concept of stress and its relevance for animal behavior. Adv. Stud. Behav., 27, 1-131.

Wasser SK, Bevis K, King G, and Hanson E. (1997). Non-invasive physiological measures of disturbance in the northern spotted owl. Cons. Biol., 11, 1019-1022.

Whitten PL, Stavisky R, Aureli F, and Russell E. (1998). Response of fecal cortisol to stress in captive chimpanzees (*Pan troglodytes*). Amer. J. Primatol., 44, 57-69.

— Section E —
Endocrine Disruption and Biotechnology

24

Endocrine Disruptors: Their Mechanism of Actions and Testing Strategies to Assess Human Hazard

*Vikas Kumar, Bhavna Sharma, Ben. M. J. Pereira and Partha Roy**

Department of Biotechnology, Indian Institute of Technology Roorkee, Roorkee – 247 667, Uttaranchal, India

ABSTRACT

It has recently been discovered that a number of synthetic chemicals and naturally occurring compounds released into the environment can influence endocrine activity. These endocrine disrupting compounds (EDCs) are highly varied in structure, provenance and mode of action. Many EDCs are anthropogenic, products of the chemical industry; others are natural compounds. Although natural hormones exist at low levels in the environment, industrial sites such as pulp and paper mills and municipal sewage treatment plants can gather large amounts of natural EDCs and release them into the environment as part of their daily operations, thus increasing background environmental concentrations. These potential consequences of endocrine disruption are serious, yet little is known about the phenomenon. The issue of endocrine disruption originally focused on chemicals that mimic the actions of the natural hormone, estrogen. However, the concern is now encompassing effects on whole endocrine system. This relates to possible effects on the environment as well as on human health for chemicals with (anti)estrogenic, (anti)androgenic or (anti)thyroid activity. In response to public awareness, regulatory agencies

**Corresponding Author.* E-mail: paroyfbs@iitr.ernet.in

(including the US EPA) and the OECD are formulating potential testing strategies and have begun the process of validating defined tests to systematically assess chemicals for their endocrine-disrupting activities. Towards this end, a large number of *in vitro* and *in vivo* assays have been identified. *In vitro* test systems offer the possibility of providing an early screen for large number of chemicals and can be useful in characterizing the mechanism of action and potency. On the other hand, the *in vivo* tests provide the evidence of metabolic activities of these chemicals in the body. Based on the mechanism of steroid receptor actions and their *in vivo* effects, *in vitro* and *in vivo* test systems are described that can be employed not only for screening but also for the elucidation of their mechanism of action.

Keywords: *Endocrine disruption, Steroid receptors, Tiered Screening.*

Introduction

During the last decade, concerns have been raised regarding the possible harmful effects of exposure to certain chemicals (natural and synthetic) that are capable of modulating or disrupting the endocrine system. In principle, an "endocrine disruptor" is defined as an exogenous substance or mixture that alters function(s) of the endocrine system and consequently causes adverse health effects in an intact organism or its progeny or (sub) population (Scientific Committee for Toxicity, 1999; International Program on Chemical Safety, 1998). This concern for EDCs is directed at observed adverse effects on different physiological changes right from fertility to cancers in both human and wild life. Initially the concern was focused around environmental chemicals (including synthetic and naturally occurring chemicals as well as environmental contaminants) that may be able to mimic the action of the female hormone estrogen and was thus termed "environmental estrogens" (Baker, 2001). However, these chemicals were further found to be not only confined to estrogens but also affected the other hormones like androgen, progesterone and thyroid hormones. These chemicals included a wide range of molecules such as organochlorine pesticides, phthalates, alkylphenols, natural hormones and pharmaceuticals (Safe and Gaido, 1998). Among these compounds natural and synthetic estrogens are shown to be most potent *in vitro* (Gutendorf and Westendorf, 2001) and *in vivo* (Laws *et al.*, 2000). Estrogenic compounds are basically prescribed for birth control and estrogen replacement therapy. Among these, the synthetic estrogen ethinylestradiol (EE2), mestranol or the estradiol valerate are widely used (Cargouët *et al.*, 2004). These compounds in addition to natural hormones, 17β-estradiol (E2), estrone (E1) and estriol (E3), are excreted by women and consequently found in wastewaters. Men also excrete estrogens but at much lower rate (Fotis and Adlercreutz, 1987; Lee and Peart, 1998).

Another major source of EDCs is the waste water treatment plants (WWTPs) which receives a large spectrum of molecules from domestic and/or industrial sources and they are not totally eliminated during the treatment process (Ternes *et al.*, 1999). At the outlets of WWTPs, a complex mixture of molecules including not only the partially eliminated wastewater molecules but also metabolites formed during treatment process are finally discharged into the rivers. In this context, WWTP discharges are considered as a major source of estrogenic surface water pollution that plays a significant role in environmental contamination. Recently, Soto *et al.* (2004) identified some androgenic and estrogenic compounds in water bodies receiving cattle feedlot effluent in certain regions of USA.

The major point of contention over the efforts to define an endocrine disruptor is in the use of the phrase 'adverse'. For a chemical to be regulated, it has to be shown to induce an adverse effect but this is a complicated issue in the endocrine disruption field due to the dual nature of the phenomenon. For example, 'estrogenecity' per se is not an adverse effect, it is a natural mechanism of the hormone action controlled via homeostatic mechanisms. However, a chemical with estrogenic properties acting out of context within the endocrine system, or to a vulnerable developmental time-point may have the potential to induce an adverse effect.

Because of these growing public and scientific concerns, epidemiological studies have been initiated to analyze the short and long term effects of EDCs. In addition, a number of assays (either *in vitro* and *in vivo*) have been developed and are undergoing validation in order to perform high throughput screening of chemical agents with possible endocrine disrupting properties, to elucidate their mechanism of action and to perform risk assessment studies.

Relationship between Diseases/Abnormalities and EDCs

Several studies have shown evidence for the adverse health effects resulting from human exposure to endocrine disruptors (Damstra *et al.*, 2002; Greim, 2004). The current human concerns about EDCs are not only based on some assumptions but on a group of observations that when considered together suggests that EDCs may have the potential to induce deleterious changes to human health (Colburn and Clement, 1992; Singleton and Khan, 2003; Fisher, 2004). In recent years there have been many reports on the increasing incidence of breast cancer in women (Davidson and Yager, 1997). There are several reviews, which debate the validity of this concept and the supporting data. Nevertheless, there is evidence of decreasing sperm counts in many countries, increase incidence of testicular germ cell cancer, higher rates of cryptochorchidism, hypospadias and nervous disorders (Toppari *et al.*, 1996, Singleton and Khan, 2003). Questions are being raised whether EDCs having estrogenic effects or those which can modulate the action of estrogens could be causal factors in breast cancer occurrences (Davies *et al.*, 1993). Recent report shows the identification of over 500 EDCs (Brody and Rudel, 2003). However, there is still the lack of data which proves conclusively whether the human exposure to these EDCs pose any risk to reproductive health (Sharpe and Irvine, 2004).

EDCs Mode of Action

The mechanism of action of the endocrine disruptors is initiated by the binding of the chemicals to the nuclear receptor. Some chemicals act with similar hormonal function as the organism and activating the signaling pathway, while other acts as antagonists. The chemicals like nonylphenol or bisphenolA which has estrogen like activity, combine with estrogen receptor (ER). Similarly vinclozolin or DDE which are examples of androgenic chemicals and are known to be the competitors for androgen receptor (AR), binds to the receptor and inhibit the androgen hormone activity (Kelce *et al.*, 1995; Sugawara *et al.*, 2002). As a whole, at the cellular level, EDC can induce its effect via a number of pathways that involve steroid receptor binding (agonist), blocking steroid receptor binding (antagonist). In addition, disruption of the biosynthesis and or metabolism of steroids could also result (Sharpe and Irvine, 2004). Recently, Fisher *et al.* (2004) reported the disruption of testosterone synthesis by certain phthalate esters and inhibition of bio-inactivation of estradiol by inhibiting the sulfotransferase enzymes by some EDCs. EDCs can also alter the expression of the receptor protein itself, increasing or decreasing the amount of receptor molecule in each cell, and thus altering cellular sensitivity to normal signalers (Francois *et al.*, 2003).

In this context, the classical view that steroids act only through the binding of the high affinity steroid receptors (Truss and Beato, 1993) has been questioned. Recently, there has been increasing evidences regarding rapid non-genomic actions of steroid hormones (Simoncini and Genazzani, 2003 and Fisher, 2004). The exact pathway (s) for this mode of steroid action is still a matter of debate. These rapid effects do not act by transcription or protein synthesis dependent pathway. Some of the evidences have shown the steroid induced modulation of cytoplasmic and cell membrane bound regulatory proteins as also the intracellular signaling cascades involving mitogen activated protein kinases, phophatidylinositol 3-OH kinase or tyrosine kinases (Fisher, 2004). The involvement of membrane bound ion channels and G protein-coupled receptors have also been reported (Simoncini and Genazzani, 2003). The net outcome of these studies indicates that the action of steroid hormones upon target cells and organs are regulated by a complex interplay of genomic and nongenomic signaling and a cross talk between them (Figure 24.1).

Another aspect of EDCs action on target tissues is the specificity of particular receptors and variation in their affinities with ligands, for example, ligand binding and affinities of many phyto-estrogens (*e.g.* coumestrol and genistein), synthetic chemicals such a bisphenol A (Matthews *et al.*, 2001) and organochlorine pesticides (*e.g.* methoxychlor and its analogue DDT) for ERα and ERβ are not the same. More precisely, ERβ is dependent on pure agonists for the activation of transcription from its target promoters, while ERα can be activated by agonists, partial agonists (such as tamixifen), and ligand independent mechanisms. Naturally appropriate test systems must be designed strategically to cover such multiple actions and interactions (Jacobs and Lewis, 2002).

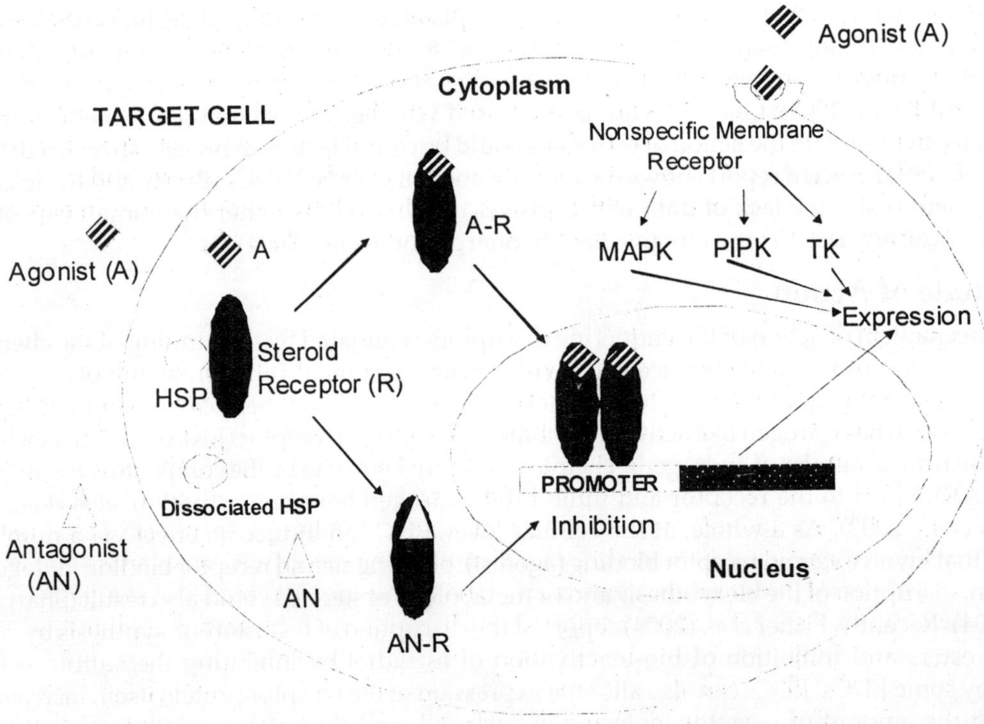

Figure 24.1: Endocrine Disruption of Steroid Hormone Action: EDCs (agonist or antagonist) bind to cytoplasmic/steroid receptor (genomic action). Agonists activate the target gene expression while the antagonists block it. Some agonists act through membrane receptor (non-genomic action) which activates target genes by MAPK, PIPK or TK pathways. HSP: Heat Shock Protein; MAPK: Mitogen Activated Protein Kinase; PIPK: Phosphatidnylinositol 3-OH kinase; TK: Tyrosine Kinase.

Tiered Bioassays for the Assessment of Endocrine Disruptors

As a result of the numerous field observations, epidemiologic reports, laboratory studies, magazine articles, books, ongoing controversy, and Congressional hearing, the issue of endocrine disruption was incorporated into the Food Quality Protection Act (FQPA) and amendments made to the Safe Water Drinking Act (SWDA) in 1996. This mandated the U.S. Environmental Protection Agency (EPA) to develop and implement a screening program to determine whether hormonal activity existed in environmental chemicals. Consequently, they established the Endocrine Disruptor Screening and Testing Advisory Committee (EDSTAC) which strongly recommended setting up a program with tiered approach to evaluate potential endocrine activity of chemicals (EDSTAC, 1998). At the same time Organization of Economic Cooperation and Development (OECD) developed the conceptual framework for the testing and assessment of endocrine disrupting chemicals consisting of several levels of screening and testing (OECD 2003). Although the mandate dealt mainly with environmental estrogens and androgens, thyroid hormone activity was also considered. The "tiered approach" proposed by EDSTAC starts with identification of substances that exhibit hormonal activity using a variety of *in vitro* and *in vivo* bioassay systems (Tier 1 Screening) and then determines which of these could evoke detrimental effects in test animals (Tier 2 Testing) (Witorsch, 2002). Marked species differences have been observed in the endocrine disruptive activities of environmental chemicals and therefore multi-tiered approaches are also being undertaken. Tier I screening battery, proposed by the US-EPA detects the hormonal activity of the compounds and Tier II characterizes and quantifies those effects (Eertmans *et al.*, 2003).

Today several *in vitro* and *in vivo* bioassays are available to screen EDC's. These assays are based on different principles right from binding assays to *in vivo* models. Of course the *in vitro* assays are sensitive, high throughput and easy to perform, but the importance of *in vivo* assays cannot be ignored since this model alone can truly assess the actual risk of the compounds. The problems with the *in vitro* assays are that the tests are carried out on an isolated system and do not take into consideration all the factors that add to the complexity of the human and animals. Many a time the parent compound may not have endocrine disruptive activity but could be metabolized by animals into compounds that disturb endocrine function. Thus, *in vitro* assays could give 'false negative' responses. It is desirable therefore that whole animal testing should follow up the results obtained with *in vitro* assays. Again, because of species differences, the results of *in vivo* bioassays used for risk assessment studies should be extrapolated to the human population with discretion. In order to give an idea of the practices, some *in vitro* and *in vivo* assays that are currently used to screen these EDCs are described.

In vitro Tests

Receptor Binding Assay

This assay is used to detect if the concerned chemical interacts directly with the steroid receptors, like ER, AR or progesterone receptor (PR) and determines the affinity of a particular chemical for the receptors. Briefly, the chemical is made to compete with the radiolabelled ligands (like estradiol, androgen) for binding a particular receptor isolated from the nuclear or cellular extract of the target tissue or cells (Wong *et al.*, 1995). Recently, there are few reports on the development of *in vitro* binding assays for estrogen (Scippo *et al.*, 2004), androgen (Scippo *et al.*, 2004; Freyberger and Ahr, 2004; Yamasaki *et al.*, 2004) and progesterone (Scippo *et al.*, 2004), where recombinant receptor proteins synthesized in bacterial systems have been used. However, the major drawback of these assays is that they do not provide information regarding hormonal activity and thus they cannot discriminate between agonistic and antagonistic actions of particular substances. Further they are also only suitable for the detection of hormone-receptor mediated effects and cannot detect pro-steroids (example, pro-estrogen) due to the absence of metabolism in the cell free system (Baker, 2001).

Cell Proliferation Assay

The E-Screen assay is best known for identifying estrogenic chemicals resulting in the proliferation of the human breast cancer cell line (MCF-7) in response to estrogen or estrogen like chemicals (Lippman and Huff, 1976; Soto *et al.*, 1995). Desaulniers *et al.* (1998) developed and optimized a MCF-7 based proliferation assay after thorough considerations of different factors involved in this assay. This assay is one of the most simple and sensitive assays available for estrogenic compounds. Rasmussen and Neilsen (2002) have described a detailed protocol including the critical parameters that should be taken into account for performing the E-Screen assay. Recently there is a report of screening androgenic compounds using the proliferation of human mammary carcinoma cell line stably transfected with human AR (Körner *et al.*, 2004). However, this type of assay is also not free of disadvantages; first, batch to batch variation of cells result in change of cell proliferation rates (Zachaarewski, 1997). Secondly, this test is difficult to perform for environmental samples containing constituents that are toxic to cells and thirdly, the studies from some laboratories have shown that the MCF-7 cells proliferate in response to a number of non-estrogenic substances like EGF, progesterone, DHT, insulin- like growth factors, lithium chloride and ethanol thus generating false positive data (Baker, 2001).

Reporter Gene Assays

This assay principally measures the level of expression of the reporter genes in response to the induction by some ligands or ligand like molecules. For these assays either the mammalian cell lines (MCF7, COS1, CHO etc.) or yeast strains such as *Saccharomyces cerevisiae* are transfected with a reporter plasmid and the receptor plasmid (in case the cell line does not have the endogenous receptor). The reporter plasmid consists of hormone response element (HRE) coupled to the reporter gene, such as luciferase (mostly in the case of mammalian cells) and β-galactosidase (Metzger *et al.*, 1998; Sohini and Sumpter, 1998; Gaido *et al.*, 1999; Hoogenboom *et al.*, 2001; Roy *et al.*, 2004). The rationale behind this assay is that exposure of the cells to a hormonally active compound, will result in the binding of this compound to its receptor. The ligand receptor complex then dimerizes and binds to the HRE resulting in the transcription and expression of the reporter genes. β-Galactosidase metabolize the substrate generating a red color that can be estimated with a spectrophotometer (Sohini and Sumpter, 1998). For luciferase as the reporter, the substrate is luciferin and this generates a flash/glow of light, which can be determined using a luminometer (Roy *et al.*, 2004). Cell based reporter gene assays are getting extremely popular day by day with its application not only in endocrine disruptor research but also in drug screening programs of pharmaceutical industries. Both yeast and mammalian cell lines can be stabilized with the transformed genes and therefore easy to use. This avoids the batch-to-batch variations as has been seen in transient expression of reporter and receptor genes in cells (Vinggaard *et al.*, 1999). Towards the development of high throughput screening system a recent report shows the use of green fluorescent protein as the reporter gene expressed in yeasts (Bovee *et al.*, 2004). This assay bypasses the cell lysis step as done in other reporter-based assays thus reducing the time and labor to a great extent. However, in spite of all these advantages there is still a question about the efficiency of this assay for screening toxic chemicals without damaging the normal physiology of the cells. Several laboratories are working on this aspect so that it can evolve into one of the best models for identification and screening of EDCs.

In vitro Test Summary

In vitro tests systems, particularly when used in combination, offer the possibility of providing an early screen for large numbers of chemicals and can be useful for understanding and analyzing their mechanism of action and potency. However, *in vitro* data always do not reflect the *in vivo* data due to lack of metabolic capacities in test systems and varied pathways involved for their mechanism of action. Finally, no single *in vitro* test is able to detect all properties of a hormonally active substance.

In vivo Tests

(anti-) Estrogenic Assays

The rodent uterotrophic assay has been done since 1930 to screen (anti-) estrogenic chemicals. This involves the *in vivo* binding of the chemicals to the estrogen receptor. The basic idea behind this assay is, the exogenous estrogen or estrogen like chemical can cause the hypertrophy of the uterus in female immature rats or ovariectomized rats. The rats are treated for 3 days with the chemicals and then they are sacrificed to collect the uterus, which is physiologically characterized. This test was evaluated by Owen and Ashby (2002) for screening some chemicals for their estrogenecity and the conclusion was that it is reliable and can detect estrogenic chemicals. In another set of assays, the abnormalities associated with the development of female sex organs and sex characters in immature animals (mainly rats) have been studied. For this, the immature female animals are exposed to the chemicals for about 20 days following which the reproductive organs are studied. Vaginal cornification assay is another assay that is also used extensively for the identification of estrogenic chemicals. In this test, the histological changes in vaginal epithelium is detected in response to different chemicals (Martin and Claringbold, 1960). These tests are yet to get the proper standardization and validation from EPA (Gray *et al.*, 2002). These tests if validated and approved finally can be used to screen compound with estrogen and thyroid hormone like activities.

(anti-) Androgenic Assay

Assays to detect androgenic properties of different chemicals have been developed since 1930s, but Hershberger *et al.* (1953) demonstrated the best among them. They analyzed in castrated rats, the response of the ventral prostrate, seminal vesicles and coagulating glands to exposure of several chemicals, mostly androgenic and some estrogenic and even progesterogenic. In this assay, among all parameters tested the capacity of the exogenously administered chemicals to restore the weight of the above-mentioned organs of the castrated male rats was determined and validated by Dorfman (1962). In a recent report, a detailed study has been done using Hershberger test and binding assay to identify thirty different chemicals for their androgenic activities (Yamasaki *et al.*, 2004). In addition to this test, another test has been developed that utilizes immature male rats for identifying and screening androgenic and thyroid properties of chemicals. This assay is based on the same principle as the rodent pubertal female assay, used for estrogenic chemicals. The assay examines abnormalities with development of the male sex organs and secondary sex characteristics. This test is also undergoing standardization and validation by the EPA (Gray *et al.*, 2002).

(anti-) Progesterogenic Assay

Not too many chemicals have yet been identified with progesterone like activities. One of the recent reports utilizes the rabbit endometrial transformation test for identifying the *in vivo* activity of progesterogenic chemicals (Tabata *et al.*, 2002). In this assay, immature female rabbits are administered with these chemicals for 5 consecutive days. The rabbits are sacrificed and uterine sections are then studied histologically for their extent of transformation in response to the chemicals.

Other *In vivo* Assays Using Non-mammalian Vertebrates

In addition to these assays some non-mammalian vertebrates are also used for identifying the EDCs. Fish is one of the popular and easily used model lists in this category. A fish reproduction assay, is used to test agents for estrogenic and androgenic effects. The adult reproductively active fishes are exposed to the chemicals for 21 days. The assay then examines abnormalities associated with survival, reproductive behavior, secondary sex characteristics and fecundity (number of spawns, number of eggs per spawn, fertility and development of offspring). This assay is being proposed to be included in the higher tier screening program (Hutchinson *et al.*, 2000). Another test for fish model is vitellogenin (vtg) protein regulation and its plasma clearance rates. Vtg is a protein expressed mostly in female fish. Male fish also possess this gene, but under normal condition they do not express this protein in a level equal or greater than that expressed in female fish. Thus, this has become an excellent marker for estrogenic chemicals in aquatic environment. A recent investigation of vtg regulation and plasma clearance in male sheepshead minnows reports that liver vtg levels returns to the base line in estradiol-exposed fish, but remains elevated in nonylphenol-treated fish (Hemmer *et al.*, 2002). Furthermore, plasma vtg clearance is both concentration and time dependent and may be detected even many months after exposure (Hemmer *et al.*, 2002).

Many man-made chemicals disrupt thyroid function either by influencing the synthesis of thyroid hormone or by adversely affecting their peripheral metabolism. Measurement of the enzyme thyroperoxidase (TPO) is considered to be a valuable endpoint to detect disruption of thyroid function (Capen 1998). Frog metamorphic assay is another assay which is performed in relation to the regression of the tadpole's tail during metamorphosis in *Xenopus laevis*. This test is based on the fact that thyroid hormone causes the metamorphosis in frogs and the assumption that exposure to the environmental chemicals would speedup the process. Along with other tests and assays this test is also undergoing validation by EPA (Yaoita and Nakajima, 1997; Tietge *et al.*, 2000).

In vivo Tests Summary

In addition to the above mentioned *in vivo* tests, OECD has made several revisions to the test methods like, 28-day repeat-dose oral toxicity study, two generation reproductive toxicity study and teratogenicity study and these are being considered for final adoption of OEDC member countries. Since these methods are not directly related to endocrine disrupting chemicals, it is apprehended that the proposed revision will improve the current guidelines by incorporating further parameters in the test system. Despite these certain advantages the *in vivo* tests also has some limitations in this method like, species variation, huge number of animals required and absence of SHBG in rodents after parturition, which makes the analysis further complicated (Baker, 2001).

Conclusion

There is increasing evidence that EDCs can disturb the optimal functioning of the reproductive system in both animals and human beings. It has been already known that the relationship between human health adverse effects and endocrine disruptors is poorly understood, this raises a controversy. In spite of these controversies there are still similarities between results obtained from animal studies and adverse effects as observed in human beings from different parts of the world, the best example being the effect of DES. Therefore, more research is needed to fully characterize the adverse effects resulting from exposure to endocrine disrupting chemicals and to understand the mechanistic basis for these effects. It is also clear that more focus is needed in considering additional endpoints, particularly those relating to the central nervous system and thyroid hormone system and to develop test methods for several potential targets. In addition, there is a need to characterize the effects of mixtures as well as resolving the possibility of low dose effects. A significant amount of work over the past few years has been focused towards the development of test methods to detect the EDCs. These tests are basically a wide range of *in vitro* assays. *In vitro* assays are definitely powerful tools for screening a large number of chemicals, particularly when used in combination, and can be useful in characterizing the mechanism of action and potency. *In vitro* assays are used for a fast, inexpensive and sensitive screening to identify and to evaluate the possible hormonal activity of a suspected compound. However, the biggest problem with these *in vitro* assays is that these data may not always correlate with the *in vivo* data due to differences in metabolic capacities in test systems and the diverse range of mechanisms by which EDCs may act. Further, a single *in vitro* test cannot predict all the properties of a hormonally active chemical. Considering all these facts it is best recommended that, a complementary battery of short and long term *in vitro* and *in vivo* assays (that assess both receptor and non-receptor mediated mechanisms of action) is the most appropriate way to comprehensively assess the potential endocrine disrupting activities of a substance or mixture of it. Test batteries which include a range of receptor and non receptor mediated endpoints (including *in vivo* elements to provide metabolic and pharmacokinetic competence) appear to provide the best opinion.

The computerized determination of Quantitative Structure-Activity Relationship (QSAR) has been used for many years in designing potentially useful chemical agents, but it is only in recent years these techniques have been used in toxicology (Sanderson *et al.*, 2004). This method can determine the structure and binding properties of a chemical computationally. But it must be borne in mind that a mere ligand receptor binding model is not enough to characterize the function of a suspected chemical. For example, two different ligands binding to the same receptor may not elicit the same result and the effects could also be cell and tissue specific. Thus, QSAR may be a fast and time saving mode of analysis but it needs further fine tuning to be used as a screening tool. In addition to QSAR some more techniques are being applied to understand the endocrine disrupting nature of these chemicals. Of these techniques, the use of DNA microarray has been recently reported by two laboratories (Francois *et al.*, 2003; Naciff and Daston, 2004). If validated properly, this technique could be useful in understanding EDCs action using the toxicogenomic approach. Hilscherova *et al.* (2004) identified the effects of some EDCs on ten different steroidogenic genes using real-time PCR technology. Availability of these advanced technologies will further improve the efficiencies of understanding the mechanism of action of these EDCs.

Recent activities globally have aimed at defining tiered testing strategies, validating screening and short term tests and evaluating the feasibility of modifying existing test guidelines to encompass the detection of EDCs. However, it is essential that a rigorous evaluation of assays is carried out prior to incorporation into testing strategy and it is important to define how the information provided by these can be used in hazard identification and possibly risk assessments.

Acknowledgements

The authors wish to thank the Council of Scientific and Industrial Research and the Department of Biotechnology, Government of India for generously funding research projects related to this review.

References

Baker VA. (2001). Endocrine disruptors–testing strategies to assess human hazard. Toxicol. *In vitro*, 15, 413-419.

Bovee TFH, Helsdingen RJR, Koks PD, Kuiper HA, Hoogenboom RLAP, and Keijer J. (2004). Development of a rapid yeast estrogen bioassay based on the expression of green fluorescent protein. Gene, 325, 187-200.

Brody JG, and Rudel RA, (2003). Environmental pollutants and breast cancer. Environ. Health Prespect., 111, 1007-1019.

Capen CC. (1998). Correlation of mechanistic data and histopathology in the evaluation of selected toxic endpoints of the endocrine system. Toxicol. Lett., 102-103, 405-409.

Cargouet M, Perdiz D, Mouatassim-Souali A, Tamisier-Korolak S, and Levi Y. (2004). Assessment of river contamination by estrogenic compounds in Paris area (France). Sci. Total Environ., 324, 55-66.

Colburn T, and Clement C. (1992). Chemically induced alterations in sexual and functional development: the wildlife/human connection. *In*: Colburn T, and Clement C. (Eds.), Adv. Modern Environ. Toxicol., Vol. 12. Princeton Scientific Publishing Co. Princeton NJ, pp. 403.

Damstra T, Barlow S, Bergman A, Kavlock R, Van Der Kraak G. (2002). International Programme on Chemical Safety (IPCS). *In*: Damstra T, Barlow S, Bergman A, Kavlock R, Van Der Kraak G. (Eds.), Global Assessment of the State-of-the–Science of Endocrine Disruptors, World Health Organization Geneva, Switzerland, WHO/PCS/EDC/02.2.

Davidson NE, and Yager JD. (1997). Pesticides and breast cancer: fact or fad? J. Natl. Cancer Inst., 89, 1743-1744.

Davis DL, Bradlow HL, Wolff M, Woodruff T, Howl DG, and Antón-Culver H. (1993). Medical hypothesis: xenoestrogens as preventable causes of breast cancer. Environ. Health Prespect., 101, 372-377.

Desaulniers D, Leingartner K, Zacharewski T, and Foster WG. (1998). Optimization of an MCF7-E3 cell proliferation assay and effects of environmental pollutants and industrial chemicals. Toxicol. in Vitro, 12, 409-422.

Dorfman R. (1962). Androgens and anabolic agents. *In:* Dorfman R. (Ed.), Methods in Hormone Research, Bioassays II, Academic Press, New York, pp. 275-313.

EDSTAC, Endocrine Disruptors Screening and Testing Advisory Committee (EDSTAC) (1998). Final report. Office pf Prevention, Pesticides and Toxic substances, United States Environmental Protection Agency, Washington, DC. http://www.epa.gov/scipoly/oscpendo/history/finalrpt.htm.

Eertmans F, Dhooge W, Stuyvaert S, and Comhaire F. (2003). Endocrine disruptors: effects on male fertility and screening tools for their assessment. Toxicol. in Vitro, 17, 515-524.

Fisher JS. (2004). Are all EDC effects mediated via steroid hormone receptors? Toxicol., 205, 33-41.

Fotis T, and Adlercreutz H. (1987). The multicomponent analysis of estrogens in urine by ion exchange chromatography and GC-MS-I. Quantitation of estrogen after initial hydrolysis of conjugates. J. Steroid Biochem., 28, 203-213.

Francois E, Wang D–Y, Fulthorpe R, Liss SN, and Edwards EA. (2003). DNA microarrays for detecting endocrine-disrupting compounds. Biotech. Advn., 22, 17-26.

Freyberger A, and Ahr H–J. (2004). Development and standardization of simple binding assay for detection of compounds with affinity for the androgen receptor. Toxicol. 195, 113-126.

Gaido KW, Leonard LS, Maness SC, Hall JM, McDonnell DP, Saville B, and Safe S. (1999). Differential interaction of the methoxychlor metabolite 2,2-bis-(p-hydroxyphenyl)-1,1,1-trichloroethane with estrogen α and β. Endocrionol. 143, 5746-5753.

Gray LEJ, Ostby J, Wlison V, Lambright C, Bonseine K, Hartig P, Hotchkiss A, Wolf C, Furr J, Price M, Parks L, Cooper RL, Stocker TE, Laws SC, Degitz SJ, Jensen KM, Khal MD, Korte JJ, Makynen EA, Tietge JE, and Ankley GT. (2002). Xenoendocrine disruptors-tiered screening and testing. Toxicol., 181-182, 371-382.

Greim HA. (2004). The Endocrine and Reproductive system: Adverse effects on hormonally active substances? Paediatrics, 113, 1070-1075.

Gutendorf B, and Westendorf J. (2001). Comparison of an array of *in vitro* assays for the assessment of the estrogenic potential of natural and synthetic estrogens, phytoestrogens and xenoetrogens. Toxicol., 166, 79-89.

Hemmer MJ, Bowman CJ, Hemmer BL, Friedman SD, Marovich D, Kroll KJ, and Denslow ND. (2002). Vitellogenin mRNA regulation and plasma clearance in male sheephead minnows, (*Cyprinodon variegates*) after cessation of exposure to 17 beta-estradiol and p-nonylphenol. Aqua. Toxicol., 58, 99-112.

Hershberger L, Shipley E, and Meyer R. (1953). Myotrophic activity of 19-nortestosterone and other steroids determined by modified levator ani muscle method. Proc. Soc. Exp. Biol. Med., 83, 175-180.

Hilscherova K, Jones PD, Gracia T, Nested JL, Zhang X, Sanderson JT, Yu RM, Wu RS, and Giesy JP. (2004). Assessment of the effects of chemicals on the expression of ten steroidogenic genes in the H295R cell line using real time PCR. Toxicol. Sci., 81, 78-89.

Hoogenboom LAP, DeHaan L, Hooijerink D, Bor G, Murk AJ, and Brouwer A. (2001). Estrogenic activity of estradiol and its metabolites in the ER-CALUX assay with human T47D breast cells. APIMS, 109, 101-107.

Hutchinson TH, Brown R, Brugger KE, Campbell PM, Holt M, Lange R, McCahon P, Tattersfield LJ, and van Egmond R. (2000). Ecological risk assessment of endocrine disruptors. Environ. Health Prespect., 108, 1007-1014.

International Programme on Chemiacl Safety (IPCS). Report of IPCS/OECD scoping meeting on endocrine disruptors (EDCs). Washington, DC; March 1998.

Jacobs MN, and Lewis DFV. (2002). Steroid hormone receptors and dietary ligands: a selected review. Proc. Nutr. Soc., 61, 105-122.

Kelce WR, Stone CR, Laws SC, Gray LE, Kemppainen JA, Wilson EM. (1995). Persistent DDT metabolite p,p'-DDE is a potent androgen receptor antagonist. Nature, 375, 581-585.

Körner W, Vinggaard AM, Térouanne BMR, Wieloch, Schlumpf M, Sultan C, and Soto AM. (2004). Interlaboratory comparison of four *in vitro* assays for assessing androgenic and antiandrogenic activity of environmental chemicals. Environ. Health Prespect., 112, 695-702.

Laws SC, Carey SA, Ferrell JM, Bodman GJ, and Cooper RL. (2000). Estrogenic activity of octylphenol, nonylphenol, bisphenol A and methoxychlor in rats. Toxicol. Sci., 54, 154-167.

Lee HB, and Peart TE. (1998). Determination of 17β estradiol and its metabolites in sewage effluent by solid-phase extraction and Gas chromatography/mass spectrometry. J. AOAC Int., 81, 1209-1215.

Lippman M, and Huff KA. (1976). A demonstration of androgen and estrogen receptors in human breast cancer using a new protamine sulfate assay. Cancer, 38, 868-874.

Martin L, and Claringbold PJ. (1960). The mitogenic action of oestrogens in the vaginal epithelium of the ovariectomized mouse. J. Endocrinol., 20, 173-186.

Matthews JB, Twomey K, and Zachareswski TR. (2001). *In vitro* and *in vivo* interactions of bisphenol A and its metabolite, bisphenol A glucuronide, with estrogen receptors α and β. Chem. Res. Toxicol., 14, 149-157.

Metzger D, White JH, and Chambon P. (1998). The human estrogen receptor functions in yeast. Nature, 334, 31-36.

Naciff JM, and Daston GP. (2004). Toxicogenomic approach to endocrine disruptors: Identification of a transcript profile characteristic of chemicals with estrogenic activity. Toxicol. Pathol., 32, 59-70.

OECD, Draft summary report of the sixth meeting of the task force on endocrine disruptors testing and assessment (EDTA 6), Draft as of March 12, 2003.

Owens JW, and Ashby J. (2002). Critical review and evaluation of the uterotrophic bioassay for the identification of possible estrogen agonists and antagonists: in support of the validation of the OECD uterotrophic protocols for the laboratory rodent. Organization for economic co-operation and development. Critical Rev. Toxicol., 32, 445-520.

Rasmussen TH, and Nielsen JB. (2002). Critical parameters in the MCF-7 cell proliferation bioassay (E-Screen). Biomarkers, 7, 322-336.

Roy P, Salminen H, Koskimies P, Simola J, Smeds A, Saukko P, and Huhtaniemi IT. (2004). Screening of some anti-androgenic endocrine disruptors using a recombinant cell-based *in vitro* bioassay. J. Steroid Biochem. Mol. Biol., 88, 157-166.

Safe SH, and Gaido K. (1998). Phytoestrogens and anthropogenic estrogenic compounds. Environ. Toxicol. Chem., 17, 119-126.

Sanderson H, Johnson DJ, Reitsma T, Brain RA, Wilson CJ, Solomon KR. (2004). Ranking and prioritization of environmental risks of pharmaceuticals in surface waters. Regul. Toxicol. Pharamcol., 39, 158-183.

Scientific Committee for Toxicity, Ecotoxicity and the environment. CSTEE Opinion on human and wildlife health effects of endocrine disrupting chemicals, with emphasis on wildlife and on ecotoxicology test methods. European Commission, DG XXIV, Consumer Policy and Consumer Health Protection; March 1999.

Scippo M–L, Argiris C, Weerdt CVD, Muller M, Willemsen P, Martial J, Rogister GM. (2004). Recombinant human estrogen, androgen and progesterone receptors for detection of potential endocrine disruptors. Anal. Bioanal. Chem., 378, 664-669.

Sharpe RM, and Irvine DS. (2004). How strong is the evidence of a link between environmental chemicals and adverse effects on reproductive health? British Med. J., 328, 447-451.

Simoncini T, and Genazzani AR. (2003). Non-genomic actions of sex steroid hormones. Eur. J. Endocrinol., 148, 281-292.

Singleton DW, and Khan SA. (2003). Xenoestrogen exposure and mechanisms of endocrine disruption. Front. Biosci., January: 1.

Sohini P, and Sumpter JP. (1998). Several environmental oestrogens are also anti-androgens. J. Endocrinol., 158, 327-339.

Soto AM, Sonneenscheein C, Chung KL, Fernandez MF, Olea N, and Serrano FO. (1995). The E-Screen assay as a tool to identify estrogens: an update on estrogenic environmental pollutants. Environ. Health Prespect., 103, 113-122.

Soto AM, Calabro JM, Prechtl NV, Yau AY, Edward FO, Daxenberger A, Kolok AS, Guillette LJJr., Bizec le B, Lange IG, and Sonnenschein C. (2004). Androgenic and estrogenic activity in water bodies receiving cattle feedlot effluent in Eastern Nebraska, USA. Environ Health Prespect 112: 346-352.

Sugawara T, Nakajima A, Nomura E. (2002). Development of a simple screening system for endocrine disruptors. Med. Sci. Monit., 8, 431-438.

Tabata Y, Iizuka Y, Masuda NT, Shinei R, Kurihara K, Okonogi T, Hoshiko S, and Kurata Y. (2002). *In vitro* and *in vivo* characterization of novel non-steroidal progesterone receptor agonists derived from the fungal metabolite PF1092C. J. Steroid Biochem. Mol. Biol., 82, 217-223.

Ternes TA, Stumf M, Mueller J, Haberer K, Wilken RD, and Servos M. (1999). Behavior and occurrence of estrogens in municipal sewage treatment plants-I. Investigations in Germany, Canada and Barzil. Sci. Total Environ., 225, 81-90.

Tietge JE, Ankley GT, and Degitz SJ. (2000). Report on the Xenopus Tail Resorption Assay as a Tier 1 Screen for Thyroid Active Chemicals, Mid-Continent Ecology Division, USEPA, Duluth MN, pp 45.

Toppari J, Larsen JC, Christiansen P, Giwercman A, Grandjean P, Guillette Jr. LJ, Jegou B, Jensen TK, Jouannet P, Keiding N, Leffers H, McLachlan JA, Meyer O, Muller J, Rajpert-de-Meyts E, Scheike T, Sharpe R, Sumpter J, and Skakkebaek NE. (1996). Male reproductive health and environmental xenoextrogens. Environ. Health Prespect., 104, 741-803.

Truss M, and Beato M. (1993). Steroid hormone receptors: interaction with deoxyribonucleic acid and transcription factors. Endocr. Rev., 14, 459-479.

Witorsch RJ. (2002). Can biological effects and environmental risks be predicted? Regul. Toxicol. Pharmacol., 36, 118-130.

Vinggaard AM, Joergensen ECB, and Latsen JC. (1999). Rapid and sensitive reporter gene assays for detection of antiandrogenic and estrogenic effects of environmental chemicals. Toxicol. Appl. Pharmacol., 155, 150-160.

Wong C, Kelce WR, Sar M, and Wilson EM. (1995). Androgen receptor antagonist versus agonist activities of the fungicide vinclozolin relative to hydroxyflutamide. J. Biol. Chem., 25, 19998-20003.

Yamasaki K, Sawaki M, Noda S, Muroi T, Takakura S, Mitoma H, Sakamoto S, Nakai M, and Yakabe Y. (2004). Comparison of the Heshberger assay and androgen receptor binding assay of twelve chemicals. Toxicol., 195, 177-186.

Yaoita Y, and Nakajima K. (1997). Induction of apoptosis and CPP32 expression by thyroid hormone in a myoblastic cell line derived from tadpole tail. J. Biol. Chem., 272, 5122-5127.

Zachaarewski TR. (1997). *In vitro* bioassay for assessing estrogenic substances. Environ. Sci. Tech., 31, 613-623.

25

Assessment of Endocrine Disruptive Activities of Xenobiotics in Teleosts

Shelley Bhattacharya and Sonali Roy*

Environmental Toxicology Laboratory, Department of Zoology, Visva Bharati University, Santiniketan – 731 235, India

ABSTRACT

The endocrine system is a highly complex and integrative system responsible for regulating various body functions including growth, development and maturation. With the phenomenal increase in the varieties, types and numbers of xenobiotic chemicals the onslaught on the teleosts is of immediate environmental concern. Endocrine defects may be addressed in three ways namely, as a) response of the endocrine sensitive tissues in an intact whole organism; b) response of the organism at multiple tiers including phenotypic expression, physiology, cell biology and molecular biology and c) by direct measurement of hormone action. Moreover, crosstalk between the different endocrine systems results in unpredictable effects in a different system altogether. Therefore one has to be cautious in making predictions of toxicity of the xenobiotics. Nevertheless some degree of specificity of xenobiotic action has been found in terms of thyrotoxicity, gonadal toxicity and interrenal toxicity. In the area of xenobiotic research there is indeed a lot of uncertainty and the problem of the endocrine disruptive substances is ominous. Further research in this area is the need of today to guarantee a cleaner healthier environmental future.

**Corresponding Author*: E-mail: bhattacharyashelley@rediffmail.com

Introduction

In recent years there has been a momentous increase in the load of xenobiotics of diverse origin in the environment. As yet there are 11 million known chemicals, of which about 100,000 are being produced currently on an industrial scale and about 1000 to 2000 new chemicals are introduced annually in the market (Younes, 1999). A large number of chemicals having deleterious effects on the endocrine systems (endocrine disrupting chemicals or endocrine disruptors) have recently emerged in the environment and studies have been focused in the fields of environmental toxicology, medicine and regulatory policy (Chen, 2001; Menditto and Turio-Baldassari,1999; Rhind, 2002). A seminal work by Sumpter and Jobling (1995) established that a large number of xenobiotics and their degradation products have appeared in the water ways. Fish being the most successful vertebrate with a high degree of adaptability in residing in freshwater, marine or brackish water systems have been found to absorb and accumulate such xenobiotics in alarmingly high concentrations. Moreover, a higher ventilation rate through the gills increases the rate of exposure to water borne toxicants including exposure to endocrine disrupting xenobiotics. It is well known that the endocrine system is a highly complex and integrative system which is responsible for regulating body functions including growth, development and maturation as well as how the different organs participate in the orchestrated phenomenon of life. The present review will deal with the impact of diverse chemical compounds or xenobiotics, arising out of human related activities, in the fish organizational system.

Definition of Endocrine Disruptor

An endocrine disruptor can be defined as a chemical present in the environment owing to human related activities in industry, agriculture, pharmaceutical operations and so on. However, the chemical nature of such compounds is widely variable from each other although dose response observations can be recorded in the endocrine system which is not the secondary effects of general systemic toxicity. These synthetic chemicals when absorbed into the body either mimic or block the hormones thereby disrupting their normal functions. These disruptions are found to occur in various ways. Xenobiotics can inhibit or stimulate the synthesis and secretion of hormones, can alter the homeostatic levels of hormones or even can influence the mechanism of action of the hormones. Any or all of these interferences lead to a disturbance in the normal function of the respective hormones. However, endocrine disruptors can also induce developmental abnormalities arising out of reproductive toxicity, teratogenicity and embryotoxicity. Several hormones have critical roles in directing the developmental process. Therefore exogenous environmental chemicals that are able to mimic, block or modulate endogenous chemical messengers can produce defects in the developmental process (Birnbaum, 1995). Moreover, reproductive endocrine system is primarily regulated by the neuroendocrine system where the EDCs can pose a potential threat. Such defects caused either directly or indirectly have also been considered as an outcome of endocrine disruption. Finally the endocrine defects may be analyzed in three ways:- i) as response of such compounds to endocrine sensitive tissues in an intact whole organism; ii) as response of such compounds at multiple levels of biological organization including phenotypic expression, physiology, cell biology and molecular biology; iii) by direct measurement of changed hormone action including gene induction or repression under controlled toxicity bioassays.

An Inventory of Endocrine Disruptors

The earliest known endocrine disruptor (ED) is diethylstilbestrol (DES) which is also cited as a classical example of delayed toxicity of a xenobiotic. Besides, other compounds which appeared in the environment over decades are dioxin, polychlorinated biphenyls or PCBs, dichlorodiethyl trichloroethane (DDT) and many other organochlorine pesticides. The number and variety of such endocrine disruptors have increased in great proportions during the last few decades (Table 25.1) and it is necessary to put on record the mechanism of action of these xenobiotic compounds in order to establish the status of biomarker of endocrine disruption in fish.

Table 1: An Inventory of Endocrine Disruptors

Endocrine Disruptor	*Reference*	*System Affected*
BKME	McMaster *et al.*, 1991; van der Kraak *et al.*, 1992; Munkittrick *et al.*, 1992	Gonadal Steroidogenesis
Cadmium	Sangalang and Freeman, 1974; Kime 1984	
PCBs	Sivarajah *et al.*, 1978; Thomas 1988	
Xenoestrogens	Sumpter and Jobling, 1995	
Carbamate and organophosphate compounds	Pundir and Saxena, 1990; Sukumar and Karpagaganapathy, 1992; Murugesan and Hanifa, 1992; Anitha Kumari and Sree Ram Kumar, 1997; Mondal 1997	
Cadmium, Malathion	Singh, 1989	
Gamma-BHC	Singh *et al.*, 1993	
Metacid 50, Carbaryl	Choudhury *et al.*, 1993	
PCBs, PB, benzo(a)pyrene	Thomas 1989, 1990a	Gonadal changes
OP and CA compounds	Bhattacharya *et al.*, 1996 Bhattacharya and Mondal 1997	
Organochlorines	Schoettger, 1972; Grant and Mehrle, 1970; Mehrle, 1971; Kumar and Bhattacharya, 1975; Singh and Singh, 1980, Sinha *et al.*, 1991a;1991b	Plasma Cortisol profile
OP and CA compounds Cd, Hg	Saxena and Mani, 1985; Guhathakurta and Bhattacharya, 1988; Ghosh *et al.*, 1989; Ghosh and Bhattacharya, 1992	Thyroid functions
Sewage Treatment Effluents	Tyler and Routledge, 1998; Kime, *et al.*,1999	Vitellogenesis
Bisphenol A	Larsson *et al.*, 1999	
BKME	Mellanen *et al.*, 1999; Trembley and Van Der Kraak, 1999	

Interference by Endocrine Disruptors

The discussion on the biological impact posed by a specific endocrine disruptor on the endocrine system is a difficult task because of the inherent interactive nature of hormone actions (Figure 25.1). Endocrine systems have a pervasive role in both short and long term manifestation of metabolic regulation. All physiological processes are intricately orchestrated by the endocrine systems giving rise to highly subtle and surprising effects due to either hypo or hyperactivity of the hormones. Endocrine disruptive compounds (EDCs) are also basically intriguing molecules because of their unknown mechanisms of action in biological systems and their actions are also guided by various factors. On the other hand, the life cycle stage of the fish can modulate the nature of response to an EDC. If the exposure takes place in the adult stage compensatory mechanisms may be triggered which mask the toxicity of an EDC. Secondly, the effect may be profound if exposure takes place when programming of the endocrine system is in progress. Thirdly, exposure to a same dose of EDC at different life cycle stages or in different seasons may produce altogether different effects. Moreover, the

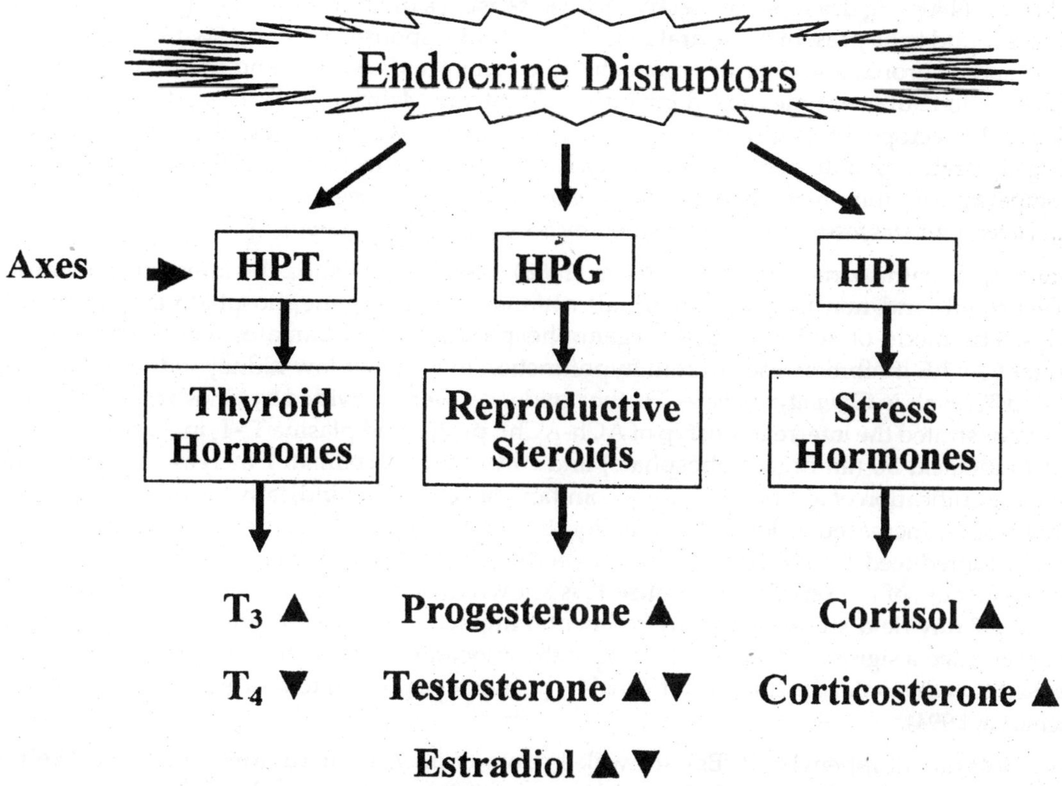

Figure 25.1: Flow Chart of Endocrine Disruptive Actions

crosstalk between different endocrine systems may result in unpredictable effects surfacing in a totally different system. Therefore considerable caution must be exercised while extrapolating results obtained from *in vitro* studies to real *in vivo* situations. In general the threat posed by the EDCs either agonistic or antagonistic is destined to be of serious concern.

Specificity of Xenobiotics

Thyrotoxicity

Thyroid hormone is an enigmatic hormone having involvement in various physiological processes. They have a major role in the metamorphosis of larval fish and amphibians into the juvenile body forms (Dickhoff *et al.*, 1990; Galton, 1992; Kikuyama *et al.*, 1993; Shi, 1994). The manifestation of thyroid hormone action is dependent upon the hypothalmus pituitary thyroid (HPT) axis in non mammals and mammals (Norris 1997) except the teleosts, where thyroid is ectopic, diffused and not encapsulated by connective tissue. In some case thyroid follicles are found in kidney, liver and also gonads. Thyroid hormones are also known to work synergistically with growth hormone (Norris, 1997) and are well known stimulators of sexual maturation (Norris, 1999).

The earliest report on the effect of organochlorines (Schoettger, 1972) on reproduction; osmoregulation and oxidative metabolism were indeed highly relevant. The work revealed that the threshold dose (480ng/g dry food) of dietary dieldrin causing a discernible physiological lesion in the thyroid was found to be present in natural foods. Thyroidal response also varied with the dose of the xenobitoic and a chronic low dose of organochlorine compounds were found to initially stimulate thyroidal function but over time caused inhibition (Grant and Mehrle, 1970). Interestingly in fish the thyroid gland is ectopic in origin and the head kidney is known to possess thyroid follicles also exhibiting high rate of iodide peroxidase activity. Kumar and Bhattacharya (1975) had demonstrated a non competitive inhibition of iodide peroxidase by endrin in *Anabas testudineus*. A similar mode of action in *Heteropneustes fossilis* was exemplified by Singh and Singh (1980).

Later on, Guhathakurta and Bhattacharya (1988) clearly established the interactive actions of target (neuronal) and non target (endocrinal) actions of both organophosphate and carbamate pesticides. The mode of action of both organophosphates and carbamates were manifested as concomitant AChE inhibition and thyroid hypofunction in terms of low kidney iodide peroxidase activity and T_4 levels in *Channa punctatus*. Further evidences were provided by Ghosh *et al.* (1989) who clearly demonstrated the interrelationship of ACh-AChE profile and plasma T_3-T_4 in *Channa punctatus* exposed to metacid 50 (an organophosphate) and carbaryl (a carbamate). Increase in T_3 indicates increased organification of iodine while a concomitant decline in T_4 is indicative of depressed release of the hormones. Increased T_3 level also ameliorates the endocrine disruptor effect because it can compensate for reduced T_4 titre. This can be described as a compensatory mechanism of counteracting the deleteriousness of the xenobiotic because T_3 is known to be the more active thyroid hormone. A comparative 7 day field study of the environmental impact of metacid 50 and carbaryl on *Channa punctatus* revealed a significant increment in T_3 with a concomitant reduction in T_4 level. Thus it may be surmised that thyroidal dysfunction effected by pesticides upon the teleost is genuine (Ghosh and Bhattacharya, 1994).

Polycholrinatedbiphenyls (PCBs) are well known endocrine disruptors which are likely to influence neuroendocrine function. Collins and Capen (1980) have reported that PCB exposure in rats resulted in fine structural lesions in thyroid in association with inhibition of proteolysis of thyroglobulin thereby decreasing the release of thyroxine. In addition thyroid hormone levels are influenced by increase in hormone metabolism by PCB (Barter and Klaassen, 1992) and ultimately reproduction (Khan and Thomas 1998). In Atlantic croaker Aroclor 1254 caused a dramatic decline in LH secretion and hypothalamic levels of 5-HT, which is stimulatory of LH secretion (Khan and Thomas, 2000) which was later found to be due to inhibition of tryptophan hydroxylase, the rate limiting enzyme in 5HT synthesis (Khan and Thomas, 2001). The decreased 5HT activity resulted in lowered LHRH content in hypopthalamus leading to down regulation of LHRH receptors on gonadotropes and decreased LH response to LHRH stimulation (Khan and Thomas, 2000). Khan and Thomas (2001) further demonstrated that parachlorine phenylalanine, a specific inhibitor of tryptophan hydroxylase, mimicked the PCB effects in causing reproductive impairment whereas co-treatment of PCB dosed fish with 5-hydroxytryptophan reversed the PCB effect. Such neuroendocrine disturbances are found to occur in the Atlantic croaker treated with lead.

Gonadal Toxicity

The gonads are the major sites of steroidogenesis of the reproductive hormones and the survivability of the species depends on successful production of the steroid hormones during the breeding season of the fish. Reduced gonadosomatic index and depressed growth of ovaries have been correlated with xenobiotic induced inhibition of estrogen synthesis (Munkittrick *et al.*, 1992; Singh, 1989; Singh *et al.*, 1993; Thomas, 1989; 1990a). Ghosh *et al.* (1990) demonstrated the impairment of the regulation of the gonadal function by non lethal or no mortality levels of metacid 50 and carbaryl in both laboratory and field conditions. Under laboratory conditions they noted a time dependent decrease in serum GtH level which was found to be significantly higher in the carbaryl treated fish. Interestingly the situation was reversed in the field where metacid 50 was found to be more inhibitive in its action. However, metacid 50 caused a higher rate of decrease in the pituitary GtH content and inhibition of GnRH activity under both laboratory and field conditions. Thus it is highly significant observation by Ghosh (1990) and Ghosh *et al.* (1990) that low doses of such pesticides cause reproductive damage in fish mediated via the hypohalamus-pituitary-gonadal (HPG). Choudhury *et al.* (1993) have demonstrated different patterns of xenobiotic action in short term and long term exposures of *Anabas testudineus* to metacid 50 and carbaryl. An initial increase in plasma and ovarian estradiol was followed by an equally dramatic fall in the hormone which was directly correlated with the GSI and decreased vitellogenic synthesis.

Integrity of the reproductive system is the key to the success of survival of any species. Fish being a very vulnerable organism to ambient pollution any interference with the normal structure and function of the endocrine systems cause a drastic reduction in fecundity and reproduction. Nowadays the use of pesticides has been maximized in India to get a good crop. Unfortunately, the agricultural run off carry all the toxic pesticides or its metabolites to the water bodies thus jeopardizing the normal life cycle of the fish. Pesticides being the major agricultural chemicals pervading the water resources the stressful events are recorded in the fish and large volume of data have been generated in this area of research. Fish are very interesting models in the study of interference posed by endocrine disruptors to the gonadal organization because reproduction is a cyclical event and functions in tune with the environmental cues. Therefore any small environmental changes can register a significant displacement of the breeding cycle. The stress effects registered can be manifested as gross changes in the structure of ovary (Sukumar and Karpagaganapathy, 1992) and testes (Pundir and Saxena, 1990). Changes are also documented as less yolky granules (Sukumar and Karpagaganapathy, 1992) ruptured oocyte walls and damaged yolky vesicles (Kulshrestha and Arora, 1984), nuclear and cytoplasmic changes (Kirubagararan and Joy, 1988, Murugesan and Haniffa, 1992). Even so called safe concentrations over a long term exposure regimen have been found to trigger atresia in *Channa punctatus* (Mani and Saxena, 1985). Dey and Bhattacharya (1989) demonstrated that during a long term exposure to phenthoate the Stage III oocytes in *Channa punctatus* are radically destroyed thus compromising the chance of successful reproduction. Similar observations have also been recorded in gamma–BHC treated *Heteropneustes fossilis* (Singh and Singh, 1980) or metacid 50 treated *Channa punctatus* and *Anabas testudineus* (Choudhury, 1991). Thus it may be surmised that the first line of attack by endocrine disruptors is ovarian damage. However, mere assessing the indicator potential do not address the exact mechanism of action of endocrine disruptors. It is well documented in fish that the steroid hormones are affected by endocrine disruptors but since it is not possible to link particular steroids to specific glands in the male or female organisms reduced GSI and inhibition of ovarian growth can be taken as a dependable index of disruptive action of chemicals. Moreover, the inhibition of ovarian growth may be considered as an expression of inhibition of estrogen synthesis by Aroclor 1254, lead and benzo(a)pyrene in Atlantic croaker (Thomas, 1989; 1990a), cadmium, malathion and 3-methycholanthrene in *Monopterus albus* (Singh, 1989), bleached kraft mill effluent (BKME) in white sucker (Munkitrick *et al.*, 1992) and gamma-BHC in *Heteropneustes fossilis* (Singh *et al.*, 1993). Different patterns of responses have been identified by Choudhury *et al.* (1993) in *Anabas testudineus* exposed for short and long term to metacid 50, a potent organophosphate pesticide in use in India. Moreover it was also revealed that an initial remarkable increase in plasma and ovarian estradiol followed by an equally drastic fall in the hormone titre was well correlated to GSI and decreased vitellogenic response.

Thomas (1988) reported an increase in the ovarian estradiol and decrease in testosterone. This finding corroborates earlier reports indicating that lowered activity of testicular steroids were more or less a general stress response as exhibited by PCB (Sivarajah, *et al.*, 1978), aldrin (Chatterjee *et al.*, 1988) or BKME (McMaster *et al.*, 1991; Munkitrick *et al.*, 1992; Van der Kraak *et al.*, 1992). *In vitro* steroidogenesis in many fish species has also attracted attention of several workers to elucidate the mechanisms of action of such chemicals causing endocrine disruption. Sangalang and O'Halloran (1973) were one of the first group of researchers to assess the damage quotient of the endocrine disruptors in both *in vitro* and *in vivo* (Sangalang and Freeman, 1974) studies with cadmium induced inhibition of 11-ketotestosterone biosynthesis. They clearly demonstrated that cadmium treatment induced elevated androgen profile in the plasma which was found to be due to impaired clearance rather than synthesis. Kime (1984) demonstrated in rainbow trout that a low dose of cadmium stimulated endogenous steroidogenesis of testosterone and 11β-hydroxytestosterone. Singh *et al.* (1994) and Kime (1995) also demonstrated in *in vitro* steroidogenetic protocols that low dose stimulated and higher doses of endocrine disruptors inhibited steroidogenesis.

Usually testosterone is the major androgen and estradiol the major estrogen. Many male teleosts, however, are found to produce 11-ketotestosterone which happens to be the predominating species of androgen. Female teleosts also may have quite a high concentration of testosterone which may be as great as estradiol. mechanism of steroid action in teleosts is similar to mammalian systems with several differences, such as in addition to ER-α and ER-β a third subtype ER-γ has been identified in teleosts (Hawkins *et al.*, 2000). Moreover, the teleost progestin receptor does not bind as many EDCs that bind to the mammalian progesterone receptor (Pinter and Thomas, 1997). Thus extreme care is necessary to extrapolate the mechanism of actions of the various EDCs across the vertebrate taxa. Endocrine disruptors that mimic estrogens or express antiandrogenic activity can indeed have a far reaching effect in the wild life exposed during development or as juveniles.

Plasma levels of hormones have sometimes been utilized as biomarkers to record rate and magnitude of steroidogenic potential of xenobiotics. Plasma estrogen level has been reliably used as a bioindicator of xenobiotic exposure by several authors (Choudhury, 1993; Kime, 1995; Thomas 1990b). Choudhury (1991) also reported increased titres of plasma progesterone and estradiol in metacid 50 and carbaryl treated *Anabas testudineus* which were corroborated by Thomas (1989). Sumpter and Jobling (1995) and Cheek *et al.* (1998) established clearly that there are several xenobiotics or their metabolites in the water ways which are weakly estrogenic and these compounds can be absorbed and bioaccumulated by the fish in sufficiently high concentrations to be deleterious in their endocrine disruptive activity. Although there are good indications that xenobiotics can disrupt gonadal activity but the mechanism of their action has not been elucidated. In an interesting study, however, the signal transduction by inorganic mercury has been successfully traced in the fish oocyte model (Mondal *et al.*, 1997). It was noted that *in vivo* and *in vitro* treatments of fish ovary with inorganic mercury caused a remarkably high rate of progesterone synthesis accompanied by a low rate of conversion to 17β-estradiol. Progesterone accumulation was suggestive of enhancement in the 3β-hydroxysteroid dehydrogenase (HSD) activity in the Hg-treated fish. It was further demonstrated that the synthesis of this specific enzyme was definitely triggered by Hg binding to oocyte plasma membrane Na^+-K^+-ATPase with a concomitant increase in the specific messenger RNA of HSD at a low level of Hg-intoxication of the fish through binding of Hg to MRE or Metal Response Element (Mondal *et al.*, 1997). The following model is suggestive of the signal transduction of inorganic mercury in fish oocytes (Figure 25.2).

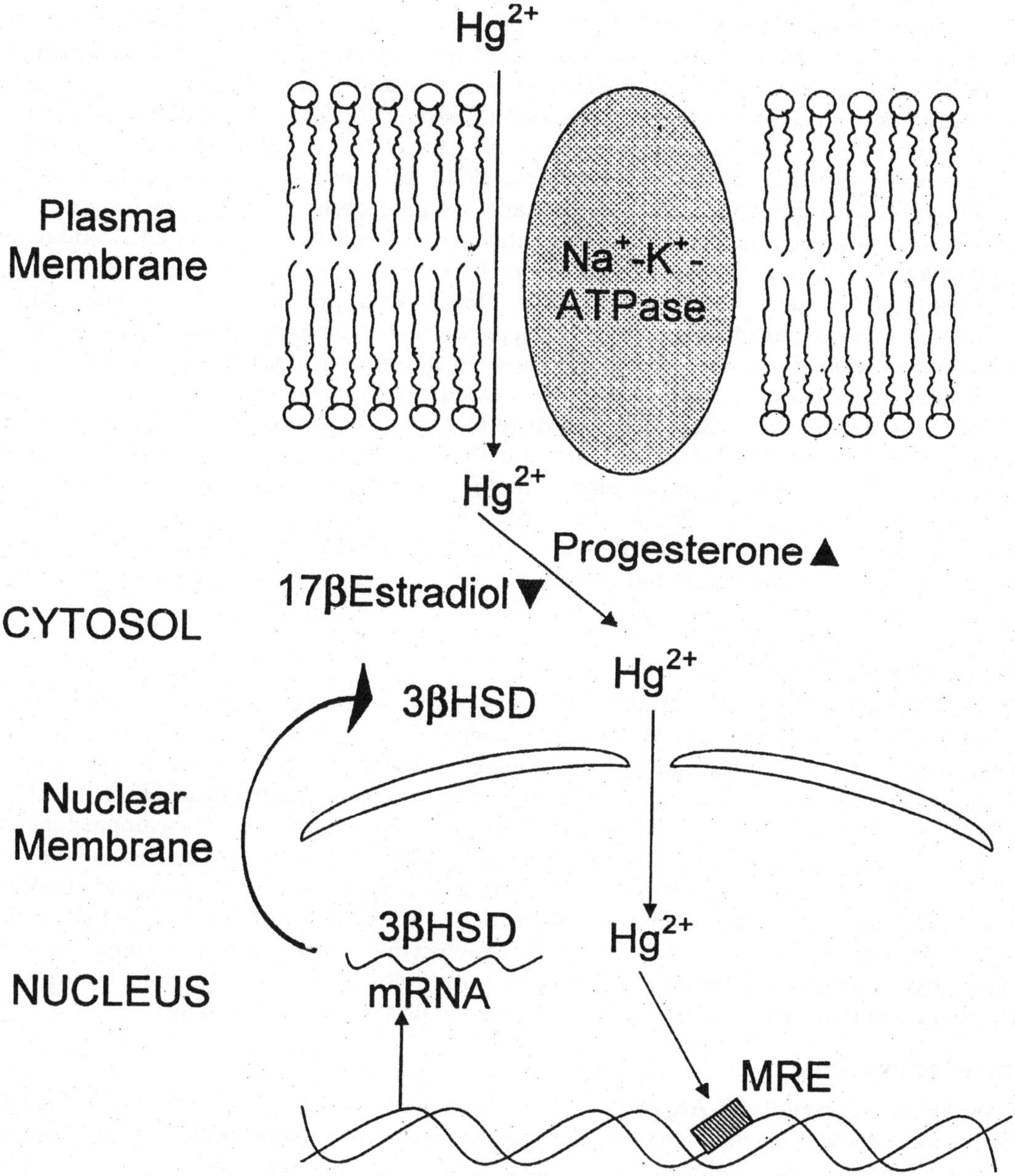

Figure 25.2: Signal Transduction by Hg in Fish Oocyte

Interrenal Toxicity

Steroidogenic tissue homologous to the mammalian adrenal cortex in non mammalian vertebrates is termed as interrenal (Bentley, 1998). The terminal neurons of the hypothalamus release corticotropin releasing hormone (CRH) which acts on corticotropes in the anterior pituitary to regulate synthesis and secretion of adrenocorticotropin hormone (ACTH) which in turn stimulates the secretion of glucocorticoids (cortisol or corticosterone) from adrenal or interrenal tissues. Besides the already known mechanisms of action of these hormones it has been increasingly realized that glucocorticoids have important "programming" effects during development and alterations in the circulating levels of these hormones can affect various endocrine axes (Philips *et al.*, 1998). The hypothalmic-pituitary-adrenal/interrenal (HPI) axis is important in regulating the stress response (Iwama *et al.*, 1997) and also demonstrates anti-immune actions (Schreck, 1996). In teleosts cortisol functions both as a mineralocorticoid controlling Na^+ and K^+ balance and as a glucocorticoid. It is on record that in some teleosts upon exposure to pesticides the plasma corticosterone level does rise (Ray and Poddar, 1983). Rise in plasma cortisol levels could also be correlated to pesticides (Bhattacharya *et al.*, 1996; Mondal, 1997) but not to heavy metals (Schreck and Lorz, 1978; Thomas, 1990b; Bhattacharya and Mondal, 1997). Therefore it may be surmised that the prototypical stress response of elevated plasma cortisol is not prominent in teleosts. On the other hand elevated plasma hormone levels may also indicate decreased rate of catabolism of steroid hormones in the liver. In a very interesting study it has been revealed by Kaltreider *et al.* (2001) that low dose of arsenic exposure affects the expression of phosphoenolpyruvate carboxy kinase gene in rat liver reducing the responsiveness of the tissue to hormonal signals by blocking the expression of the glucocorticoid receptor (GR). Thus arsenic has been found to be a new type of endocrine disruptor which mediates its action via specific hormone receptors even at non toxic doses. Although comparable reports are yet to be generated in teleosts considering the environmental importance of arsenic in the present day scenario this information is of importance and future studies in the fish system would definitely fill the lacunae.

Future Research Needs

The problem of the EDCs has not been addressed adequately in the developing countries. Therefore more research is warranted in this area to derive general conclusions of the effects of the EDCs. However, in spite of the insufficiency ample evidence is there about the ominous nature of these xenobiotics with adverse endocrine related effects posing serious threat to wildlife. In the wildlife the problem of bioaccumulation of the EDCs is acute and with little information on hormonal defects due to low dose chronic exposures it is indeed difficult to identify specific cause-effect relationships. Further research is necessary to address the uncertainties in this area of study. It would also be very challenging to look at the chronic low dose effects of the EDCs which do not kill the fish but cause irreparable harm in their functional organization. Only when the cause effect relationship is understood to some degree of certainty, abatement and prevention protocols may be designed to tackle the problems of endocrine disruptive activities of the xenobiotics generated in the environment.

Acknowledgements

The authors are grateful to ICAR, DST and CSIR for project grants and to UGC for the DSA(COSIST) support to the department which enabled the present study. We are also grateful to Miss Anamika Priyadarshini and Dr Sudipta Maitra for their help in preparing the manuscript.

References

Anitha Kumari S, and Sree Ram Kumar N. (1997). Histopathological alterations induced by aquatic pollutants in *Channa punctatus* from Husainsagar Lake (A.P.). J. Environ. Biol., 18, 1-16.

Barter RA, and Klaassen CD. (1992). UDP-grucuronosyltransferase inducers reduce thyroid hormone levels in rats by an extrathyroidal mechanism. Toxicol. Appl. Pharmacol., 113, 36-42.

Bentley PJ. (1998). *Comparative Vertebrate Endocrinology*. Third Edition. Cambridge University Press, Cambridge.

Bhattacharya S, Majumder C, and Mondal S. (1996). Xenobiotic action on gonadal and interrenal steroidogenesis in *Channa punctatus*. Symp. Environ. Endocrine Disruptors. 3rd AOSCE Congress, Sydney, Australia, S-14.

Bhattacharya S, and Mondal S. (1997). Disruptions of endocrine functions by environmental contaminants. *In:* Kawashima S, and Kikuyama S. (Eds.), *Advances in Comparative Endocrinology*. Monduzzi Editore, Bologna, Italy, pp. 1729-1732.

Birnbaum LS. (1995). Developmental effects of dioxins and related endocrine disrupting chemicals. Toxicol. Lett., 82-83, 743-750.

Chatterjee S, Roy A, Bagchi P, and Deb C. (1988). Suppression of testicular steroidogenesis in rats by organochlorine insecticide Aldrin. Environ. Pollut., 2, 35-40.

Cheek AO, Vonier PM, Oberdorster E, Burow BC, and McLachlan JA. (1998). Environmental signaling: a biological context for endocrine disruption. Environ. Health Perspec. 106, Suppl. 1, 5-10.

Chen CW. (2001). Assessment of endocrine disrupters: Approaches, issues and uncertainties. Folia Histochem. Cytobiol., 39 (Suppl.), 20-23.

Choudhuri C, Ray AK, Bhattacharya S, Bhattacharya S. (1993). Nonlethal concentrations of pesticides impair ovarian function in the freshwater perch. Env. Biol. Fishes., 36, 319-324.

Choudhury C. (1991). Impact of metacid 50 and carbaryl on the ovarian maturation in *Anabas testudineus* (Bloch) and *Channa punctatus* (Bloch). Ph.D Dissertation, Visva Bharati University, Santiniketan, India.

Collins WT, and Capen CC. (1980). Fine structural lesions and hormonal alterations in thyroid glands of perinatal rats exposed *in vitro* and by milk to polychlorinated biphenyls. Am. J. Pathol., 99, 125-142.

Dey S, and Bhattacharya S. (1989). Ovarian damage to *Channa punctatus* after chronic exposure to low concentrations of elsan, mercury and ammonia. Ecotoxicol. Environ. Safety., 17, 247-257.

Dickhoff W, Brown CL, Sullivan CV, Bern HA. (1990). Fish and amphibian models for developmental endocrinology. J. Exp. Zool., 12 (Suppl), 4, 90-97.

Galton VA. (1992). The role of thyroid hormones in amphibian metamorphosis. Trends Endocrinol. Metab., 3, 96-100.

Ghosh P. (1990). Interrelationship of acetylcholinesterase-acetylcholine, triiodothyronine–thyroxine and gonadotropin-gonadotropin releasing hormone in pesticide treated murrel, *Channa punctatus* (Bloch). PhD dissertation, Visva Bharati University, Santiniketan, India.

Ghosh P, and Bhattacharya S. (1992). *In vivo* and *in vitro* acetylcholinesterase inhibition by metacid 50 and carbaryl in *Channa punctatus* under natural field condition. Biomed. Environ. Sci., 5, 18-24.

Ghosh P, and Bhattacharya S. (1994). Impact of environmental pollutants on fish thyroid physiology: A field study. *In*: Roy M. (Eds.), *Recent Researches in Ecology, Environment and Pollution*. Vol 9, Today and Tomorrow's Printer and Publishers, New Delhi, India, pp. 19-23.

Ghosh P, Bhattacharya S, and Bhattacharya S. (1989). Impact of non lethal levels of metacid 50 and carbaryl on thyroid function and cholinergic system of *Channa punctatus*. Biomed. Environ. Sci., 2, 92-97.

Ghosh P, Bhattacharya S, and Bhattacharya S. (1990). Impairment of regulation of gonadal function in *Channa punctatus* by metacid 50 and carbaryl under laboratory and field conditions. Biomed. Environ. Sci., 3, 106-112.

Grant BF, and Mehrle PM. (1970). Pesticide effects on fish endocrine function. Progress in Sport Fishery Research 1969. Bur. Sport Fish Wildl. Res., 88, 13-15.

Guhathakurta S, and Bhattacharya S. (1988). Target and non target actions of phenthoate and carbofuran: brain acetylcholinesterase, kidney iodide peroxidase and blood thyroxine profiles in *Channa punctatus*. Biomed. Environ. Sci., 1, 59-63.

Hawkins MB, Thornton JW, Crews D, Skipper JK, Dotte A, and Thomas P. (2000). Identification of a third distinct estrogen receptor and reclassification of estrogen receptors in teleosts, Proc. Natl. Acad. Sci., USA, 97, 10751-10756.

Iwama GK, Pickering AD, Sumpter JP, and Schreck CB. (1997). *Fish Stress and Health in Aquaculture.* Cambridge University Press. Cambridge.

Kaltreider RC, Davis AM, Lariviere JP, and Hamilton JW. (2001). Arsenic alters the function of the glucocorticoid receptor as a transcription factor. Environ. Helth Perspec., 109, 245- 251.

Khan IA, and Thomas P. (1998). Estradiol 17β and *o,p2* -DDT stimulate gonadotropin release in Atlantic croaker. Mar. Environ. Res., 46, 149-152.

Khan IA, and Thomas P. (2000). Lead and Aroclor 1254 disrupt reproductive neuroendocrine function in Atlantic croaker. Mar. Environ. Res., 50, 119-123.

Khan IA, and Thomas P. (2001). Disruption of neuroendocrine control of luteinizing hormone secretion by Aroclor 1254 involves inhibition of hypothalamic tryptophan hydroxylase activity. Biol. Reprod., 64, 955-964.

Kikuyama S, Kawamura K, Tanaka S, and Yamamoto K. (1993). Aspects of amphibian metamorphosis: Hormonal control. Int. Rev. Cytol., 145, 105-148.

Kime D. (1984). The effect of cadmium on Steroidogenesis by the testes of the rainbow trout, *Salmo gairdneri*. Toxicol. Letters, 22, 83-88.

Kime DE. (1995). The effect of pollution on reproduction in fish. Rev. Fish Biol. Fish., 5, 92-96.

Kime DE, Nash JP, and Scott AP. (1999). Vitellogenesis as a biomarker of reproductive disruption by xenobiotics. Aquacult., 177, 345-352.

Kirubagaran R, and Joy KP. (1988). Toxic effects of mercuric chloride and emisan 6 (an organic mercurial fungicide) on ovarian recrudescence in the catfish *Clarias batrachus* (L). Bull. Environ. Contam. Toxicol., 41, 902-909.

Kulshrestha SK, and Arora N. (1984). Impairments induced by sublethal doses of two pesticides in the ovaries of a freshwater teleosts *Channa striatus* Bloch. Toxicol. Letters, 20, 93-98.

Kumar D, and Bhattacharya S. (1975). Effect of endrin on the head kidney peroxidase activity of a fish, *Anabas testudineus*. Indian Biologist, 7, 45-51.

Larsson DGJ, Adolfsson-Erici M, Parkkonen J, Petersson M, Berg AH, Olsson P-E, and Förlin L. (1999). Ethinylestradiol–an undesired fish contraceptive? Aquat. Toxicol., 45, 91-97.

Mani M, and Saxena PK. (1985). Effect of safe concentration of some pesticides on ovarian recrudescence in the freshwater murrel, *Channa punctatus* (Bloch): a quantitative study. Ecotoxicol. Environ. Safety, 9, 241-249.

McMaster ME, Van der Kraak GJ, Portt CB, Munkittrick KR, Sibley PK, Smith IR, and Dixon DG. (1991). Changes in hepatic mixed function oxygenase (MFO) activity, plasma steroid levels and age at maturity of a white sucker (*Catastomus commersoni*) population exposed to bleached kraft paper mill effluent. Aquat. Toxicol., 21, 199-218.

Mehrle PM. (1971). Amino acid metabolism in rainbow trout, *Salmo gairdneri* is affected by chronic dieldrin exposure. PhD Dissertation, University of Missouri, Columbia, USA.

Mellanen P, Soimasuo M, Holmbom B, Oikari A, and Santti R. (1999). Expression of the vitellogenin gene in the liver of juvenile whitefish (*Coregonus lavaretus* L. *s.l.*) exposed to effluents from pulp and paper mills. Ecotoxicol. Environ. Safety, 43, 133-137.

Menditto A, and Turio-Baldassari L. (1999). Environmental and biological monitoring of endocrine disrupting chemicals. Chemosphere, 39, 1301-1307.

Mondal S. (1997). Pesticides and heavy metals influence steroidogenic activity in fish gonad and interrenal. PhD Dissertation, Visva Bharati University, Santiniketan, India

Mondal S, Mukhopadhyay B, and Bhattacharya S. (1997). Inorganic mercury binding to fish oocyte plasma membrane induces steroidogenesis and translatable messenger RNA synthesis. Biometals, 10, 285-290.

Munkittrick KR, Mcmaster ME, Portt CB, Smith IR, and Dixon DG. (1992). Changes in maturity, plasma sex steroid levels, hepatic mixed function oxygenase activity and the presence of external lesions in Lake Whitefish (*Coregonus clupeaformis*) exposed to blached kraft mill effluent. Can. J. Fish. Aquat. Sci., 49, 1560-1569.

Murugesan AG, and Hanifa MA. (1992). Histopathological and histochemical changes in the oocytes of the air breathing fish *Heteropneustes fossilis* (Bloch) exposed to textile mill effluent. Bull. Environ. Contam. Toxicol., 48, 929-936.

Norris DO. (1997). Vertebrate Endocrinology, Third edition. Pergamon Press, SanDiego, CA.

Norris DO. (1999). Thyroid hormones in subavian vertebrates. *In*: *Encyclopedia of Reproduction*, Vol. 4. Academic Press, San Diego, CA. pp. 807-812.

Philips DIW, Barker DJP, Fall CHD, Seckl JR, Whorwood CB, Wood PH. (1998). Elevated plasma cortisol concentrations: an expression for the relationship between low birth weight and adult cardiovascular risk factors. J. Clin. Endocrinol. Metab., 83, 757-760.

Pinter J, and Thomas P. (1997). The ovarian pregestogen receptor in the spotted seatrout, *Cynoscion nebulosus*, demonstrates steroid specificity intermediate between progesterone and glucocorticoid receptors in other vertebrates. J. Steroid Biochem. Mol. Biol., 60(1-2), 113-119.

Pundir R, and Saxena AB. (1990). Seasonal changes in the testis of fish, *Puntius ticto*, and their relation to heavy metal toxicity. Bull. Environ. Contam. Toxicol., 45, 288-293.

Ray SK, and Poddar MK. (1983). Carbaryl induced elevation of corticosterone level and cholinergic mechanism. Bio. Sci. Rep., 3, 973-977.

Rhind SM. (2002). Endocrine disrupting chemicals and farm animals: Their properties, actions and routes of exposure. Domest. Animal Endocrinol., 23, 179-187.

Sangalang GB, and Freeman HC. (1974). Effects of sublethal cadmium on maturation and testosterone and 11-ketotestosterone production *in vivo* in brook trout. Biol. Reprod., 11, 429-435.

Sanglang GB, and O'Halloran MJ. (1973). Adverse effects of cadmium on brook trout testis and on *in vitro* testicular androgen synthesis. Biol. Reprod., 9, 394-403.

Saxena PK, and Mani K. (1985). Protein bound iodine levels in the blood plasma of freshwater teleost, *Channa punctatus* (Bloch) exposed to subtoxic pesticide concentrations. Toxicol. Letters., 24, 33-36.

Schoettger R. (1972). Progress in Sport Fishery Research 1971. Bur. Sport Fish. Wildl. Res. Pub., 90, 2-17.

Schreck CB. (1996). Immunomodulation: Endogenous factors. *In*: Iwama G, and Nakanishi T. (Eds.), *The Fish Immune System: Organism, Pathogen, and Environment*. Academic Press, Inc. San Diego, CA, pp. 118-146.

Schreck CB, and Lorz HH. (1978). Stress response of coho salmon (*Oncorhynchus kisutch*) elicited by cadmium and copper and potential use of crotisol as and indicator of stress. J. Fish. Res. Bd. Canada, 35, 1124-1129.

Shi YB. (1994). Molecular biology of amphibian metamorphosis: A new approach to an old problem. Trends Endocrinol. Metab., 5,14-20.

Singh H. (1989). Interaction of xenobiotics with reproductive endocrine functions in a protogynous teleost, *Monopterus albus*. Mar. Env. Res., 28, 285-289.

Singh H, Kime D, and Singh TP. (1993). Modulatory actions of *Mystus* gonadotropin on gamma-BHC induced histological changes, cholesterol, and sex steroid levels in *Heteropneustes fossilis*. Ecotoxicol. Env. Safety, 25, 141-145.

Singh H, and Singh TP. (1980). Thyroid activity and TSH potency of the pituitary gland and blood serum in response to Cythion and Hexadrin treatment in the freshwater catfish, *Hetropneustes fossilis* (Bloch). Env. Res., 22, 184-189.

Singh PB, Kime DE, Epler P, and Chyb I. (1994). Impact of gamma-hexachlorocyclohexane exposure on plasma gonadotropin levels and *in vitro* stimulation of gonadal steroid production by carp hypophyseal homogenate in *Carassius auratus*. J. Fish Biol., 44, 195-204.

Sinha N, Lal B, and Singh TP. (1991a). Carbaryl-induced thyroid dysfunction in the freshwater catfish *Clarias batrachus*. Ecotoxicol. Environ. Safety, 21, 240-247.

Sinha N, Lal B, and Singh TP. (1991b). Effect of endosulfan on thyroid physiology in the freshwater catfish *Clarias batrachus*. Toxicology, 67, 187-197.

Sivarajah K, Franklin CS, and Williams WP. (1978). The effect of polychlorinated biphenyls on the plasma steroid levels and hepatic microsomal enzymes in fish. J. Fish Biol., 13, 401-409.

Sukumar A, and Karpagaganapathy PR. (1992). Pesticide-induced atresia in ovary of a freshwater fish, *Colisa ialia* (Hamilton-Buchanan). Bull. Environ. Contam. Toxicol., 48, 457-462.

Sumpter JP, and Jobling S. (1995). Vitellogenesis as a biomarker for estrogenic contamination of the aquatic environment. Environ. Health Perspec., 103 Suppl. 7, 173-178.

Thomas P. (1988). Reproductive endocrine function in female Atlantic croaker exposed to pollutants. Mar. Env. Res., 24, 179-183.

Thomas P. (1989). Effects of arochlor 1254 and cadmium on reproductive endocrine function and ovarian growth in Atlantic croaker. Mar. Env. Res., 28, 499-503.

Thomas P. (1990a). Teleost model for studying the effects of chemicals on female reproductive function. J. Exp. Zool. Suppl. 4, 126-128.

Thomas P. (1990b). Molecular and biochemical responses of fish to stressors and their potential use in environmental monitoring. Am. Fish Soc. Symp., 8, 9-28.

Tremblay L, and Van Der Kraak G. (1999). Comparison between the effects of the phytosterol β-sitosterol and pulp and paper mill effluents on sexually immature rainbow trout. Environ. Toxicol. Chem., 18, 329-336.

Tyler CR, and Routledge EJ. (1998). Oestrogenic effects in fish in English rivers with evidence of their causation. Pure Appl. Chem., 70, 1795-1804.

Van der Kraak GJ, Munkittrick KR, McMaster ME, Portt CB, and Chang JP. (1992). Exposure to bleached kraft pulp mill effluents disrupts the pituitary-gonadal axis of white sucker at multiple sites. Toxicol. Appl. Pharmacol., 115, 224-233.

Younes M. (1999). Specific issues in health risk assessment of endocrine disrupting chemicals and international activities. Chemosphere, 39, 1253-1257.

26

Ovarian Dysfunction Under Galactose Toxicity

Syed N. Kabir* and Soma Bandyopadhyay

Reproductive Biology Research Division, Indian Institute of Chemical Biology, 4, Raja S. C. Mullick Road, Jadavpur, Kolkata – 700 032, India

ABSTRACT

Galactose, an elementary structural component of a number of important biomolecules, holds a critical place in the intermediary metabolism of carbohydrates. Some inborn errors of galactose metabolism threat the female reproductive system to a considerable extent and induce premature gonadal failure. The pathophysiological attributes of gonadal failure under galactose toxicity, however, remain elusive. This review presents an overall perspective of ovarian dysfunction pertaining to galactose toxicity and evaluates the suitability of experimental galactose toxicity as a model to unveil the pathological mechanisms underlying ovarian dysfunction.

Introduction

Lactose (milk sugar), a disaccharide of glucose and galactose, is the main carbohydrate in the diet of almost all young mammals. Adults, too, ingest galactose, since this sugar is found in glycoproteins and glycolipids; and considerable amounts of free galactose are present in some fruits, vegetables,

---

**Corresponding Author*: E-mail: snkabir@iicb.res.in; Phone: 033 24733491/24730492; Fax: 91 33 24730284/24735197

nuts, legumes, and beverages. Humans also synthesize substantial quantities of galactose *de novo* from glucose and also from the reservoir of galactose within glycoproteins and mucopolysaccharides. This is important for the maintenance of galactose and galactose metabolites needed for the synthesis of galactose-containing macromolecules. While glucose, the primary energy source for most cells, is metabolised directly after phosphorylation, higher organisms are apparently unable to metabolise galactose without first transforming it into glucose.

The Metabolism of Galactose

In humans, ingested lactose is hydrolysed to glucose and galactose by the intestinal enzyme called lactase. The further metabolism of galactose to uridine diphospho (UDP)-glucose involves three major enzymes–galactokinase, galactose-1-phosphate uridyltransferase (GALT) and UDP-galactose-4/–epimerase. The pathways for galactose metabolism and associated enzyme reactions are shown in Figure 26.1. Galactose can be converted into energy by entering the glycolytic pathway, which occurs through a series of steps. First, galactose is phosphorylated by galactokinase to yield galactose-1-phosphate. Epimerisation of galactose-1-phosphate to glucose-1-phosphate requires the transfer of UDP from UDP-glucose catalysed by GALT. This generates UDP-galactose and glucose-1-phosphate. UDP-galactose is then epimerised to UDP-glucose by UDP-galactose-4/–epimerase. The UDP portion is exchanged for phosphate generating glucose-1-phosphate, which then is converted to glucose-6-phosphate by phosphoglucomutase. UDP-galactose holds an important place in the intermediary metabolism by serving the role of galactose donor in the process of synthesis of glycoproteins, glycolipids and complex carbohydrates (Segal and Berry, 1995).

Biochemical Abnormalities of Galactose Metabolism

There are three clinically significant inborn errors of galactose metabolism, each of which is attributed to deficiency of one of the three major enzymes as referred above (Holton, 1990). The most common error, however, is classical galactosaemia, a disorder with autosomal recessive inheritance and is caused by the deficiency of GALT. It is biochemically characterized by elevated blood galactose, increased galactose-1-phosphate in various tissues and decreased UDP-galactose in red blood cells. Untreated newborns with this disorder develop early organ damage as a result of accumulation of galactose and galactose metabolites in the liver, kidneys, lens and nervous system. In addition, ovarian failure occurs in almost all female patients with galactosaemia, despite adequate treatment. The age of onset of ovarian insufficiency has been variable, occurring as early as 1 or 2 years and as late as late 20's, and even after normal pregnancy.

Ovarian Dysfunction in Galactosaemia

The mammalian ovary is particularly susceptible to damage from the accumulation of galactose and galactose metabolites; and germ cells, follicular (granulosa) cells, stromal cells and thecal cells are all affected in galactosaemia-related ovarian dysfunction (Kaufman *et al.*, 1987; Kaufman *et al.*, 1989; Gibson, 1995). The increased galactose load possibly functions as a reproductive toxicant. Ovarian dysfunction in galactosaemia was first recognised by Kaufman *et al.* (1979) and Hoefnagel *et al.* (1979), and further confirmed by later studies (Kaufman *et al.*, 1981, 1986, 1988).

Retrospective surveys suggest that up to 75-96 per cent of all females who are homozygous for galactosaemia, suffer from ovarian dysfunction (Waggoner *et al.*, 1990; Schweitzer *et al.*, 1993), while those who are heterozygous for the condition do not suffer (Sardharwalla and Wraith, 1987; Hagenfeldt *et al.*, 1989). According to Waggoner *et al.* (1990), 70-80 per cent galactosaemic women eventually develop premature ovarian failure (POF). The consequences of this ovarian dysfunction include failure

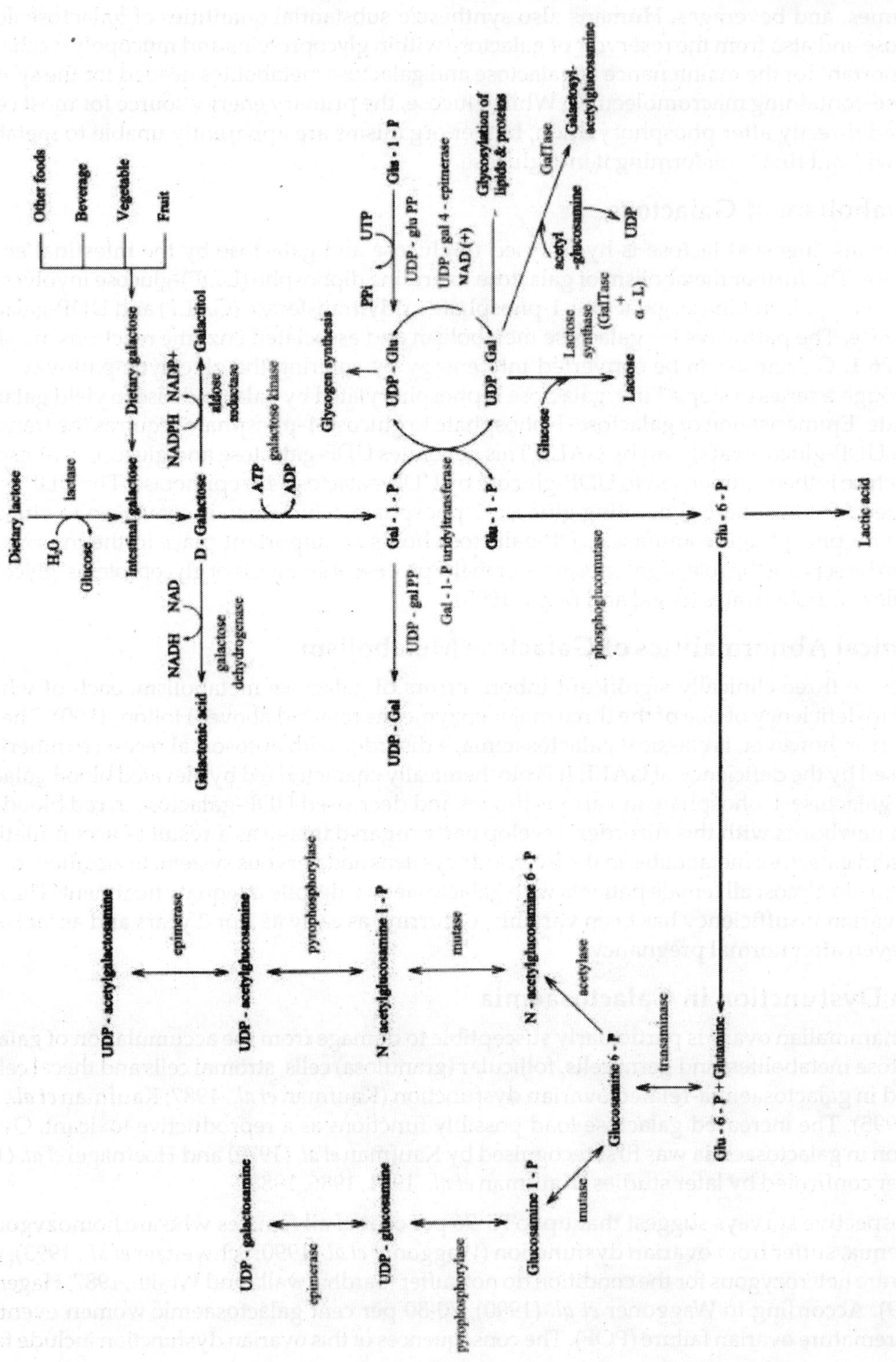

Figure 26.1: Metabolic Pathways of Galactose

of pubertal development, primary amenorrhea, secondary amenorrhea and premature menopause. The ovarian failure, which can occur at any age during the reproductive period (Coulam, 1981; Levy *et al.*, 1984), may also typically manifest in reduced follicular sensitivity to gonadotrophin (Resistant Ovary Syndrome) with an eventual hypergonadotrophic hypogonadal state (Fraser *et al.*, 1986). The ovary's susceptibility to galactose toxicity may in part be explained by unusually high local concentrations as well as activities of enzymes responsible for the metabolism of galactose (Xu *et al.*, 1989). The conc entrations of GALT mRNA in various adult organs from highest to lowest were liver > ovary ~ cerebellum ~ kidney ~ heart > cerebrum > lung > testis > skeletal muscle, as measured by Rogers *et al.* (1992) and Heidenreich *et al.* (1993). Interestingly, these organs are affected in this order from greatest to least in patients with galactosaemia. In normal ovarian tissue, UDP-glucose pyrophosphorylase activity is about 50 times that of transferase. This high phosphorylase activity, supposed to be important in nucleotide sugar production, plays a critical role in the maintenance of germ cell function, maturation of follicles and steroidogenesis leading to normal development and functioning of the ovary. Any interference with this balance of galactose metabolism understandably could lead to germ and follicular cell dysfunction. It is therefore hypothesised that POF in galactosaemic patients occurs due to interference with nucleotide sugar metabolism and defective synthesis of galactose-containing macromolecules consequent to the enzymatic defect in the major pathway of galactose metabolism (Xu *et al.*, 1989).

Experimental Galactose Toxicity as a Model for Ovarian Dysfunction

The exact mechanisms of ovarian failure in galactosaemic females remain obscure, however, increased levels of galactose and its metabolites are held responsible for most of the toxic manifestations of galactosaemia. Resort has therefore been made to develop an animal model of galactose toxicity by experimental manipulation of galactose metabolism simply by way of feeding large amounts of galactose that overpowers the normal capacity of the animal to metabolise this sugar (Gibson, 1995; Segal, 1995b). Administration of a 30 per cent to 50 per cent galactose diet to rats results in the development of cataracts, aminoaciduria and liver abnormalities resembling the human galactosaemic syndrome. Thus, the rodent provides an excellent model system to study the effects of galactose toxicity on different body systems. These models are not directly comparable to humans with enzyme deficiencies, nevertheless, important information about the mechanism of galactose toxicity has been uncovered. When this strategy of feeding high galactose diet is extended to the pregnant rat, there is evidence that galactose readily crosses the placenta (Segal, 1995b), and some abnormal metabolites found in galactosaemia are present in the foetuses and newborn rat pups. The offspring exhibits increased blood levels of galactose and galactose-1-phosphate (Chen *et al.*, 1984), and produces a sequel of abnormalities that characterise human galactosaemia (Segal and Bernstein, 1963; Spatz and Segal, 1983). This experimental condition is easy to acquire, maintain and modulate, and therefore has been the usual model for the study of gonadal dysfunction in galactosaemia.

Chen and colleagues (1981) demonstrated that female rat foetuses exposed to high levels of galactose (the result of a 50 per cent maternal diet) throughout gestation and continuing till weaning had significantly decreased number of oocytes in the ovarian pool. Postnatal galactose exposure, however, exerted no such effects on the oocyte number.

Swartz and Mattison (1988) fed adult mice a 40 per cent galactose diet. Inhibition of oocyte maturation occurred after 2 weeks as shown by decreased numbers of corpora lutea and attenuated response to superovulatory stimuli. This effect was reversible with restoration of a normal carbohydrate diet. Meyer *et al.* (1992) had also reported similar effects in prepubertal rats fed a high galactose diet, however, an aldose reductase inhibitor lessened the ovarian abnormalities.

The findings of Liu *et al.* (2000) were also consistent with those from the rodent model experiments. They observed a decrease in egg production and an increase in both mortality and mental depression on feeding a 10 per cent galactose diet to quails. We adopted a similar strategy of galactose treatment in which pregnant rats were fed with 35 per cent galactose diet from day 3 of conception and continuing through weaning of the litters at postnatal day 21. This treatment produced a sequel of ovarian dysfunction that characterised the basic tenets of diverse phenotypes of POF including hypergonadotrophic hypoestrogenism with reduced follicular reserve or optimum follicular reserve with decreased sensitivity to gonadotrophins (Bandyopadhyay *et al.*, 2003a, 2003b). This strong analogy in respects of ovarian dysfunction between experimental galactose toxicity and classical galactosaemia proved it to be a suitable model to explore the cause and effect relationship between galactosaemia and ovarian dysfunction.

Pathogenesis of Ovarian Dysfunction in Galactose Toxicity

Onset of Puberty

Delayed puberty is a frequent finding in women with galactosaemia (Fraser *et al.*, 1986). Chen *et al.* (1981) have reported late onset of puberty under experimental galactose toxicity. We made a number of pertinent observations in this regard. We demonstrated that the timing of vaginal opening (VO), a reproductive tract response to ovarian steroids and one of the key indices for puberty in the rodents, was significantly delayed under galactose toxicity. In prepubertal female rats, the hypothalamo-pituitary-ovarian axis undergoes a maturation process that results in onset of puberty, which is marked by the first preovulatory surges of oestrogen and LH followed by ovulation. During this prepubertal period, oestrogen and inhibin are secreted from the ovary and the pituitary attains the capability of responding to and integrating these signals that allow GnRH to be secreted in discrete pulses (Ojeda and Urbanski, 1994).

In immature female rats, FSH secretion remains high until postnatal day (PND) 18 and decreases thereafter (Thyssen and Libertun, 1996). A subsequent rise in FSH level occurs at the time of VO. This sequential drop and rise of FSH secretion is likely to be of phenomenal importance in the process of onset of puberty. Drop in FSH during PND18 has been suggested to be a steroid-independent event of cerebral origin, while the rise of gonadotrophins that immediately precedes VO is brought about by pre–pubertal gradual rise of oestrogen secretion that acts at the CNS-pituitary level to trigger a preovulatory pro–oestrus–like surge of gonadotrophins (Ojeda *et al.*, 1976). We demonstrated that the galactose-exposed litters had elevated FSH levels both at weaning as well as at the stipulated peri-pubertal phase; and E2 levels remained significantly low throughout. It is therefore tempting to speculate that the absence of stimuli in the form of E2 perhaps blunts the puberty-specific sensitisation of the CNS-pituitary unit and inhibits the maturation of GnRH pulse generator.

Ovarian Architecture

Galactosaemia encompasses a broad spectrum of ovarian pathology. While some patients clearly demonstrate ovarian failure related to follicle depletion, others have primordial follicles appearing normal in ovarian biopsy but lacking growth and development (Fraser *et al.*, 1986). In order to typify the ovarian effects of galactose toxicity, we looked into the ovarian follicular reserve and also evaluated its response to gonadotrophins. Chen *et al.* (1981) reported that prenatal galactose toxicity led to a significant reduction in the number of small follicles. Our histopathological examination, however, did not reveal such generalised effect. While many rats suffered from deficient follicular reserve (Figure 26.2B), others exhibited apparently normal follicular complement, but with arrested growth and

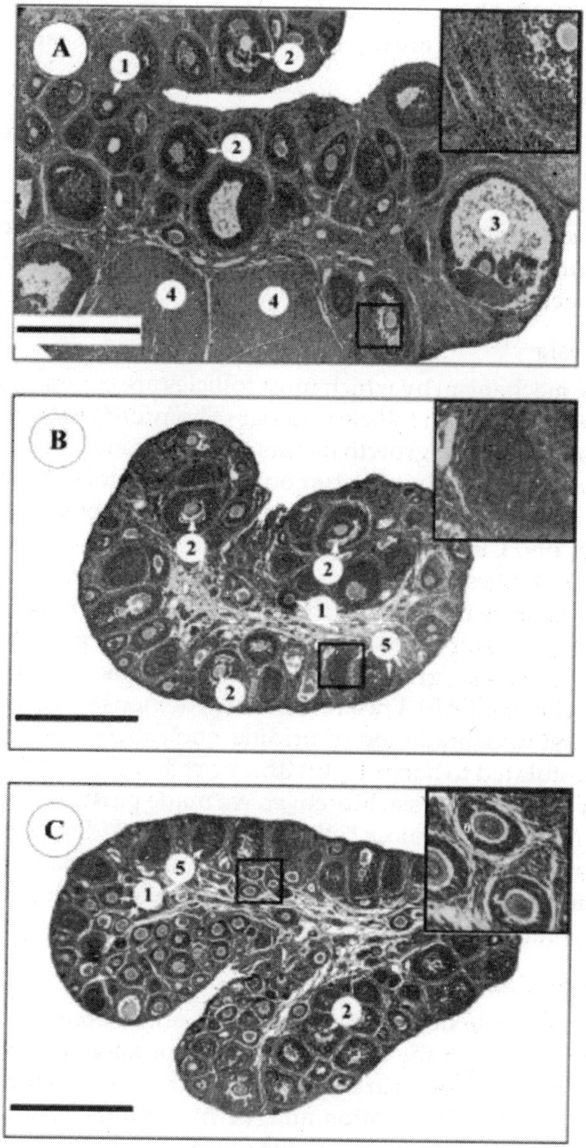

Figure 26.2: Histological Sections Through Ovaries of 55–60-day Old Control (A) and Galactose-exposed Rats (B and C).

The control ovary exhibits the presence of follicles at different stages of maturity and a functional corpus luteum (CL). The galactose-exposed ovaries exhibit an infantile appearance with a deficient state of follicular reserve (B), or the presence of apparently normal complement of unstimulated follicles arrested mostly at the pre-antral stage and a preponderance of atretic follicles (C). The absence of hcraafian follicles and CL was the most salient feature. 1, pre–antral or secondary follicles; 2, early antral follicles or vesicular follicles; 3, hcraafian follicle; 4, corpus luteum; 5, atretic follicles. Bars = 50 mm. (*Reproduced from Bandyopadhyay et al.,* 2003b).

maturation possibly due to follicular resistance to gonadotrophins (Figure 26.2C). Thus the basic tenets of the diverse phenotypes of POF are conserved in experimental galactose toxicity.

Deficient Follicular Reserve

The size of a follicular complement at any point of reproductive life represents the balance between the numbers of follicles endowed in the initial ovarian pool and those expended in between. A deficient follicular reserve may therefore ensue either from an accelerated rate of atresia and/or an under-endowment of follicles in the ovarian pool early in life due to impaired germ cell migration–the major form of follicular expenditure. We explored both of these possibilities in connection with deficient follicular reserve under galactose toxicity.

Accelerated Follicular Atresia

Follicular atresia is the mechanism by which most follicles are expended, and a small increase in atresia rate can have dramatic effects on follicle number (Thomford *et al.*, 1987). A number of factors including gonadotrophins, oestrogens, growth factors, cytokines and nitric oxide (Amsterdam *et al.*, 1997) rescue oocytes from apoptosis. In contrast, tumour necrosis factor (*TNF*)-a, Fas ligand, androgens, *IL*-6, and gonadal gonadotrophin-releasing hormone (GnRH)-like proteins promote programmed cell suicide (Kaipia and Hsueh, 1997). Evidence suggests that these diverse factors converge on selective pathways including the gene families of *Bcl-2* and *ICE* to regulate apoptosis (Billig *et al.*, 1996). Galactose toxicity perhaps alters this balance in favour of the apoptosis-inducing milieu. As discussed earlier, within the reproductive system, ovary is one of the major targets of galactose metabolites that have detrimental effects on the germ cells, granulosa cells, stromal cells and thecal cells (Kaufman *et al.*, 1987; Kaufman *et al.*, 1989; Gibson, 1995). Fraser *et al.* (1986) demonstrated that ovarian accumulation of galactose-1-phosphate led to degradation of uridine nucleotides and induced follicular atresia. Galactose has also been postulated to perturb glutathione redox cycle and provoke accumulation of methylglyoxal and trigger follicular atresia. Moreover, we made particular observation on attenuated biopotency of FSH and prohibited granulosa cell production of E2 (discussed in detail in subsequent sections), which might have contributed to the process of ensuing apoptotic process. A number of these mechanisms, in isolation or concerted fashion, might have contributed to the loss of follicles as evidenced by general preponderance of atretic secondary follicles in the galactose-exposed group.

Impaired Germ Cell Locomotion

Females receive a finite follicle store during foetal life, and thereafter no additional follicles are formed. So, the size of germ cells in the initial gonadal pool may have an impact on the ovarian follicular reserve. The lower the number is in the initial pool, the earlier is the depletion of the follicular reserve. Abnormalities in germ cell formation and/or migration could therefore be mechanisms favouring a deficient initial pool of germ cells in the ovary. We therefore investigated the effects of galactose toxicity on the migration of primordial germ cells (PGCs).

During early development, the PGCs migrate from the developing hindgut to the site of gonadal development. A unique glycoconjugate with terminal N-acetylgalactosamine (GalNAc) is selectively expressed on the surface of PGCs only transiently during the course of active migration in the rat (Fazel *et al.*, 1987). Because of binding specificity of *Dolichos biflorus* agglutinin (DBA) for GalNAc, we used DBA-HRP conjugate as a histochemical probe to identify and localise the PGC in the migratory path. We demonstrated that the number of PGC at the day-specific sites on the path of migration was significantly lower (Figure 26.3) and only fewer cells finally colonised the gonadal stroma of the embryos exposed to high galactose. It was possible that the formation of germ cells and/or their

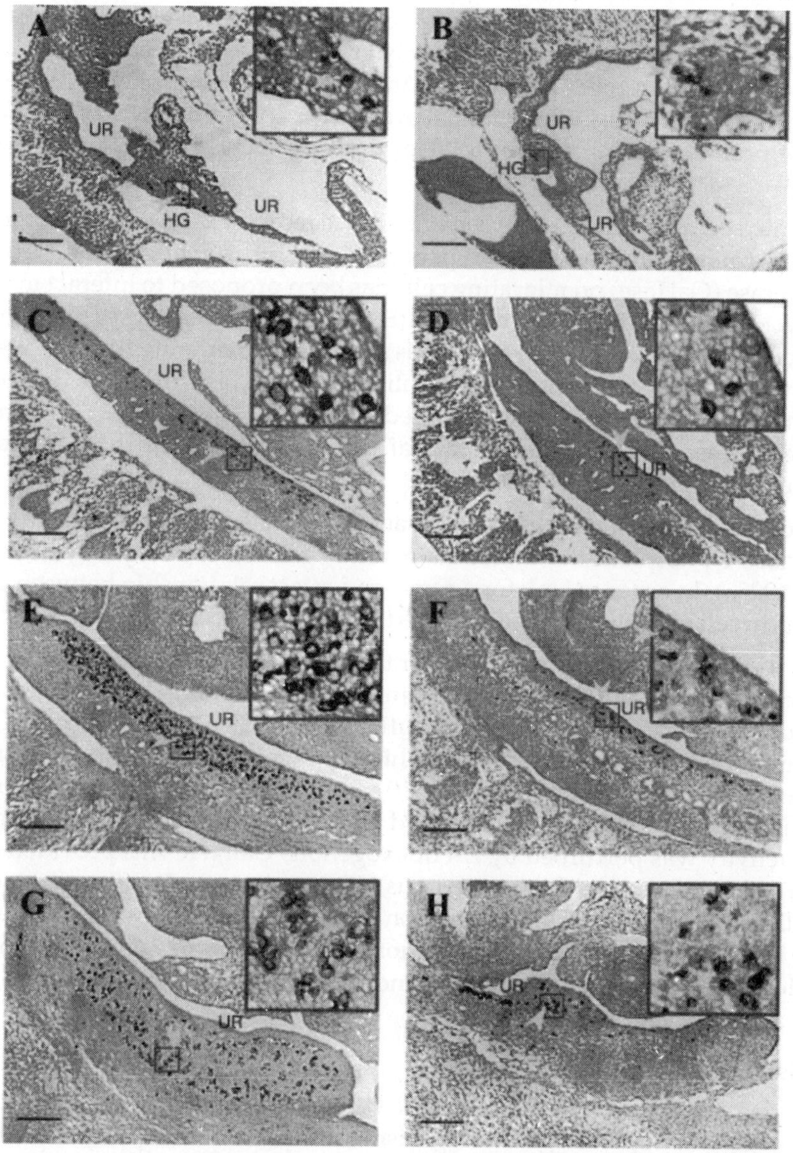

Figure 26.3: Photomicrographs (Plates A–H) Show DBA-reactive PGC (arrows) on their Path of Migration in 12-15 Day Old Rat Embryos

PGC are found scattered posterior to the developing hindgut in day 12 control (plate A) and galactose-exposed (plate B) embryos. Plates C and D demonstrate the distribution of PGC along the coelomic epithelium of the dorsal mesentery in 13-day old control and galactose-exposed embryos, respectively. In 14-day old control (E) and galactose-exposed (F) embryos, PGC are located in the epithelium and mesenchyme of the developing gonad. On day 15, gonads are infiltrated and colonized by the PGC in both control (G) and galactose-exposed (H) groups. Note an appreciably thinner population of PGC in galactose-exposed embryos on all days when compared with the corresponding controls. UR = urogenital ridge; HG = hindgut. Bar = 20 mm. Inset in each plate presents a part of the corresponding photograph (red-bordered) at higher magnification. (Reproduced from *Bandyopadhyay et al.,* 2003a).

mitotic proliferation was impaired (for some reasons not yet known) and therefore the numbers of PGC at the corresponding sites were proportionately lower, or the germ cells were produced optimally, but their locomotion was perturbed. Whatever be the reason, the net result was development of gonad with low initial pool of germ cells. Deficient initial ovarian pool of germ cells may therefore be implicated in the process of premature depletion of follicles under galactose toxicity (Bandyopadhyay *et al.*, 2003a).

The coordinated communication between cell surface receptors and the extracellular matrix (ECM) plays important roles in many cellular processes including migration. Outer cell surface galactosyltransferase (GalTase) on migrating cells has been proposed to interact in a receptor-ligand fashion with glycosyl acceptors located on ECM (Shur, 1993), and catalyses the transfer of galactose from UDP-galactose onto it. Primordial germ cells are known to migrate from the wall of the hindgut through the ECM of the dorsal mesentery and localisation of GalTase in the PGC has been documented (Shur and Roth, 1975). It has been demonstrated that while leaving the dorsal mesentery, PGCs accumulate on a ribbon of laminin, a principal GalTase substrate that lines the migratory pathway in the basal lamina (Runyan *et al.*, 1986).

A number of reports document that galactosaemic subjects suffer from dearth of UDP-galactose. Moreover, galactose-1-phosphate that accumulates under galactose toxicity may also attenuate GalTase activity (Segal, 1995a). In order to look for a cause and effect relationship between galactose toxicity and inhibited germ cell migration, we adjudicated if GalTase serves any role in oogonial locomotion.

Embryonic GalTase activity was experimentally modulated by three sets of GalTase activity modifiers: a-lactalbumin (LA), N-acetylglucosamine (GlcNAc) and UDP-galactose. LA modulates GalTase function by altering its substrate specificity from GlcNAc to glucose, GlcNAc acts as a competitive inhibitor of endogenous substrate, while UDP-galactose plays the ro le of galactose donor. These test agents were made available to the day-10 embryo through microinjections under laparotomy and germ cell migration was studied. We observed that germ cell migration was inhibited in situations when GalTase activity was perturbed by LA or exogenous GalNAc was provided as an acceptor of galactose from UDP-galactose, while exogenous UDP- galactose significantly increased the PGC population at all day-specific sites. This observation lends additional support in favour of a hypothetical role of GalTase in the process of oogonial migration and tempts us to suggest that impaired germ cell migration under galactose toxicity was a secondary consequence of attenuated GalTase activity (Bandyopadhyay *et al.*, 2004).

Follicular Resistance to Gonadotrophins

Clinical evidence suggests that galactose toxicity renders the follicles refractory to gonadotrophins. The state of hypergonadotrophic hypo-oestrogenism in the presence of apparently normal complement of follicles under galactose toxicity attests to the state of follicular refractoriness to gonadotrophins. There could be two possible factors that might have entailed this sort of ovarian effects: either galactose toxicity had a deleterious effect at the gonadotrophin receptor/post-receptor levels, or the gonadotrophins produced under toxic state were biologically not active. The involvement of the first possibility was, however, ruled out by indirect evaluation of the receptor potential to respond to gonadotrophins *in vitro*. No difference in the rates of E2 and androstenedione production was observed following culture of granulosa and theca cells from control as well as galactose-exposed rats. But when the controlled target cells were cultured in the presence of sera from control and study rats, a graded increment in the volume of sera was responded by a gradually increased thecal cell production of androstenedione in both groups. The granulose cell, however, responded differentially to sera from control and study group. There was gradually increased granulosa cell production of E2 in response to control sera, while the experimental sera exerted practically little or no effect. The results of these *in-vitro* studies may collectively be interpreted to mean that follicles remained refractory to gonadotrophin stimulation because of lack of biological activity of FSH. This interpretation also gets further support from the earlier suggestions that in galactosaemic women some aberrant forms of gonadotrophins may be produced that are biologically inactive (Prestoz *et al.*, 1997). In order to replace such putative defective gonadotrophins, we induced superovulation in rats by exogenous gonadotrophins. Ovarian response, expressed in terms of the number of rats exhibiting ovulation or the number of oocyte ovulated per rat, was significantly poorer in the study group. It was not surprising that ovaries with deficient follicles responded poorly to gonadotrophins. But it was significant that post-autopsy histological analysis of the gonadotrophin-exposed ovaries showed those to possess apparently normal complement of unstimulated follicles. If reduced biopotency of the endogenous gonadotrophins were the sole factor responsible for follicular refractoriness, these follicles should have responded to the exogenous gonadotrophins. In order to look for additional factor(s), we carried out the induction of superovulation in the absence of or diminished *in vivo* endogenous gonadotrophin milieu. The objective was to assess the ovarian response to exogenous gonadotrophins in the absence of intervening effects of endogenous gonadotrophins, if any.

We induced experimental down-regulation of GnRH receptor by chronic administration of GnRH to achieve a gonadotrophin-deficient milieu. A significant increase in ovarian response was observed in this GnRH down-regulated group. This raises the possibility that endogenous gonadotrophins were not only biologically inactive, but possibly also had exerted anti-gonadotrophic activity, presumably at the receptor level. This possibility gets further support from the observation that FSH-stimulated production of E2 was further increased in the presence of control sera, while the study sera suppressed the production of E2, rather than increasing it. In view of these findings, we hypothesised that some altered forms of gonadotrophins might be produced under galactose toxicity that lack biological activity and/or act-as anti-gonadotrophins.

Effects on Galactosylation Reactions

There is compelling evidence to suggest that galactosaemia interferes with galactosylation process and provokes defective synthesis of carbohydrate–containing macromolecules, and this could be a contributing factor in the aetiology of the long-term complications observed in this disorder. The first suggestion of a glycosylation defect in galactosaemia was based on the observation of abnormal glycoprotein pattern in the brain of a galactosaemic patient (Haberland *et al.*, 1971). The analysis of the glycolipid content in the brain of another galactosaemic infant also exhibited reduced production of galactose or GalNAc-containing glycosphingolipids in parallel with accumulation of their precursors (Petry *et al.*, 1991). Fibroblasts collected from galactosaemic subjects showed impaired incorporation of [35 S]-sulfate (Tedesco and Miller, 1979), possibly as a consequence of a deficient incorporation of galactose into glycoproteins. The idea of defective galactosylation has been also supported by the work of Dobbie *et al.* (1990) who found that the galactose to mannose ratio was abnormally low in a hydrolysate of the cultured fibroblasts of galactosaemic patients. These findings suggest that an abnormality in galactosylation of complex carbohydrates is also a likely phenomenon in galactose toxicity.

The enzyme GalTase catalyses transfer of galactose from UDP-galactose to the terminal GlcNAc residues on the oligosaccharide complex. A number of reports document that galactosaemic subjects suffer from dearth of UDP-galactose. Moreover, galactose-1-phosphate that accumulates under galactose toxicity may also attenuate GalTase activity (Segal, 1995a). Both of these consequences may rate limit the process of galactosylation. We therefore looked into the possibility if galactose toxicity had any impact on these galactosylation-modulating parameters. This was addressed by direct assessment of GalTase activity in the sera of galactose-exposed rats and indirect assessment of UDP-galactose milieu by measuring the activity of epimerase, the enzyme involved at multiple steps of galactose metabolism including formation of UDP-galactose by catalysing the freely reversible interconversion of UDP-galactose and UDP-glucose. We observed reduced GalTase activity as well as significantly decreased epimerase activity, which indirectly support the conceivability of deficiency of UDP-galactose in experimental subjects.

We also studied the degree of galactosylation of the glycoproteins in experimental rat sera to determine the extent to which glycoproteins would serve as galactose acceptors when radiolabeled UDP-galactose and purified milk GalTase were made available *in vitro*. The rationale for this approach was that an impairment of galactosylation would result in a greater number of glycoprotein oligosaccharides being terminated in GlcNAc. Normally, galactose is linked to GlcNAc in the biosynthesis of the carbohydrate moieties of glycoproteins by a GalTase. We reasoned that if sera samples were incubated with radiolabeled UDP-galactose and purified GalTase, radioactive galactose would be transferred to terminal Glc NAc sites. McGuire *et al.* (1989) have shown that a GalTase purified from human milk transfers radiolabeled galactose from UDP-galactose to free GlcNAc or into various glycoproteins from which galactose has been removed. Our results revealed the existence of a greater number of vacant galactose acceptor sites, thus supporting the idea that glycoconjugates are produced in hypo-galactosylated states under galactose toxicity.

Effects on Gonadotrophin Structure/Function

Gonadotrophins are the members of glycoprotein family of hormones. Carbohydrates attached to the protein core of these hormones influence a number of intracellular and extracellular processes including activation of the respective receptors and efficient triggering of signal transduction. As with other glycoprotein hormones, FSH is produced and released as an array of isoforms that differ from each other in respect of their oligosaccharide attachments. Individual FSH isoforms, while displaying identical amino acid sequences, differ in their extent of post-translational modification and carbohydrate make up. As a result of these variations, the FSH isoforms exhibit differences in overall charge, degree of sialic acid or sulphate incorporation, receptor binding affinity and plasma half-life. The content of sialic acid in FSH and the variations in the content or arrangement of the carbohydrate residues internal to sialic acid are closely correlated with the isoelectric point (pI) (Sherins *et al.*, 1973) and the functional characteristics of this glycoprotein hormone (Bousfield and Ward, 1994). FSH isoforms have been determined with pIs 3.7–6.7 in female rats and women (Ozawa and Wakabayashi, 1988; Ulloa-Aguirre *et al.*, 1988). The most abundant FSH isoforms in the rats and women are found with pIs 4-5 and pIs 4-6, respectively.

Our preliminary *in-vitro* studies reflected a possible lack of biological activity of FSH. To look for probable production of FSH with altered carbohydrate make up, we assessed the relative concentrations of different isoforms of FSH by immunoblot after separation of the sera proteins by isoelectric focusing (IEF). We noted that in addition to the commonly found FSH species with a median pI of 4-5, the sera of the treated animals contained qualitatively abnormal FSH isoforms with a pI close to neutral (6.4-7.0) (unpublished observation). The altered FSH isoforms can be explained by the deficiency of galactose in the carbohydrate moiety of FSH to which sialic acid is linked. Sialic acid bears a negative charge and the degree of sialic acid content in FSH is in linear relationship with the pH of FSH (Sherins *et al.*, 1973). Deficiency of sialic acid due to the deficiency of galactose explains the cathodal shift of FSH isoforms.

This kind of a partial deficiency of the terminal disaccharide content, galactose and sialic acid, has been demonstrated by IEF for serum lysosomal enzymes of galactosaemic patients (Jaeken *et al.*, 1992; Prestoz *et al.*, 1997). The relatively neutral FSH isoforms indicate that possibly because of GALT deficiency, abnormal forms of gonadotrophins may be produced. In fact, deglycosylated forms of FSH act as anti-hormones, producing antagonistic effects at the FSH receptor levels (Kessel *et al.*, 1988). *In vitro* and *in vivo* experiments have shown that the deglycosylated FSH has a higher binding affinity to its receptor than the glycosylated form, but is unable to activate cAMP.

Onset of Ovarian Damage: Prenatal or Postnatal ?

Galactosaemia is associated with acquired ovarian failure in virtually all female subjects. The age at which ovarian damage occurs is variable and apparently preceded by a progressive rise in gonadotrophin levels and an abnormal response to GnRH, which are indicative of ovarian insufficiency. There is little doubt that galactose or galactose-1-phosphate accumulates in the gonad and exerts direct toxic effects on the ovarian parenchyma. But there exists no uniformity of opinion on whether the toxic effects are prenatal or postnatal events. Chen and colleagues (1981) demonstrated that offspring of rats fed a high galactose diet during pregnancy had significantly reduced number of oocytes at birth, suggesting that prenatal toxicity is responsible for ovarian dysfunction in galactosaemia. On the contrary, Levy and colleagues (1984) reported that a histologically normal ovary with abundant oocytes in a 5-day old infant with galactosaemia, who died with *Escherichia coli* sepsis, indicating that although damage from galactose or galactose-1-phosphate in the foetus is initiated *in-utero*, it requires continued accumulation of metabolites in postnatal life to cause ovarian atrophy. This is also consistent with the observation of normal-appearing ovaries in childhood but streak ovaries after puberty in a female patient with galactosaemia (Kaufman *et al.*, 1981).

Taking all these reports into consideration, we postulate that perhaps the toxic effects of galactose are exerted at the pre- as well as postnatal life, however, the mode of effects differ. Since the effects at embryonic stage jeopardize the process of oogonial migration with ultimate development of ovaries having sub-optimal number of germ cells, it leaves an irreversible imprint on ovarian functions; while the effects during postnatal stage is rather transitional. If the effects of galactose exerted during postnatal stage are superimposed on the damage already ensued following prenatal exposure, the overall damaging effect is compounded. In such situation, the fewer numbers of follicles present in the ovary are further exposed to atretic influence of accumulated ovarian galactose metabolites and also suffers from deficiency of gonadotrophic support. But when exposed only during postnatal life, the follicles are rendered refractory to gonadotrophins. Execution of the toxic manifestations perhaps occurs at the ovary as well as pituitary levels. Accumulation of galactose and galactose-1-phosphate exerts direct toxic effects on ovarian parenchyma and induce follicular atresia. This effect is accentuated further because of production of galactose-deficient isoforms of FSH to which the follicles are resistant. Execution of these effects, however, needs continued exposure to galactose; and restoration of ovarian function therefore ensues following withdrawal of galactose support. We presume that as the galactose metabolites wane below the toxic levels, the ovary gets rid of the atretic influence and the pituitary starts secreting optimally galactosylated forms of FSH.

Conclusion

Galactose plays an intricate role in the promotion of normal ovarian functions. Inimically altered galactose metabolism jeopardises ovarian functions at multiple steps. Wide spectra of ovarian dysfunction under experimental galactose toxicity prove its credentials as a model to explore many basic tenets of the multifaceted phenotypes of ovarian pathology.

Acknowledgements

This study was supported by grants from the Department of Science and Technology (SP/SO/B31/97), New Delhi, India.

References

Amsterdam A, Dantes A, Selvaraj N, and Aharoni D. (1997). Apoptosis in steroidogenic cells: structure-function analysis. Steroids, 62, 207-211.

Bandyopadhyay S, Banerjee S, Pal AK, Goswami SK, Chakravarty BN, and Kabir SN. (2004). Primordial germ cell migration in the rat: preliminary evidence for a role of galactosyltransferase. Biol. Reprod., 71, 1822-1827.

Bandyopadhyay S, Chakrabarti J, Banerjee S, Pal AK, Bhattacharyya D, Goswami SK. *et al.* (2003a). Prenatal exposure to high galactose adversely affects initial gonadal pool of germ cells in rats. Hum. Reprod., 18, 276-282.

Bandyopadhyay S, Chakrabarti J, Banerjee S, Pal AK, Goswami SK, Chakravarty BN. *et al.* (2003b). Galactose toxicity in the rat as a model for premature ovarian failure: an experimental approach readdressed. Hum. Reprod., 18, 2031-2038.

Billig H, Chun SY, Eisenhauer K, and Hsueh AJ. (1996). Gonadal cell apoptosis: hormone-regulated cell demise. Hum. Reprod. Update, 2, 103-117.

Bousfield GR, and Ward DN. (1994). Evidence for two folding domains in glycoprotein hormone alpha-subunits. Endocrinol., 135, 624-635.

Chen YT, Mattison DR, Bercu BB, and Schulman JD. (1984). Resistance of the male gonad to a high galactose diet. Pediatr. Res., 18, 345-348.

Chen YT, Mattison DR, Feigenbaum L, Fukui H, and Schulman JD. (1981b). Reduction in oocyte number following prenatal exposure to a diet high in galactose. Science, 214, 1145-1147.

Coulam CB. (1981). Hormonal status of galactosaemia homozygotes. Lancet, 1, 165.

Dobbie JA, Holton JB, and Clamp JR. (1990). Defective galactosylation of proteins in cultured skin fibroblasts from galactosaemic patients. Ann. Clin. Biochem., 27, 274-275.

Fazel AR, Schulte BA, Thompson RP, and Spicer SS. (1987). Presence of a unique glycoconjugate on the surface of rat primordial germ cells during migration. Cell Differ., 21, 199-211.

Fraser IS, Russell P, Greco S, and Robertson DM. (1986). Resistant ovary syndrome and premature ovarian failure in young women with galactosaemia. Clin. Reprod. Fertil., 4, 133-138.

Gibson JB. (1995). Gonadal function in galactosaemics and in galactose-intoxicated animals. Eur. J. Pediatr., 154 (Suppl 2), S14-S20.

Haberland C, Perou M, Brunngraber EG, and Hof H. (1971). The neuropathology of galactosaemia. A histopathological and biochemical study. J. Neuropathol. Exp. Neurol., 30, 431-447.

Hagenfeldt, K., von Dobeln, U. and Hagenfeldt, L. (1989) Gonadal failure in young women and galactose-1-phosphate uridyl transferase activity. Fertil. Steril., 51, 177-178.

Heidenreich RA, Mallee J, Rogers S, and Segal S. (1993). Developmental and tissue-specific modulation of rat galactose-1-phosphate uridyltransferase steady state messenger RNA and specific activity levels. Pediatr. Res., 34, 416-419.

Hoefnagel D, Wurster-Hill D, and Child EL. (1979). Ovarian failure in galactosaemia. Lancet, 2, 1197.

Holton JB. (1990). Galactose disorders: an overview. J. Inherit. Metab. Dis., 13, 476-486.

Jaeken J, Kint J, and Spaapen L. (1992). Serum lysosomal enzyme abnormalities in galactosaemia. Lancet, 340, 1472-1473.

Kaipia A, and Hsueh AJ. (1997). Regulation of ovarian follicle atresia. Annu. Rev. Physiol., 59, 349-363.

Kaufman F, Kogut MD, Donnell GN, Koch R, and Goebelsmann U. (1979). Ovarian failure in galactosaemia. Lancet, 2, 737-738.

Kaufman FR, Donnell GN, and Lobo RA. (1987). Ovarian androgen secretion in patients with galactosaemia and premature ovarian failure. Fertil. Steril., 47, 1033-1034.

Kaufman FR, Donnell GN, Roe TF, and Kogut MD. (1986). Gonadal function in patients with galactosaemia. J. Inherit. Metab. Dis., 9, 140–146.

Kaufman FR, Kogut MD, Donnell GN, Goebelsmann U, March C, and Koch R. (1981). Hypergonadotrophic hypogonadism in female patients with galactosaemia. N. Engl. J. Med., 304, 994-998.

Kaufman FR, Xu YK, Ng WG, and Donnell GN. (1988). Correlation of ovarian function with galactose-1-phosphate uridyl transferase levels in galactosaemia. J. Pediatr., 112, 754-756.

Kaufman FR, Xu YK, Ng WG, Silva PD, Lobo RA, and Donnell GN. (1989). Gonadal function and ovarian galactose metabolism in classic galactosaemia. Acta Endocrinol. (Copenh.)., 120, 129-133.

Kessel B, Dahl KD, Kazer RR, Liu CH, Rivier J, Vale W. *et al.* (1988). The dependency of bioactive follicle-stimulating hormone secretion on gonadotrophin–releasing hormone in hypogonadal and cycling women. J. Clin. Endocrinol. Metab., 66, 361-66.

Levy HL, Driscoll SG, Porensky RS, and Wender DF. (1984). Ovarian failure in galactosaemia. N. Engl. J. Med., 310, 50-50.

Liu G, Hale GE, and Hughes CL. (2000). Galactose metabolism and ovarian toxicity. Reprod. Toxicol., 14, 377-384.

McGuire EJ, Kerlin R, Cebra JJ, and Roth S. (1989). A human milk galactosyltransferase is specific for secreted, but not plasma, IgA. J. Immunol., 143, 2933-2938.

Meyer WR, Doyle MB, Grifo JA, Lipetz KJ, Oates PJ, DeCherney AH. *et al.* (1992). Aldose reductase inhibition prevents galactose-induced ovarian dysfunction in the Sprague-Dawley rat. Am. J. Obstet. Gynecol., 167, 1837-1843.

Ojeda SR, Wheaton JE, Jameson HE, and McCann SM. (1976). The onset of puberty in the female rat: changes in plasma prolactin, gonadotrophins, luteinizing hormone-releasing hormone (LHRH), and hypothalamic LHRH content. Endocrinol., 98, 630-638.

Ojeda SR, and Urbanski HF. (1994). Puberty in the rat. *In*: Knobil E, and Neill J. (Eds.), *The Physiology of Reproduction*, 2nd ed, Raven Press Ltd., New York, pp. 363-410.

Ozawa K, and Wakabayashi K. (1988). Dynamic change in charge heterogeneity of pituitary FSH throughout the estrous cycle in female rats. Endocrinol. (Jpn.)., 35, 321-332.

Petry K, Greinix HT, Nudelman E, Eisen H, Hakomori S, Levy HL. *et al.* (1991). Characterization of a novel biochemical abnormality in galactosaemia: deficiency of glycolipids containing galactose or N-acetylgalactosamine and accumulation of precursors in brain and lymphocytes. Biochem. Med. Metab. Biol., 46, 93-104.

Prestoz LLC, Couto AS, Shin YS, and Petry KG. (1997). Altered follicle stimulating hormone isoforms in female galactosaemia patients. Eur. J.Pediatr., 156, 116-120.

Rogers S, Heidenreich R, Mallee J, and Segal S. (1992). Regional activity of galactose-1-phosphate uridyltransferase in rat brain. Pediatr. Res., 31, 512-515.

Runyan RB, Maxwell GD, and Shur BD. (1986). Evidence for a novel enzymatic mechanism of neural crest cell migration on extracellular glycoconjugate matrices. J. Cell Biol., 102, 432-441.

Sardharwalla IB, and Wraith JE. (1987). Galactosaemia. Nutr. Health, 5, 175-188.

Schweitzer S, Shin Y, Jakobs C, and Brodehl J. (1993). Long-term outcome in 134 patients with galactosaemia. Eur. J. Pediatr., 152, 36-43.

Segal S. (1995a). Defective galactosylation in galactosaemia: is low cell UDPgalactose an explanation? Eur. J. Pediatr., 154 (Suppl 2), S65-S71.

Segal S. (1995b). In-utero galactose intoxication in animals. Eur. J. Pediatr., 154 (Suppl. 2), S82-S86.

Segal S, and Bernstein H. (1963). Observations on cataract formation in the newborn offspring of rats fed a high-galactose diet. J. Pediatr., 62, 363-370.

Segal S, and Berry GT. (1995). Disorders of galactose metabolism. *In*: Scriver C, Beauder A, Sly W. *et al.* (Eds.), *The metabolic and molecular basis of inherited disease*. 7th edn., Mcgraw-Hill, New York, pp.996-1000.

Sherins RJ, Vaitukaitis JL, and Chambach A. (1973). Physical characterization of hFSH and its desialylation products by isoelectric focusing and electrophoresis in polyacrylamide gel. Endocrinol., 92, 1135-1141.

Shur BD. (1993). Glycosyltransferases as cell adhesion molecules. Curr. Opin. Cell Biol., 5, 854-863.

Shur BD, and Roth S. (1975). Cell surface glycosyltransferases. Biochim. Biophys. Acta, 415, 473-512.

Spatz M, and Segal S. (1983). Transplacental galactose toxicity in rats. J. Pediatr., 67, 438-446.

Swartz WJ, and Mattison DR. (1988). Galactose inhibition of ovulation in mice. Fertil. Steril., 49, 522-526.

Tedesco TA, and Miller KL. (1979). Galactosaemia: alterations in sulfate metabolism secondary to galactose-1-phosphate uridyltransferase deficiency. Science, 205, 1395-1397.

Thomford PJ, Jelovsek FR, and Mattison DR. (1987). Effect of oocyte number and rate of atresia on the age of menopause. Reprod. Toxicol., 1, 41-51.

Thyssen SM, and Libertun C. (1996). Alpha-difluoromethylornithine modifies FSH secretion and puberty onset in the female rat. Proc. Soc. Exp. Biol. Med., 211, 76-80.

Ulloa-Aguirre A, Espinoza R, Damian-Matsumura P, Larrea F, Flores A, Morales L. *et al.* (1988). Studies on the microheterogeneity of anterior pituitary follicle-stimulating hormone in the female rat. Isoelectric focusing pattern throughout the estrous cycle. Biol. Reprod., 38, 70-78.

Waggoner DD, Buist NR, and Donnell GN. (1990). Long-term prognosis in galactosaemia: results of a survey of 350 cases. J. Inherit. Metab. Dis., 13, 802-818.

Xu YK, Ng WG, Kaufman FR, Lobo RA, and Donnell GN. (1989). Galactose metabolism in human ovarian tissue. Pediatr. Res., 25, 151-155.

27

Regulation of Avian Growth Plate Development and its Interruption by Thiram, A Dithiocarbamate Pesticide

Narayan C. Rath, William E. Huff, Geraldine R. Huff*

U.S. Department of Agriculture, Agricultural Research Service, Poultry Science Center, University of Arkansas, Fayetteville, AR 72701, USA

ABSTRACT

Long bone growth is achieved by a process called 'endochondral ossification' taking place in the growth plates located at the end of these bones. The growth plate development is a complex process which encompasses chondrocyte proliferation, maturation related differentiation, and the eventual replacement of the cartilage scaffold by bone. This process is regulated by a number of hormones, cytokines, and growth factors the interference of which can lead to growth plate abnormality and consequent bone problems. We have been studying the mechanisms underlying a growth plate abnormality in poultry called "tibial dyschondroplasia (TD)" where parts of the cartilage scaffold fail to form bone and are retained as avascular plugs of tissues that cause lameness. Using thiram, a thiocarbamate fungicide and an endocrine disruptor, to induce TD in broiler chickens, we studied its effects on the metabolism, proliferation, differentiation associated gene expression, and cell death in the growth plate to understand the mechanisms associated with this defect. The results show that thiram did not produce chondrocyte hyperplasia or changes in maturation associated gene expression that could explain an unusual

**Corresponding Author*: E-mail: nrath@uark.edu

increase in the growth plate size. On the other hand, it reduced cartilage glutathione levels, caused endothelial cell death and prevented angiogenesis and induced subsequent death of chondrocytes in the maturing zone cartilage. Thiram treatment also increased the blood corticosterone levels. It is concluded that thiram selectively prevents angiogenesis and induces premature chondrocyte death preventing subsequent bone formation that results in growth plate dysplasia.

Abbreviations Used

BMP: Bone morphogenetic protein; Cbf1: Core binding factor 1; ECM: Extracellular matrix; EDC: Endocrine disrupting chemical; FGF: Fibroblast growth factor; IGF: Insulin-like growth factor; IHH: Indian hedgehog; GH: Growth hormone; MMP: Matrix metalloproteinase; NF-kB: Nuclear factor kappa-B; PHC: Prehypertrophic chondrocyte; PTHrP: Parathyroid related hormone; RA: Retinoic acid; RUNX2: Runt-related gene; TD: Tibial dyschondroplasia; TGF-β: Transforming growth factor β; TUNEL: Terminal deoxynucleotidyl transferase-mediated dUTP nick end labeling; VEGF: Vascular endothelial growth factor.

Introduction

Postnatal development is largely a growth and maturation process of organ systems during which they achieve their full functional competency. Exposure to adverse environmental influences such as chemicals and pollution, during this process can cause epigenetic changes that result in organ dysfunctions, carcinogenesis, phenotypic anomalies, or maladaptive behavior. Several pesticides cause such problems by virtue of their ability to interfere with physiological homeostasis either by disrupting basic cellular functions, or signaling processes. Therefore, some environmental chemicals have been qualified as 'endocrine disrupting chemicals (EDC)' because they interfere with neuro-hormonal signaling processes. While the mechanisms of action of many EDC are recognizable by their selective effects on certain target organs, at other times it may be difficult to recognize their effects if their modes of action are diverse and the organs systems depend upon multiple signaling processes for their growth and development. The skeletal tissues form one such complexity whose development is multifactorial. The current review is based on studies which show that certain dithiocarbamate pesticides can affect the development of the cartilaginous growth plate of post-hatch chickens inducing growth plate dysplasia leading to the failure of bone development.

Growth Plate Development

Long bones develop by a process called "endochondral ossification" where bones form by the replacement of cartilaginous growth plates, located at the ends of these bones beneath articular cartilage. The avian growth plate, like its mammalian counterpart, consists of a hyaline cartilage which undergoes a complex series of development beginning with the proliferation of the germinal layer of mesenchymal cells to form proliferative zone cartilage which differentiates to form a 'maturation zone' of prehypertrophic chondrocytes (PHC). The PHC undergo hypertrophy under a complex series of chemical signaling regulated by different physiological factors that lead to chondrolysis and their replacement by osteoblasts which form bone (Howlett, 1979; White and Wallis, 2001; Provot and Schipani, 2005). Accompanying these morphological transitions there are extracellular matrix synthesis and turnover, calcification, and angiogenesis which are pivotal to successful bone formation. The development of growth plate entails a number of hormones, growth, and nutritional factors which facilitate different aspects of these transitions. The avian growth plate is much thicker than a typical mammalian growth plate and lacks a strict columnar organization of the chondrocytes in the maturing zone cartilage, and is relatively better vascularized (Leach and Gay, 1987; Loveridge *et al.*, 1992; Barreto and Wilsman, 1993). In post hatch chicks, the undifferentiated reserve zone chondrocytes are contiguous with articular cartilage and typically consists of several layers of cells that are flat or spindle shaped and proliferate as discoid layers of cells, and transform to round PHC. The PHC synthesize abundant extracellular matrix (ECM) molecules predominantly consisting of type II collagen and proteoglycans such as aggrecan and perlecan. Types IX and XI collagens along with other minor ECM proteins such as cartilage oligomeric matrix protein and matrilins are also coexpressed in this region. As PHC progress to hypertrophy, the cells synthesize different types of collagen protein most prominently type X collagen which characterizes chondrocyte maturation. Type X collagen in conjunction with several other factors facilitates calcification of the growth plate and provides signals for angiogenesis, beginning the process called chondrolysis and matrix modeling, which precede bone formation (Linsenmayer *et al.*, 1998). The calcification of ECM is facilitated by alkaline phosphatase whereas remodeling is facilitated by a number of proteases such as matrix metalloproteinases (MMP) and cathepsins that are produced by hypertrophic chondrocytes (Carlevaro *et al.*, 2000; Kronenberg, 2003; Ortega *et al.*, 2004). These developments which embody cell division, differentiation, and cell death, are regulated by multiple genes involving numerous growth and nutritional factors that harmoniously control this process the failing of which evidently lead to growth plate disease and bone problems.

Regulation of Chondrocyte Development

A number of local and systemic factors coordinate both development and differentiation of growth plate chondrocytes (Leach and Twal, 1994; Farquharson, 2002; Karsenty, 2003; Kronenberg, 2003; van der Eerden *et al.*, 2003; Ionescu *et al.*, 2004). Among the major locally acting autocrine and paracrine factors which play a crucial part in the development and differentiation of chondocytes are the Indian hedgehog protein (IHH), parathyroid hormone related protein (PTHrP), bone morphogenetic proteins (BMP), transforming growth factor-β (TGF-β), fibroblast growth factors (FGF), and vascular endothelial growth factors (VEGF). Several transcription factors, most prominently, SOX9 and Cbf1 or RUNX2 play important roles in growth plate differentiation. While many of these factors have been identified to play a definite role in local regulation of growth plate development there are many others whose functions are still not understood. Systemic factors known to regulate growth plate chondrocyte development are growth hormone (GH), thyroid hormone, insulin, insulin-like growth factor (IGF), retinoic acid, vitamin D metabolites, and glucocorticoids which interactively contribute to the development of growth plate cartilage (Leach and Twal, 1994; van der Eerden *et al.*, 2003; Kronenberg, 2003). Despite significant advance in chondrocyte cell biology there are major gaps in the understanding of how these factors are coordinated.

Proliferation and hypertrophy of chondrocytes are important aspects of longitudinal bone growth (Hunziker, 1994). Once the germinal zones of mesenchymal cells divide, they differentiate to form PHC. The chondrocyte proliferation and the transition of PHC to the hypertrophic state are governed by several autocrine and paracrine factors. The PHC produce a signaling protein called Indian Hedgehog which through a paracrine loop induces the production of PTHrP by periarticular cells. PTHrP at downstream activates its own receptors and promotes chondrocyte proliferation while preventing hypertrophic differentiation of PHC through a negative feedback loop (Vortkamp *et al.*, 1996; Kronenberg, 2003; Ionescu *et al.*, 2004; Provot and Schipani, 2005). PTHrP production appears to be mediated by TGF-β that also locally regulates chondrocyte proliferation and prevents chondrocyte hypertrophy (Ionescu *et al.*, 2004). Several forms of BMPs and their receptors also play significant roles in controlling chondrogenesis through cooperative interactions with FGF and Wnt signaling processes (Reddi, 1994; Leboy *et al.*, 2002; Yoon and Lyon, 2004; Wan and Cao, 2005). The BMPs are the members of TGF-β superfamily which are differentially expressed by proliferating and hypertrophic chondrocytes and they favor chondrocyte proliferation (Wan and Cao, 2005). Different forms of FGF along with their promiscuous receptors regulate both chondrocyte proliferation and hypertrophy (Di Nino and Linsenmayer, 2004). FGF, IGF, and GH support chondrocyte proliferation in culture (Rosselot *et al.*, 1994). The SOX family transcription factors regulate chondrocyte proliferation and suppress hypertrophy (Bi *et al.*, 1999; Wan and Cao, 2005). Pituitary growth hormone (GH) is thought to contribute to longitudinal growth of bone (Lindah *et al.*, 1986) but its effect on avian growth plate is not known although avian chondrocytes have been shown to express GH receptors (Monsonego *et al.*, 1995). Glucocorticoids on the other hand, inhibit chondrogenesis preventing both proliferation and differentiation and possibly promote chondrocyte death and impair bone growth (van der Eerden *et al.*, 2003; Chrousos, 2004).

Hypertrophic differentiation of chondrocytes leads to chondrolysis and ultimately cartilage remodeling in the growth plate that precedes angiogenesis and bone formation. Though the avian growth plate is relatively better vascularized compared to a mammalian growth plate, the general scheme of development follows the same conservative pattern and involves many similar growth factors for its development. The maturation process is identified by the deposition of type X collagen and increased alkaline phosphatase activities (Pines *et al.*, 1998). Retinoic acid (RA), thyroid hormone, glucocorticoid, and vitamin D metabolites all signal hypertrophic transformation (Quarto *et al.*, 1992; Ishikawa *et al.*, 1998; Nie *et al.*, 1998; Krohn *et al.*, 2003; Ionescu *et al.*, 2004). Pharmacological interference with RA signaling blocks maturation (Pacifici *et al.*, 2004). RUNX2, also called Cbf1, is a transcription factor produced by chondrocytes undergoing hypertrophy which appears to be essential for skeletal development and promotes production of type X collagen by the maturing chondrocytes (Stricker *et al.*, 2002; Karsenty, 2003; Kronenberg, 2003; Ionescu *et al.*, 2004). RUNX2 also regulates vascular invasion of hypertrophic chondrocytes which is mediated through VEGF signaling. The calcification of matrix provides a signal for angiogenesis that is facilitated by the production of VEGF and several forms of MMP. Additionally angiogenesis also facilitates chondroclast invasion of matrix that helps turnover of ECM (Baron, 2003). Matrix metalloproteinases (MMP) are Zn dependent endoproteinases that are produced by the hypertrophic chondrocytes that induce ECM degradation, activate growth factors such as TGF-β, and release angiogenic molecules to facilitate vascularization (Gerber *et al.*, 1999; Carlevaro *et al.*, 2000; Vu and Werb, 2000; Ortega *et al.*, 2004). The chicken growth plate contains several different types of MMP which play a major role in growth plate development and modeling (Wardale and Duance, 1996; Rath *et al.*, 1997; D'Angelo *et al.*, 2000; Tong *et al.*, 2003). Chondrocyte hypertrophy leads to changes in the chondrocyte volume (Hunziker, 1994; White and Wallis, 2001) which also contributes to chondrolysis, the regulation of which is little understood. Whether chondrocyte hypertrophy leads to a change in membrane fluidity and permeability along with down regulation of mitochondrial activities is not known but chondrolysis presumably accompanies apoptotic death of the hypertrophied chondrocytes that are replaced by osteoblasts. On the contrary, some believe that terminally differentiated chondrocytes transdifferentiate to osteoblasts (Roach, 1997). In normal avian growth plates, we have not found much apoptosis at the osteochondral junction of metaphysis (Rath *et al.*, unpublished).

Avian Growth Plate Abnormalities

Because of the interplay of different genetic, nutritional, and environmental factors such as instituted for intensive poultry production, growth plate and bone abnormalities in chickens and turkeys are often encountered. Genetic abnormalities affecting collagen or other ECM proteins do bring about a variety of skeletal abnormalities such as dwarfism, osteochondrodysplasia and compromise skeletal function in mammals. Congenital abnormalities such as dwarfism-related growth plate defects and malformatiom are not well studied in birds although such possibilities exist. On the other hand, nutritional deficiencies have been known to cause growth plate abnormalities in birds. Vitamin D deficiency and Ca or P metabolism disorders produces a rachitic growth plate leading to the accumulation of hypertrophic cartilage which does not calcify (Klasing and Austic, 2003). These defects are reversed once the nutritional deficiencies are corrected. Tibial dyschondroplasia (TD) is a frequently encountered growth plate defect in meat-type poultry that is characterized by partial failure of growth plate to form an endochondral bone (Figure 27.1; Leach and Nesheim, 1965; Edwards, 1984; Rath *et al.*, 1994), the etiology of which is largely unknown. The failure of growth plate to form bone causes lameness in broiler chickens and turkeys undergoing rapid growth. The pathophysiological studies of TD have shown reduction in the levels of various maturation and calcification-related parameters including growth factors and enzymes such as the vitamin D receptors, TGF-β, alkaline phosphatase, and osteopontin, and PTHrP (Pines *et al.*, 1998; Praul *et al.*, 2000; Farquharson, 2002). It has been postulated that genetic predisposition and the deficiency of some growth or nutritional factors may be responsible for this problem in broiler chickens and turkeys. The absence of adequate physical activity of birds reared under environmentally controlled and crowded conditions has been

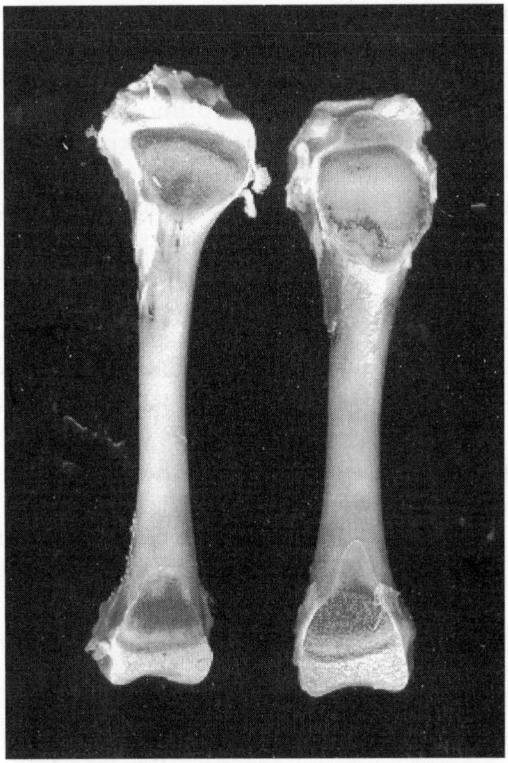

Figure 27.1: Growth Plate Dysplasia in the Proximal Tibia of 4 Week-old Broiler Chicken Showing the Presence of a Plug of Avascular Cartilage (Right) as Compared to a Sickle Shaped Normal Growth Plate Underlying Articular Cartilage (Left)

suggested as a factor for tibial growth plate abnormality. Using naturally occurring disease our studies have determined that TD was related to abnormal chondrocyte death in the growth plate (Rath *et al.*, 1998).

Growth Plate Dysplasia Induced by Dithiocarbamate pesticide

Dithiocarbamates are organic pesticides, used as fungicides, pest repellents, and wood preservatives in a variety of agricultural and household applications (Edwards *et al.*, 1991; U. S. EPA, 2001). Several dithiocarbamates are classified as endocrine disrupting chemicals (EDC) since they interfere with hormonal signaling processes although the mechanisms of their action are far from clear (Colborn, 1998; McLachlan, 2001). Tetramethyl thiuram disulfide (thiram, Figure 27.2) is used to prevent crop damage and as a seed disinfectant, and accelerator in the vulcanization of rubber. Its ethyl analogue disulfiram, is used for alcohol aversion therapy since these chemicals inhibit aldehyde dehydrogenase (Edwards *et al.*, 1991). Dithiocarbamates also act as metal chelators (Sharma *et al.*, 2003; Shian *et al.*, 2003). Endocrine disruption is affected by the environmental chemicals if they modify the metabolism of natural hormones by their interaction with hormone receptors, or blocking the pathways of hormone action, or induce hormone breakdown leading to persistent physiological changes resulting in carcinogenesis, mutagenesis, toxicity, and organ dysfunctions (McLachlan, 2001). Although much less is known about the mode of action of dithiocarbamates as EDC, there is some information to show that they interfere with gonadotropin secretion and ovulation (Stoker *et al.*, 1993; Cooper *et al.*, 1999). Thiram also interferes with corticosteroid metabolism inhibiting 11 β hydroxysteroid dehydrogenase-2 that irreversibly transforms cortisol and corticosterone to their biologically inactive metabolites, cortisone and 11-dehydrocorticosterone, respectively (Atanasov *et al.*, 2003; Mazancova *et al.*, 2005). Metabolism of some dithiocarbamates generates thiourea products that have potential to interfere with thyroid hormone synthesis metabolism (Brucker-Davis, 1998; U. S. EPA, 2001). Dithiocarbamates also interfere with the metabolism of glutathione which is pivotal in balancing the oxido-reductive status of cells thereby preserving mitochondrial function and membrane integrity of cells. Similarly, they can interact and inhibit sulfhydryl containing enzymes. Several dithiocarbamates also inhibit nuclear factor kappa B (NF-kB), a transcription factor that regulates a plethora of biological activities, the inhibition of which can lead to cellular apotosis (Schreck *et al.*, 1992; Orrenius *et al.*, 1996; Nobel *et al.*, 1997; Wang *et al.*, 2003). Both thiram and disulfiram have been shown to be angiogenesis inhibitors (Marilkovsky, 2002; Shian *et al.*, 2003; Sook Han *et al.*, 2003). Overall, it appears that dithiocarbamates are capable of interfering with cellular functions at multiple levels.

Based on correlation between the use of thiram and the outbreak of lameness in chickens, it was shown that both thiram and disulfiram when fed continuously to broiler chickens cause growth plate dysplasia (Waibel *et al.*, 1957; Vargas *et al.*, 1983; Guitart *et al.*, 1996; Wickramanayake, 1996). We have shown that inclusion of several different dithiocarbamates including thiram, disulfiram, pyrrolidine

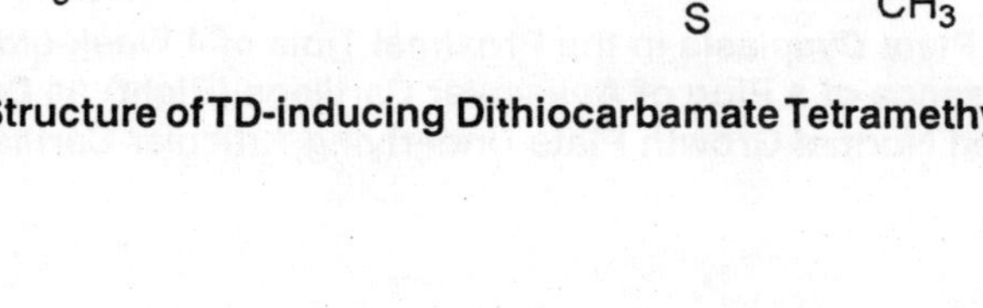

Figure 27.2: Chemical Structure of TD-inducing Dithiocarbamate Tetramethyl Thiuram Disulfide (thiram)

dithiocarbamate, ferbam, maneb, and ziram at concentration of 100 ppm in feed for one to two days was sufficient to induce growth plate dysplasia in > 90 per cent of chickens (Rath et. al., 2004; Rath unpublished observation). A concentration dependent increase in the incidence and severity of TD was observed at 10-80 ppm of thiram (Rath *et al.*, 2004). However, this effect was not seen using the monomer dimethyldithiocarbamate or certain other metabolites including propylenethiourea and methylamine (unpublished). To understand how dithiocarbamates induce growth plate dyschondroplasia, we fed 7-day-old broiler chickens thiram at 100 ppm concentration for 48 h and analyzed growth plate cartilage tissues starting from 2 h after the end of feeding to 7 days after feeding for their abilities to divide, undergo changes in metabolism and gene expressions, or cell death that may be consequential to their phenotypic abnormalities (Rath *et al.*, 2005; Figure 27.3a,b). Using reverse transcription-polymerase chain reaction (RT-PCR), there was no change in the transcription profiles of either type II or type X collagen, aggrecan, MMP gene expression indicating that thiram did not affect PHC function nor precluded their transition to maturity (Rath *et al.*, 2005). Similar results were also observed in naturally occurring tibial dyschondroplasia where type X collagen was present in the dyschondroplastic lesions (Rath *et al.*, 1994). Although, the broadening of growth plate was associated with the accumulation of cartilage cells in the maturing zone as was the absence of remodeling and vascularization, it was not reflected in transcription profiles of VEGF, MMP-13, and RUNX2 genes of growth plate during the first 48-50 h of its action. This work shows that these key regulators of chondrocyte hypertrophy are possibly not affected by thiram treatment. Additionally, the accumulation of chondrocyte in the cartilage plug seems not to be due to an increase in proliferation of PHC because bromodeoxyuridine labeling failed to show the presence of any cell division in the maturing region of cartilage in the growth plate derived from either control or thiram-treated birds. On the other hand, the proliferative regions showed incorporation of bromodeoxyuridine in both control and thiram treated birds (Rath *et al.*, 2005). Since thiram had no effect on cell division, we asked if the accumulation of chondrocytes was possible due to cell death in the maturing zone cartilage precluding their further development and differentiation to bone. The results showed that as early as 24 h after the end of thiram feeding, apoptosis of capillary endothelial cells was evident by teminal deoxynucleotidyl transferase-mediated dUTP nick end labeling (TUNEL) which at later time points showed extensive labeling of capillary vessels and chondrocytes in the surrounding regions. This was confirmed by various other morphological criteria such as nuclear shrinkage and blebbing, and DNA fragmentation (Figure 27.3c). These results indicate that induction of cell death is probably the primary event for the accumulation of chondrocytes in the growth plate which do not resorb and are not removed because of their inaccessibility to phagocytes. However, it is intriguing as to why differential cell death takes place in the maturing zone chondrocytes while thiram was unable to mortally affect the proliferative or reserve zone chondrocytes and the bone always grew in length albeit at a lower rate compared to control birds. While the exact molecular mechanism of action of dithiocarbamates on hypertrophying chondrocytes are not clear, It is possible that cell volume expansion during hypertrophy (Barreto and Wilsman, 1994; White and Wallis, 2001) changes the membrane fluidity permitting the incorporation of lipophilic compounds such as thiram into the cells to interfere with cellular function. Our studies with chondrocytes in culture as well as others using different types of cells have shown that chronic exposure to thiram or disulfiram can induce apoptosis (Rath *et al.*, 1995; Orrenius *et al.*, 1996; Sook Han *et al.*, 2003). Additionally, we have shown that growth plates of chickens treated with thiram had lower glutathione concentrations, indicating that mitochondrial activities of the cells are probably impaired (Rath *et al.*, 2005). Since the thiol reserves in chondrocytes undergoing hypertrophy are already reduced (Adams *et al.*, 2002) an additional burden with thiol-active dithiocarbamates may significantly impair mitochondrial activities (Balakirev and Zimmer, 2001) of hypertrophying cells

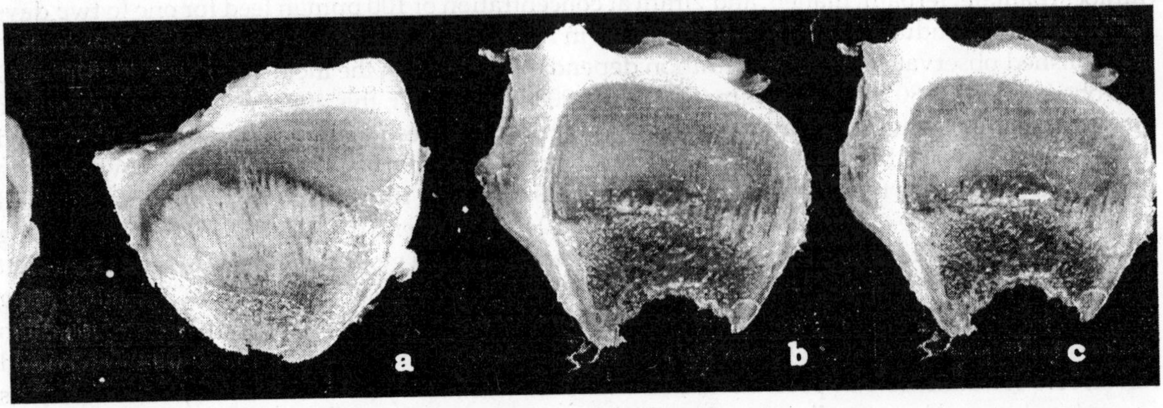

Figure 27.3: Tibial Dyschondroplasia (b) Induced by Feeding Thiram for 48 Hours as Compared to Control Fed Chickens (a); Arrows Indicate Chondrocytes in the Growth Plates of thiram-treated Birds Exhibiting Apoptosis (c)

pushing them to apoptosis. Targeted apoptosis of maturing region chondrocytes possibly leads to dysplasia. There are other possible mechanisms of action of dithiocarbamates that may lead to endothelial cell and chondrocyte apoptosis. As stated earlier, dithiocarbamates impair NF-kB activation which has been suggested to induce apoptosis (Schreck *et al.*, 1992; Nobel *et al.*, 1997; Wang *et al.*, 2003). Retinoic acid (RA) synthesis which is catalyzed by the enzyme aldehyde dehydrogenase in its final step is inhibited by thiram. Since, RA is a critical factor for hypertrophic maturation of chondrocytes, the cartilage maturation could be impaired by thiram although our data indicate that the initial step of maturation is not be affected by it (Rath *et al.*, 2005). Increased tissue concentrations of glucocorticoids due to 11-β hydroxysteroid dehydrogenase-2 inhibition could also lead to the accumulation of glucocorticoids in the cartilage tissues prompting apoptosis of maturing chondrocytes. We have observed that thiram treatment increases serum corticosterone level from 6.3 •} 1.4 ng/ml to 12.2 •} 1.8 ng/ml although there was no change in the serum thyroxin levels (Rath *et al.*, 2004). While the death of maturing chondrocytes explains the origin of growth plate dysplasia induced by thiram its exact mode of action may extend to many levels affecting different growth, endocrine, and nutritional factors as well as basic cellular functions.

Conclusion

Conventionally, EDC work as hormone mimics thereby disrupting the functions of natural hormones; however, it is likely that some chemicals, such as dithiocarbamates can have effects at multiple levels by virtue of their ability to act on different aspects of metabolizing systems either by themselves or through their metabolites. The avian growth plate development, which is regulated by the actions of multiple growth, endocrine, and nutritional factors, provides a unique model to study how these chemicals affect connective tissue development and differentiation.

References

Adams CS, Mansfield KD, Rajpurohit R, Tachibana H, Teixeira CM, and Shapiro IM. (2002). Components of cartilage matrix regulate chondrocyte apoptosis. *In*: Shapiro IM, Boyan B, Clark-Anderson HC. (Eds.), *The Growth Plate*, IOS Press, Amsterdam, pp 63-76.

Atanasov AG, Tam S, Rocken JM, Baker ME, and Odermatt A. (2003). Inhibition of 11 beta-hydroxysteroid dehydrogenase type 2 by dithiocarbamates. Biochem. Biophys. Res. Commun., 308, 257-262.

Balakirev MY, and Zimmer G. (2001). Mitochondrial injury by disulfiram: two different mechanisms of the mitochondrial permeability transition. Chem. Biol. Interact., 138, 299-311.

Baron R. (2003). General principles of bone biology. *In:* Favus MJ. (Ed), *Primers on the Metabolic Bone Diseases and Disorders of Mineral Metabolism.* 5th Edition. American Society of Bone and Mineral Research, Washington, D.C. pp1-8.

Barreto C, and Wilsman NJ. (1994). Hypertrophic chondrocyte volume and growth rates in avian growth plates. Res. Vet. Sci., 56, 53-61.

Bi W, Deng JM, Zhang Z, Behringer RR, and de Crombrugghe B. (1999). Sox9 is required for cartilage formation. Nature Genet., 22, 85-89.

Brucker-Davis F. (1998). Effects of environmental synthetic chemicals on thyroid function. Thyroid, 8, 827-856.

Carlevaro MF, Cermelli S, Cancedda R, Descalzi Cancedda F. (2000). Vascular endothelial growth factor (VEGF) in cartilage neovascularization and chondrocyte differentiation: auto-paracrine role during endochondral bone formation. J. Cell Sci., 113, 59-69.

Chrousos GP. (2004). Is 11beta-hydroxysteroid dehydrogenase type 1 a good therapeutic target for blockade of glucocorticoid actions? Proc. Natl. Acad. Sci. USA, 101, 6329-630.

Colborn T. (1998). Endocrine disruption from environmental toxicant. *In*: Rom WN. (Ed.), *Environmental and Occupational Medicine*, Third edition, Lippincott-Raven Publishers, Philadelphia.

Cooper RL, Goldman JM, and Stoker TE. (1999). Neuroendocrine and reproductive effects of contemporary-use pesticides. Toxicol. Ind. Health, 15, 26-36.

D'Angelo M, Yan Z, Nooreyazdan M, Pacifici M, Sarment DS, Billings PC, and Leboy PS. (2000). MMP-13 is induced during chondrocyte hypertrophy. J. Cell Biochem., 77, 678-693.

Di Nino D, and Linsenmayer TF. (2004). Perichondrial and periosteal regulation of endochondral growth. *In*: Massaro EJ, Rogers JM. (Eds.), *The Skeleton: Biochemical, Genetic, and Molecular Interactions in Development and Homeostasis.* Humana Press, Totowa, N. J., pp 229-247.

Edwards HM Jr. (1984). Studies on the etiology of tibial dyschondroplasia in chickens. J. Nutr., 114, 1001-1013.

Edwards IR, Ferry DH, and Temple WA. (1991). Fungicides and related compounds. *In*: Hayes Jr. WJ, and Laws ER. (Eds.), *Handbook of Pesticide Toxicology*, Vol. 3. Academic Press, N. Y., pp1436-1469.

Farquharson C. (2002). Tibial dyschondroplasia: a growth plate abnormality caused by delayed terminal differentiation. *In*: Shapiro IM, Boyan B, Clark-Anderson HC. (Eds.), *The Growth Plate*, IOS Press, Amsterdam, pp 201-211.

Gerber HP, Vu TH, Ryan AM, Kowalski J, Werb Z, and Ferrara N. (1999). VEGF couples hypertrophic cartilage remodeling, ossification and angiogenesis during endochondral bone formation. Nature Med., 5, 623-628.

Guitart R, Mateo R, Gutierrez JM, and To-Figueras J. (1996). An outbreak of thiram poisoning on Spanish poultry farms. Vet. Hum. Toxicol., 38, 287-288.

Howlett CR. (1979). The fine structure of the proximal growth plate of the avian tibia. J. Anat., 128, 377-399.

Hunziker EB. (1994). Mechanism of longitudinal bone growth and its regulation by growth plate chondrocytes. Microsc. Res. Tech., 28, 505-519.

Ionescu AM, Drissi MH, and O'Keefe RJ. (2004). Regulation of chondrocyte differentiation. *In*: Massaro EJ, Rogers JM. (Eds), *The Skeleton: Biochemical, Genetic, and Molecular Interactions in Development and Homeostasis*. Humana Press, Totowa, N. J., pp 43-53.

Ishikawa Y, Genge BR, Wuthier RE, and Wu LN. (1998). Thyroid hormone inhibits growth and stimulates terminal differentiation of epiphyseal growth plate chondrocytes. J. Bone Miner. Res., 13, 1398-1411.

Karsenty G. (2003). The complexities of skeletal biology. Nature, 423, 316-318.

Klasing K, Austic RE. (2003). Nutritional diseases. *In*: Saif YM. (Ed.), *Diseases of Poultry*. 11th edition, Iowa state Press, Ames, IA, pp 1027-1054.

Kronenberg HM. (2003). Developmental regulation of the growth plate. Nature, 423, 332-336.

Krohn K, Haffner D, Hugel U, Himmele R, Klaus G, Mehls O, and Schaefer F. (2003). 1, 25$(OH)_2$D3 and dihydrotestosterone interact to regulate proliferation and differentiation of epiphyseal chondrocytes. Calcif. Tiss. Int., 73, 400-410.

Leach RM, and Nesheim MC. (1965). Nutritional, genetic, and morphological studies of an abnormal cartilage formation in young chickens. J. Nutr., 86, 236-244.

Leach RM Jr, and Gay CV. (1987). Role of epiphyseal cartilage in endochondral bone formation. J. Nutr., 117, 784-790.

Leach RM, and Twal WO. (1994). Autocrine, paracrine, and hormonal signals involved in growth plate chondrocyte differentiation. Poult. Sci., 73, 883-888.

Leboy PS, Grasso-Knight G, D'Angelo M, and Adams S. (2002). BMP-regulated chondrocyte hypertrophy. *In*: Shapiro IM, Boyan B, and Clark-Anderson HC. (Eds.), The Growth Plate, IOS Press, Amsterdam, pp 223-234.

Lindah A, Isgaard J, Nilsson A, and Isaksson O. (1986). Growth hormone potentiates colony formation of epiphyseal chondrocytes in suspension culture. Endocrinology, 118, 1843-1848.

Linsenmayer TF, Long F, Nurminskaya M, Chen Q, and Schmid TM. (1998). Type X collagen and other up-regulated components of the avian hypertrophic cartilage program. Prog. Nucleic Acid Res. Molec. Biol., 60, 79-109.

Loveridge N, Thomson BM, and Farquharson C. (1992). Bone growth and turnover. In: Whitehead CC. (Ed), Bone Biology and Skeletal Disorders in Poultry. Carfax Publishing Co., Abingdon, England, pp 3-17.

Mazancova K, Kucka M, Miksik I, and Pacha J. (2005). Glucocorticoid metabolism and Na+ transport in chicken intestine. J. Exp. Zoology. A, Comp. Exp. Biol., 303, 113-122.

Marilkovsky M. (2002). Thiram inhibits angiogenesis and slows the development of experimental tumors in mice. Br. J. Canc., 86, 779-787.

McLachlan JA. (2001). Environmental signaling: what embryos and evolution teach us about endocrine disrupting chemicals. Endocr. Rev., 22, 319-341.

Monsonego E, Halevy O, Gertler A, Hurwitz S, and Pines M. (1995). Growth hormone inhibits differentiation of avian epiphyseal growth-plate chondrocytes. Mol. Cell Endocrinol., 114, 35-42.

Nie D, Ishikawa Y, Yoshimori T, Wuthier RE, and Wu LN. (1998). Retinoic acid treatment elevates matrix metalloproteinase-2 protein and mRNA levels in avian growth plate chondrocyte cultures. J. Cell. Biochem., 68, 90-99.

Nobel CS, Burgess DH, Zhivotovsky B, Burkitt MJ, Orrenius S, Slater AF. (1997). Mechanism of dithiocarbamate inhibition of apoptosis: thiol oxidation by dithiocarbamate disulfides directly inhibits processing of the caspase-3 proenzyme. Chem. Res. Toxicol., 10, 636-643.

Orrenius S, Nobel CSI, van den Dobbelsteen DJ, Burkitt MJ, and Slater AFG. (1996). Dithiocarbamates and redox regulation of cell death. Biochem. Soc. Trans. 24,1-3.

Ortega N, Behonick DJ, and Werb Z. (2004). Matrix remodeling during endochondral ossification. Trends Cell. Biol., 14, 86-93.

Pacifici M, Gentili C, Golden E, and Koyama E. (2004). Retinoids and Indian hedgehog orchestrate long bone development. *In*: Massaro EJ, and Rogers JM. (Eds.), *The Skeleton: Biochemical, Genetic, and Molecular Interactions in Development and Homeostasis.* Humana Press, Totowa, N. J., pp. 159-170.

Pines M, Knopov V, Genina O, Hurwitz S, Faerman A, Gerstenfeld LC, Leach RM. (1998). Development of avian tibial dyschondroplasia: gene expression and protein synthesis. Calcif. Tissue Int., 63, 521-527.

Praul CA, Ford BC, Gay CV, Pines M, and Leach RM. (2000). Gene expression and tibial dyschondroplasia. Poult. Sci. 79, 1009-1013.

Provot S, and Schipani E. (2005). Molecular mechanisms of endochondral bone development. Biochem. Biophys. Res. Commun., 328, 658-665.

Quarto R, Campanile G, Cancedda R, and Dozin B. (1992). Thyroid hormone, insulin, and glucocorticoids are sufficient to support chondrocyte differentiation to hypertrophy: a serum-free analysis. J. Cell Biol., 119, 989-995.

Rath NC, Bayyari GR, Balog JM, Huff WE. (1994). Physiological studies of turkey tibial dyschondroplasia. Poult. Sci., 73, 416-424.

Rath NC, Huff WE, Bayyari GR, and Balog JM. (1995). Effect of thiram on chick chondrocytes in culture. J. Toxicol. Environ. Health, 44, 369-376.

Rath NC, Huff WE, Bayyari GR, and Balog JM. (1997). Matrix metalloproteinase activities in avian tibial dyschondroplasia. Poult. Sci., 76, 501-505.

Rath NC, Huff WE, Bayyari GR, and Balog JM. (1998). Cell death in avian tibial dyschondroplasia. Avian Dis., 42, 72-79.

Rath NC, Huff WE, Balog JM, and Huff GR. (2004). Comparative efficacy of different dithiocarbamates to induce tibial dyschondroplasia in poultry. Poult. Sci., 83, 266-274.

Rath NC, Richards MP, Huff WE, Huff GR, and Balog JM. (2005). Changes in the tibial growth plates during thiram induced tibial dyschondroplasia in chickens. J. Comp. Pathol. (in Press)

Reddi AH. (1994). Cartilage morphogenesis: role of bone and cartilage morphogenetic proteins, homeobox genes and extracellular matrix. Matrix Biol., 14, 599-606.

Roach HI. (1997). New aspects of endochondral ossification in the chick: chondrocyte apoptosis, bone formation by former chondrocytes, and acid phosphatase activity in the endochondral bone matrix. J. Bone Miner. Res., 12, 795-805.

Rosselot G, Vasilatos-Younken R, Leach RM. (1994). Effect of growth hormone, insulin-like growth factor I, basic fibroblast growth factor, and transforming growth factor beta on cell proliferation and proteoglycan synthesis by avian postembryonic growth plate chondrocytes. J. Bone Miner. Res., 9, 431-439.

Schreck R, Meier B, Mannel DN, Droge W, and Baeuerle PA. (1992). Dithiocarbamates as potent inhibitors of nuclear factor kappa B activation in intact cells. J. Exp. Med., 175, 1181-1194.

Sharma VK, Aulak JS, Malik AK. (2003). Thiram: degradation, applications and analytical methods. J. Environ. Monit., 5, 717-723.

Shian SG, Kao YR, Wu FY, Wu CW. (2003). Inhibition of invasion and angiogenesis by zinc-chelating agent disulfiram. Mol. Pharmacol., 64, 1076-1084.

Sook Han M, Shin KJ, Kim YH, Kim SH, Lee T, Kim E, Ho RS, and Suh PG. (2003). Thiram and ziram stimulate non-selective cation channel and induce apoptosis in PC12 cells. Neurotoxicol., 24, 425-434.

Stoker TE, Goldman JM, and Cooper RL. (1993). The dithiocarbamate fungicide thiram disrupts the hormonal control of ovulation in the female rat. Reprod. Toxicol. 7, 211-218.

Stricker S, Fundele R, Vortkamp A, and Mundlos S. (2002). Role of Runx genes in chondrocyte differentiation. Dev. Biol., 245, 95-108.

Tong A, Reich A, Genin O, Pines M, Monsonego-Ornan E. (2003). Expression of chicken 75-kDa gelatinase B-like enzyme in perivascular chondrocytes suggests its role in vascularization of the growth plate. J. Bone Miner. Res., 18, 1443-1452.

U.S. EPA, (2001). http://www.epa.gov/scipoly/sap/2001/september7/dithiofinal_aug17.pdf

van der Eerden BC, Karperien M, and Wit JM. (2003). Systemic and local regulation of the growth plate. Endocrine Rev., 24, 782-801.

Vargas MI, Lamas JM, and Alvarenga V. (1983). Tibial dyschondroplasia in growing chickens experimentally intoxicated with tetramethylthiuram disulfide. Poult. Sci., 62, 1195-1200.

Vortkamp A, Lee K, Lanske B, Segre GV, Kronenberg HM, and Tabin CJ. (1996). Regulation of rate of cartilage differentiation by Indian hedgehog and PTH-related protein. Science, 273, 613-622.

Vu TH, and Werb Z. (2000). Matrix metalloproteinases: effectors of development and normal physiology. Genes Dev., 14, 2123-2133.

Waibel PE, Johnson EL, Pomeroy BS, and Howard LB. (1957). Toxicity of tetramethyl thiuram disulfide. Poult. Sci., 36, 697-703.

Wardale RJ, Duance VC. (1996). Collagen expression in chicken tibial dyschondroplasia. J. Cell Sci., 109, 1119-1131.

Wan M, and Cao X. (2005). BMP signaling in skeletal development. Biochem. Biophys. Res. Commun., 328, 651-657.

Wickramanayake SD. (1996). Thiram poisons India's poultry. World Poultry, 12, 47.

Wang W, McLeod HL, and Cassidy J. (2003). Disulfiram-mediated inhibition of NF-kappaB activity enhances cytotoxicity of 5-fluorouracil in human colorectal cancer cell lines. Int. J. Canc., 104, 504-511.

White A, and Wallis G. (2001). Endochondral ossification: A delicate balance between growth and mineralization. Current Biol., 11, R589-R591.

Yoon BS, Lyon KM. (2004). Multiple functions of BMPs in chondrogenesis. J. Cell Biochem., 93, 93-103.

Index